The design of experiments

The design of experiments

STATISTICAL PRINCIPLES FOR PRACTICAL APPLICATIONS

R. MEAD

Department of Applied Statistics, University of Reading

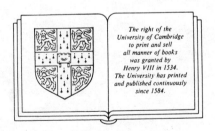

The right of the
University of Cambridge
to print and sell
all manner of books
was granted by
Henry VIII in 1534.
The University has printed
and published continuously
since 1584.

CAMBRIDGE UNIVERSITY PRESS

Cambridge
New York   Port Chester   Melbourne   Sydney

Published by the Press Syndicate of the University of Cambridge
The Pitt Building, Trumpington Street, Cambridge CB2 1RP
40 West 20th Street, New York, NY 10011–4211, USA
10 Stamford Road, Oakleigh, Victoria 3166, Australia

First published 1988
First paperback edition (with corrections) 1990
Reprinted 1991

Printed in Great Britain at the University Press, Cambridge

British Library cataloguing in publication data
Mead, R.
  The design of experiments: statistical
  principles for practical applications.
  1. Experimental design
  I. Title
  001.4'34'028   QA279

Library of Congress cataloguing in publication data
Mead, R. (Roger)
  The design of experiments

Bibliography
  Includes index.
  1. Experimental design.   2. Mathematical statistics.
I.   Title
QA279.M388   1988   001.4'34   87-24236

ISBN 0 521 24512 5 hardback
ISBN 0 521 28762 6 paperback

DE

CONTENTS

PREFACE

My aim in this book is to explain the fundamental statistical concepts required for designing efficient experiments to answer real questions. The book has grown out of 25 years experience of designing experiments for research scientists, and of teaching the concepts of statistical design both to statisticians and to experimenters. This experience has convinced me that the whole subject of statistical design needs to be reassessed in the context of modern statistical computing facilities. The development of statistical philosophy about the design of experiments has been dominated during the last 30 years by mathematical theory. The influence of the availability of vastly improved computing facilities on teaching, textbooks, and, most crucially, practical experimentation appears to have been slight.

The existence of statistical programs capable of analysing the results from any experimental design does not imply any changes in the main statistical concepts of design. However these concepts have become restricted by the earlier need to develop mathematical theory for design in such a way that the results from the designs can be analysed without recourse to computers. The fundamental concepts now require re-examination and re-interpretation outside the limits implied by classical mathematical theory so that the full range of design possibilities may be considered. The result of the revolution in computing facilities is that experimental design should become a much wider and more exciting subject. I hope that this book will display that breadth and excitement.

The development of this book has been particularly motivated by teaching postgraduate students specialising in statistics. However the intention of the book encompasses a wider audience. Understanding the fundamental concepts of design is essential for all research scientists involved in programmes of experimental work. In addition to this general need for an understanding of the philosophy of experimental design there are particular aspects of design, such as repeated measurements (Chapter 14) or levels of replication (Chapter 6) which are relevant to virtually all research disciplines in which experimental work is required. Because of the concentration on basic concepts and their implications the book could be used for courses for final-year undergraduates provided such courses allow sufficient time for the concepts to be thoroughly discussed.

I have tried in writing this book to concentrate on the ideas of design rather than those of analysis, on the statistical concepts rather than the mathematical theory, and on practically useful experiments rather than classes of possible experimental designs. Obviously it is necessary to know how to analyse data from experiments and the philosophy and methods of analysis are discussed in the introductory first part of the book. Also of course, examples of analysis punctuate many later sections. However, after the first part of the book it is assumed that the analysis of experimental data is not difficult when modern computing facilities are used. Consequently ideas of analysis are introduced only when they illumine or motivate the design concepts.

The formal language of statistics is mathematical. It is not possible to discuss design without some mathematically complex formulation of models and ideas. Some of the mathematical language used in the book requires a sound mathematical background beyond school level. However in all parts of the book it is the statistical concepts which are important, and the structure of the book hopefully allows the less-mathematical reader to bypass the more complex mathematical details. Throughout the book the development of concepts relies on many examples. I hope that readers will consider the detailed arguments of these examples. By trying to solve the problems which underly the examples before reading through the explanation of the solutions, I believe that readers will start to develop the intuitive understanding of design concepts which is essential to good design. For the mathematically sophisticated reader the mathematical details provide additional support for the statistical concepts.

Most importantly, the book is intended to show how practical problems of designing real experiments should be solved. To stimulate this practical emphasis real examples of design problems are described at the beginning of each chapter from Chapter 6 onwards (except Chapter 9). The final chapter of the book attempts an overall view of the problem-solving aspects of design. The bias of the areas of application discussed in this book inevitably reflects my personal experience. Much of my experience has been concerned with agricultural experimentation and to write with anything like a comprehensive knowledge about the practical application of design concepts I believe that a concentration on agricultural applications is inevitable. However all the examples are intended to illustrate particular forms of problem that will be relevant in many other fields of application. I hope that statisticians and research scientists in a wide range of experimental disciplies will be able to interpret and adapt the concepts discussed in the book to their own requirements through the use of analogy when the examples discussed are not directly relevant to their discipline.

The book is divided into an overture and two main subjects with a final coda to bring together all the previous material. Chapters 1 to 5 constitute the overture, providing a general introduction and the basic theory necessary for analysis of experimental data. Chapters 1, 2 and 3 should be familiar to readers who have taken an elementary course in experimental design. Alternatively, for those readers without any previous training in design, these three chapters provide an introductory presentation of the two most important ideas of design, blocking and factorial structure. Chapter 4 is the mathematically heavy chapter, providing the necessary theory for general linear models and the analysis of data from designed experiments, with an initial explanation of the important results at a rather simpler level. Chapter 5 ventures briefly, and rashly, into an inevitably prejudiced view of computing needs and implications.

The first main subject is unit variation and control. The fundamental concepts of replication, blocking (with either one or two systems of control) and randomisation are each examined separately. My aim is to distinguish the purposes and practical relevance of each concept and to eliminate the confusion about these concepts which seems to be common in the minds of users of experimental designs. The other two chapters in this subject, on covariance (for control) and on assumptions (to express variation realistically) are biased towards analysis rather than design but their implications for design and for the choice of measurement variables are important.

The second main subject is treatment questions and structure. Chapter 12 presents a broad view of the need for statisticians to be involved in all stages of discussion about the choice of treatments and the interpretation of results. The classical ideas of factorial structure and single and fractional replicates are presented in Chapter 13. Important consequences of particular practical requirements for factorial structure are described in Chapters 14 (split levels of information) and 15 (avoiding confounding). Some necessary mathematical theory for confounding is included in Chapter 16. The choice of experimental treatments for the investigation of the response to quantitative factors is discussed in Chapters 17 and 18.

Finally in the coda, Chapter 0 seeks to draw the concepts of the two main subjects together to provide guidance on choosing experimental designs to satisfy particular practical requirements. A book on practical experimental design should start with the approach of Chapter 0 but this chapter requires knowledge from the previous chapters before it can be read and understood. Hence the number and position of this chapter.

I owe a considerable debt to many consultees and collaborators both for the stimulus to consider why the problems they presented should be covered by a book on design and also for the many examples they have provided. There are too many for me to thank them individually here for their stimulating requests and they therefore remain anonymous (some should prefer it that way, and others are too far into the mists of time for anything else). I have also drawn exercises at the end of the chapters from Chapter 6 onwards from many examination papers.

I have also benefitted from many discussions with colleagues at Reading and wish to thank Richard Coe, Robert Curnow, Derek Pike and Roger Stern, without whom this book would have been a more stunted growth. I am particularly grateful to Richard Jarrett for a most stimulating dinner conversation in Melborne, from which Chapter 0 was born.

Finally I record my gratitude to those who have made the production of this book possible. Clive Bowman and Marilyn Allum manipulated GLIM with great skill to provide many analyses and information about precision for possible designs. Audrey Wakefield has worked devotedly through thousands of pages of manuscript, second thoughts, third thoughts, total reorganisations..., and has endured the slings and arrows of an early generation word-processor. Rosemary Stern has used her art to convert ideas into figures. And David Tranah and Martin Gilchrist at Cambridge University Press have been most patient with a tardy author.

PART I

OVERTURE

1

Introduction

1.1 Why a statistical theory of design?

The need to develop statistical theory for designing experiments stems, like the need for statistical analysis of numerical information, from the inherent variability of experimental results. In the physical sciences, this variability is frequently small and, when thinking of experiments at school in physics and chemistry, it is usual to think of 'the correct result' from an experiment. However, practical experience of such experiments makes it obvious that the results are, to a limited extent, variable, this variation arising as much from the complexities of the measurement procedure as from the inherent variability of experimental material. As the complexity of the experiment increases, and the differences of interest become relatively smaller, then the precision of the experiment becomes more important. An important area of experimentation within the physical sciences, where precision of results and hence the statistical design of experiments are important, is the control of industrial chemical processes.

Whereas the physical sciences are thought of as exact, it is quite obvious that biological sciences are not. Most experiments on plants or animals use many plants or animals because it is clear that the variation between two plants, or between two animals, is very large. It is impossible, for example, to predict quantitatively the characteristics of one plant from the corresponding characteristics.of another plant of the same species, age and origin.

Thus, no medical research worker would make confident claims for the efficacy of a new drug merely because a single patient responded well to the drug. In the field of market research, no newspaper would publish an opinion poll based on interviews with only two people, but would require a sample of at least 500, together with information about the method of selection of the sample. In a drug trial, the sample of patients would normally be less than 500, possibly between 20 and 100. In psychological experiments, the number of subjects used might be only eight to 12. In agricultural experiments, there may be 20 to 100 plots of land, each with a crop grown on it. In a laboratory experiment hundreds of plants may be treated and examined individually. Or just six cows may be examined while undergoing various diets, with measurements taken frequently and in great detail.

The size of an experiment will vary according to the type of experimental method and the objective of the experiment. One of the important statistical ideas of experimental design is the choice of the size of an experiment. Another is the control of the use of experimental material. It is of little value to use large numbers of patients in the comparison of two drugs, if all the patients given one drug are male, aged between 20 and 30, and all the patients given the other drug are female, aged 50 to 65. Any reasonably sceptical person would doubt claims made about the relative merits of the two drugs from such a trial. This example may seem trivially obvious, but the scientific literature in medicine and many other disciplines shows that many examples of badly planned (or unplanned) experiments occur.

And this is just the beginning of statistical design theory. From avoiding foolish experiment, we can go on to plan improvements in precision for experiments. We can consider the choice of experiments as part of research strategy and can, for example, discuss the relative merits of many small experiments or a few large experiments. We can consider how to design experiments when our experimental material is generally heterogeneous, but includes groups of similar experimental units. Thus, if we are considering the effects of applying different chemicals on the properties of different geological materials, then these may be influenced by the environment from which they are taken, as well as by the chemical treatment applied. However, we may have only two or three samples from some environments, but as many as ten samples from other environments; how then do we decide which chemicals to apply to different samples so that we can compare six different chemical treatments?

1.2   History, computers and mathematics

If we consider the history of experimental design, then most of the developments have been in biological disciplines, in particular in agriculture, and also in medicine and psychology. There is therefore an inevitable agricultural bias to any discussion of experimental design. Most of the important principles of experimental design were developed in the 1920s and 1930s by R. A. Fisher. The practical manifestation of these principles was very much influenced by the calculating capacity then available. Had the computational facilities which we now enjoy been available when the main theory of experimental design was being developed then, I believe, the whole subject of design would have developed very differently. Whether or not this belief is valid, it is certainly true that a view of experimental design today must differ from that of 50, or even 30, years ago. The principles have not changed, but the principles are often forgotten, and only the practical manifestation of the principles retained; these practical applications do require rethinking.

The influence of the computer is one stimulus to reassessing experimental design. Another cause for concern in the development of experimental design is the tendency for increasingly formal mathematical ideas to supplant the statistical ideas. Thus the fact that a particularly elegant piece of mathematics can be used to demonstrate the existence of groups of designs, allocating treatments to blocks of units in a particular way, begs the statistical question of whether such designs would ever be practically useful.

Although myself originally a mathematician, I believe that the presentation of statistical design theory has been quite unnecessarily mathematical, and I shall hope to demonstrate the important ideas of statistical design without excessive mathematical encumbrance. The language of statistical theory, like that of physics, is mathematical and there will be sections of the book where those with a mathematical education beyond school level will find a use for their mathematical expertise. However, even in these sections, which I believe should be included because they will improve the understanding of statistical theory of the readers able to appreciate the mathematical demonstrations, there are intuitive explanations of the theory at a less advanced mathematical level.

1.3   The influence of analysis on design

To write a book solely about the theory of experimental design, excluding all mention of the analysis of data, would be impossible. Any experimenter must know how he intends to analyse his experimental data before he designs his experiment to yield the data. If not, how can he know whether the form of information which he collects can be used to answer the questions which prompted him to do an experiment?

Thus, consider again the medical trial to compare two drugs. Suppose the experimenter failed to think about the analysis and argued that one of the drugs was well known, while the other was not; in the controlled experiment to compare them, there are available 40 patients. Since a lot is already known about drug A, let it be given to one patient, and let drug B be given to the other 39. When the data on the response to the drugs is obtained, the natural analysis is to compare the mean responses to the drugs, and to consider the difference $(\bar{x}_A - \bar{x}_B)$. To test the strength of evidence for a real difference in the effects of the two drugs, we need the standard error of $(\bar{x}_A - \bar{x}_B)$, which will be:

$$\sigma(1/1 + 1/39)^{1/2} = 1.013\sigma.$$

However, if the experimenter had considered this analysis of data before designing the experiment, he would have realised that the effectiveness of his experiment depended on making the standard error of $(\bar{x}_A - \bar{x}_B)$ within his experiment as small as possible (to use the previous knowledge about the effects of drug A requires that the experimental conditions are identical to those in which the previous experience was gained). This is achieved by allocating 20 patients to drug A and 20 to drug B. This gives a standard error of

$$\sigma(1/20 + 1/20)^{1/2} = 0.316\sigma,$$

giving an improved precision by a factor of more than three. Of course, it would not be necessary to consider the standard error formally to guess that equal allocation of patients gives the most precise answer. In a sense, it would be intuitively surprising if any unequal division were to be more efficient than the equal division. Prior thought of what is to be done with the results of the experiment should lead to avoiding a foolish design.

Nevertheless, although analysis is integral to any consideration of design, this book is about the design of experiments, and will not be concerned with the analysis of data, except insofar as this is essential to the understanding of the design principles discussed. The general theory of the analysis of data from designed experiments is developed in Chapter 4, and particular examples of this general theory will appear in later chapters for some examples of experimental designs. In Chapter 5, the developments in computing which allow the analysis by a general computer program of any designed experiment are described. The joint implication of these two chapters is clear; not only is it possible to derive the algebraic formulae necessary to define the calculations required for the analysis of data from any experiment, but the computer programs should make it possible for an experimenter, without the mathematical skills necessary to derive the algebraic formulae, to understand the statistical basis for the interpretation of the calculations. The need, in the earlier development of statistical methods, to be able to derive the algebraic form of analysis has made it easier to discuss some of the important design principles we shall discuss later. In this way, the earlier lack of advanced computing power may be regarded as a blessing, though there must be many erstwhile technical assistants to statisticians who would find that hard to accept! A caveat to the advice to use the general computer programs to analyse all experimental data must be to remember that much of our understanding of design theory was stimulated by algebraic necessity because of the lack of computers. Consequently, we should search for increased understanding both through the use of computers and through algebraic manipulation.

1.4   Separate consideration of units and treatments

Throughout this book, I shall emphasise the need to think separately about the properties of the experimental units available for the experiment, and about the choice of experimental treatments. Far too often, the two aspects are confused at an early stage, and an inefficient or useless experiment results. Thus, an experimenter considering a comparison of the effects of different diets on the growth of rats may observe that the rats available for experimentation come from litters with between five and ten animals per litter. Having heard about the randomised block design, he decides that he must use litters as blocks, and therefore will have blocks of five units. He further decides that he should consequently have five treatments, and chooses his treatments on this premise, rather than considering how many treatments are required for the objectives of his experiments.

Similarly, a microbiologist investigating the growth of salmonella as affected by five different temperatures, four different media and two independent different chemical additives, may decide that a factorial set of 80 treatments is necessary. Recognising the possibility of day-to-day variation and also knowing about the randomised block design he tries to squeeze the preparation of 80 units into a single day, where only 40 can be efficiently managed, and consequently suffers a much higher error variance for his observations than would normally be expected.

In both of these examples, which though plausible are fictitious so far as my personal experience extends, I believe that separate consideration of properties of units and choice of treatments, together with an appreciation of the possibilities of modern statistical design, would lead to better experiments. Thus, in the rat growth experiment, on considering the experimental material available, the experimenter might recognise that each litter forms a natural block, and that he has one block of five units, one of six units, two of seven units, and one of ten units; each block could be reduced in size while still retaining its natural block quality. For treatments, he has a control and three additions to the basic diet, each of which he wishes to try at two concentrations; a total of seven treatments. I believe that it is quite simple to construct a sensible design to compare seven treatments in blocks of five, six, seven, seven and ten units, respectively, and this problem will be discussed further in Chapter 7.

For the microbiologist, again the solution of his problem requires only the application of standard statistical design theory, though because the problem does not fit into any neat mathematical classification, the reader of most books on experimental design might be forgiven for thinking that the problem was impossible or at best required very complex mathematical theory. The problem of using blocks of 40 units with a set of 80 treatments in a $5 \times 4 \times 2 \times 2$ factorial structure is a very simple example of confounding to be met again in Chapter 15.

In summary, to achieve good experimental design, the experimenter should think first about the experimental units, and should make sure that he understands the likely patterns of variation. He should also consider the set of treatments appropriate to the objectives of his experiment. If the set of treatments fits simply with the structure of units using a standard design, then the experiment is already designed. If not, then a design should be constructed to fit the requirements of units and treatments, either by the experimenter alone or with the assistance of a statistician. The natural pattern of units and treatments should not be deformed in a Procrustean fashion to fit a standard design. It may perhaps reassure the reader to point out that, for most experiments (80 or 90%), a standard design will be appropriate. To counterbalance the note of reassurance, it is obvious from the experimental scientific literature that frequently an inappropriate simple standard design has been used when an appropriate but slightly more complex standard design exists and should have been used.

2

Elementary ideas of blocking: the randomised block design

2.1   Controlling variation in experimental units

In an experiment to compare different treatments, each treatment must be applied to several different units. This is because the response from different units varies, even if the units are treated identically. If each treatment is applied to only a single unit, the results will be ambiguous, in that we will not be able to distinguish whether a difference between the response from two units is caused by the different treatments, or is simply due to the inherent differences between the units.

The simplest experimental design is that in which, in order to compare t treatments, treatment 1 is applied to n_1 units, treatment 2 to n_2 units, and treatment t to n_t units. In many experiments the numbers of units per treatment, $n_1, n_2, \ldots, n_t$, will be equal, but this is not necessary, or even always desirable. Some treatments may be of greater importance than others, in which case more information will be needed about them, and this will be achieved by the use of more units for these treatments. This design, in which the only recognisable difference between units is the treatments which are applied to those units, is called a *completely randomised design*.

With several units for each treatment the ambiguity which occurs when each treatment has only a single unit is not eliminated. One treatment might be 'lucky' in the selection of units to which it is to be applied. There is no foolproof method of overcoming the vagaries of allocating treatments to units, since the responses from the units if the same treatment were applied to all units are not known before the experiment. However, most experimenters have some idea about which units are likely to behave similarly, and such ideas can be used to control the allocation of treatments to units in an attempt to make the allocation more 'fair'. Essentially each group of 'similar' units should include roughly equal numbers of units for each treatment. This control is called *blocking*, and the simplest, and by far the most frequently used, design resulting from the idea of blocking is the randomised block design, sometimes called the *randomised complete block design*.

The term blocking arises from the introduction of the idea of control in agricultural crop experiments. Here, the experimental unit is a small area of land, or plot, typically between $2 \, \text{m}^2$ and $20 \, \text{m}^2$, depending on the crop and on the treatments. The set of units or plots for an experiment might be situated as in Figure 2.1(a). In general, we might expect that adjacent plots

Figure 2.1. Possible blocking plans (b) and (c) for a set of 60 plots.

(a)

(b)

(c)

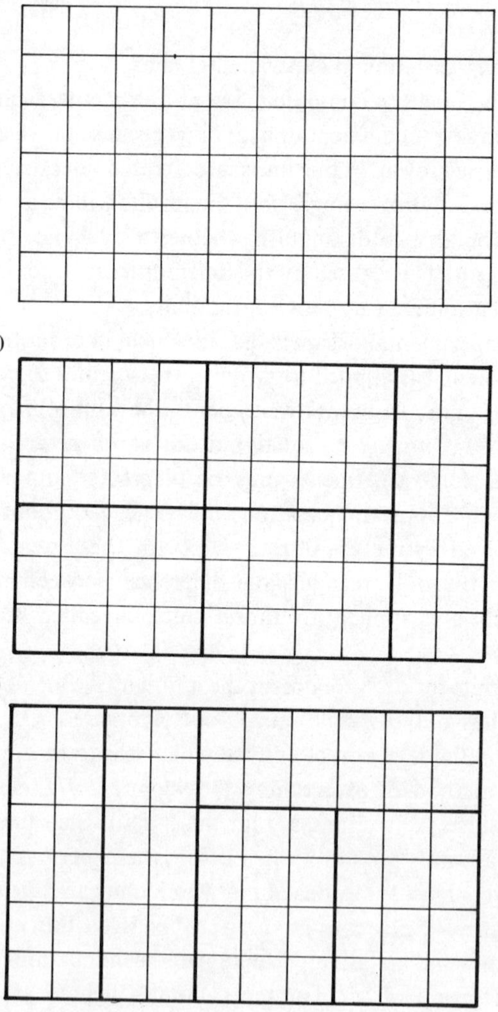

would behave more similarly than those further apart. Of course, the experimenter might have more specific knowledge about the plots than this. For example, there might be a slope from east to west, so that the plots down the east side might be expected to behave similarly, and differently from those on the west side. However, either from general or particular considerations, if the plots are grouped together in sets of, say, 12, so that the plots in a set might be expected to behave similarly, then the resulting grouping will usually look like Figure 2.1(b) or (c). Such groupings are naturally called blocks of plots, and the term has become standard for groupings of units in disciplines other than agronomy.

The patterns of variation on which blocking systems can be based are very numerous, and always the specialist knowledge of the experimenter about his experimental material must be the most important source of information in determining blocks of units. Some examples of blocking systems are discussed here, to help the experimenter to think about the choice of blocks.

In experiments with plants in pots in glasshouses or laboratories, where different treatments are applied to different pots, geographical blocks of similarly situated pots will usually prove useful, the major controllable source of variation between yields from different pots being nearness to the edge of the room or to the door or to other features. In experiments with perennial crops, such as fruit trees, the experimental unit will often be an individual tree, and geographical blocks may again be useful. Possibly more useful blocks can be based on yields of the trees in previous years, or more generally based on the previous histories of the trees. When the experimental unit is an animal, then blocks can be based on genetic similarity, leading to the use of complete litters as blocks, or on the weights of the animals prior to the experiment, or on previous history of the animals.

In medical trials, where the experimental unit is usually an individual person, the number of classifications which can be used as the basis for blocking is large. These include the age of the person, sex, height or weight, social class, medical history, racial characteristics or time at which the person becomes available for the trial. Many of the same classifications are relevant for psychological experiments. Often, in these experiments, the experimental unit will be a short time period in a person's life. If several observations are made for different periods for the same person, with different stress treatments being applied during the different periods, then the experimental unit may be defined to be a short period within a person's life, and the set of units for one person constitutes a sensible block. A possible complication is that time might be thought to be an important factor influencing response *within* the total experimental period, in which case the experimental design requires two forms of blocking for persons and for times.

In industrial experiments, a small number of machines may each be observed at different times, working under different conditions. The set of observations for a single machine may be a sensible block. Alternatively, if the environment changes diurnally or between days the set of observations at the same time within a day, or within a week, may be a useful block. In industrial or other experiments, there may be suspected to be consistent differences between different machine operators, or different observers or recorders, and the sets of observations for each operator or observer may be regarded as blocks.

In many experiments, there may be several different ways of blocking the experimental units, and we shall return to this problem later in Chapters 7, 8 and 10. For the present, we shall assume that the experimenter has considered the available units and, for the set of *t* treatments, has divided the units into blocks so that each block contains *t* units, which are expected to behave similarly, and has then, in each block, allocated one unit to each treatment. The resulting design in which there is exactly one observation for each treatment in each block is the randomised block design. The analysis of data from this design, and from the completely randomised design, are now considered in some detail at a fairly simple level.

2.2   The analysis of variance identity

The initial technique in the analysis of most experimental data is the *analysis of variance*. This has two purposes. First, it provides a subdivision of the total variation between the experimental units into separate components, each component representing a different source of variation, so that the relative importance of the different sources can be assessed. Second, and more important, it gives an estimate of the underlying variation between units which provides a basis for inferences about the effects of the applied treatments. We develop the principles of the analysis of variance for the two designs described in the previous section, using two examples.

First we consider a laboratory experiment in which three different chemicals (A, B and C) were added to cultures of a virus. The purpose of the experiment was to compare the effects of the chemicals on the growth of the virus, measured by the diameter of the affected area. The design of the experiment was a completely randomised design. The total number of units was 25, divided unequally (10, 9, 6) between the treatments, and the data are given in Table 2.1.

Table 2.1. *Diameter (mm) of each affected area.*

A	B	C
17.1	24.3	14.3
14.9	21.7	14.5
23.6	22.3	17.1
16.0	26.4	13.2
18.7	23.2	15.6
21.7	20.7	10.3
21.4	22.9	
23.0	19.6	
19.0	22.4	
18.6		

The second data set, in Table 2.2, is from a randomised block design. Ten spacing treatments for rice were compared in four randomised blocks of ten plots each. The ten spacings were all possible combinations of pairs of inter-seedling distances (15 cm, 20 cm, 24 cm and 30 cm), and the general purpose of the experiment was to investigate both the effect of changing density of seedlings and the effect of spatial arrangement. The area available for the experiment was sufficient for 40 experimental plots, each measuring 2.4 m × 3.6 m. The area was divided into four quarters, each containing ten plots, so that the ten plots in each quarter were hopefully similar, and then the ten spacings were allocated, one to each plot in each

Table 2.2. *Yields of rice grain (kg per plot).*

	Block			
Spacing	I	II	III	IV
30 cm × 30 cm	5.95	5.30	6.50	6.35
30 cm × 24 cm	7.10	6.45	6.60	5.75
30 cm × 20 cm	7.00	6.50	6.35	8.90
30 cm × 15 cm	8.10	5.50	6.60	7.50
24 cm × 24 cm	8.85	7.65	7.00	7.90
24 cm × 20 cm	7.65	6.90	8.25	8.30
24 cm × 15 cm	7.80	6.75	8.20	7.25
20 cm × 20 cm	8.05	6.65	8.10	8.05
20 cm × 15 cm	9.30	8.75	8.75	8.00
15 cm × 15 cm	9.35	8.10	7.60	7.75

of the four blocks, or quarters. The experiment, at Belle Vue Farm, Mauritius, was conducted during the 1970–71 season, using variety Taichung Native No. 1.

The usual philosophy of analysis is to assume that each unit has an inherent yield which is modified by the effect of the particular treatment applied to the unit. For the completely randomised design, this is expressed by writing the yield for the kth unit to which treatment j is applied:

$$y_{jk} = \mu + t_j + \varepsilon_{jk}. \tag{2.1}$$

Hence, μ represents an average yield for the whole set of experimental units, and ε_{jk} represent the deviations of the particular units from that average. The treatment effects, t_j, represent deviations, for particular treatments, from the average of the set of treatments included in the experiment.

If the units are grouped in blocks, then an additional set of terms is needed to represent the average differences between blocks. Hence, for the randomised block design, the yield of treatment j in block i, y_{ij}, is written as

$$y_{ij} = \mu + b_i + t_j + \varepsilon_{ij}. \tag{2.2}$$

Here μ again represents an average yield for the whole set of experimental units, b_i is the additional term representing the average deviation from μ of the set of units in block i, and ε_{ij} represent the deviations of the particular units from the average yield, $\mu + b_i$, of the units in block i. The terminology used here will be used throughout the book. To complete the definition for the completely randomised design and the randomised block design we assume that, in the completely randomised design, there are t treatments and n_j units for treatment j, with a total of $n = \sum n_j$ units; and in the randomised block design there are b blocks and t treatments, and a total of $n = bt$ experimental units. The definitions of b_i and t_j imply that:

$$\sum_{j=1}^{t} t_j = 0 \qquad \sum_{i=1}^{b} b_i = 0.$$

The algebraic statements (2.1) and (2.2), with the accompanying verbal explanations, constitute models for the sets of yields. Two basic assumptions implied by the use of these models must be remembered when calculating an analysis. The first is that the individual unit variation, ε, is not affected by the particular treatment applied to the unit, and, second, that the effect of each treatment is additive. A consequence of these assumptions is that, if two treatments applied to a pair of units had been swapped and applied each to the other unit of the pair, then the total yield from the pair of units would have been unaltered, the loss of one yield being exactly counterbalanced by the gain of the other. This is, of course, a purely hypothetical concept, since each unit can have only one treatment applied to it, but the consequence of the assumption is a powerful one which the experimenter must be prepared to accept as reasonable if the standard form of analysis is to be used. We return to further consideration of the assumptions implied by these models in Chapter 11.

The analysis of variance of data from an experimental design is simply the division of the total variation between the n observations into sensibly distinguishable components, and the subsequent interpretation of the relative size of the components.

In the completely randomised design, there are two distinguishable components of variation, the variation between the units receiving the same treatment, and the variation between units receiving different treatments. The variation between the yields for units receiving treatment j is most simply written

$$\sum_k (y_{jk} - y_{j.})^2, \tag{2.3}$$

where the use of a dot replacing a suffix indicates the mean over all possible values of the suffix j:

$$y_{j.} = \sum_k y_{jk}/n_j.$$

The total variation between all n observations is

$$\sum_{jk} (y_{jk} - y_{..})^2, \tag{2.4}$$

where $y_{..}$ is the mean of all n observations.

Now, both expressions (2.3) and (2.4) can be written in extended form as

$$\sum_k (y_{jk} - y_{j.})^2 = \sum_k y_{jk}^2 - \left(\sum_k y_{jk}\right)^2 \bigg/ n_j$$

and

$$\sum_{jk} (y_{jk} - y_{..})^2 = \sum_{jk} y_{jk}^2 - \left(\sum_{jk} y_{jk}\right)^2 \bigg/ n,$$

from which we may deduce the analysis of variance identity

$$\sum_{jk} (y_{jk} - y_{..})^2 \equiv \sum_j \left\{ \sum_k (y_{jk} - y_{j.})^2 \right\} + \sum_j \sum_k (y_{j.} - y_{..})^2, \tag{2.5}$$

where the last term may also be written

$$\sum_j \sum_k (y_{j.} - y_{..})^2 = \sum_j \left(\sum_k y_{jk}\right)^2 \bigg/ n_j - \left(\sum_{jk} y_{jk}\right)^2 \bigg/ n.$$

The three terms in the identity (2.5) represent the total yield variation over all n observations, the sum of within treatment variations and the variation between treatment means. For the calculation of sample variances (2.3) would be divided by (n_j-1) for each treatment variance, and (2.4) by $(n-1)$ for the variance from the total sample. An analogous identity for these divisors, which are called degrees of freedom (df), is

$$(n-1)=\sum_j (n_j-1)+(t-1) \tag{2.6}$$

and the divisor $(t-1)$ is used to calculate a between treatment variance.

The analysis of variance structure is normally presented in tabular form (see Table 2.3). For the virus growth data, the calculations are:

$$\sum_{jk} y_{jk}^2 = 9705.41$$

$$\sum_j \left(\sum_k y_{jk}\right)^2 \Big/ n_j = 9569.13$$

$$\left(\sum_{jk} y_{jk}\right)^2 \Big/ n = 9312.25,$$

and the results are summarised in Table 2.4. Clearly the major component of the total variation is attributable to the treatment effects, almost two-

Table 2.3.

Source	Sum of squares (SS)	df
Between treatments	$\sum_j \left(\sum_k y_{jk}\right)^2 \Big/ n_j - \left(\sum_{jk} y_{jk}\right)^2 \Big/ n$	$t-1$
Within treatments	$\sum_j \left\{\sum_k y_{jk}^2 - \left(\sum_k y_{jk}\right)^2 \Big/ n_j\right\}$	$\sum_j (n_j-1)$
Total	$\sum_{jk} y_{jk}^2 - \left(\sum_{jk} y_{jk}\right)^2 \Big/ n$	$n-1$

Table 2.4.

Source	SS	df
Between treatments	256.88	2
Within treatments	136.28	22
Total	393.16	24

thirds of the total variation representing between treatment variation. The comparison of variances, calculated by dividing sums of squares (SS) by degrees of freedom, is even more dramatic, the variances being 128, 6 and 16 for between treatments, within treatments and total, respectively.

The analysis of variance of data from a randomised block design is also based on an algebraic identity, but with four component terms. The extra term, compared with the completely randomised analysis of variance, is required to allow for the variation between blocks. The analysis of variance identity is

$$\sum_{ij} (y_{ij} - y_{..})^2 \equiv \sum_{ij} (y_{i.} - y_{..})^2 + \sum_{ij} (y_{.j} - y_{..})^2 + \sum_{ij} (y_{ij} - y_{i.} - y_{.j} + y_{..})^2$$

total SS $\equiv$ block SS $+$ treatment SS $+$ error SS.

The second and third terms in the identity represent the variation between the block averages, $y_{i.}$, and between the treatment averages, $y_{.j}$, respectively. By analogy with the between treatment variation in the previous analysis of variance, the corresponding df are $(b-1)$ and $(t-1)$, respectively. The final term in the identity, the error SS corresponds to the ε terms in the model, and represents the inconsistency of treatment differences over the different blocks. The label 'error' is unfortunate but traditional.

The truth of the randomised block analysis of variance identity may be verified simply by rewriting the various terms in extended form:

$$\sum_{ij} (y_{ij} - y_{..})^2 = \sum_{ij} y_{ij}^2 - \left(\sum_{ij} y_{ij}\right)^2 \Big/ n = \sum_{ij} y_{ij}^2 - Y_{..}^2/n,$$

where we have introduced an extension of the dot suffix notation, the dot suffix with a capital letter representing the total yield summing over the dotted suffix or suffices;

$$\sum_{ij} (y_{i.} - y_{..})^2 = \sum_{i} Y_{i.}^2/t - Y_{..}^2/n$$

$$\sum_{ij} (y_{.j} - y_{..})^2 = \sum_{j} Y_{.j}^2/b - Y_{..}^2/n$$

$$\sum_{ij} (y_{ij} - y_{i.} - y_{.j} + y_{..})^2 = \sum_{ij} y_{ij}^2 - \sum_{i} Y_{i.}^2/t - \sum_{j} Y_{.j}^2/b - Y_{..}^2/n.$$

The term $Y_{..}^2/n$, which occurs in each expression, is often called the *correction factor*, since it modifies, or 'corrects', simple sums of squared quantities to be sums of squared deviations.

The randomised block analysis of variance structure is normally presented in tabular form (see Table 2.5). For the rice spacing example given earlier, the calculations are as follows:

$$\text{total SS} = 2252.61 - 2211.17 = 41.44$$
$$\text{block SS} = 2217.05 - 2211.17 = 5.88$$
$$\text{treatment SS} = 2234.31 - 2211.17 = 23.14$$

and the analysis of variance is shown in Table 2.6. For the final column in Table 2.6, each mean square is calculated by dividing the SS by the corresponding df. Each mean square is a variance, the method of calculation being such that the mean squares are directly comparable. Thus the treatment mean square (TMS) is:

$$\text{TMS} = \left(\sum_j Y_{.j}^2 \Big/ b - Y_{..}^2/n \right) \Big/ (t-1)$$

$$= 1/b \left[\left\{ \sum_j Y_{.j}^2 - \left(\sum_j Y_{.j} \right)^2 \Big/ t \right\} \Big/ (t-1) \right]$$

$$= 1/b \text{ (variance of the } Y_{.j}\text{'s)},$$

which, since $Y_{.j}$ is the sum of b observations, is on the same scale as $\text{Var}(y_{ij})$. Thus the block, treatment and error mean squares may be

Table 2.5.

Source	SS	df
Blocks	$\sum_i Y_{i.}^2/t - Y_{..}^2/n$	$b-1$
Treatments	$\sum_j Y_{.j}^2/b - Y_{..}^2/n$	$t-1$
Error	by subtraction	$(b-1)(t-1)$ by subtraction
Total	$\sum_{ij} y_{ij}^2 - Y_{..}^2/n$	$n-1$

Table 2.6.

Source	SS	df	Mean square (MS)
Blocks	5.88	3	1.96
Treatments	23.14	9	2.57
Error	12.42	27	0.46
Total	41.44		

compared to assess the relative magnitude of the between block variance, the between treatment variance and the between unit variance or error variance.

For the rice spacing experiment, the block variance appears to be rather larger than the error variance, and the treatment variance to be considerably larger. Later, in Chapter 4, we show formally that the mean squares can be compared using F tests if we assume normally distributed unit variation. For the present, our purpose is to consider the analysis of variance as an arithmetical subdivision of variance. We may reasonably conclude, however, that the different spacings lead to different yields.

2.3 Estimation of variance and the comparison of treatment means

In the construction of the models (2.1) and (2.2), the terms ε_{jk} or ε_{ij} represented the deviations of unit yields from the average yield or from the average yield of units in block i. The precision of comparisons between the mean yields for different treatments will depend only on the variability of these ε_{ij}, since in the completely randomised design there are no other effects, and in the randomised block design block differences have been eliminated from treatment comparisons, if the model assumptions are correct. To assess the precision of treatment comparisons, we require explicit assumptions about the mean and variance of the ε's. Formally, we assume that the units, and hence the ε's, are a random sample from a population, such that the expectation of ε, $E(\varepsilon)$, equals zero, and the variance $\mathrm{Var}(\varepsilon) = \sigma^2$.

In the completely randomised design, the variance from the sample of observations for each treatment provides an estimate of σ^2. The pooled within treatment SS provides a better estimate of σ^2 than any single treatment sample, since the variance of observations is assumed to be unaffected by the particular treatment considered. Hence, the estimate of σ^2 for a completely randomised design is the within treatment MS:

$$s^2 = \sum_j \left\{ \sum_k y_{jk}^2 - \left(\sum_k y_{jk} \right)^2 \bigg/ n_j \right\} \bigg/ \sum_j (n_j - 1).$$

The variance of the difference between two treatment means is

$$\mathrm{Var}(y_{.j} - y_{.j'}) = \sigma^2(1/n_j + 1/n_{j'})$$

and is estimated by $s^2(1/n_j + 1/n_{j'})$.

In the randomised block design model (2.2), the b_i sum to zero and the expected value of a treatment mean yield is

$$E(Y_{.j}/b) = (\mu + t_j).$$

Also, since

$$Y_{.j} = b\mu + bt_j + \sum_i \varepsilon_{ij}$$

the variance of a treatment mean is

$$\text{Var}(Y_{.j}/b) = \text{Var}(\mu + t_j + \sum_i \varepsilon_{ij}/b) = \text{Var}\left(\sum_i \varepsilon_{ij}/b\right) = \sigma^2/b.$$

Similarly,

$$\text{Var}(y_{.j} - y_{.j'}) = 2\sigma^2/b.$$

The definition of σ^2 is clear, but it is not immediately clear how to estimate σ^2, since the observations, y_{ij}, all contain components due to block and treatment differences. However, it is clear from the analysis of variance identity that the error SS is not affected by block or treatment differences, and in fact the error MS provides an unbiassed estimate of σ^2. To see that this is algebraically correct, consider the error SS:

$$\sum_{ij} (y_{ij} - y_{i.} - y_{.j} + y_{..})^2$$

$$= \sum_{ij} (y_{ij}^2 + y_{i.}^2 + y_{.j}^2 + y_{..}^2 - 2y_{ij}y_{i.} - 2y_{ij}y_{.j} + 2y_{ij}y_{..}$$

$$\qquad + 2y_{i.}y_{.j} - 2y_{i.}y_{..} - 2y_{.j}y_{..})$$

$$= \sum_{ij} y_{ij}^2 + t \sum_i y_{i.}^2 + b \sum_j y_{.j}^2 + ny_{..}^2 - 2b \sum_i y_{i.}^2 - 2b \sum_j y_{.j}^2 + 2ny_{..}^2$$

$$\qquad + 2ny_{..}^2 - 2ny_{..}^2 - 2ny_{..}^2$$

$$= \sum_{ij} y_{ij}^2 - t \sum_i y_{i.}^2 + b \sum_j y_{.j}^2 + ny_{..}^2.$$

But, from model (2.1)

$$y_{..} = \mu + \varepsilon_{..}$$
$$y_{i.} = \mu + b_i + \varepsilon_{i.}$$
$$y_{.j} = \mu + t_j + \varepsilon_{.j},$$

and, since $E(\varepsilon_{ij}) = 0$, $\text{Var}(\varepsilon_{ij}) = \sigma^2$,

$$E\left\{\sum_{ij} (y_{ij} - y_{i.} - y_{.j} + y_{..})^2\right\} = E\left(\sum_{ij} y_{ij}^2 - t \sum_i y_{i.}^2 - b \sum_j y_{.j}^2 + ny_{..}^2\right)$$

$$= \left(n\mu^2 + t \sum_i b_i^2 + b \sum_j t_j^2 + n\sigma^2\right) - t\left\{b\mu^2 + \sum_i b_i^2 + b(\sigma^2/t)\right\}$$

$$\qquad - b\left\{t\mu^2 + \sum_j t_j^2 + t(\sigma^2/b)\right\} + n(\mu^2 + \sigma^2/n)$$

$$= (n - t - b + 1)\sigma^2 = (b - 1)(t - 1)\sigma^2.$$

Hence the expected value of the error SS is $(b-1)(t-1)\sigma^2$, and the expected value of the error MS is σ^2. The error MS is therefore called s^2 and provides an unbiassed estimator of σ^2 and of variances required for comparisons of treatment means. In this chapter only simple comparisons of treatments are considered. The philosophy of choosing treatment comparisons to answer the questions which provoked the experiment is considered in Chapter 12.

The obvious treatment comparisons are between two particular treatments, for example $(t_1 - t_2)$, estimated by $(y_{.1} - y_{.2})$ for which the standard error is $(2\sigma^2/b)^{1/2}$. An alternative comparison is between one treatment and the average of two other treatments, $\{t_1 - (t_2 + t_3)/2\}$; the relevant standard error is

$$\{\sigma^2/b + \sigma^2/(2b)\}^{1/2} = (3\sigma^2/2b)^{1/2}.$$

The problem with making many comparisons is two-fold. First, the more significance tests we make of treatment comparisons, the more likely we are to find a 5% significance result from the 1 in 20 chance inherent in the concept of a significance test. In addition, it is natural, when making comparisons, to concentrate on those comparisons which look larger, which will be between the treatments giving the more extreme results. These problems are discussed in Chapter 12, but both these dangers must be borne in mind when comparing treatment means after any analysis of variance.

In the experiment on viral growth the three treatment means are

A	B	C
19.4	22.6	14.2,

and the standard errors of the difference between two means are

    1.14 for A − B

    1.29 for A − C

    1.31 for B − C.

Clearly chemical C reduces the growth of the virus compared with either A or B. The difference between A and B is the smallest of the three differences but even this is quite large compared with the standard error of the difference, and we may reasonably conclude that the growth rate is different for all three chemicals, with B giving the greatest growth and C the smallest.

In the rice spacing example, there are various obvious simple comparisons, for example:

(i) between spacings with one dimension in common $t_1 - t_2$, $t_1 - t_3$, $t_9 - t_{10}$;

(ii) between spacings which are square $t_1 - t_5$, $t_5 - t_8$, $t_8 - t_{10}$;

(iii) between spacings at a similar overall density $t_4 - t_8$.

In this instance, however, as in many experiments with quantitative treatments, a graphical presentation of mean is best. Figure 2.2 shows mean yield plotted against area per plant. The values of mean yield and area per plant for the ten treatments are given in Table 2.7. The regression of mean yield on area per plant gives

$$\text{mean yield} = 9.11 - 0.0034 \text{ area}$$

$$\text{regression } SS = 4.20 \quad \text{on 1 df}$$

$$\text{residual } SS = 1.59 \quad \text{on 8 df}$$

$$\text{total } SS = 5.79 \quad \text{on 9 df.}$$

Clearly the regression accounts for most of the treatment variation, and the residual treatment variation is relatively small, suggesting that the interpretation of results should concentrate on the effect of space per plant. The analysis of this set of data is considered further in Chapter 12.

Figure 2.2. Relationship between mean yield and area per plant.

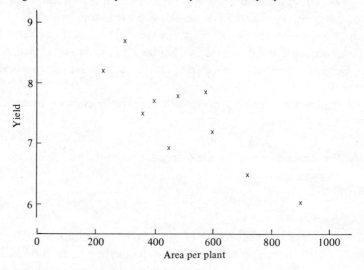

Table 2.7.

Mean yield (y)	Area per plant (x)
6.02	900
6.48	720
7.19	600
6.92	450
7.85	576
7.78	480
7.50	360
7.71	400
8.70	300
8.20	225

2.4 Residuals and the meaning of error

The error SS in the analysis of variance is calculated from the quantities

$$r_{ij} = y_{ij} - y_{i.} - y_{.j} + y_{..}.$$

These are called *residuals* because they represent the unexplained, or residual, variation for each observation after the block and treatment differences have been allowed for. The residuals for the rice spacing data are listed in Table 2.8 and show the patterns imposed on the residuals by the elimination of block and treatment differences. The residuals contain no element due to treatments and so the total of residuals is zero for each treatment; similarly, the block totals of residuals are all zero. The pattern of the residuals displays the consistency of treatment (or block) effects. A treatment with very small residuals, such as treatment 8 (20×20), shows similar effects in all blocks, whereas one with large residuals, such as treatment 3 (30×20), has rather discrepant effects in the different blocks.

The interpretation of error as consistency can be seen more clearly in the analysis of a subset of the data with only two treatments. For treatments 1 and 2, the yields and analysis are shown in Table 2.9(a) and (b), respectively. Using the treatment differences from the four blocks and calculating the mean difference gives the same estimate for the difference between the two treatments: -0.45. Using the variance of the sample of four differences,

$$s_d^2 = \{(1.15)^2 + (1.15)^2 + (0.10)^2 + (0.60)^2 - (1.80)^2/4\}/3 = 0.735,$$

the standard error of this mean difference is

$$(s_d^2/4)^{1/2} = 0.43,$$

Table 2.8. *Residuals for rice spacing experiment.*

	Block			
Spacing	I	II	III	IV
30×30	-0.55	-0.14	$+0.52$	$+0.19$
30×24	$+0.14$	$+0.55$	$+0.16$	-0.87
30×20	-0.66	-0.10	-0.79	$+1.58$
30×15	$+0.70$	-0.84	-0.28	$+0.44$
24×24	$+0.52$	$+0.38$	-0.81	-0.09
24×20	-0.61	-0.30	$+0.51$	$+0.38$
24×15	-0.18	-0.17	$+0.74$	-0.39
20×20	-0.14	-0.48	$+0.43$	$+0.20$
20×15	$+0.12$	$+0.63$	$+0.09$	-0.84
15×15	$+0.67$	$+0.48$	-0.56	-0.59

Table 2.9(a). *Yields.*

Block	Treatment 1	Treatment 2	Total	Difference (1 − 2)
I	5.95	7.10	13.05	−1.15
II	5.30	6.45	11.75	−1.15
III	6.50	6.60	13.10	−0.10
IV	6.35	5.75	12.10	+0.60
Total	24.10	25.90	50.00	−1.80

(b). *Analysis.*

Source	SS	df	MS
Blocks	0.69	3	
Treatments	0.40	1	
Error	1.11	3	$s^2 = 0.37$
Total	2.20	7	

which is the same as the standard error of the difference between the two spacing means from the analysis of variance,

$$(2s^2/4)^{1/2} = (0.74/4)^{1/2} = 0.43.$$

Thus, when only two treatments are considered, the error MS from the analysis of variance is a direct measure of the consistency of the difference between the two treatments, over the different blocks.

2.5   The random allocation of treatment to units

We have discussed the randomised block design in terms of blocking to control the effects of patterns of variation among the experimental units. The other aspect of this design is that, within each block, the treatments are allocated to units by a formal process of randomisation. This process ensures that each treatment is equally likely to be allocated to any unit within the block. This random allocation has two important consequences, the more important of which is discussed in detail in Chapter 9. The other consequence is quite simply that it removes the subjective element of choosing how to allocate treatments.

Subjective allocation is extremely difficult because it inevitably involves simultaneous avoidance of any tendency to include or exclude any particular patterns of allocation. If anyone is asked to allocate six treatments, A, B, C, D, E, F, to six units, 1, 2, 3, 4, 5, 6, in each of a number of blocks, it is commonly found that because of the subjective wish to avoid identical or similar patterns each letter tends to be spread evenly over the positions 1 to 6, which in itself produces a systematic pattern which may be very desirable but is also clearly a pattern and therefore not what is assumed. More simple failures of subjective allocations have occurred when doctors, having to allocate two drugs to pairs of patients, show a tendency to allocate one treatment to those patients more obviously in need.

Application of a random allocation procedure allows all patterns of treatment allocation to occur. For example, in a randomised block design, with four blocks of five units each, it allows the design shown in Figure 2.3. If this design seems undesirable in prospect, then this implies that the blocking system used is insufficient, and suggests that a further blocking restriction be introduced. It is important to distinguish blocking, which controls the extent to which treatments may be allocated to similar units, from randomisation. Randomisation offers no control and allows all possible allocations to occur with equal probability, but it avoids subjective allocation with its inevitable, but unspecified, tendency towards particular patterns.

The random allocation of treatments to units in each block may be

Figure 2.3. Random allocation of five treatments in five randomised blocks.

C	E	A	D	B

D	A	E	C	B

A	E	C	B	D

E	C	A	D	B

A	E	D	C	B

achieved through the use of tables of pseudo-random numbers, either in printed versions to be found in many statistical textbooks and tables, or from a computer printout (or from final digits of telephone numbers in a telephone directory). For a set of ten treatments, each unit in the block is numbered 0 to 9. The first treatment is allocated to the plot corresponding to the first digit in the random number sequence. Suppose the random number sequence is:

3 9 3 8 4 3 1 9 4 3 7 7 1 5 5 4 0 0 4 7 7 0 5 3 2 1.

Then treatment A goes to plot 3, B to 9, C to 8 (not 3 which is already occupied), D to 4, E to 1, F to 7, G to 5, H to 0, I to 2 and, by elimination, J to 6:

0   1   2   3   4   5   6   7   8   9
H   E   I   A   D   G   J   F   C   B.

If there are less than ten treatments, then digits 9, 8, 7 . . . are ignored. If there are more than ten treatments, then two digit numbers can be used, but usually there are many fewer than 100 treatments, and to avoid 'wasting' digits each two-digit number may be replaced by the remainder after dividing by the number of treatments. To maintain equal probabilities for each treatment those two digit numbers greater than or equal to the largest multiple of the number of treatments less than 100 are ignored. To illustrate this, suppose that, in the earlier example, there were 12 treatments. The two digit numbers formed from the sequence are

39   38   43   19   43   77   15   54   00   47   70   53   21.

After division by 12 these yield remainders

3   2   7   7   7   5   3   6   0   11   10   5   9.

The treatment allocation would now be A to unit 3, B to unit 2, C to 7, D to 5, E to 6, F to 0, G to 11, H to 10, with the remaining allocation of I, J, K and L to 1, 4, 8 and 9, determined by the continuation of the sequence.

The method of using the random numbers can be reversed, so that, instead of the digits representing units, they can represent treatments: A=0, B=1, C=2, . . . up to J=9. Now, starting at the beginning of the random number sequence and reverting to the ten treatment case, the treatment corresponding to the first digit is allocated to the first plot, the next to the second plot, and so on:

0   1   2   3   4   5   6   7   8   9
D   J   I   E   B   H   F   A   C   G.

Each block must be randomised separately. For other designs involving blocking, the randomisation procedures will be discussed in Chapter 9. For the completely randomised design, there is no blocking, but simply a set of treatments, A, B, . . . , to be allocated, each to a number of units. This may be achieved by numbering all units and selecting the set of units for each treatment in turn randomly from the complete set. Suppose six treatments are to be allocated to four units each; the units are numbered 1 to 24 and the previous sequence of two-digit random numbers is used. The remainders after division by 24, ignoring 96, 97, 98 and 99, are

$$15 \quad 14 \quad 19 \quad 19 \quad 19 \quad 5 \quad 6 \quad 0 \quad 23 \quad 22 \quad 5 \quad 21,$$

so treatment A is allocated to units 5, 14, 15 and 19, treatment B to units 0, 6, 22 and 23 and treatment C to unit 21, further numbers being required to complete the randomisation.

2.6   Practical choices of blocking patterns

In some situations, the choice of a blocking system for experimental units may be obvious. In others, there may be two or more patterns of likely variation between units, which could usefully be identified with blocks. In yet other situations, there may be no obvious pattern on which to base a blocking system, but there may be advantages in using blocks as 'insurance' against possible patterns not yet identified.

When there are alternative patterns on which to base a blocking system then the experimenter can adopt one of two approaches. Either he may select one of the patterns, usually that suspected of causing more variation, and record for each unit information relevant to the other patterns, hoping to be able to use it later, perhaps, by the method of covariance discussed in Chapter 10. Or, he may attempt to use two or more blocking patterns, classifying each unit by two or more blocking factors and arranging his treatments so that differences between the blocks of either system may be eliminated from treatment comparisons; this is the idea of multiple blocking systems, which are discussed in Chapter 8. An example where the first method might be appropriate is in experiments on animals, where both litter effects and initial size effects could be used as a basis for blocks. Using litters as blocks and adjusting for initial weight by covariance is a standard technique. An example of the second method would occur in industrial or laboratory experiments, where there are a limited number of machines or pieces of equipment which may be expected to show systematic differences, and where observations cannot all be taken at the same time, and time differences must be allowed for.

The most common situation where blocks are used without any dominant pattern to define them is in field crop experiments. Typically, the experimenter is allocated an area of land, apparently fairly homogeneous, and has to choose the size and shape of his plots. Let us suppose that the experiment involves comparison of 12 treatments, and it is decided to have four plots for each treatment. Although the area is apparently uniform, it is a good precaution to decide to use four randomised blocks of 12 plots each, because there is likely to be some variation within the area, and plots close together are, in general, more likely to behave similarly. The experimenter is, however, left with several choices. Blocks could be defined in various ways, some of which are shown in Figure 2.4; plots can be arranged within each block in many patterns, some of which are shown in Figure 2.5. How should the choice be made? Plots within each block should be as similar as possible. Hence, blocks and plots must be chosen to maximise the differences between plots in different blocks, and minimise the differences between plots within blocks.

All other things being equal, and they rarely are, these criteria suggest that blocks should tend to be nearly square in shape, and that within each block the plots should be long and thin, thus sampling the remaining differences in the block equally. But this suggestion must be considered afresh in each situation. Suppose that there is suspected to be a fertility gradient from left to right within each part of Figure 2.5; then it may well be more effective to use Figure 2.5(*b*) rather than (*a*). And long, thin plots may be quite inappropriate if only the centre portion of the plot is to be harvested as representative of plants in a large field, because the discarded edge area will be a very large proportion of the total plot. So ignore the rules, but remember the principles.

Figure 2.4. Alternative divisions of area into blocks.

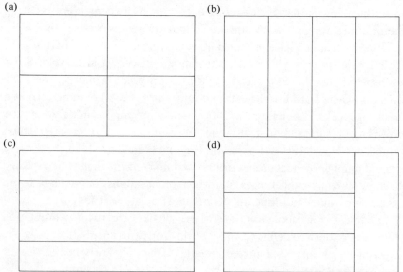

A more general example of the problem of choosing plots and blocks is illustrated in Figure 2.6. This shows a plan of about 200 rubber trees planted along the contours on a steeply sloping hill. Further difficulties are caused by the occurrence of drains, roads and rocks, all of which may

Figure 2.5. Alternative divisions of blocks into plots.

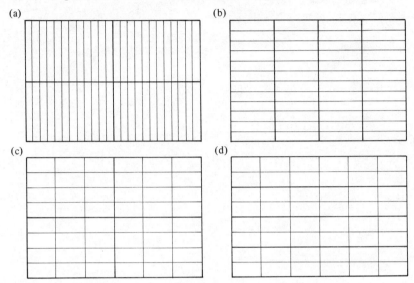

Figure 2.6. Map of 200 rubber trees from Sri Lanka, for a future experiment.

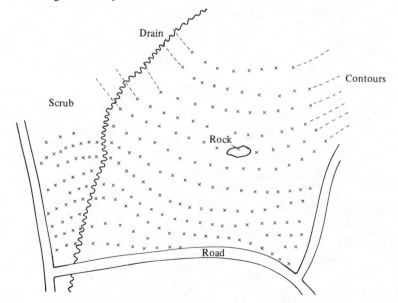

influence production of neighbouring trees. At the beginning of the experiment some trees will have 'dried up' and cannot be used. Others will dry up during the experiment. The minimum plot size is a set of four adjacent trees. How does the experimenter define blocks as groups of plots, each plot containing four trees, so that the plots within blocks will be as homogeneous as possible?

3

Elementary ideas of treatment structure

3.1 Choice of treatments

It might seem that the choice of treatments is not a statistical matter, being a question for the experimenter alone. However, there are very many examples of research programmes where the objectives of the programme either have been, or could have been, achieved much more efficiently by using statistical theory. In some situations, there are several intuitively appealing alternative sets of treatments, and the choice between these alternatives can be assisted by a statistical assessment of the efficiency with which the alternative sets of treatments satisfy the objectives of the experiment. In other situations statistical methods can allow more objectives to be achieved from the same set of experimental resources. In yet other situations statistical arguments can lead to adding experimental treatments which avoid the necessity of making possibly unjustifiable assumptions in the interpretation of results from the experimental data. Classical theory of the design of experiments has been less concerned with the choice of treatments than with the control of variation between units, but there is a large body of knowledge about treatment structure. More recently much of the statistical theory of experimental design has been concerned with the optimal choice of treatments, and while some of this recent theory is very mathematical and has little relevance to real experiments, the general principles are important and will be considered in Chapter 17.

The first point to clarify is that there are many different forms of objective for an experiment. Some experiments are concerned with the determination of optimal operating conditions for some chemical or biological process. Others are intended to make very precise comparisons between two or more methods of controlling or encouraging the growth of a biological organism. Some experiments are intended to provide information about the effects of increasing amounts of some stimulus, on the output of the experimental units to which the stimulus is applied. And other experiments are planned to screen large numbers of possible contestants for a job, for example drugs for treating cancer or athletes for a place in the Olympics. And most experiments have not one but several objectives.

In this chapter we consider some basic aspects of a particular pattern of treatments, namely factorial structure. Factorial structure is the single most important concept of treatment design, and much of the third part of the book is devoted to this concept.

3.2   Factorial structure

Suppose we are investigating the effects on the growth of young chicks of adding small levels of copper to their diet, and wish to assess and compare these effects for two basic diets, maize (M) and wheat (W). To find out whether the effects of the copper additive are similar for both wheat and maize based diets it is necessary to use as experimental treatments the same set of different copper levels, both for wheat based diets and for maize based diets. We might decide to use three levels of copper (0, 50 and 100 units), both with a wheat based diet and with a maize based diet. If the set of six experimental treatments is reduced then it is not possible to assess whether the effects of copper on chick growth are different for the two basic diets. Consider the difficulties using the set of four treatments:

(1) W + 0 units copper

(2) W + 50 units copper

(3) W + 100 units copper

(4) M + 0 units copper.

We can assess the maize–wheat difference from a comparison of treatments (1) and (4); we can find something about the effect of copper from treatments (1), (2) and (3). But there is no information on whether the copper effects will be the same for a maize based diet, and to predict the chick growth for M + 100, for example, we have to make strong assumptions which may well be very wrong.

Other benefits of factorial structure, which are perhaps more surprising and important even than these, are examined in the final section of this chapter. Before looking at those advantages, we examine the standard form of analysis for factorial structure.

First some terminology: a *factor* is a set of treatments which can be applied to experimental units. A *level* of a factor is a particular treatment from the set of treatments which constitute the factor. An *experimental treatment* is the description of the way in which a particular unit is treated, and comprises one level from each factor.

In the chick growth example, there were two factors (copper, basic diet): there were three levels of the copper factor (0, 50, 100) and two levels of the basic diet factor (maize, wheat). The set of experimental treatments is $3 \times 2 = 6$, being $M + 0$, $M + 50$, $M + 100$, $W + 0$, $W + 50$, $W + 100$.

Other examples of factorial structure can be found in all areas of experimentation. In medical research, different drugs for treatment of a disorder could be one factor, different concentrations of the drug could be a second factor, and different frequencies of administration a third factor. In experiments on metal fractures the thickness of material could be one factor, the form of alloy a second factor, and the force of the stress a third factor. In microbiological investigations, different strains of a virus could be one factor, and different media in which the virus is immersed a second factor. In the chemical industry, the number of constituents for a process is large, and variation of each constituent would define a separate factor; for many processes it would be easy to think of a factorial structure with eight or more factors, and this produces different problems of design discussed in Chapter 13. In crop experiments the spacing of the crop, the crop variety, the nutrients added to the soil, or the timing of sowing offer a wide range of possible factors.

3.3   Models for main effects and interactions

Given a set of treatments with factorial structure, such as those for the chick growth experiment, there are several different approaches to interpreting the results. Consider the following results, set out in a form which emphasises the structure:

	copper		
	0	50	100
wheat	20.6	24.1	25.6
maize	21.8	26.7	25.9

(i) We could consider the response to copper for each basal diet in turn. For a wheat diet, the effect of increasing copper appears to be to produce an increased weight for copper increasing both from 0 to 50 and from 50 to 100, but with the second increase smaller than the first. In contrast, for a maize diet, increasing copper produces first an increase and then a decrease in weight. To complete the summary, we might note that weights are, on average, 1.37 greater for maize than for wheat.

(ii) Alternatively, and in this example rather less convincingly, we could consider the maize–wheat yield difference for each copper level; these are 1.2 at zero copper, 2.6 at 50 and 0.3 at 100. In addition, we should consider the average response to copper:

0	50	100
21.2	25.4	25.75

showing an initial increase disappearing almost totally between 50 and 100.

(iii) The third approach, which is that conventionally adopted for factorial structure, is to consider first the average maize–wheat difference, 1.37, then the average response to copper, 21.2 to 25.4 to 25.75, and then the way in which the overall pattern differs from a combination of these two effects. Qualitatively, this last component of interpretation could be expressed by saying either that the maize–wheat difference was largest for the middle copper level, or that the response to increasing copper declines more rapidly for maize than for wheat.

The philosophy of this third approach can be expressed in an algebraic model

$$t_{jk} = d_j + c_k + (dc)_{jk}, \tag{3.1}$$

where

t_{jk} is the treatment effect for basal diet j, and copper level k; d_j is the average treatment effect for basal diet j; c_k is the average treatment effect for copper level k; and $(dc)_{jk}$ is the difference between t_{jk} and $(d_j + c_k)$.

The model (3.1) is apparently very similar to (2.2) for the randomised block design in the previous chapter:

$$y_{ij} = \mu + b_i + t_j + \varepsilon_{ij},$$

but the interpretation of the terms is different, and the two models should not be thought of as being two particular cases of the same model. In particular, whereas ε_{ij} is a characteristic of the experimental units representing variation between the units within each block, and is quite independent of the treatment effects, t_j, the $(dc)_{jk}$ term represents deviations of the treatment effects relative to both d_j and c_k and is only meaningful in the context of the d_j and c_k effects.

Effects which involve comparisons between levels of only one factor are called *main effects* of that factor, and effects which involve comparisons for more than a single factor are called *interactions*. To make these concepts more precise, we define effects as follows.

Main effect of a factor: a comparison between the expected yields for different levels of one factor, averaging over all levels of all other factors. Algebraically, this is written

$$\sum_j l_j t_{j.},$$

where $\sum_j l_j = 0$ and $t_{j.}$ represents the average of t_{jk} over all possible levels of k.

Interaction between two factors: a comparison, of an effect comparison between levels of one factor, between levels of the second factor. Algebraically, and more simply than in words, this is written

$$\sum_k m_k \left(\sum_j l_j t_{jk} \right),$$

where $\sum_k m_k = 0$. Although defined non-symmetrically, the interaction effect is symmetric, as is apparent if the definition is expanded:

$$\sum_k m_k \left(\sum_j l_j t_{jk} \right) = \sum_j \sum_k l_j m_k t_{jk}$$

The effects d_j, c_k and $(dc)_{jk}$ in model (3.1) can now be recognised as main effects and interactions if we define them as follows:

$$d_j = t_{j.} - t_{..}$$
$$c_k = t_{.k} - t_{..}$$
$$(dc)_{jk} = (t_{jk} - t_{j.}) - (t_{.k} - t_{..}) = (t_{jk} - t_{.k}) - (t_{j.} - t_{..})$$

For the numerical values used earlier, the treatment effects t_{jk} are estimated by the deviations of treatment mean yields from the overall average 24.12:

$$-3.52 \qquad -0.02 \qquad +1.48$$
$$-2.32 \qquad +2.58 \qquad +1.78$$

and the estimates of effects are:

$$d_1 = (-3.52 - 0.02 + 1.48)/3 = -0.68$$
$$d_2 = (-2.32 + 2.58 + 1.78)/3 = +0.68$$
$$c_1 = (-3.52 - 2.32)/2 = -2.92$$
$$c_2 = (-0.02 + 2.58)/2 = +1.28$$
$$c_3 = (+1.48 + 1.78)/2 = +1.63$$

$(dc)_{11} = (-3.52) - (-0.68) - (2.92) = +0.08$		$(dc)_{21} = -0.08$
$(dc)_{12} = \quad \cdots \quad = -0.62$		$(dc)_{22} = +0.62$
$(dc)_{13} = \quad \cdots \quad = +0.53$		$(dc)_{23} = -0.53.$

These numerical values confirm the general pattern of conclusions discussed earlier. The average maize–wheat difference is 1.37. The major effect for the copper levels is the lower weight for zero copper; the difference between 50 and 100 is small. And the interaction pattern occurs mainly between the second and third copper levels, with wheat producing a positive additional difference: $(dc)_{13} - (dc)_{12} = +1.21$, to add to the main effect difference $c_3 - c_2 = +0.35$, giving $+1.56$, whereas maize produces a negative difference: $+0.35 + (-1.21) = -0.86$.

3.4   The analysis of variance identity

We return now to the formal representation of the models on which the analysis of data from designed experiments may be based. The model for the randomised block design,

$$y_{ij} = \mu + b_i + t_j + \varepsilon_{ij},$$

and the associated analysis of variance were discussed in Chapter 2. Now suppose that the treatments have a factorial structure so that the treatment effects are classified by two factors, and can be expressed in terms of two main effects and an interaction:

$$t_{jk} = p_j + q_k + (pq)_{jk}.$$

Here, p_j and q_k represent the main effects of factors P and Q and $(pq)_{jk}$ represents the interaction effects. As in the earlier definitions of b_i and t_j, the p_j and q_k are regarded as deviations from the average yield of all treatment combinations for factors P and Q, respectively. In addition, each $(pq)_{jk}$ is regarded as the deviation of the yield of the treatment combination for levels j and k of factors P and Q from the sum $(p_j + q_k)$. Consequently, there are implied relations within each set of effects:

$$\sum_{j=1}^{p} p_j = 0, \ \sum_{k=1}^{q} q_k = 0, \ \sum_{j=1}^{p} (pq)_{jk} = 0, \ \sum_{k=1}^{q} (pq)_{jk} = 0,$$

where p, q are the numbers of levels for factors P and Q.

The definition of a model for the treatment effects for a factorial structure is reflected in a second part of the analysis of variance, dividing the treatment SS into constituent parts. Since two factors are used to classify the treatment effects, the experimental observations are classified by three suffixes giving an overall model:

$$y_{ijk} = \mu + b_i + p_j + q_k + (pq)_{jk} + \varepsilon_{ijk}.$$

The treatment SS becomes

$$\sum_{ijk} (y_{.jk} - y_{...})^2$$

and the analysis of variance identity for this sum of squares is

$$\sum_{ijk} (y_{.jk}-y_{...})^2 \equiv \sum_{ijk} (y_{.j.}-y_{...})^2 + \sum_{ijk} (y_{..k}-y_{...})^2$$

$$+ \sum_{ijk} (y_{.jk}-y_{.j.}-y_{..k}+y_{...})^2.$$

The three terms on the right hand side may be identified as:
(i) the main effect SS for factor P

$$\sum_{ijk} (y_{.j.}-y_{...})^2 = \sum_j Y_{.j.}^2/(bq) - Y_{...}^2/n;$$

(ii) the main effect SS for factor Q

$$\sum_{ijk} (y_{..k}-y_{...})^2 = \sum_k Y_{..k}^2/(bp) - Y_{...}^2/n;$$

(iii) the interaction SS for factors P and Q

$$\sum_{ijk} (y_{.jk}-y_{.j.}-y_{..k}+y_{...})^2 = \sum_{jk} Y_{.jk}^2/b - Y_{...}^2/n - SS(P) - SS(Q).$$

Combined with the earlier analysis of variance identity, this gives a full analysis of variance structure for a randomised block design with factorial treatment structure (see Table 3.1). The similarity of the structure of the analysis of variance identity for treatments to that for the randomised block design is clear. In each, we isolate two constituents, each representing variation over the different classes of one of the two classifications, and then consider the remaining variation as a third component. However, as with the similarity between models (2.2) and (3.1), this analogy should not be pursued too enthusiastically. In terms of designed experiments, block

Table 3.1.

Source	SS	df
Blocks	$\sum_i Y_{i..}^2/(pq) - Y_{2..}/n$	$b-1$
Main effect P	$\sum_j Y_{.j.}^2/(bq) - Y_{...}^2/n$	$p-1$
Main effect Q	$\sum_k Y_{..k}^2/(bp) - Y_{...}^2/n$	$q-1$
Interaction PQ	$\sum_{jk} Y_{.jk}^2/(b) - \sum_j Y_{.j.}^2/(bq) - \sum_k Y_{..k}^2/(bp) + Y_{...}^2/n$	$(p-1)(q-1)$
Error	by subtraction	$(b-1)(pq-1)$
Total	$\sum_{ijk} Y_{ijk}^2 - Y_{...}^2/n$	$n-1$

and treatment effects have quite different interpretations, and the error SS, although arithmetically equivalent to the block × treatment interaction, is more appropriately viewed as representing variation between units within a block.

Nonetheless, the similarity of the two analysis of variance identities does suggest a method of extending the analysis of variance to include the case of three factors in the treatment structure. The second factor Q may be replaced by two factors Q and R. The main effect SS for the earlier factor, Q, may be divided into constituent SS for the main effects of Q and of R and for the QR interaction. Similarly, the original PQ interaction SS can be divided into the interaction SS for PQ and PR and the remainder which can be interpreted as due to the interaction of all three factors. The resulting extended analysis of variance identity for the complete three-factor treatment structure becomes:

$$\sum_{ijk} (y_{.jkl} - y_{...})^2 \equiv \sum_{ijkl} (y_{.j..} - y_{....})^2 + \sum_{ijkl} (y_{..k.} - y_{....})^2$$

$$+ \sum_{ijkl} (y_{...l} - y_{....})^2 + \sum_{ijkl} (y_{.jk.} - y_{.j...} - y_{..k.} + y_{...})^2$$

$$+ \sum_{ijkl} (y_{.j.l} - y_{.j..} - y_{...l} + y_{....})^2$$

$$+ \sum_{ijkl} (y_{..kl} - y_{...l} - y_{..k.} + y_{....})^2$$

$$+ \sum_{ijkl} (y_{.jkl} - y_{.jk.} - y_{..kl} - y_{.j.l}$$

$$+ y_{.j..} + y_{..k.} + y_{...l} - y_{....})^2.$$

The further extension to four or more factors is conceptually simple though even more typographically painful!

Example 3.1

The analysis of variance calculation for a design with factorial treatment structure is illustrated for data from an experiment on the water uptake of amphibia. Frogs and toads were kept in moist or dry conditions prior to the experiment. Half of the animals were injected with a mammalian water balance hormone. There were thus three treatment factors: species (S), pre-experiment moisture condition (M) and hormone (H)–each with two levels. Two animals were observed for each of the eight treatment combinations, but there was no blocking of the 16 animals. The model for the analysis of variance is therefore an extension of that derived above but with no block effect term:

$$y_{jkli} = \mu + s_j + m_k + h_l + (sm)_{jk} + (sh)_{jl} + (mh)_{kl} + (smh)_{jkl} + \varepsilon_{jkli}$$

Table 3.2.

Treatment	Results		Total
Toad wet control	+ 2.31	− 1.59	+ 0.72
Toad dry control	+17.68	+25.23	+42.91
Toad wet hormone	+28.37	+14.16	+42.53
Toad dry hormone	+28.39	+27.94	+56.33
Frog wet control	+0.85	+ 2.90	+ 3.75
Frog dry control	+ 2.47	+17.72	+20.19
Frog wet hormone	+ 3.82	+ 2.86	+ 6.68
Frog dry hormone	+13.71	+ 7.38	+21.09

Table 3.3.

	Wet	Dry	Total	Control	Hormone
Toad	43.25	99.24	142.49	43.63	98.86
Frog	10.43	41.28	51.71	23.94	27.77
Total	53.68	140.52	194.20	67.57	126.63
Control	4.47	63.10	67.57		
Hormone	49.21	77.42	126.63		

The variable measured was the percentage increase in weight after immersion in water for two hours (see Table 3.2). The totals for the eight combinations of treatment levels can be further totalled in a set of two-way tables as in Table 3.3. The method of manual calculation of the sums of squares is shown below:

$$\text{SS (species)} = (142.49^2 + 51.71^2)/8 - 194.20^2/16 = 515.06$$
$$\text{SS (moisture)} = (53.68^2 + 140.52^2)/8 - 194.20^2/16 = 471.33$$
$$\text{SS (SM)} = (43.25^2 + 99.24^2 + 10.43^2 + 41.28^2)/4$$
$$- 194.20^2/16 - 515.06 - 471.33 = 39.50$$
$$\text{SS (SMH)} = (0.72^2 + 42.91^2 + 42.53^2 + 56.33^2 + 3.75^2 + 20.19^2$$
$$+ 6.68^2 + 21.09^2)/2 - 194.20^2/16 - \text{SS(S)} - \text{SS(M)}$$
$$- \text{SS(H)} - \text{SS(SM)} - \text{SS(SH)} - \text{SS(MH)} = 43.53.$$

The analysis of variance is given in Table 3.4. Note that there is no sum of squares for blocks, since the animals were not blocked. The error SS could therefore be calculated alternatively from the within treatment variation:

$$\text{error SS} = (2.31^2 + 1.59^2 - 0.72^2/2) + (17.68^2 + 25.23^2 - 42.91^2/2 +$$
$$\ldots + (13.71^2 + 7.38^2 - 21.09^2/2) = 276.05$$

Table 3.4.

Source	SS	df	MS
Species	515.06	1	
Moisture	471.33	1	
Hormone	218.01	1	
SM	39.50	1	
SH	165.12	1	
MH	57.73	1	
SMH	43.53	1	
Error	276.05	8	$s^2 = 34.5$
Total	1786.33	15	

3.5   Interpretation of main effects and interactions

The analysis of variance includes a subdivision of the treatment variation into various components, and this division can be used to provide a structure for the interpretation of the results of the experiment. The relative sizes of the sums of squares show the relative importance of the different sets of effects. As a further aid to interpretation, each mean square = sum of squares/degrees of freedom can be compared with the error MS, s^2, to assess which effects are important. A formal derivation of the procedure for assessing the significance of the component MS will be developed in Chapter 4. This procedure involves the calculation of the ratio of each mean square to the error MS and the comparison of the ratio with the F distribution for the degrees of freedom of the two mean squares.

The possible patterns of large or small mean squares (and the corresponding F ratios) are many, and rules for the interpretation of sets of effects are vulnerable to counterexamples. However, the following principles will be useful in most situations:

(i) The definition of main effects and interactions implies a natural ordering of the effects for interpretation. Main effects should be examined first, then the two-factor interactions, then three-factor interactions, and so on. This is because interactions essentially represent modifications of the main effects, and have no sensible interpretation without consideration, at the same time, of the corresponding main effects.

(ii) If the two-factor, and higher order interactions, appear negligible, then the results of the experiment should be interpreted in terms of the main effect mean yields only, ignoring the mean yields for the combinations of levels. If the three-factor and higher order interactions appear negligible but the two-factor interactions are not negligible then the mean yields for combinations of levels from pairs of factors should provide the basis for interpretation.

(iii) If a two-factor interaction is clearly important, then the interpretation of the effects of these two factors should normally be based on the mean yields for the combinations of levels for those two factors. If the main effect MS are of the same order of magnitude as the interaction MS, then the main effect yields will add little to an interpretation based on the mean yields for the two-factor combinations of levels. If, on the other hand, one or both of the main effect MS are large compared with the inter-action MS, then the main effect comparison of mean yields will be meaningful, and the interpretation should be in terms of the main effect, or main effects, modified by the interaction effects. In other words mean yields should be presented for the two-way combinations of levels and also for the levels of the factor whose main effect is larger than the interaction.

For the example of the toads and frogs, the mean squares for species and moisture are large (515 and 471), that for the hormone treatment is not so large (218), but still substantially bigger than the error MS (34.5) and the species × hormone interaction (165) is certainly big enough for that two-factor interaction to be investigated. The other three interaction MS are small (39, 58 and 44) and those effects can be ignored. In interpreting the results, the first thing to notice is that the moisture effect appears not to be influenced by other factors, so that for the moisture factor only the main effect mean yields need to be presented. The species main effect MS is quite a bit larger than the species × hormone interaction MS and therefore we should present the species main effect yields as well as the species–hormone combination mean yields. However, the hormone main effect mean yields are unlikely to be helpful in arriving at an interpretation because the mean squares for hormone main effect and species × hormone interaction are similar in size. The relevant mean yields are therefore:

moisture

	wet	dry	difference (D -- W)
	6.72	17.56	10.84

$$\text{SE(difference } (D - W)) = \{2(34.5)/8\}^{1/2} = 2.94.$$

species + hormone

	toads	frogs	
control	10.91	5.98	
hormone	24.72	6.94	difference (T − F)
mean	17.82	6.46	11.36 SE = 2.94

			difference of differences (H − C)
differences (H − C) 13.81	0.96		12.85

$$\text{SE(differences } (H - C)) = \{2(34.5)/4\}^{1/2} = 4.15$$

$$\text{SE(difference of differences)} = \{4(34.5)/4\}^{1/2} = 5.87.$$

The interpretation is clear. Toads show a greater increase in weight after immersion in water than frogs. The effect of the hormone is to raise the weight increase for toads from 11% to 25%, but for frogs the hormone appears to have little effect. Finally, keeping the animals in dry conditions before the experiment leads to an additional 10% increase in weight, as compared with animals kept in moist conditions before the experiment.

This example, with two levels for each factor, is not difficult to interpret. The use of more levels for a factor allows a wider range of possible patterns of results, and hence of interpretations. Many such examples occur later in the book. However, the general principles enunciated in this section will usually provide a sound basis for interpretation.

3.6   Advantages of factorial structure

In introducing the ideas of factorial structure, we have emphasised the advantage of being able to examine interactions between factors. There are two other advantages. First, the conclusions about the effects of a particular factor have a broader validity because of the range of conditions under which that factor has been investigated. Thus, in the example of frogs and toads, the effect of the different pre-experimental conditions has been assessed for both frogs and toads, and with and without a hormone treatment. The estimate of the extra 10% weight increase might be expected to apply in other conditions as well, because of its consistency over levels of two other factors within this experiment.

The other advantage of factorial experiments is still more important, and is essentially that a factorial experiment allows several experiments to be done simultaneously. This advantage is seen most clearly in the situation where there are assumed to be no interactions between factors. Consider an experimenter who wants to investigate the effects of three factors, each at two levels, and who has the resources sufficient for 24 observations. Let the factors be P, Q and R with levels p_0 and p_1, q_0 and q_1, r_0 and r_1. Three different designs of experiment are considered:

(a) Three separate experiments, one for each factor, eight observations per experiment:

   (i) $(p_0q_0r_0, p_1q_0r_0)$ four observations each,

   (ii) $(p_0q_0r_0, p_0q_1r_0)$ four observations each,

   (iii) $(p_0q_0r_0, p_0q_0r_1)$ four observations each.

This is the classical scientific experiment, isolating the effect of each factor in turn by controlling all other factors.

(b) Instead of 'wasting' resources by using $(p_0q_0r_0)$ in each sub-experiment, the four distinct treatments may be replicated equally:

$(p_0q_0r_0)$, $(p_1q_0r_0)$, $(p_0q_1r_0)$, $(p_0q_0r_1)$ with six observations each.

(c) The factorial experiment with eight treatments:

$(p_0q_0r_0)$, $(p_0q_0r_1)$, $(p_0q_1r_0)$, $(p_0q_1r_1)$, $(p_1q_0r_0)$, $(p_1q_0r_1)$, $(p_1q_1r_0)$, $(p_1q_1r_1)$, each having three observations.

To compare the three designs, consider the variance of the comparison of mean yields for p_0 and p_1; obviously the comparison of q_0 and q_1, or of r_0 and r_1 will be equivalent. The three experiments give variances for $(p_1 - p_0)$ as follows:

(a) $2\sigma^2/4 = 0.5\sigma^2$

(b) $2\sigma^2/6 = 0.33\sigma^2$

(c) $2\sigma^2/12 = 0.17\sigma^2$.

The third experiment gives the smallest variance, and is therefore the most efficient design, because in the absence of interactions the difference

$$(p_1q_kr_l) - (p_0q_kr_l)$$

has the same expectation for all (k, l) pairs. Thus, the effective replication of the $(p_1 - p_0)$ comparisons is 12. Looked at from another angle, in comparing p_1 with p_0 in experiment (iii), all 24 observations are used, whereas in (i) and (ii) only 8 and 12 observations, respectively, are used. The factorial experiment is said to have 'hidden replication' for the $p_1 - p_0$ comparison.

If there are interactions between factors, then some of this hidden replication may disappear, and this advantage of factorial experiments is diminished. However, if there are interactions, then the other two forms of design, (i) and (ii), are still inferior to the factorial, because they do not permit recognition that the size of the $(p_1 - p_0)$ effect depends on the particular combination of factor Q and R levels, and hence results from (i) and (ii) may not be reproducible if the levels of Q and R are changed.

Only some of the hidden replication is lost when there are interactions between factors, and it is important to realise that hidden replication is still a benefit when some, but not all, interactions exist. Consider again the experiment for the frogs and toads. Because the moisture factor does not interact with the other factors, the effective replication for the wet–dry comparison is eight – two explicit replications times four hidden replications. The species and hormone interaction requires consideration of the mean yields for the four combinations. Each of these mean yields is based on four replications – two explicit replications times two hidden replications. So, the factorial structure has enabled us to discover the interaction effect of species and hormone, but also gives the benefit of some hidden replication for all the comparisons of means which are of importance.

4

General principles of linear models for the analysis of experimental data

4.1 Introduction and some examples

In this chapter I shall consider the principles on which the classical analysis of experimental data is based. The explanation of these principles necessarily involves mathematical terminology, but in the main part of the chapter this will be reduced as far as possible so that the results will hopefully be appreciated by readers without a sophisticated mathematical background. The formal mathematical derivation of results is included in the appendix to this chapter, the numbering of sections in the appendix matching the section numbering in the main body of the chapter. The discussion of general principles will be illustrated by reference to three examples of data.

Two of the examples have already been seen in the two previous chapters. *Example A* is the randomised block design with ten spacing treatments for rice in four blocks of ten plots each, for which the yield data is given in Section 2.2. The interest in this example is in the relative magnitude of the treatment effects, the block effects and the residual variation, and also in the relationship between yield and spacing.

Example B is the experiment on the water uptake of amphibia, described in Section 3.4, in which eight treatment combinations were replicated twice each. The interest in this example is in the separation of the different components of treatment variation, using the factorial structure of the eight treatments.

Example C is an experiment on tomatoes in which five spray treatments using a chemical growth regulator were compared. Thirty plots were used, arranged in six rows and five columns. The results were as shown in Table 4.1. The analysis must allow for systematic differences between rows and between columns, while being principally concerned with the differences between the five treatments for which the questions are concerned with the

Table 4.1.

A 3.72	B 3.39	C 2.95	D 2.92	E 1.68
C 3.50	D 2.73	E 2.99	A 3.08	B 1.72
D 2.30	A 4.36	B 4.18	E 1.27	C 0.81
E 3.98	C 5.45	D 5.48	B 5.26	A 4.85
B 6.40	E 3.25	A 8.44	C 4.90	D 2.53
B 7.69	C 5.97	E 6.91	D 5.41	A 6.92

The treatment definitions were as follows:
A: early spray 75 p.p.m.
B: early spray 150 p.p.m.
C: late spray 75 p.p.m.
D: late spray 150 p.p.m.
E: control.

general effect of spraying with the growth regulator, and the particular factorial comparisons of the first four treatments.

4.2   The principle of least squares and least squares estimators

A general model for the linear dependence of yields, from a set of experimental units, on a set of parameters which may be expected to cause some of the variation observed in the yields may be expressed as follows:

$$y_i = \sum_{j=1}^{p} a_{ij}\theta_j + \varepsilon_i. \tag{4.1}$$

In this formulation y_i are the yields, θ_j the parameters, a_{ij} represent the structure of dependence of yields on parameters, and ε_i represent the random variation resulting from inherent variation between individual units. The models we have already used for examples A and B can be recognised as fitting within this general structure. For example A the model was

$$y_{ij} = \mu + b_i + t_j + \varepsilon_{ij}. \tag{4.2}$$

An alternative model representing only block averages and a dependence of yield on area per plant would be

$$y_{ij} = b_i + \beta a_j + \varepsilon_{ij}. \tag{4.3}$$

For example B a model for the observations on the eight treatments (t_1 to t_8) could be written

$$y_{jk} = t_j + \varepsilon_{jk}. \tag{4.4}$$

In the general model (4.1) the parameters θ_j may represent block or treatment effects (b_i, t_j) or a mean effect (μ); in each case the a_{ij} term takes values 1 or 0 according to whether the particular effect (j) is relevant or not to yield i, respectively. For quantitative relationships such as that between yield and area per plant the θ_j parameter becomes the regression coefficient, β, in model (4.4) and the a_{ij} become the values of the quantitative variable a_j in model (4.4).

To obtain estimates of the parameters some assumptions are necessary about the properties of the variation among experimental units, represented by the ε_{ij}. We must also define the principle on which the estimation of effects is to be based. For the ε_{ij} the simple assumption is that the observations from different units are independent (apart from the patterns represented by the θ_j) and that the variability of the observations is homogeneous (unaffected by θ_j). Formally these assumptions may be written

$$E(e_{ij}) = 0$$
$$\text{Var}(\varepsilon_{ij}) = \sigma^2$$
$$\text{Cov}(\varepsilon_{ij}, \varepsilon_{i'j'}) = 0 \text{ unless } (i, j) = (i', j').$$

The principle on which the estimation of parameters is based is the *least squares principle*. This requires that the estimates of parameters $(\hat{\theta})$ are chosen so that the deviations of the observed yields from the fitted values $(\hat{y})$, where

$$\hat{y}_i = \sum_{j=1}^{p} a_{ij}\hat{\theta}_j,$$

are minimised, in the sense of minimising the sum of squares of deviations

$$S = \sum_{i=1}^{n} \left(y_i - \sum_{j=1}^{p} a_{ij}\theta_j \right)^2. \tag{4.5}$$

The minimisation of S with respect to any particular parameter, θ_k, involves setting

$$\partial S/\partial \theta_k = 0,$$

which leads to an equation

$$\sum_{i=1}^{n} \left\{ \left(y_i - \sum_{j=1}^{p} a_{ij}\theta_j \right) a_{ik} \right\} = 0,$$

which may be written

$$\sum_{j=1}^{p} \hat{\theta}_j \left(\sum_{i=1}^{n} a_{ik}a_{ij} \right) = \sum_{i=1}^{n} a_{ik} y_i, \tag{4.6}$$

where the 'hat' symbol represents an estimate $(\hat{\theta})$ of the parameter θ.

For a set of p parameters a set of p equations of the form (4.6) is obtained and the parameter estimates, $\hat{\theta}_j$, satisfying these equations are termed *least squares estimates*.

We illustrate these equations for models (4.3) and (4.4). First for (4.3) in which, with parameters representing the four average block values (b_i) and the regression coefficient (β), there are five parameters and 40 observations. The individual model equations are typically

$$y_{11} = b_1 + \beta(900) + \varepsilon_{11}$$
$$y_{12} = b_1 + \beta(720) + \varepsilon_{12}$$
$$\vdots \qquad \vdots$$
$$y_{410} = b_4 + \beta(225) + \varepsilon_{410},$$

and the least squares equations are

$$10\hat{b}_1 + 5011\hat{\beta} = 79.15$$
$$10\hat{b}_2 + 5011\hat{\beta} = 68.55$$
$$10\hat{b}_3 + 5011\hat{\beta} = 73.95$$
$$10\hat{b}_4 + 5011\hat{\beta} = 75.75$$
$$5011(\hat{b}_1 + \hat{b}_2 + \hat{b}_3 + \hat{b}_4) + 2883301\hat{\beta} = 144847.9,$$

which may be solved to give

$$\hat{\beta} = -0.0034, \ \hat{b}_1 = 9.62, \ \hat{b}_2 = 8.56, \ \hat{b}_3 = 9.10, \ \hat{b}_4 = 9.29.$$

For model (4.4) there are eight parameters and 16 observations. The individual equations are

$$y_{11} = t_1 + \varepsilon_{11}$$
$$y_{12} = t_1 + \varepsilon_{12}$$
$$\vdots \qquad \vdots$$
$$y_{82} = t_8 + \varepsilon_{82}$$

and the least squares equations are

$$2\hat{t}_1 = 0.72, \ 2\hat{t}_2 = 42.91, \ 2\hat{t}_3 = 42.53, \ 2\hat{t}_4 = 56.33,$$
$$2\hat{t}_5 = 3.75, \ 2\hat{t}_6 = 20.19, \ 2\hat{t}_7 = 6.68, \ 2\hat{t}_8 = 21.09,$$

from which the solutions are obtained immediately.

For most of the models considered in this book the least squares equations have a particularly simple form. Whenever the coefficients a_{ij} for a parameter θ_j are all 0 or 1, as in the case for b_i in (4.3) or t_j in (4.4) the corresponding least squares equation has on the right hand side the total of observations for which θ_j is part of the yield expression. The left hand side is the expected value of the total on the right hand side, written in terms of parameter estimates $\hat{\theta}$, instead of parameters θ.

4.3   Properties of least squares estimators

In this section we consider the properties of least squares estimators, and to do so we assume that the set of least squares equations (4.6) can be solved uniquely. In fact, for most of the experimental design models considered in this book, the equations do not have unique solutions and we consider how to obtain solutions in these cases in the next section. The properties of the modified least squares estimators of the next section are simply related to those of the simple least squares estimators for the case where (4.6) does yield unique solutions, and it is convenient to derive these properties for the simple case.

The most important properties for any estimator are the expected value of the estimator and the variance of the estimator. Since the variance of the estimator will inevitably be related to the variance of the original observations, σ^2, it is necessary to be able to obtain an estimate of σ^2 in order to make practical use of the information about the variance of estimators.

To demonstrate the argument required to derive the properties of least squares estimators it is helpful to re-express the least squares equations in terms of matrices. The set of design coefficients (a_{ij}) are defined for n observations, referenced by suffix i, and p parameters, referenced by suffix j. The set can be considered in the form of a two-way array, or design matrix:

$$\mathbf{A} = \begin{bmatrix} a_{11} & a_{12} & \cdots & a_{1p} \\ a_{21} & a_{22} & \cdots & a_{2p} \\ \vdots & \vdots & \vdots & \vdots \\ a_{n1} & a_{n2} & \cdots & a_{np} \end{bmatrix}.$$

It is convenient to define

$$c_{jk} = \sum_{i=1}^{n} a_{ij} a_{ik}.$$

Essentially c_{jk} is the sum of products of pairs of values in columns j and k for each row of the matrix $\mathbf{A}$. The set $\{c_{jk}\}$ may also be written as a matrix:

$$\mathbf{C} = \begin{bmatrix} c_{11} & c_{12} & \cdots & c_{1p} \\ c_{21} & c_{22} & \cdots & c_{2p} \\ \vdots & \vdots & \vdots & \vdots \\ c_{p1} & c_{p2} & \cdots & c_{pp} \end{bmatrix}.$$

Notice that the definition of c_{jk} means that $c_{kj} = c_{jk}$, and the matrix $\mathbf{C}$ is said to be symmetric. Equation (4.6) may be rewritten in terms of the c_{jk} in the form

$$\sum_{j=1}^{p} c_{jk} \theta_j = \sum_{i=1}^{n} a_{ik} y_i.$$

There is an equation of this form for each of the p values of k, and the set of such equations,

$$\sum_{j=1}^{p} c_{j1}\hat\theta_j = \sum_{i=1}^{n} a_{i1} y_i$$
$$\vdots \qquad\qquad \vdots$$
$$\sum_{j=1}^{p} c_{jp}\hat\theta_j = \sum_{i=1}^{n} a_{ip} y_i,$$

(4.7)

may be written in matrix form

$$\mathbf{C}\hat\theta = \mathbf{A}'\mathbf{y}.$$

(4.8)

If the matrices $\mathbf{C}$, θ, $\mathbf{A}$, $\mathbf{y}$ are written as arrays

$$\begin{bmatrix} c_{11} & c_{12} & \cdots & c_{1p} \\ c_{21} & c_{22} & \cdots & c_{2p} \\ \vdots & \vdots & & \vdots \\ c_{p1} & c_{p2} & \cdots & c_{pp} \end{bmatrix} \begin{bmatrix} \hat\theta_1 \\ \hat\theta_2 \\ \vdots \\ \hat\theta_p \end{bmatrix} \begin{bmatrix} a_{11} & a_{12} & \cdots & a_{1p} \\ a_{21} & a_{22} & \cdots & a_{2p} \\ \vdots & \vdots & & \vdots \\ a_{n1} & a_{n2} & \cdots & a_{np} \end{bmatrix} \begin{bmatrix} y_1 \\ y_2 \\ \vdots \\ y_n \end{bmatrix}$$

then each sum of terms in (4.7) is a sum, over rows, of products of terms in $\hat\theta$ and a column of $\mathbf{C}$, or in $\mathbf{y}$ and a column of $\mathbf{A}$.

Now associated with the matrix $\mathbf{C}$, for the situation when (4.8) has a unique solution there is a matrix $\mathbf{C}^{-1}$, the inverse matrix of $\mathbf{C}$, such that the solutions to (4.8) are given by

$$\hat\theta = \mathbf{C}^{-1}(\mathbf{A}'\mathbf{y}),$$

(4.9)

where the multiplication of the matrix $\mathbf{C}^{-1}$ by $(\mathbf{A}'\mathbf{y})$ is again a set of products obtained by multiplying a column of $\mathbf{C}^{-1}$ by the column of values represented by $\mathbf{A}'\mathbf{y}$.

The crucial characteristic about least squares estimators which may be seen directly from (4.9) or may be inferred from (4.6) is that least squares estimators are linear combinations of the original observations, y_i. From this general characteristic and the particular form of (4.9) the following results may be deduced (proofs in appendix to this chapter).

(1) The expected value of a least squares estimator of a parameter in a general linear model is the true value of the parameter; that is, least squares estimators are unbiased.

(2) The variance of a least squares estimator of a parameter is a simple multiple of σ^2, the multiplying factor being the corresponding diagonal element of the matrix $\mathbf{C}^{-1}$.

(3) The covariance of least squares estimators for a pair of parameters is also a simple multiple of σ^2, the multiplying factor being a nondiagonal element of $\mathbf{C}^{-1}$.

(4) For any linear comparison or combination of parameters, the corresponding contrast of the least squares estimators of the

parameters provides an unbiassed estimate, and this estimate is the most precise estimate that can be obtained using an unbiassed linear combination of observations.

The implications of these results are that if we are considering linear models, that is models in which the parameters appear additively, and intend to use linear combinations of the observations to estimate parameters or combinations of parameters, then least squares estimates are the best available.

The other important question is how to estimate σ^2? The least squares principle requires that the parameter estimates minimise the sum of squared deviations (4.5). If we consider the resulting minimised SS

$$S_r = \sum_{i=1}^{n} \left(y_i - \sum_{j=1}^{p} a_{ij}\theta_j \right)^2$$

then it can be shown that the expected value of S_r is $(n-p)\sigma^2$ (see the appendix).

The calculation of S_r is straightforward following an expansion of the squared terms of the summation as follows:

$$
\begin{aligned}
S_r &= \sum_{i=1}^{n} \left\{ y_i^2 - \sum_{k=1}^{p} \hat{\theta}_k a_{ik} \left(2y_i - \sum_{j=1}^{p} a_{ij}\theta_j \right) \right\} \\
&= \sum_{i=1}^{n} y_i^2 - \sum_{k=1}^{p} \hat{\theta}_k \left(2 \sum_{i=1}^{n} a_{ik} y_i - \sum_{j=1}^{p} \hat{\theta}_j \sum_{i=1}^{n} a_{ij}a_{ik} \right) \\
&= \sum_{i=1}^{n} y_i^2 - \sum_{k=1}^{p} \hat{\theta}_k \left(\sum_{i=1}^{n} a_{ik} y_i \right).
\end{aligned}
$$

The first term is the sum of squares of the original observations. The second is the sum of products of each $\hat{\theta}_k$ estimate and the corresponding right hand side total in the least squares equations (4.6).

In practice therefore we set up and solve the least squares equations and calculate the residual sum of squares S_r by subtracting from the total sum of squares the fitting sum of squares, which is calculated as the sum of products of each least squares estimate times the corresponding total.

For the model (4.3) for example A, the total sum of squares is

$$\text{total SS} = 5.95^2 + \cdots + 7.75^2 = 2252.61$$

$$\text{fitting SS} = (9.62)(79.15) + (8.56)(68.55) + (9.10)(73.95) + (9.29)(75.85$$

$$+ (-0.0034)(144847.9) = 2233.85.$$

Hence the residual SS is

$$S_r = 2252.61 - 2233.85 = 18.76,$$

and the estimate of σ^2 is

18.76/(40   5) = 0.536

For the model (4.4) for example B the sums of squares are

$$\text{total SS} = 2.31^2 + \cdots + 7.38^2 = 17821.49$$
$$\text{fitting SS} = (0.36)(0.72) + (21.46)(42.91) + \cdots$$
$$+ (10.54)(21.09) = 12545.44.$$

Hence the residual SS is

$$S_r = 17821.49 - 17545.44 = 276.05,$$

and the estimate of σ^2 is 34.5, as in Section 3.4.

4.4   Overparameterisation, constraints and practical solution of least squares equations

The natural form in which most models for data from designed experiments are expressed includes more parameters than can be estimated from the information available. The cause of the overparameterisation is almost always the desire to retain symmetry for sets of effects. The most obvious example is the randomised block design model such as (4.2) where we include a complete set of block effects (b_i) and a complete set of treatment effects (t_j) in addition to the overall mean effect (μ). When this model was first developed in Chapter 2 the b_i represented deviations of each block from the experiment average, and the t_j represented the deviation for a specific treatment from the average of all treatments. It was recognised that this interpretation implied two restrictions

$$\sum_i b_i = 0 \quad \text{and} \quad \sum_j t_j = 0.$$

If we consider the predicted values from the fitted model

$$\hat{y}_{ij} = \hat{\mu} + \hat{b}_i + \hat{t}_j$$

then it is clear that the model contains superfluous parameters. Suppose that we have a set of parameter estimates satisfying the least squares equations. Then clearly we would get the same fitted values, and therefore the same residual SS, if the value for $\hat{\mu}$ was increased by 1 and the values for $\hat{b}_i$ were each reduced by 1. Alternatively, we could decrease $\hat{\mu}$ by 2 and increase each $\hat{t}_j$ by 2. We cannot simply identify μ as superfluous and consider the model

$$y_{ij} = b_i + t_j + \varepsilon_{ij}, \tag{4.10}$$

since if each of the fitted $\hat{b}_i$ values is increased by 5 and each $\hat{t}_j$ decreased by 5 then again the fitted $\hat{y}$ values will be unchanged.

For the model (4.10) we could think of the b_i as representing average block yields and the t_j as deviations from the average of all treatments, implying $\sum_j t_j = 0$. Or, by the symmetry of the data structure, we could think of the t_j as representing average treatment yields and the b_i as deviations from the average of all blocks, implying $\sum_i b_i = 0$.

I believe that, for a practical approach to models for experimental data, it is sensible to accept that models will be overparameterised and that we should obtain least squares estimates by recognising the inevitability of overparameterisation and using methods which allow us to deal with the resulting difficulty simply and directly. The theory of a general method of modified least squares estimation for overparameterised models is developed in the appendix to this chapter. We shall concentrate here on a practical approach.

Usually consideration of the practical interpretation of the terms in the model will identify sensible restrictions, or constraints, on the parameter values, and use of these constraints with the least squares equations will lead directly to parameter estimates. Consider the set of least squares equations for the model (4.2) for the randomised block rice spacing example

$$40\hat{\mu} + 10 \sum_i \hat{b}_i + 4 \sum_j \hat{t}_j = \sum_i \sum_j y_{ij} = 297.40$$

$$10\hat{\mu} + 10\hat{b}_1 + \sum_j \hat{t}_j = \sum_j y_{1j} = 79.15$$

$$\vdots \qquad\qquad \vdots \quad \vdots$$

$$10\hat{\mu} + 10\hat{b}_4 + \sum_j \hat{t}_j = \sum_j y_{4j} = 75.75$$

$$4\hat{\mu} + \sum_i \hat{b}_i + 4\hat{t}_1 = \sum_i y_{i1} = 24.10$$

$$\vdots \qquad\qquad \vdots \quad \vdots$$

$$4\hat{\mu} + \sum_i \hat{b}_i + 4\hat{t}_{10} = \sum_i y_{i10} = 32.80.$$

If the restrictions $\sum_i b_i = 0$ and $\sum_j t_j = 0$ are imposed then obviously equivalent restrictions will apply to the parameter estimates, $\sum_i \hat{b}_i = 0$ and $\sum_j \hat{t}_j = 0$, and all summations on the left hand side of the equations disappear leaving

$$40\hat{\mu} \qquad\quad = 297.40$$

$$10\hat{\mu} + 10\hat{b}_1 = 79.15$$

$$\vdots$$

$$4\hat{\mu} + 4\hat{t}_{10} = 32.80$$

from which least squares estimates can be obtained

$$\hat{\mu} = 7.435$$
$$\hat{b}_1 = 0.480$$
$$\vdots$$
$$\hat{t}_{10} = 0.765.$$

We now consider briefly the effect of alternative constraints. We have already mentioned the possibility of setting $\mu = 0$ and $\sum_j t_j = 0$. The least squares estimates are then

$$\hat{b}_1 = 7.915$$
$$\vdots$$
$$\hat{b}_4 = 7.575$$

and the $\hat{t}_j$ remain unaltered.

If we consider all block effects relative to block 1 and all treatment effects relative to treatment 1 then we could use the constraints $b_1 = 0$, $t_1 = 0$. The solution of the least squares equations is more complex but can be achieved quite quickly to give

$$\hat{\mu} = 6.51 \quad \hat{b}_2 = -1.06 \quad \hat{t}_2 = 0.45$$
$$\hat{b}_3 = -0.52 \quad \hat{t}_3 = 1.16$$
$$\hat{b}_4 = -0.34 \quad \vdots$$
$$\hat{t}_{10} = 2.18.$$

The fitted values are

$$\hat{y}_{11} = 6.51 \quad \hat{y}_{21} = 5.45 \quad \hat{y}_{31} = 5.99 \quad \hat{y}_{41} = 6.17$$
$$\hat{y}_{21} = 6.96 \quad \hat{y}_{22} = 5.90 \quad \hat{y}_{32} = 6.44 \quad \hat{y}_{42} = 6.62$$
$$\vdots \qquad\qquad \vdots \qquad\qquad \vdots$$
$$\hat{y}_{110} = 8.69 \quad \hat{y}_{210} = 7.63 \quad \hat{y}_{310} = 8.17 \quad \hat{y}_{410} = 8.35,$$

exactly the same as for any other set of constraints.

Because the fitted values are unchanged by the use of alternative constraint systems it follows that the residual SS is invariant over constraint systems. In particular if constraints were all of the form $\theta_j = 0$ for sufficient different j we can recognise that the number of parameters would be reduced to the number that can actually be estimated from the data. If the model is rewritten to exclude the parameters which are set to zero by the constraints, then the effective number of parameters, p', may be identified. From the results of the previous section the expected value of S_r is known to be $(n - p')\sigma^2$. This allows us to estimate σ^2, when using constraints of any form, p' being the number of parameters minus the number of constraints.

For the usual randomised block model (4.2) using constraints $\sum_i b_i = 0$, $\sum_j t_j = 0$ the least squares estimates have been obtained earlier in this section. The fitted SS is calculated as

$$\text{fitted SS} = \hat{\mu}\left(\sum_{ij} y_{ij}\right) + \hat{b}_1\left(\sum_j y_{1j}\right) + \cdots + \hat{t}_{10}\left(\sum_i y_{i10}\right)$$

$$= (7.435)(297.40) + (0.480)(79.15) + \cdots + (0.765)(32.80)$$

$$= 2240.19.$$

The residual SS is $2252.61 - 2240.19 = 12.42$. This residual SS is the same as would have been obtained if b_1 and t_1 had been set to zero, in which case the number of parameters in the model would have been

$$p' = 1 + 3 + 9 = 13.$$

Therefore the estimate of σ^2 is

$$S_r/(n - p') = 12.42/27 = 0.46.$$

Variances for parameters when constraints have been used can be obtained quite easily, but some care is necessary in interpreting variance information about parameters subject to constraints. If the model is

$$y_{ij} = \mu + b_i + t_j + \varepsilon_{ij}$$

and constraints on the b_i and on the t_j are used then because there is an arbitrariness in the choice of constraints there is a corresponding arbitrariness about individual parameter values. Thus, as argued previously, the value for $\hat{b}_1$ is dependent on the values for $\hat{\mu}$ and $\hat{t}_j$, and therefore discussion about the absolute variance of $\hat{b}_1$ is meaningless. However, differences between the parameters within the set of block parameters or between treatment parameters are not affected by the choice of constraint.

To derive the variance of any comparison of parameters the estimate of the comparison must be expressed as a linear combination of the observations, y_{ij}. The variance of a linear combination of y values can then be evaluated directly. Thus for the rice spacing data the difference $(t_1 - t_2)$ is estimated by

$$\left(\sum_i y_{i1} - \sum_i y_{i2}\right)\bigg/ 4,$$

and the variance is obviously $2\sigma^2/4$.

For a final illustration of the use of constraints and the calculation of least squares estimates we consider example C. The model must allow for row effects, column effects and treatment effects,

$$y_{ij} = \mu + r_i + c_j + y_{k(ij)} + \varepsilon_{ij},$$

where the actual treatment k occurring for the combination of row i and column j is determined by the design

$$30\hat{\mu}+5\sum_i \hat{r}_i+6\sum_j \hat{c}_j+6\sum_k \hat{t}_k \qquad =\sum_{ij} y_{ij}=125.04$$

$$5\hat{\mu}+5\hat{r}_1+\sum_j \hat{c}_j+\sum_k \hat{t}_k \qquad =\sum_j y_{1j}=14.66$$

$$\vdots \qquad\qquad \vdots \qquad \vdots$$

$$5\hat{\mu}+5\hat{r}_6+\sum_j \hat{c}_j+\sum_k \hat{t}_k \qquad =\sum_j y_{6j}=32.90$$

$$6\hat{\mu}+\sum_i \hat{r}_i+6\hat{c}_1+\sum_k \hat{t}_k+\hat{t}_B \qquad =\sum_i y_{i1}=27.59$$

$$\vdots \qquad\qquad \vdots \qquad \vdots$$

$$6\hat{\mu}+\sum_i \hat{r}_i+6\hat{c}_5+\sum_k \hat{t}_k+\hat{t}_A \qquad =\sum_i y_{i5}=18.51$$

$$6\hat{\mu}+\sum_i \hat{r}_i+\sum_j \hat{c}_j+\hat{c}_5+6\hat{t}_A \qquad =\sum_A y_{ij}=31.37$$

$$\vdots \qquad\qquad \vdots \qquad \vdots$$

$$6\hat{\mu}+\sum_i \hat{r}_i+\sum_j \hat{c}_j+\hat{c}_3+6\hat{t}_E \qquad =\sum_E y_{ij}=20.08.$$

The natural constraints are $\sum_i r_i=0$, $\sum_j c_j=0$, $\sum_k t_k=0$ and applying these gives simplified equations:

$$30\hat{\mu} \qquad\qquad = 125.04$$

$$5\hat{\mu}+5\hat{r}_1 \qquad = 14.66$$

$$\vdots \qquad \vdots$$

$$5\hat{\mu}+5\hat{r}_6 \qquad = 32.90$$

$$6\hat{\mu}+6\hat{c}_1+\hat{t}_B \;\; = 27.59$$

$$\vdots \qquad \vdots$$

$$6\hat{\mu}+6\hat{c}_5+\hat{t}_A \;\; = 18.51$$

$$6\hat{\mu}+\hat{c}_5+6\hat{t}_A \;\; = 31.37$$

$$\vdots \qquad \vdots$$

$$6\hat{\mu}+\hat{c}_3+6\hat{t}_E \;\; = 20.08.$$

The estimates for $\hat{\mu}$ and $\hat{r}_i$ are simple. Those for $\hat{c}_j$ and $\hat{t}_k$ occur in pairs of equations, of which one pair, by chance, is the two equations for $\hat{c}_5$ and $\hat{t}_A$ shown adjacently above. After calculating $\hat{\mu}=4.168$ and eliminating $\hat{\mu}$ from the $\hat{c}_5$ and $\hat{t}_A$ equations we have

$$6\hat{c}_5+\hat{t}_A = -6.50$$

$$\hat{c}_5+6\hat{t}_A = +6.36$$

and hence

$$\hat{c}_5 = -1.296 \quad \hat{t}_A = +1.276.$$

Similarly

$$\hat{c}_1 = +0.339 \quad \hat{t}_B = +0.549$$
$$\hat{c}_2 = +0.065 \quad \hat{t}_C = -0.249$$
$$\hat{c}_4 = -0.268 \quad \hat{t}_D = -0.562$$
$$\hat{c}_3 = +1.159 \quad \hat{t}_E = -1.015.$$

For completeness the other estimated effects are

$$\hat{\mu} = 4.168$$
$$\hat{r}_1 = -1.234, \hat{r}_2 = -1.362, \hat{r}_3 = -1.585, \hat{r}_4 = +0.838, \hat{r}_5 = +0.938,$$
$$\hat{r}_6 = +2.414.$$

To determine the variance of a difference between two treatments, for example t_A and t_E, we have to express $\hat{t}_A - \hat{t}_E$ as a linear combination of the y values. To do this we note that if we consider the simple differences $c_5 - c_3$ and $t_A - t_E$ then the equations can be written

$$6(\hat{c}_5 - \hat{c}_3) + (\hat{t}_A - \hat{t}_E) = 18.51 - 30.95 = C_5 - C_3$$
$$(\hat{c}_5 - \hat{c}_3) + 6(\hat{t}_A - \hat{t}_E) = 31.37 - 20.09 = T_A - T_E,$$

where capital letters are used to denote yield totals. We can now solve for $(\hat{t}_A - \hat{t}_E)$:

$$35(\hat{t}_A - \hat{t}_E) = 6(T_A - T_E) - (C_5 - C_3)$$
$$= 6y_{11} + y_{13} - 7y_{15} - 5y_{23} + 6y_{24} - y_{25}$$
$$+ 6y_{32} + y_{33} - 6y_{34} - y_{35} - 6y_{41} + y_{43}$$
$$+ 5y_{45} - 6y_{52} + 7y_{53} - y_{55} - 5y_{63} + 5y_{65}.$$

A simple diagrammatic method for writing out such linear combinations with minimum error is described in Section 7.1. The variance of $(\hat{t}_A - \hat{t}_E)$ can now be calculated as

$$\text{Var}(\hat{t}_A - \hat{t}_E) = \sigma^2 \{6(36) + 2(49) + 4(25) + 6(1)\}/(35)^2$$
$$= 12\sigma^2/35.$$

This is, interestingly, very close to $2\sigma^2/6$ which would be the variance for a simple design with six blocks of five treatments. We consider designs of this pattern further in Chapters 7 and 8.

Finally for this example note that the residual SS can be calculated by first finding the total SS and fitting SS:

$$\text{total SS} = 3.72^2 + 3.39^2 + \cdots + 6.92^2 = 630.12,$$
$$\text{fitting SS} = (4.168)(125.04) + (-1.234)(14.66) + \cdots$$
$$+ (-1.015)(20.08)$$
$$= 622.00.$$

Hence, residual SS $= 630.12 - 622.07 = 8.05$

$$s^2 = S_r/(35 - 14) = 8.05/21 = 0.39.$$

4.5   Subdividing the parameters and the extra sum of squares

For many models the set of parameters consists of several subsets, and the interest in the parameters may be restricted to only some of the subsets. More generally we may think of the model as being built in stages by

(i) starting with parameters which are consequences of the initial structure of the experimental units,

(ii) adding parameters to represent the treatment effects,

(iii) trying, parsimoniously, to reduce the set of parameters required to represent the effects of the treatments, and possibly

(iv) adding further parameters to represent modification of the treatment effects by the environment.

For example in a randomised block design there are three subsets of parameters, μ, $\{b_i\}$ and $\{t_j\}$. The model must include μ and $\{b_i\}$ because they are implied by the structure of the units in the design. However, the real interest is in the $\{t_j\}$ and consequently in their estimates, standard errors and the sum of squares for fitting the treatment effects. In the rice spacing example we have used two models representing the effects of treatments, first the full model (4.2)

$$y_{ij} = \mu + b_i + t_j + \varepsilon_{ij},$$

and secondly a model where the treatment effects are represented by a linear dependence on area per plant, model (4.3)

$$y_{ij} = \mu + b_i + \beta a_j + \varepsilon_{ij}.$$

The latter is a special case of the former assuming a special pattern among the t_j (and using one parameter instead of ten).

Factorial structure also provides a situation in which different subsets of treatments may be identified as being of different *a priori* importance. Main effects will almost certainly be important, two-factor interactions may be, but three-factor interactions are unlikely to be important, though we may wish to check this last presumption.

The fundamental results which allow us to analyse these situations are derived by considering a simple split of the vector of parameters θ into two subsets θ_1 and θ_2. There is a corresponding split of the design matrix A into those columns related to θ_1, denoted by A_1, and the remainder related to θ_2, denoted by A_2. There is a further, corresponding, split of the matrices C and C^{-1} so that C_{11} is the submatrix obtained from pairs of columns both in A_1, C_{22} from pairs of columns both in A_2, and C_{12} from pairs of columns, one from A_1 and one from A_2.

We assume that the interest is in the second set of parameters θ_2, while θ_1 represents parameters inevitably included in the model but not of direct

Table 4.2.

Source	SS	df
Fitting θ_1 ignoring θ_2	S_{p-q}	$p-q$
Fitting θ_2 allowing for θ_1	S_q	q
Residual	S_r	$n-p$

interest. Explicit expressions may be obtained for estimates of θ_2, allowing for θ_1, without actually calculating $\hat{\theta}_1$, and these are derived in Appendix 4.A5. It is also important to consider the sum of squares, which is the difference between the sum of squares for fitting the full model with θ_1 and θ_2 and the sum of squares for fitting the reduced model including only θ_1. If the complete set θ includes p parameters, and the subset of interest θ_2 includes q parameters, then we can identify an analysis of variance structure with three components as shown in Table 4.2. We have noted that it can be shown that the expected value of S_r is $(n-p)\sigma^2$. It can also be shown that if all the parameters in the set θ_2 are zero then the expected value of S_q is $q\sigma^2$, making no assumptions about the values of the parameters in the set θ_1. In other words we can use S_q as a basis for assessing the evidence about the parameters θ_2, while allowing for the effects of the parameters θ_1. The sum of squares, S_q, is termed the 'extra sum of squares' attributable to θ_2.

There is one further important concept and this is whether S_{p-q} may be used to assess the importance of the parameters θ_1. Clearly the roles of θ_1 and θ_2 could be reversed and the model fitted first with θ_2 only so that the fitting SS for fitting θ_1, allowing for θ_2, may be calculated. This 'extra' SS for θ_1, allowing for θ_2, provides an assessment of θ_1 and may sometimes be the same as the SS for θ_1 ignoring θ_2. It can be shown (Appendix 4.A6) that, for the SS for fitting θ_1 ignoring θ_2 and the SS for fitting θ_1 allowing for θ_2 to be identical, it is necessary that the design matrix columns for θ_1 shall be orthogonal to those for θ_2. This orthogonality may be expressed in terms of the columns of the design matrix A in the following form: the sum (over rows) of products of pairs of values in columns j (from A_1) and k (from A_2) for each row must be zero for each possible pairing (j, k). Any constraints on parameters must be allowed for by rewriting the parameters when applying this concept of orthogonality.

We can interpret orthogonality practically as requiring that the estimates of comparisons between the parameters θ_2 are unaffected by the need to allow for the parameters θ_1. In this practical sense comparisons between treatments in randomised complete block design are clearly unaffected by differences between blocks because all block–treatment combinations occur exactly once, and hence treatments are orthogonal to blocks. However, in example C the occurrence of treatment A twice in column 5 and treatment E twice in column 3 means that the apparent $(A - E)$ difference could be partially caused by differences between columns 5 and 3. Therefore in example C treatments are not orthogonal to columns.

The subdivision of parameters into two sets θ_1 and θ_2 generalises to more than two sets, and analyses of variance can be constructed with components:

(i) fitting θ_1, ignoring $\theta_2, \ldots, \theta_k$;

(ii) fitting θ_2, allowing for θ_1, ignoring $\theta_3, \ldots, \theta_k$;

(iii) fitting θ_3, allowing for θ_1 and θ_2, ignoring $\theta_4, \ldots, \theta_k$;

$$\vdots$$

(k) fitting θ_k, allowing for $\theta_1, \theta_2, \ldots, \theta_{k-1}$.

Unless all sets of parameters are mutually orthogonal, the SS for fitting a set of parameters will depend on the order of fitting which defines the other parameter sets allowed for or ignored, and the interpretation of each SS is specific to this context.

Consider again example A. For the model (4.2)

$$y_{ij} = \mu + b_i + t_j + \varepsilon_{ij}$$

we can define three sets of parameters $\theta_1 = \mu$, $\theta_2 = \{b_i\}$ and $\theta_3 = \{t_j\}$. Fitting $\theta_1 = \mu$ is only sensible as the first step since the interpretation of $\{b_i\}$ and $\{t_j\}$ as deviations from an overall average yield would not be valid if μ is not included in the model. We might consider therefore two orders of fitting terms for the analysis of variance, $\theta_1 \rightarrow \theta_2 \rightarrow \theta_3$ or $\theta_1 \rightarrow \theta_3 \rightarrow \theta_2$. In fact, block effects and treatment effects are orthogonal so the order of fitting is irrelevant and the resulting analysis of variance shown in Table 4.3, is essentially that calculated in Chapter 2. It is clear from the practical form of calculation of block and treatment SS in Chapter 2 that the SS are unaffected by the order of fitting. We could

Table 4.3.

Source	SS	df
μ	2211.17	1
b_i	5.88	3
t_j	23.14	9
Residual	12.42	27
Total	2252.61	40

reinforce this by noting that the block effects are
$$\hat{b}_1 = +0.480, \ \hat{b}_2 = -0.580, \ \hat{b}_3 = -0.040, \ \hat{b}_4 = +0.140,$$
as calculated in the previous section whether or not the $\{t_i\}$ are included in the model.

Now consider example C with the model
$$y_{ij} = \mu + r_i + c_j + t_{k(ij)} + \varepsilon_{ij}$$
for which we can identify four sets of parameters $\theta_1 = \mu, \theta_2 = \{r_i\}, \theta_3 = \{c_j\},$ $\theta_4 = \{t_k\}$. Sets θ_2 and θ_3 are orthogonal since each row 'occurs' in each column and vice versa, and also θ_2 and θ_4 are orthogonal since each treatment occurs in each row. However, the pattern for θ_3 and θ_4 is not orthogonal, since the replication pattern of the five treatments varies between columns. The implications of this non-orthogonality can be identified by considering the situation where there are no difference effects between the columns. If the treatments give different values then there will appear to be differences between the columns reflecting the treatments which appear in the columns. Thus suppose we assume for the model
$$y_{ij} = \mu + r_i + c_j + t_{k(ij)} + \varepsilon_{ij}$$
that all c_j are zero and, to make the pattern simple, that $\mu = 20$, all r_i and ε_{ij} are zero and the t_k are $(+5, +3, +1, -1, -8)$. Then the observed yields would be as in Table 4.4. Column 3 appears to give lower values than column 5, if possible treatment effects are ignored. That this effect does not disappear if the row effects or random effects are made non-zero can be confirmed by adding any set of numerical values for such effects to the observed yields.

In the previous section the least squares equations for example C allowed us to estimate the column and treatment sets of parameters, fitted together, giving
$$\hat{c}_1 = +0.339, \hat{c}_2 = +0.065, \hat{c}_3 = +1.159, \hat{c}_4 = -0.268, \hat{c}_5 = -1.296;$$
$$\hat{t}_A = +1.276, \hat{t}_B = +0.549, \hat{t}_C = -0.249, \hat{t}_D = -0.562, \hat{t}_E = -1.015.$$

Table 4.4.

					Row totals
A 25	B 23	C 21	D 19	E 12	100
C 21	D 19	E 12	A 25	B 23	100
D 19	A 25	B 23	E 12	C 21	100
E 12	C 21	D 19	B 23	A 25	100
B 23	E 12	A 25	C 21	D 19	100
B 23	C 21	E 12	D 19	A 25	100
Column totals 123	121	112	119	125	

If treatment effects are ignored the estimates for column effects are

$$\hat{c}_1 = +0.430, \hat{c}_2 = +0.024, \hat{c}_3 = +0.990, \hat{c}_4 = -0.361, \hat{c}_5 = -1.083.$$

Column 3 appears less high and column 5 less low because the extra occurrence in column 3 of the low value treatment E and the extra occurrence of the high value treatment A in column 5 are ignored.

Similarly if we ignore the column effects the treatment effect estimates are

$$\hat{t}_A = +1.060, \hat{t}_B = +0.605, \hat{t}_C = -0.238, \hat{t}_D = -0.606, \hat{t}_E = -0.821.$$

The SS for fitting column effects, treatment effects, and both sets of effects are:

$$\text{SS (columns only)} = (0.430)(27.59) + \cdots = 14.82$$
$$\text{SS (treatments only)} = (1.060)(31.37) + \cdots = 15.53$$
$$\text{SS (both)} = (0.339)(27.59) + \cdots + (1.276)(31.37) + \cdots$$
$$= 34.24.$$

The two analyses of variance for the two orders $\theta_1 \rightarrow \theta_2 \rightarrow \theta_3 \rightarrow \theta_4$ and $\theta_1 \rightarrow \theta_2 \rightarrow \theta_4 \rightarrow \theta_3$ are given in Table 4.5(a) and (b).

The SS for column effects and for treatment effects are broadly similar in the two analyses, and in each case are markedly larger than the residual, so that the interpretation is going to be similar for treatment effects whether or not we allow for column effects. Nevertheless the change of 3.89 in the SS is sufficiently large to indicate that where an effect was less clearly substantial, or where the non-orthogonality or degree of interference between treatment and column effects was larger, then the conclusions

Table 4.5(a).

Source	SS	df
μ (ignoring all other terms)	521.17	1
r_i (ignoring c_j and t_k)	66.66	5
c_j (allowing for r_i, ignoring t_k)	14.82	4
t_k (allowing for r_i and c_j)	19.42	4
Residual	8.05	21

(b).

Source	SS	df
μ (ignoring all other terms)	521.17	1
r_i (ignoring c_j and t_k)	66.66	5
t_k (allowing for r_i, ignoring c_j)	15.53	4
c_j (allowing for r_i and t_k)	18.71	4
Residual	8.05	21

from an analysis which failed to allow for other effects when assessing treatment effects could be severely misleading.

One further example of the use of the extra SS principle concerns example A again and the comparison of the two models considered for the treatment effects. In one form of model the ten separate treatment effects are included giving a SS of 23.14. Alternatively the effects of treatments are represented by a regression on area per plant. The full analysis of variance for this alternative model is given in Table 4.6. Now the regression model for treatment effects is a special case of the more general model allowing for ten different treatment effects, and we could write an extended model to include both regression and separate treatment effects:

$$y_{ij} = \mu + b_i + \beta a_j + t'_j + \varepsilon_{ij}.$$

We can calculate SS for fitting four sets of parameters $\theta_1 = \mu$, $\theta_2 = \{b_i\}$, $\theta_3 = \beta$, $\theta_4 = \{t'_j\}$. However, since the above model is exactly equivalent to the earlier model which allowed for possible differences between all ten treatments

$$y_{ij} = \mu + b_i + t_j + \varepsilon_{ij},$$

we already have the SS for all sub-models and for the full model and can construct the analyses of variance given in Table 4.7. We may now observe that the treatment SS after allowing for the regression effect is not very large so that we may be prepared to accept the regression model as an adequate summary of the treatments.

Table 4.6.

Source	SS	df
μ	2211.17	1
Block effect	5.88	3
Regression on area per plant	16.80	1
Residual	18.76	35

Table 4.7.

Source	SS	df
μ	2211.17	1
Blocks	5.88	3
Regression	16.80	1
Remaining treatment effects (t'_j)	6.34	8
Residual	12.42	27

4.6   Distributional assumptions and inferences

All the results discussed so far in this chapter have ignored assumptions about the sampling distributions of the observations. It is important to notice that much of the analysis of experimental data does not rely on particular distributional assumptions. However, if we assume that the observations are taken from a normally distributed population, then we can develop methods for more detailed inferences.

The first consequence of the normal distribution assumption is that, since least squares estimates are linear combinations of the y's and since linear combinations of normally distributed variables are themselves normally distributed, then the least squares estimates are normally distributed. Least squares estimates are unbiassed and variances of least squares estimates were derived in Section 4.3, so the information on the distribution of least squares estimates is complete. In particular, confidence intervals for parameters may be constructed, and tests of significance calculated using the t distribution.

The second important consequence of the normal distribution assumption is that the distributional properties of SS in the analysis of variance are all described in terms of χ^2 distributions. First, quite generally, the residual SS, S_r, is distributed as σ^2 times a χ^2 distribution on $(n-p')$ df. Second, where the parameters are partitioned into two sets, θ_1 and θ_2, then, whenever the null hypothesis (that the parameters in the second set are zero) is true, the SS due to fitting θ_2 allowing for θ_1 is distributed as σ^2 times a χ^2 distribution on q df, where q is the number of effective parameters in θ_2. Hence, the null hypothesis that the parameters θ_2 are zero may be tested by calculating the ratio of the mean square for θ_2 to the residual mean square, and comparing that ratio with the F distribution on q and $(n-p')$ df. Note that for the fitting order of θ_1 followed by θ_2 the F distribution applies only to the mean square for θ_2, and not to that for θ_1, unless θ_1 and θ_2 are orthogonal.

To illustrate the application of these distributional results we repeat several previous analyses of variance with added mean squares and F statistics.

Example A

First the simple analysis for block and treatment effects (see Table 4.8). The treatment mean square with an F statistic on 9 and 27 df is clearly significant so that the differences between treatments are well substantiated. The F statistic for blocks could also be tested because of the orthogonality between blocks and treatments, but the formal hypothesis that $b_i = 0$ for all i is neither credible nor important.

Table 4.8.

Source	SS	df	MS	F
μ	2211.17	1		
Blocks	5.88	3	1.96	4.26
Treatments	23.14	9	2.57	5.59
Residual	12.42	27	0.46	

Table 4.9.

Source	SS	df	MS	F
μ	2211.17	1		
Blocks	5.88	3	1.96	
Regression	16.80	1	16.80	36.52
Remaining treatment effects	6.34	8	0.79	1.72
Residual	12.42	27	0.46	

The secondary analysis with the regression effect is given in Table 4.9. The F tests provide confirmation, if any was needed, that there is strong regression effect. More importantly we can examine the remaining treatment variation not explained by the regression. The F statistic of 1.72 indicates that there may be other important effects amongst the treatments but the evidence is not significant at the 5% level. We return to the interpretation of this data in Chapter 12.

Example C
The analysis of variance (Table 4.10) includes the SS for fitting treatment effects allowing for column effects, and it is that SS for treatments that we should use to test the significance of the treatment effects. Plainly the treatment effects are very significant. Although we should not wish to test the significance of the mean squares for columns or rows, since the

Table 4.10.

Source	SS	df	MS	F
μ	521.17	1		
Rows	66.66	5	13.33	
Columns (ignoring treatments)	14.82	4	3.70	
Treatments (allowing for columns)	19.42	4	4.86	12.67
Residuals	8.05	21	0.38	

hypothesis of no row or column effects is not important to the objectives of the experiment, there are clearly substantial differences between rows and, to a lesser extent, between columns. The design of the experiment has thus been clearly successful in providing more precise information about treatments, and the degree of that success may be assessed by estimating the error mean square that might have been obtained if the treatments had been allocated without restriction. The error mean square might then have been expected to be about

$$(66.66 + 14.82 + 8.05)/(5 + 4 + 21) = 2.98$$

instead of 0.38.

4.7 Contrasts, treatment comparisons and component sums of squares

The final set of theoretical concepts relevant to the analysis of experimental data is concerned with formally defined contrasts between observations. The main use of these from a practical viewpoint is in the detailed examination of treatment comparisons. However, the definition and identification of contrasts between individual observations is also useful, particularly in understanding the concepts of confounding in Chapter 15. We consider contrasts first in the context of treatment comparisons.

Assume a set of t treatments, and for simplicity assume initially that there are n observations for each treatment, and that each treatment effect estimate is orthogonal to all block effects and is in the form

$$\hat{t}_j = y_{.j} - y_{..}.$$

A general treatment comparison, L_k, is defined as a linear contrast of treatment effects

$$L_k = \sum_j l_{kj} t_j,$$

where $\sum_j l_{kj} = 0$. The least squares estimate of L_k is

$$\hat{L}_k = \sum_j l_{kj} \hat{t}_j$$
$$= \sum_j l_{kj} (y_{.j} - y_{..})$$
$$= \sum_j l_{kj} y_{.j}$$

since $\sum_j l_{kj} = 0$. The variance of $\hat{L}_k$ is

$$\mathrm{Var}(\hat{L}_k) = \sigma^2 \sum_j l_{kj}^2 / n,$$

which may be estimated, using the residual mean square, s^2, as

$$s^2 \sum_j l_{kj}^2 / n.$$

The estimate of a contrast and the estimated variance of that estimate provide all the necessary information about the contrast, but it is useful to develop the concept further to consider the relationship between different contrasts and the relationship between individual contrasts and the total treatment variation. First, two treatment comparisons, L_k and $L_{k'}$, are defined to be orthogonal if

$$\sum_j l_{kj} l_{k'j} = 0 \quad \text{if } k \neq k'.$$

It can be shown that we can define a component SS corresponding to a contrast which can be identified in the form

$$SS(L_k) = n \left(\sum_j l_{kj} y_{.j} \right)^2 \bigg/ \left(\sum_j l_{kj}^2 \right)$$

(see the appendix to this chapter). Note that the divisor in this expression is a scaling factor. There are no restrictions on the size of coefficients in the definition of a comparison so that

$$t_1 - t_2 \text{ and } 20t_1 - 20t_2$$

are both valid forms of comparison. In calculating component SS the arbitrariness of scale inherent in the definition of L_k must be allowed for.

These component SS for comparisons L_k have two important properties.

(i) If $L_k = 0$, that is the treatment comparison has no effect, the SS is distributed as σ^2 times a χ^2 distribution on 1 df.

(ii) The total of the SS for $(t-1)$ orthogonal comparisons, from t treatments, is the total treatment SS.

The implications are that we can examine $(t-1)$ comparisons between the treatments independently provided the comparisons are constructed so as to be orthogonal, and together such a set of $(t-1)$ orthogonal comparisons represents the total variation between the treatments.

The practical use of treatment comparisons can be illustrated for all three of our examples. For example B on water uptake of amphibia the analysis of variance for main effects and interactions was considered in Chapter 3. The same analysis may be constructed in terms of treatment contrasts. Seven contrasts corresponding to the main effects and interactions may be defined as shown in Table 4.11. Contrasts (1), (2) and (3) are main effect contrasts between the two levels of each factor. Contrasts (4), (5) and (6) are two-factor interactions, in the form of differences of differences, and contrast (7) is the three-factor interaction, a difference of differences of differences. It can be verified by inspection that any two contrasts are orthogonal. The component SS are calculated, using $n = 2$;

Table 4.11.

Treatment	L_1	L_2	L_3	L_4	L_5	L_6	L_7	Treatment mean
Toad wet control	+1	+1	+1	+1	+1	+1	+1	0.360
Toad wet hormone	+1	+1	−1	+1	−1	−1	−1	21.265
Toad dry control	+1	−1	+1	−1	+1	−1	−1	21.455
Toad dry hormone	+1	−1	−1	−1	−1	+1	+1	28.165
Frog wet control	−1	+1	+1	−1	−1	+1	−1	1.875
Frog wet hormone	−1	+1	−1	−1	+1	−1	+1	10.095
Frog dry control	−1	−1	+1	+1	−1	−1	+1	3.340
Frog dry hormone	−1	−1	−1	+1	+1	+1	−1	10.545

for example,

$$SS(L_1) = 2(0.360 + 21.265 + \cdots - 10.545)^2 / (1^2 + 1^2 + \cdots + 1^2)$$
$$= 2(45.39)^2 / 8 = 515.06,$$

agreeing with the value for the species SS calculated in Chapter 3.

For example C there are five treatments and various alternative sets of contrasts may be considered. Meaningful contrasts that might be considered are defined in Table 4.12. The set (L_1, L_2, L_3, L_4) are orthogonal, and the contrasts represent first the effect of spray compared with control and then the main effects and interaction for the factorial structure of the four spray treatments. Alternatively the control may be viewed as a zero level spray and (L_5, L_6) defined to represent two orthogonal contrasts between the three levels of spray, L_5 comparing the two extremes and representing linear trend while L_6 compares the middle level with the extremes, representing the curvature of the trend. A second complete set of four orthogonal contrasts is $(L_2, L_4, L_5$ and $L_6)$ but neither L_1 nor L_3 is orthogonal to either L_5 or L_6.

The calculation of component SS for these contrasts introduces a problem not considered in the development of the component SS because

Table 4.12.

Spray treatment	L_1	L_2	L_3	L_4	L_5	L_6	$\hat{t}_k$
Early 75 p.p.m.	+1	−1	−1	+1	0	−3	+1.276
Early 150 p.p.m.	+1	−1	+1	−1	+1	+2	+0.549
Late 75 p.p.m.	+1	+1	−1	−1	0	−3	−0.249
Late 150 p.p.m.	+1	+1	+1	+1	+1	+2	−0.562
Control	−4	0	0	0	−2	+2	−1.015

the treatment effects are not orthogonal to the column effects. Consequently the treatment effects are not in the simple form specified earlier in this section. We can now extend the results for component SS for the situation where, as in example C, treatment effects are not orthogonal to block effects but the precision of treatment differences is identical for all treatment differences. We shall return to this concept of equality of precision (balance) in Chapters 7 and 8. For the present it is sufficient to note that the symmetry of the design for example C, with each treatment occurring a second time in just one column, implies that the variances of all simple treatment difference should be equal.

In Section 4.4 the variance of a simple treatment difference was derived as $12\sigma^2/35$, which is larger than the variance for a treatment difference in a randomised block design with the same six-fold replication, by a factor of $36/35$. It can be shown that the variance for any treatment contrast is similarly inflated by the factor $36/35$, and that correspondingly the component SS must be reduced by the reciprocal factor $35/36$. Essentially the non-orthogonality reduces the information on treatment differences by a factor $35/36$.

The SS for the six components are therefore

$$SS(L_1) = (35/36)6[1.276 + 0.549 + 0.249 + 0.562 - 4(1.015)]^2/20 = \quad 7.51$$
$$SS(L_2) = (35/36)6[-1.276 - 0.549 + 0.249 + 0.562]^2/4 \qquad = 10.13$$
$$SS(L_3) = (35/36)6[-1.276 + 0.549 - 0.249 + 0.562]^2/4 \qquad = \quad 1.58$$
$$SS(L_4) = (35/36)6[1.276 - 0.549 - 0.249 + 0.562]^2/4 \qquad = \quad 0.25$$
$$SS(L_5) = (35/36)6[0.549 + 0.562 - 2(1.015)]^2/6 \qquad = \quad 3.96$$
$$SS(L_6) = (35/36)6[-3(1.276) + 2(0.549) - 3(0.249) + 2(0.562) + 2(1.015)]^2/30$$
$$= 5.13.$$

And the two alternative analyses of treatment variation are given in Table 4.13(a) and (b). Clearly the timing of the spray (L_2) has the largest single effect but there is no evidence of an interaction (L_4) between timing and concentration. The most striking pattern of the effect of concentration is apparently provided by L_1, the simple contrast of spray with no spray. Alternatively the mean yields at the three concentrations should be presented

control   75p.p.m.   150p.p.m.
3.15   4.68   4.16.

As a final illustration of the use of contrasts note that the regression effect for example A could have been expressed as a treatment comparison. If the regression is written in the form

$$E(y_{.j}) = \alpha + \beta(a_j - \bar{a})$$

Table 4.13(a).

Source	MS	df	F
L_1	7.51	1	19.59
L_2	10.13	1	26.43
L_3	1.58	1	4.12
L_4	0.25	1	0.65
Residual	0.38	21	

(b).

Source	MS	df	F
L_2	10.13	1	26.43
L_4	0.25	1	0.65
L_5	3.96	1	10.33
L_6	5.13	1	13.38
Residual	0.38	21	

then a regression comparison with coefficients $(a_j - \bar{a})$ may be defined as shown in Table 4.14. The component SS for the regression effect is therefore

$$\text{SS(regression)} = 4\{(398.9)(6.02) + \cdots$$
$$- (276.1)18.20\}^2/(398.94^2 + \cdots + 276.1^2)$$
$$= 4(-1250.185)^2/372289 = 16.79.$$

At the beginning of this section the comment was made that the concept of contrasts may be applied more widely than within the set of treatments.

Table 4.14.

Area/plant for each treatment	Treatment contrast $(a_j - \bar{a})$	Treatment mean
900	$+398.9$	6.02
720	$+218.9$	6.48
600	$+98.9$	7.19
450	-51.1	6.92
576	$+74.9$	7.85
480	-21.1	7.78
360	-141.1	7.50
400	-101.1	7.71
300	-201.1	8.70
225	-276.1	8.20
mean = 501.1		

We may more generally define contrasts within the complete set of n observations. A complete set of contrasts would lead to an analysis of variance in which each SS was based on a single df and would, of course, be quite useless. Nevertheless, the identification and interpretation of individual contrasts may sometimes be helpful in understanding the properties of designs. To illustrate this wider concept of contrasts we consider a trivial example of a design for four treatments in three randomised blocks. A set of possible contrasts is defined in Table 4.15.

Contrasts L_1 and L_2 are contrasts between blocks; contrasts L_3, L_4 and L_5 are contrasts between treatments; the remaining contrasts are for residual variation. Rather laboriously it may be verified that all pairs of contrasts are orthogonal. In fact the residual contrasts are formed as products of block and treatment contrasts ($L_6 = L_1 L_3$, $L_7 = L_1 L_4, \dots,$ $L_{11} = L_2 L_5$), which emphasises the interpretation of the residual variation both as block $\times$ treatment interaction and as a measure of the consistency of treatment effects over the different blocks.

Table 4.15.

Block	Treatment	L_1	L_2	L_3	L_4	L_5	L_6	L_7	L_8	L_9	L_{10}	L_{11}
1	1	-1	-1	-1	-1	$+1$	$+1$	$+1$	-1	$+1$	$+1$	-1
1	2	-1	-1	-1	$+1$	-1	$+1$	-1	$+1$	$+1$	-1	$+1$
1	3	-1	-1	$+1$	-1	-1	-1	$+1$	$+1$	-1	$+1$	$+1$
1	4	-1	-1	$+1$	$+1$	$+1$	-1	-1	-1	-1	-1	-1
2	1	$+1$	-1	-1	-1	$+1$	-1	-1	$+1$	$+1$	$+1$	-1
2	2	$+1$	-1	-1	$+1$	-1	-1	$+1$	-1	$+1$	-1	$+1$
2	3	$+1$	-1	$+1$	-1	-1	$+1$	-1	-1	-1	$+1$	$+1$
2	4	$+1$	-1	$+1$	$+1$	$+1$	$+1$	$+1$	$+1$	-1	-1	-1
3	1	0	$+2$	-1	-1	$+1$	0	0	0	-2	-2	$+2$
3	2	0	$+2$	-1	$+1$	-1	0	0	0	-2	$+2$	-2
3	3	0	$+2$	$+1$	-1	-1	0	0	0	$+2$	-2	-2
3	4	0	$+2$	$+1$	$+1$	$+1$	0	0	0	$+2$	$+2$	$+2$

APPENDIX TO CHAPTER 4

4.A2   Least squares estimators for linear models
The general linear model, relating n observations to p parameters, may be written

$$E(\mathbf{y}) = \mathbf{A}\boldsymbol{\theta} \qquad D(\mathbf{y}) = \sigma^2 \mathbf{I}.$$

The SS of deviations of y_i from its expected value is

$$S = (\mathbf{y} - \mathbf{A}\boldsymbol{\theta})'(\mathbf{y} - \mathbf{A}\boldsymbol{\theta})$$

and is minimised by

$$\frac{\partial S}{\partial \theta} = 2\mathbf{A}'(\mathbf{y} - \mathbf{A}\boldsymbol{\theta}) = 0,$$

giving least squares equations

$$\mathbf{A}'\mathbf{A}\hat{\boldsymbol{\theta}} = \mathbf{A}'\mathbf{y}$$

or

$$\mathbf{C}\hat{\boldsymbol{\theta}} = \mathbf{A}'\mathbf{y}$$

where

$$\mathbf{C} = \mathbf{A}'\mathbf{A}.$$

If the rank of $\mathbf{C}$ is p, $\mathbf{C}^{-1}$ exists and

$$\hat{\boldsymbol{\theta}} = \mathbf{C}^{-1}\mathbf{A}'\mathbf{y}.$$

4.A3   Properties of least squares estimators
The expected value and mean of $\boldsymbol{\theta}$ may be obtained as follows:

$$
\begin{aligned}
E(\hat{\boldsymbol{\theta}}) &= E(\mathbf{C}^{-1}\mathbf{A}'\mathbf{y}) \\
&= \mathbf{C}^{-1}\mathbf{A}'E(\mathbf{y}) \\
&= \mathbf{C}^{-1}\mathbf{A}'(\mathbf{A}\boldsymbol{\theta}) \\
&= (\mathbf{A}'\mathbf{A})^{-1}(\mathbf{A}'\mathbf{A}\boldsymbol{\theta}) \\
&= \boldsymbol{\theta}.
\end{aligned}
$$

Hence $\hat{\boldsymbol{\theta}}$ is unbiassed. Also

$$\hat{\boldsymbol{\theta}} - E(\hat{\boldsymbol{\theta}}) = \hat{\boldsymbol{\theta}} - \boldsymbol{\theta} = \mathbf{C}^{-1}\mathbf{A}'\mathbf{y} - \mathbf{C}^{-1}\mathbf{A}'\mathbf{A}\boldsymbol{\theta}$$
$$= \mathbf{C}^{-1}\mathbf{A}'(\mathbf{y} - \mathbf{A}\boldsymbol{\theta}).$$

Hence

$$\begin{aligned}
\text{Var}(\hat{\boldsymbol{\theta}}) &= E\{\hat{\boldsymbol{\theta}} - E(\hat{\boldsymbol{\theta}})\}\{\hat{\boldsymbol{\theta}} - E(\hat{\boldsymbol{\theta}})\}' \\
&= E\{\mathbf{C}^{-1}\mathbf{A}'(\mathbf{y} - \mathbf{A}\boldsymbol{\theta})(\mathbf{y} - \mathbf{A}\boldsymbol{\theta})'\mathbf{A}\mathbf{C}^{-1}\} \\
&= \mathbf{C}^{-1}\mathbf{A}'E\{(\mathbf{y} - \mathbf{A}\boldsymbol{\theta})(\mathbf{y} - \mathbf{A}\boldsymbol{\theta})'\}\mathbf{A}\mathbf{C}^{-1} \\
&= \sigma^2\mathbf{C}^{-1}\mathbf{A}'\mathbf{A}\mathbf{C}^{-1} \\
&= \sigma^2\mathbf{C}^{-1}
\end{aligned}$$

since

$$\mathbf{A}'\mathbf{A} = \mathbf{C}.$$

An important result about least squares estimators is contained in the following theorem proved originally by Gauss.

Theorem Consider estimation of a linear contrast $\mathbf{l}'\boldsymbol{\theta} = \sum_j l_j \theta_j$ by a linear combination of the observations $\mathbf{m}'\mathbf{y} = \sum_i m_i y_i$. Then the unbiassed estimator with the minimum variance is the corresponding contrast of the least squares estimates $\mathbf{l}'\hat{\boldsymbol{\theta}}$.

Proof The first stage of the proof concerns the requirement that the estimator be unbiassed, for which we require the condition

$$E(\mathbf{m}'\mathbf{y}) = \mathbf{l}'\boldsymbol{\theta} \text{ for all } \boldsymbol{\theta},$$

or equivalently

$$\mathbf{m}'\mathbf{A}\boldsymbol{\theta} = \mathbf{l}'\boldsymbol{\theta} \text{ for all } \boldsymbol{\theta}$$

whence

$$\mathbf{m}'\mathbf{A} = \mathbf{l}'.$$

The variance of the estimator is

$$\text{Var}(\mathbf{m}'\mathbf{y}) = \mathbf{m}' \text{ Var}(\mathbf{y})\mathbf{m} = \mathbf{m}'\mathbf{m}\sigma^2.$$

Now

$$\begin{aligned}
\mathbf{m}'\mathbf{m} &= (\mathbf{m}' - \mathbf{l}'\mathbf{C}^{-1}\mathbf{A}' + \mathbf{l}'\mathbf{C}^{-1}\mathbf{A}')(\mathbf{m} - \mathbf{A}\mathbf{C}^{-1}\mathbf{l} + \mathbf{A}\mathbf{C}^{-1}\mathbf{l}) \\
&= (\mathbf{m}' - \mathbf{l}'\mathbf{C}^{-1}\mathbf{A}')(\mathbf{m} - \mathbf{A}\mathbf{C}^{-1}\mathbf{l}) + \mathbf{l}'\mathbf{C}^{-1}\mathbf{A}'\mathbf{A}\mathbf{C}^{-1}\mathbf{l} \\
&\quad + \mathbf{l}'\mathbf{C}^{-1}\mathbf{A}'(\mathbf{m} - \mathbf{A}\mathbf{C}^{-1}\mathbf{l}) + (\mathbf{m}' - \mathbf{l}'\mathbf{C}^{-1}\mathbf{A}')\mathbf{A}\mathbf{C}^{-1}\mathbf{l} \\
&= (\mathbf{m}' - \mathbf{l}'\mathbf{C}^{-1}\mathbf{A}')(\mathbf{m} - \mathbf{A}\mathbf{C}^{-1}\mathbf{l}) + \mathbf{l}'\mathbf{C}^{-1}\mathbf{A}'\mathbf{A}\mathbf{C}^{-1}\mathbf{l}
\end{aligned}$$

since

$$\mathbf{m}'\mathbf{A} = \mathbf{l}'\mathbf{C}^{-1}\mathbf{A}'\mathbf{A} = \mathbf{l}'.$$

And hence $\mathbf{m}'\mathbf{m}$ is minimised when $\mathbf{m}' = \mathbf{l}'\mathbf{C}^{-1}\mathbf{A}'$, which is the required result. Further the minimised form is $\mathbf{l}'\mathbf{C}^{-1}\mathbf{l}$, which could of course have been deduced from the form of the variance–covariance matrix of $\boldsymbol{\theta}$.

This completes the properties of $\hat{\boldsymbol{\theta}}$. Now consider the estimation of σ^2 and the minimised or residual SS:

$$S_r = (\mathbf{y} - \mathbf{A}\hat{\boldsymbol{\theta}})'(\mathbf{y} - \mathbf{A}\hat{\boldsymbol{\theta}}).$$

Theorem $E(S_r) = (n-p)\sigma^2$, and hence σ^2 may be estimated by $S_r/(n-p)$.

Proof First note that

$$(\mathbf{y} - \mathbf{A}\boldsymbol{\theta})'(\mathbf{y} - \mathbf{A}\boldsymbol{\theta}) = \{\mathbf{y} - \mathbf{A}\hat{\boldsymbol{\theta}} + \mathbf{A}(\hat{\boldsymbol{\theta}} - \boldsymbol{\theta})\}'\{\mathbf{y} - \mathbf{A}\hat{\boldsymbol{\theta}} + \mathbf{A}(\hat{\boldsymbol{\theta}} - \boldsymbol{\theta})\}$$
$$= (\mathbf{y} - \mathbf{A}\hat{\boldsymbol{\theta}})'(\mathbf{y} - \mathbf{A}\hat{\boldsymbol{\theta}}) + (\hat{\boldsymbol{\theta}} - \boldsymbol{\theta})'\mathbf{A}'\mathbf{A}(\hat{\boldsymbol{\theta}} - \boldsymbol{\theta}),$$

the other terms being zero since $\mathbf{A}'\mathbf{y} = \mathbf{A}'\mathbf{A}\hat{\boldsymbol{\theta}}$. Now

$$E\{(\mathbf{y} - \mathbf{A}\boldsymbol{\theta})'(\mathbf{y} - \mathbf{A}\boldsymbol{\theta})\} = \sum_i E\{y_i - E(y_i)\}^2 = n\sigma^2$$

and

$$E\{(\hat{\boldsymbol{\theta}} - \boldsymbol{\theta})'\mathbf{A}'\mathbf{A}(\hat{\boldsymbol{\theta}} - \boldsymbol{\theta})\} = \sum_{jk} c_{jk}E\{(\hat{\theta}_j - \theta_j)(\hat{\theta}_k - \theta_k)\}.$$

But

$$E\{(\hat{\theta}_j - \theta_j)(\hat{\theta}_k - \theta_k)\} = \text{Cov}(\hat{\theta}_j, \hat{\theta}_k) = (\mathbf{C}^{-1})_{jk}\sigma^2.$$

Hence

$$E\{(\hat{\theta}_j - \theta_j)(\hat{\theta}_k - \theta_k) = \sum_{jk} c_{jk}(\mathbf{C}^{-1})_{jk}\sigma^2$$
$$= \sum_{jk} (\mathbf{C})_{kj}(\mathbf{C}^{-1})_{jk}\sigma^2$$

since $\mathbf{C}$ is symmetric and $c_{jk} = c_{kj} = (\mathbf{C})_{kj}$. Hence,

$$E\{(\hat{\theta}_j - \theta_j)(\hat{\theta}_k - \theta_k)\} = \sum_k \left\{\sum_j (\mathbf{C})_{kj}(\mathbf{C}^{-1})_{jk}\right\}\sigma^2$$
$$= \sigma^2 \text{ trace } (\mathbf{C}\mathbf{C}^{-1})$$
$$= \sigma^2 p.$$

Hence

$$E(S_r) = E\{(\mathbf{y} - \mathbf{A}\boldsymbol{\theta})'(\mathbf{y} - \mathbf{A}\boldsymbol{\theta})\} - E\{(\hat{\boldsymbol{\theta}} - \boldsymbol{\theta})\mathbf{A}'\mathbf{A}(\hat{\boldsymbol{\theta}} - \hat{\boldsymbol{\theta}})\}$$
$$= n\sigma^2 - p\sigma^2$$

as required. Finally, the practically most useful form of S_r is

$$S_r = (\mathbf{y} - \mathbf{A}\boldsymbol{\theta})'(\mathbf{y} - \mathbf{A}\hat{\boldsymbol{\theta}})$$
$$= \mathbf{y}'\mathbf{y} - \mathbf{y}'\mathbf{A}\hat{\boldsymbol{\theta}}$$

since

$$\mathbf{A}'\mathbf{y} = \mathbf{A}'\mathbf{A}\hat{\boldsymbol{\theta}}.$$

This relationship is expressed verbally as

residual SS = total SS − fitting SS,

where the fitting SS, $\mathbf{y}'\mathbf{A}\hat{\boldsymbol{\theta}}$ or $(\mathbf{A}'\mathbf{y})'\hat{\boldsymbol{\theta}}$, is the SS attributable to the fitted parameters, $\hat{\boldsymbol{\theta}}$.

4.A4 Overparameterisation and constraints

Now consider the general model

$$y = A\theta + \epsilon,$$

where the model contains more parameters than are necessary. Formally this is expressed by the statements that

(i) the matrix A is singular, or

(ii) the p columns of A are linearly dependent, or

(iii) the rank of A is p', less than p.

It follows that the rank of $C = A'A$ is less than p and that the least squares equations $A'A\hat{\theta} = A'y$ cannot be directly solved. To solve the equations formally, additional equations defining the relationships between the parameters, θ_j, are needed. These additional equations are called *constraints*, and because there are many possible sets of constraint equations it is necessary to determine the conditions which the constraint equations must satisfy so that there shall be unique solutions to the least squares equations. The properties of the resulting modified least squares parameter estimates must also be examined.

Theorem If the rank of A is p' and the least squares equations are considered without additional constraint equations then

(i) there is no set of estimates My of the set of parameters, θ, such that the set of estimates are all unbiassed for all possible values of θ, but

(ii) an unbiassed estimate $m'y$ of a linear combination of parameters $l'\theta$ does exist provided the set of coefficients l satisfies the same linear dependence relation as the columns of the matrix A.

Proof (i) The linear relationship between the columns of A can be expressed in the form

$$AD = 0,$$

where D is a $p \times (p - p')$ matrix of rank $(p - p')$. Now

$$E(My) = MA\theta$$

and hence for My to be a set of unbiassed estimators for θ we require

$$\theta = MA\theta \text{ for all } \theta$$

which would require

$$MA = I.$$

Consequently

$$MAD = D$$

and hence

$$D = 0$$

since $\mathbf{AD} = \mathbf{0}$. But as $\mathbf{D}$ is of rank $(p - p') > 0$ this cannot be true, and therefore there is no unbiassed set of estimates of $\boldsymbol{\theta}$.

(ii) For a linear combination $\mathbf{m'y}$ to provide an unbiassed estimate of $\mathbf{l'\theta}$ we require

$$\mathbf{l'\theta} = \mathbf{m'A\theta} \text{ for all } \boldsymbol{\theta},$$

whence

$$\mathbf{l'} = \mathbf{m'A}$$

and

$$\mathbf{l'D} = \mathbf{m'AD} = \mathbf{0}.$$

Therefore, if $\mathbf{l'D} = \mathbf{0}$, or equivalently if the linear relations between the columns of $\mathbf{A}$ hold also for the coefficients of the linear combination, $\mathbf{l}$, then there is an unbiassed estimator of $\mathbf{l'\theta}$.

Theorem If a set of $(p - p')$ constraints $\mathbf{B\theta} = \mathbf{0}$ is imposed, subject to the condition that $\mathbf{BD}$ has full rank $(p - p')$ then the modified least squares estimates are given by the least squares equations and the constraint equations, and the minimised SS, S_r is independent of the particular form of constraint.

Proof The constraints $\mathbf{B\theta} = \mathbf{0}$ satisfy the condition that $\mathbf{BD}$ has full rank, which is equivalent to the requirement that the linear dependence relations for the columns of $\mathbf{A}$ do not hold for the columns of $\mathbf{B}$.

Modified least squares estimators are those which minimise

$$S = (\mathbf{y} - \mathbf{A\theta})'(\mathbf{y} - \mathbf{A\theta})$$

subject to the constraint

$$\mathbf{B\theta} = \mathbf{0}.$$

By introducing a vector, $\boldsymbol{\lambda}$, of undetermined multipliers, the least squares estimators of $\boldsymbol{\theta}$ must minimise

$$S' = (\mathbf{y} - \mathbf{A\theta})'(\mathbf{y} - \mathbf{A\theta}) + \boldsymbol{\lambda}\mathbf{B\theta}.$$

Differentiating with respect to $\boldsymbol{\theta}$

$$2(\mathbf{y} - \mathbf{A\hat{\theta}})'\mathbf{A} - \boldsymbol{\lambda}\mathbf{B} = \mathbf{0}.$$

Postmultiplying by $\mathbf{D}$ gives

$$\boldsymbol{\lambda}\mathbf{BD} = \mathbf{0} \text{ since } \mathbf{AD} = \mathbf{0}$$

whence

$$\boldsymbol{\lambda} = \mathbf{0}$$

since $\mathbf{BD}$ is of full rank. Hence the absolute minimum of S is the same as the minimum conditional on $\mathbf{B\theta} = \mathbf{0}$ for any set of constraints for which $\mathbf{BD}$ has full rank. And the equations satisfied by the modified least squares estimates are

$$A'A\hat{\theta} = A'y$$
$$B\hat{\theta} = 0.$$

General solutions may be obtained by using the partitioned matrix

$$\begin{bmatrix} A \\ B \end{bmatrix} = A^*,$$

which is of full rank because the linear dependencies for the columns of **A** do not apply to the columns of **B**. Hence,

$$A'A + B'B = A^{*'}A^*$$

is non-singular. Hence the modified least squares equations can be written

$$A^{*'}A^*\hat{\theta} = A'y$$

with the solution

$$\hat{\theta} = (A^{*'}A^*)^{-1}A'y = (A'A + B'B)^{-1}A'y.$$

Theorem (This theorem extends the earlier Gauss' theorem.) The minimum-variance unbiassed linear estimator of a linear contrast $l'\theta$ is the corresponding contrast of the modified least squares estimators $l'\hat{\theta} = l'(A'A + B'B)^{-1}A'y$.

Proof A linear estimator, $m'y$, will be unbiassed if

$$l'\theta = m'A\theta$$

whenenever $B\theta = 0$ or equivalently

$$l'\theta = m'A\theta + n'B\theta \text{ for all } \theta,$$

which requires

$$l' = m'A + n'B.$$

Here **n** is an arbitrary column vector, which may be eliminated by postmultiplying by **D**

$$l'D = m'AD + n'BD = n'BD$$

whence

$$n' = n'D(BD)^{-1}$$

and the required condition for $m'y$ to be unbiassed becomes

$$m'A = l' - l'D(BD)^{-1}B.$$

But

$$(A'A + B'B)D = B'BD$$

since $AD = 0$ and hence

$$D(BD)^{-1} = (A'A + B'B)^{-1}B'$$

giving the final form of the unbiassedness condition

$$\mathbf{m'A} = \mathbf{l'} - \mathbf{l'(A'A + B'B)^{-1}B'B}$$
$$= \mathbf{l'(A'A + B'B)^{-1}A'A}.$$

Now consider the variance of $\mathbf{m'y}$,

$$\begin{aligned} \mathrm{Var}(\mathbf{m'y}) &= \sigma^2 \mathbf{m'm} \\ &= \sigma^2 \{\mathbf{m'} - \mathbf{l'(A'A + B'B)^{-1}A'} + \mathbf{l'(A'A + B'B)^{-1}A'}\} \\ &\quad \times \{\mathbf{m} - \mathbf{A(A'A + B'B)^{-1}l} + \mathbf{A(A'A + B'B)^{-1}l}\} \\ &= \sigma^2 [\{\mathbf{m'} - \mathbf{l'(A'A + B'B)^{-1}A'}\}\{\mathbf{m} - \mathbf{A(A'A + B'B)^{-1}l}\} \\ &\quad + \mathbf{l'(A'A + B'B)^{-1}A'A(A'A + B'B)^{-1}l}]. \end{aligned}$$

Hence $\mathbf{m'y}$ has minimum variance when $\mathbf{m'} = \mathbf{l'(A'A + B'B)^{-1}A'}$, and

$$\mathrm{Var}(\hat{\boldsymbol{\theta}}) = \sigma^2 \mathbf{(A'A + B'B)^{-1}A'A(A'A + B'B)^{-1}}. \tag{4.11}$$

Since it is still true that $\mathbf{A'A\hat{\boldsymbol{\theta}} = A'y}$ the expression for the minimised residual SS is unchanged

$$S_r = \mathbf{y'y} - \mathbf{y'A\hat{\boldsymbol{\theta}}}.$$

Further, since the above results hold for any set of constraints for which **BD** is of rank $(p - p')$ they hold for the special set of constraints which set $(p - p')$ of the parameters of $\boldsymbol{\theta}$ equal to zero. In this case the model has no surplus parameters and the results of the previous section hold. In particular

$$\mathrm{E}(S_r) = (n - p')\sigma^2$$

since the reduced model has precisely p' parameters.

Before leaving the subject of constraints for overparameterised models it must be emphasised that the theory of this section does not give the usual form of calculations for practical analysis of data. For most analyses of experimental data, the sensible approach is to write down the least squares equations, use sensible constraint equations to simplify the least squares equations as far as possible and then solve the least squares equations directly with the aid of the constraints. The benefit of the theory is in the assurance that it provides that modified least squares estimators can be obtained and that the form of constraint does not affect any practically important estimator. In particular, the variances of estimators are, in practice, not calculated from (4.11) but through considering each estimator as a linear combination of y values.

4.A5   Partitioning the parameter vector and the extra SS principle

Now consider the partition of the p parameter vector $\boldsymbol{\theta}$ into two parameter vectors:

$\boldsymbol{\theta}_1$ containing $p - q$ parameters

and

$\boldsymbol{\theta}_2$ containing q parameters.

Our primary interest will be in $\boldsymbol{\theta}_2$. The model for the observations, $\mathbf{y}$, will be

$$E(\mathbf{y}) = \mathbf{A}_1\boldsymbol{\theta}_1 + \mathbf{A}_2\boldsymbol{\theta}_2$$

where $\mathbf{A}$ is partitioned into two matrices $\mathbf{A}_1\{n \times (p-q)\}$ and $\mathbf{A}_2(n \times q)$. The matrix $\mathbf{C} = \mathbf{A}'\mathbf{A}$ is similarly partitioned:

$$\mathbf{C} = \begin{bmatrix} \mathbf{C}_{11} & \mathbf{C}_{12} \\ \mathbf{C}_{21} & \mathbf{C}_{22} \end{bmatrix},$$

where $\mathbf{C}_{22}$ is $q \times q$, $\mathbf{C}_{11}$ is $(p-q) \times (p-q)$.

From Section 4.4 (and 4.A4) any model for which the rank of $\mathbf{A}$ is less than p, may be reparameterised to produce an effective model for which the least squares matrix, $\mathbf{C}$, can be inverted. Without losing any generality therefore we discuss the theory for partitioned parameter vectors for a model in which the rank of $\mathbf{A}$ is p. Hence $\mathbf{C}^{-1}$ exists and is partitioned according to the same pattern as $\mathbf{C}$,

$$\mathbf{C}^{-1} = \begin{bmatrix} (\mathbf{C}^{-1})_{11} & (\mathbf{C}^{-1})_{12} \\ (\mathbf{C}^{-1})_{21} & (\mathbf{C}^{-1})_{22} \end{bmatrix},$$

where $(\mathbf{C}^{-1})_{11}$ is $(p-q) \times (p-q)$ and $(\mathbf{C}^{-1})_{22}$ is $q \times q$. The least squares equations are, in partitioned form,

$$\mathbf{C}_{11}\hat{\boldsymbol{\theta}}_1 + \mathbf{C}_{12}\hat{\boldsymbol{\theta}}_2 = \mathbf{A}_1'\mathbf{y}$$
$$\mathbf{C}_{21}\hat{\boldsymbol{\theta}}_1 + \mathbf{C}_{22}\hat{\boldsymbol{\theta}}_2 = \mathbf{A}_2'\mathbf{y}.$$

The fitting procedure is taken in two stages. First ignore $\boldsymbol{\theta}_2$ and fit only $\boldsymbol{\theta}_1$, obtaining

$$\boldsymbol{\theta}_1^* = (\mathbf{C}_{11})^{-1}\mathbf{A}_1'\mathbf{y}.$$

Second, fit $\boldsymbol{\theta}_2$ allowing for $\boldsymbol{\theta}_1$ by eliminating $\boldsymbol{\theta}_1$ from the partitioned equations

$$\hat{\boldsymbol{\theta}}_1 = \mathbf{C}_{11}^{-1}\mathbf{A}_1'\mathbf{y} - \mathbf{C}_{11}^{-1}\mathbf{C}_{12}\hat{\boldsymbol{\theta}}_2,$$

whence the second set of least squares equations becomes

$$(\mathbf{C}_{22} - \mathbf{C}_{21}\mathbf{C}_{11}^{-1}\mathbf{C}_{12})\hat{\boldsymbol{\theta}}_2 = (\mathbf{A}_2' - \mathbf{C}_{21}\mathbf{C}_{11}^{-1}\mathbf{A}_1')\mathbf{y}.$$

By consideration of the similar equations arising from the equation $\mathbf{C}\mathbf{C}^{-1} = \mathbf{I}$

$$\mathbf{C}_{11}(\mathbf{C}^{-1})_{12} + \mathbf{C}_{12}(\mathbf{C}^{-1})_{22} = 0$$
$$\mathbf{C}_{21}(\mathbf{C}^{-1})_{12} + \mathbf{C}_{22}(\mathbf{C}^{-1})_{22} = \mathbf{I}$$

we can deduce that $(\mathbf{C}_{22} - \mathbf{C}_{21}\mathbf{C}_{11}^{-1}\mathbf{C}_{12})(\mathbf{C}^{-1})_{22} = \mathbf{I}$, and hence

$$\hat{\boldsymbol{\theta}}_2 = (\mathbf{C}^{-1})_{22}(\mathbf{A}_2' - \mathbf{C}_{21}\mathbf{C}_{11}^{-1}\mathbf{A}_1')\mathbf{y}.$$

This result for $\hat{\boldsymbol{\theta}}_2$ is not in any way different from that obtained directly from the original least squares equations. However, use of this form to consider the SS for fitting sets of parameters leads to a particularly neat form for the SS.

Theorem If the SS for fitting the full parameter vector is $S_p = y'A\hat{\theta}$, and the SS for fitting only the subset of parameters θ_1 is $S_{p-q} = y'A_1'\theta_1$ then the difference $S_q = S_p - S_{p-q}$ is the fitting SS for θ_2 allowing for the effects of θ_1 and is

$$S_q = \hat{\theta}_2'(C^{-1})_{22}^{-1}\hat{\theta}_2.$$

Proof

$$\begin{aligned}
S_q &= S_p - S_{p-q} \\
&= y'A\hat{\theta} - y'A_1\theta_1^* \\
&= y'(A_1\hat{\theta}_1 + A_2\hat{\theta}_2) - y'A_1\theta_1^* \\
&= y'A_1(C_{11}^{-1}A_1'y - C_{11}^{-1}C_{12}\hat{\theta}_2) + y'A_2\hat{\theta}_2 - y'A_1C_{11}^{-1}A_1'y \\
&= y'(A_2 - A_1C_{11}^{-1}C_{12})\hat{\theta}_2 \\
&= \hat{\theta}_2'(C^{-1})_{22}^{-1}\hat{\theta}_2.
\end{aligned}$$

This SS for fitting θ_2, allowing for the effects of θ_1, is known as 'the extra SS' and the extra SS principle is fundamental to almost all linear model analysis.

Note that the SS for fitting θ_2, ignoring θ_1, is not usually the same as the extra SS for θ_2 after fitting θ_1. The reason for this can be seen simply as follows. The SS for fitting θ_2, without θ_1, is

$$y'A_2(C_{22})^{-1}A_2'y,$$

whereas the extra SS for θ_2, allowing for θ_1, is

$$y'(A_2 - A_1C_{11}^{-1}C_{12})(C_{22} - C_{21}C_{11}^{-1}C_{12})^{-1}(A_2' - C_{21}C_{11}^{-1}A_1')y.$$

It can be seen that a sufficient condition for the equality of these two expressions is $C_{12} = 0$, implying also $C_{21} = 0$ by the symmetry of C.

The condition $C_{12} = 0$ implies that the product of any two column vectors from A, one from A_1 and the other from A_2, is zero. This is simply expressed by saying that the column vectors of A_1 are orthogonal to those of A_2, or equivalently the estimation of the parameters θ_1 and θ_2 is orthogonal.

4.A6   Distributional assumptions and inferences

Theorem Assuming that the ε are normally distributed the residual SS, S_r, is distributed as $\sigma^2\chi_{n-p}^2$, independently of $S_p = y'A\hat{\theta}$, the fitted SS. Further, if $\theta = 0$ then S_p is itself distributed as $\sigma^2\chi_p^2$.

Proof This can be developed from a purely algebraic approach. There is also a geometric interpretation which may assist understanding.

Consider the model

$$y = A\theta + \varepsilon,$$

and an orthogonal transformation,

$$z = Ly,$$

where L is an orthogonal matrix such that

$$H = LA = \begin{bmatrix} t_{11} & t_{12} & t_{13} & t_{14} & \cdots & t_{1p} \\ 0 & t_{22} & t_{23} & & \cdots & t_{2p} \\ 0 & 0 & t_{33} & & \cdots & t_{3p} \\ \vdots & \vdots & & & & \vdots \\ 0 & 0 & & & 0 & t_{pp} \\ 0 & 0 & & & 0 & 0 \end{bmatrix} = \begin{bmatrix} T \\ 0 \end{bmatrix},$$

where T is an upper triangular matrix $(p \times p)$ and 0 is a $(n-p) \times p$ matrix of zero elements.

Define also $\eta = L\varepsilon$ so that the original model

$$y = A\theta + \varepsilon$$

transforms to

$$z = H\theta + \eta.$$

(The geometric interpretation thus far is as follows. The design matrix A is considered as a set of vectors $a_1, \ldots, a_p$, where $A = \{a_1, a_2, \ldots, a_p\}$, and the vectors y, a_1, a_2, $\ldots$, a_p are considered in the n-dimensional space. Since the rank of A is p, the vectors a_1 to a_p span a subspace of dimension p. The orthogonal transformation $z = Ly$ effectively rotates the axes so that the first axis lies along a_1, the second in the plane $\{a_1, a_2\}$, the third in the hyperplane $\{a_1, a_2, a_3\}$ and so on.)

Since L is orthogonal

$$\eta'\eta = \varepsilon'L'L\varepsilon = \varepsilon'\varepsilon.$$

Also $\eta = L\varepsilon$ so that the η are normally distributed with common variance σ^2. Also

$$\hat{\theta} = (H'H)^{-1}H'z$$
$$= (A'L'LA)^{-1}A'L'y$$
$$= (A'A)^{-1}A'y$$

and

$$S_p = z'H\hat{\theta} = y'L'LA\hat{\theta} = y'A\hat{\theta}$$

and finally

$$\text{total SS} = z'z = y'L'Ly = y'y.$$

Hence $\hat{\theta}$, S_r and S_p are unchanged by the transformation $\mathbf{z} = \mathbf{Ly}$. (Because the transformation is simply a rotation of axes the SS, which are distances in the geometrical interpretation, will inevitably be unchanged. The relationship of the fitted model and the data, through the fitted parameter values, also must be unchanged by a rotation of axes.) Now

$$(\mathbf{z} - \mathbf{H}\theta)'(\mathbf{z} - \mathbf{H}\theta) = \sum_{i=1}^{p} \left(z_i - \sum_{j=1}^{p} t_{ij}\theta_j \right)^2 + \sum_{i=p+1}^{n} z_i^2.$$

The parameter estimates are given uniquely by the equations

$$z_i = \sum_{j=1}^{p} t_{ij}\hat{\theta}_j, \quad \text{or} \quad \mathbf{z}_1 = \mathbf{T}\hat{\theta},$$

where $\mathbf{z}_1$ includes the first p elements of $\mathbf{z}$. These equations can be solved directly and uniquely. But

$$E(z_i) = 0, \quad i > p \text{ since } z_i = \eta_i.$$

Hence S_r is the sum of $(n - p)$ squared normal variables, η_i, each with zero expectation and variance σ^2 and therefore S_r is $\sigma^2 \chi_{n-p}^2$, independent of $S_p = \sum_{i=1}^{p} z_i^2$.

Further, if $\theta < 0$, $E(z_i) = 0$, $i \leqslant p$ and so S_p is $\sigma^2 \chi_p^2$.

One direct consequence of the theorem is that to test the rather useless hypothesis that $\theta = 0$ the appropriate statistic is,

$$F = \frac{S_p/p}{S_r/(n-p)},$$

which is compared with the F distribution for p and $(n - p)$ df.

One further result for the entire parameter vector of parameters relates to the joint distribution of the parameter estimates. Since

$$(\mathbf{y} - \mathbf{A}\theta)'(\mathbf{y} - \mathbf{A}\theta) \text{ is } \sigma^2 \chi_n^2$$

and

$$(\mathbf{y} - \mathbf{A}\hat{\theta})'(\mathbf{y} - \mathbf{A}\hat{\theta}) \text{ is } \sigma^2 \chi_{n-p}^2,$$

and from Section 4.2

$$(\mathbf{y} - \mathbf{A}\theta)'(\mathbf{y} - \mathbf{A}\theta) = (\mathbf{y} - \mathbf{A}\hat{\theta})'(\mathbf{y} - \mathbf{A}\hat{\theta}) + (\hat{\theta} - \theta)'\mathbf{A}'(\hat{\theta} - \theta).$$

Then

$$(\hat{\theta} - \theta)'\mathbf{A}'\mathbf{A}(\hat{\theta} - \theta) \text{ is } \sigma^2 \chi_p^2,$$

independent of the true values of θ.

Hence, confidence intervals and confidence regions for the parameters can be constructed.

These distributional results extend to the partitioned parameter vector as follows.

Theorem If S_q is the fitting SS for θ_2, the extra SS, for which the algebraic expression derived earlier was

$$S_q = \hat{\theta}_2'(C^{-1})_{22}^{-1}\hat{\theta}_2,$$

then on the hypothesis that $\theta_2 = 0$, S_q is distributed as $\sigma^2\chi_q^2$, and is independent of S_r.

Proof Again, use the orthogonal transformation, **L**, such that

$$LA = \begin{pmatrix} T \\ 0 \end{pmatrix}$$

and consider $W = T^{-1}$ and the partitioned forms of **T** and **W**.

$$T = \begin{bmatrix} T_{11} & T_{12} \\ 0 & T_{22} \end{bmatrix} \quad W = \begin{bmatrix} W_{11} & W_{12} \\ 0 & W_{22} \end{bmatrix}$$

Now

$$C^{-1} = (A'A)^{-1} = (A'L'LA)^{-1} = (T'T)^{-1} = WW'.$$

Therefore

$$(C^{-1})_{22} = W_{22}W_{22}'.$$

Since

$$T_{22}W_{22} = I$$
$$(C^{-1})_{22}^{-1} = T_{22}'T_{22}$$

and therefore

$$\hat{\theta}_2'(C^{-1})_{22}^{-1}\hat{\theta}_2 = \theta_2'T_{22}'T_{22}\hat{\theta}_2$$

$$= \sum_{i=p-q+1}^{p} \left(\sum_{j=1}^{p} t_{ij}\theta_j \right)^2$$

$$= \sum_{i=p-q+1}^{p} z_i^2$$

and, on the hypothesis $\theta_2 = 0$, $E(z_i) = 0$, $p - q + 1 \leqslant i \leqslant p$ and S_q is $\sigma^2\chi_{n-p}^2$.

This result provides the basic theory for interpreting analyses of variance. The first consequence is that, on the hypothesis that $\theta_2 = 0$,

$$\frac{S_q/q}{S_r/(n-p)}$$

is distributed as F for q and $n - p$ df and so the extra SS can be used formally to assess the parameters θ_2.

The χ^2 distributional results also allow dfs to be added to the analysis of variance structure derived in Sections 4.5 and 4.A5 (see Table 4.16).

It is important to emphasise that, when θ is partitioned into θ_1 and θ_2, then, in the second theorem of the previous section, the two subsets of

Table 4.16.

Source	SS	df
Due to θ_1 (ignoring θ_2)	S_{p-q}	$p-q$
Due to θ_2 (allowing for θ_1)	S_q	q
Residual	S_r	$n-p$

parameters have quite distinct interpretations. We cannot simply reverse S_q and S_{p-q} and use S_{p-q} to test the hypothesis that $\theta_1 = 0$. In most practical situations, this would not even be a practically desirable procedure. Thus, if θ_1 consists of the set of block effects in a randomised block design and θ_2 is the set of treatment effects, we should not be interested in testing whether $\theta_1 = 0$ because the experiment is designed to allow for block differences, and therefore the analysis of experimental data must reflect this aspect of design. In other words, we should not consider obtaining estimates of treatment effects without simultaneously allowing for the block effects.

However, suppose that we were interested in inferences about θ_1 in addition to those about θ_2 and that we have fitted θ_1 first ignoring θ_2 and subsequently fitted θ_2 allowing for the effects of θ_1. Consider again the expected values of the z_i,

$$
\begin{aligned}
E(z_i) &= \sum_{j=1}^{p} t_{ij}\theta_j \\
&= \sum_{j=1}^{p-q} (t_{ij}\theta_j) + \sum_{j=p-q+1}^{p} (t_{ij}\theta_j).
\end{aligned}
$$

The crucial difference between the expressions for the expected values for the first $(p-q)$ z_i's and for the next q is that, for the latter z_i's, the expected values do not involve θ_1 because of the triangular form of T, whereas the first $(p-q)$ z_i's do involve θ_2.

The necessary conditions for $S_{p-q} = \sum_{i=1}^{p-q} z_i^2$ to be distributed as σ^2 times χ^2_{p-q} are not only that $\theta_1 = 0$, which would set the first sum equal to zero, but also that

$$
t_{ij} = 0 \text{ for } 1 < i < p-q, \; p-q+1 < j < p.
$$

If the t_{ij} coefficients of θ_2 in the expected values of the first $(p-q)$ z_i's are zero, then hypotheses about the θ_1 parameters may be tested independently of the values of θ_2. The condition can be written

$$
\mathbf{T}_{12} = \mathbf{0}.
$$

This is equivalent to $\mathbf{C}_{12} = 0$, so that $\mathbf{C}$ partitions in the pattern

$$\mathbf{C} = \begin{bmatrix} \mathbf{C}_{11} & 0 \\ 0 & \mathbf{C}_{22} \end{bmatrix}$$

and the first $p-q$ columns of the design matrix $\mathbf{A}$ are orthogonal to the last q columns. Thus the condition $\mathbf{C}_{12} = 0$ shown in Section 4.A5 to be sufficient for independent interpretation of the two SS S_{p-q} and S_q is here shown to be necessary also.

4.A7 Treatment comparisons and component sums of squares

A general treatment comparison L_k is defined as a linear contrast of treatment effects

$$L_k = \sum_j l_{kj} t_j,$$

where

$$\sum_j l_{kj} = 0.$$

Two treatment comparisons, L_k, $L_{k'}$, are orthogonal if

$$\sum_j l_{kj} l_{k'j} = 0$$

if $k \neq k'$. A scale-standardised form of comparison is defined as

$$c_{kj} = l_{kj} \Big/ \left(\sum_j l_{kj}^2 \right)^{1/2}$$

Now the coefficients c_{kj} satisfy

$$\sum_j c_{kj} = 0$$

$$\sum_j c_{kj}^2 = 1$$

$$\sum_j c_{kj} c_{k'j} = 0,$$

$k' \neq k$. Consider a set of $(t-1)$ orthogonal comparisons. We can write

$$\mathbf{x} = \mathbf{C}\hat{\mathbf{t}},$$

where the matrix $\mathbf{C}$ has elements c_{kj} ($k = 1, 2, \ldots, t-1, j = 1, 2, \ldots, t$), and the vector $\mathbf{x}$ is the set of estimates of the $(t-1)$ orthogonal standardised contrasts for $L_1, L_2, \ldots, L_{t-1}$. The orthogonality of the contrasts means that the x_k are independent. If the observations are assumed to be normally distributed then $\mathbf{x}$ is a vector of normally and independently distributed variables, with each element having zero mean and unit variance.

Define $\mathbf{C}^*$ to be the matrix $\mathbf{C}$ with an added row vector consisting of

$t^{-1/2}$ in each column, so that

$$C^* = \left[\begin{array}{c} C \\ t^{-1/2}, t^{-1/2}, \ldots, t^{-1/2} \end{array} \right]$$

then $C^* C^{*\prime} = I$ and C^* is orthogonal. Now

$$x_k^2 = \left(\sum_j c_{kj} \hat{t}_j \right)^2$$

$$= \left(\sum_j l_{kj} y_{\cdot j} \right)^2 \Big/ \left(\sum_j l_{kj}^2 \right)$$

$$= \left[\sum_j l_{kj} \left(\mu + t_j + \sum_i \varepsilon_{ij} \Big/ n \right)^2 \right] \Big/ \left(\sum_j l_{kj}^2 \right),$$

where n is the number of observations per treatment (or the effective replication when treatments are not orthogonal to blocks but treatment differences are all equally precise). Since

$$\sum_j l_{kj} = 0$$

$$x_k^2 = \left\{ \sum_j l_{kj} \left(t_j + \sum_i \varepsilon_{ij} \Big/ n \right)^2 \right\} \Big/ \left(\sum_j l_{kj}^2 \right).$$

Hence

$$E(x_k^2) = \left(\sum_j l_{kj} t_j \right)^2 \Big/ \left(\sum_j l_{kj}^2 \right) + \left(\sum_j l_{kj}^2 \sigma^2 \right) \Big/ \left(\sum_j l_{kj}^2 n \right)$$

$$= \left(\sum_j l_{kj} t_j \right)^2 \Big/ \left(\sum_j l_{kj}^2 \right) + \sigma^2 / n.$$

But if $L_k = 0$, $E(x_k) = 0$ and, since x_k is normally distributed, then on the hypothesis $H_k: L_k = 0$,

$$n x_k^2 \text{ is } \sigma^2 \chi_1^2.$$

Now consider $x'x = \hat{t}'C'C\hat{t}$. Since

$$C^{*\prime}C^* = C'C + J(1/t),$$

where J is the matrix whose elements are all 1, then

$$\hat{t}'C^{*\prime}C^*\hat{t} = \hat{t}'C'C\hat{t} + (1/t)\hat{t}'J\hat{t}.$$

However, C^* is orthogonal and $J\hat{t} = 0$, because of the constraints $\sum_j t_j = 0$. Hence

$$\hat{t}'C'C\hat{t} = \hat{t}'\hat{t}.$$

Consequently

$$x'x = \hat{t}'\hat{t}$$

$$= \sum_j (y_{\cdot j} - y_{\cdot \cdot})^2$$

$$= \sum_j \{ Y_{\cdot j}/n - Y_{\cdot \cdot}/(nt) \}^2$$

$$= (\text{treatment SS})/n.$$

Hence the SS for comparison L_k is identified in the form

$$SS(L_k) = nx_k^2 = \left(\sum_j l_{kj} Y_{.j}\right)^2 \bigg/ \left(n \sum_j l_{kj}^2\right),$$

with the properties that

   (i) If $L_k = 0$, the SS is σ^2 times a χ^2 variable on 1 df, and

   (ii) the sum of the SS for $(t-1)$ orthogonal comparisons is the total treatment SS.

5

Computers for analysing experimental data

5.1 Introduction

There are at the present time a wide range of computer programs and packages for the analysis of data from designed experiments (a package is essentially a larger program with a well-defined internal language which allows the user to determine the particular use of the package), and it is clear that the number and diversity of programs and packages will increase. It would be foolish in any book to give precise advice on which programs and packages to use. Such advice would rapidly become out of date, might be irrelevant to many readers immediately because of a lack of availability of some programs, and would be clearly biassed by the personal preference of the writer. Instead I shall explain the principles on which different types of program for the analysis of experimental data may be based and will attempt to advise on the characteristics which should be sought in selecting appropriate computer packages.

It is informative to trace some of the developments of computer programs for the analysis of experimental data. At the time of writing the entire history is a mere third of a century. When computers first became generally available for statistical use, the methods for analysing experimental data were already well developed. There was a wide range of designs for which the algebraic definition for the calculation of the analysis of variance was well defined. The natural way to attempt to develop statistical programs was to aim to provide programs to perform the calculations, which were then laboriously achieved by manual calculation. There could, at that time, be no concept of the increases in computer speed and size that would become available so that thoughts about the possible generality of computer programs were inevitably not considered.

Initially, then, many statisticians wrote programs to analyse randomised complete block designs, latin square designs and factorial designs, and the complexity of design catered for by these programs gradually increased. Typically, institutions with large experimental programmes developed suites of programs for different designs. The different programs at a single institution tended to have some standardisation of input and output. New programs were developed mainly through modification of existing programs. Special routines for missing values or for covariance were devised and added to many programs in each suite.

Gradually aspirations to generality led to attempts to replace the suites of programs by general packages which would provide analyses for most of the frequently used designs. Initially the dominant characteristic of designs which the general programs were constructed to accept was orthogonality. Because orthogonality allowed the separate calculation of sum of squares, and treatment means for orthogonal 'factors', general programs could allow completely unrestricted orders for the introduction and definition of the factors to be included in the model. Programs used two basic principles. Either the sums of squares and effects were calculated directly from the original yields, or the calculations for each set of levels were performed on the residuals from the previous sets of calculations using the philosophy of the sweeping methods developed by Kuipers (1952), Corsten (1958) and Wilkinson (1970). The latter approach was one of the first indications that computers would use different forms of calculations from those in general use before computers.

The major change which encouraged thoughts of computer programs which could handle a very much wider range of designs, far wider than was conceivable in pre-computer days, was the development of computer methods for handling matrices. After the introduction of analysis programs and packages based on the matrix algebra of the previous chapter, the scope of the programs and packages that could be made available was determined by the objectives of the programmer or the 'package philosopher'. In attempting to classify the resultant spectrum of programs and packages three major dimensions may be recognised:

    (i) the (specific–general) dimension,
   (ii) the (basic principle–user friendly) dimension,
  (iii) the (simple–complex) provision of information dimension.

To these can be added the fourth, methodological, dimension contrasting the factor and regression model approaches to the analysis.

In the next two sections we examine the range of possibilities for the first three main dimensions and will try to specify requirements that programs should satisfy. The philosophies of the factor and regression approaches will be discussed in Sections 5.4 and 5.5. Finally the implications of the computer revolution for the design of experiments, and for the remainder of this book, will be discussed. Because the computer packages include the possibility of analysing data from any of the designs discussed in this book, this chapter includes references to concepts which are introduced formally in chapters after this one. I hope this will cause at most minor disturbance to the reader.

5.2   How general, how friendly

The programs now available range from the simple program written by an experimenter to analyse an experiment with four randomised complete blocks for five treatments, to packages which can handle almost any designed experiment. How should the user choose the programs that should be available in the institution where he or she works, and how should the program to be used in any particular situation be chosen?

The distinction between specific and general programs is not absolute. A specific program could provide facilities sufficiently extensive to analyse:
   (i) randomised block designs with one, two, three or four treatment factors,
   (ii) latin square designs with three treatment factors,
   (iii) simple split plot designs, and
   (iv) balanced incomplete block designs,
whereas a general package might aim to allow the analysis of any orthogonal multiple factor block–treatment design with only a single error term. Some general programs will not provide some of the facilities offered by some specific programs.

There is one very important benefit which is implied by the decision to implement a general program. The limits on available experimental designs have been primarily imposed by the available computational facilities and the mathematical knowledge necessary for devising more complex methods of analysis within the limitation of those facilities. The history of the development of new experimental designs has reflected the increasing mathematical ingenuity of statisticians in finding design patterns for which the analysis of data could be performed using only the limited computing facilities offered by manually operated calculating machines. Now that packages exist which can provide the analysis for any experiment, including experimental structures deformed by accidents, experimenters and statisticians should avoid choosing to equip themselves only with programs which perform, much more quickly, the analysis of designs available in the pre-computer era. Correspondingly the limits on available experimental designs must change.

There are, of course, advantages in using specific programs. If the data for analysis come from a randomised block design with three blocks of a 2^3 factorial set of treatment combinations and a program is available for just that design then the program will require nothing more than the set of data, perhaps with identification of the factor names, and a definition of the order in which the data is to be presented. A more general program may require preliminary information about the numbers of blocks and of factors and factor levels, and a definition of the pattern of treatment allocation to blocks. It may be tempting to retain some of the simpler, more specific, programs and thereby reduce the need to come to terms with the more general programs. I do not believe that research resources will be used with maximum efficiency if we voluntarily prescribe the scope of experimental designs by relying on specific programs.

One problem with general programs, already implied in the previous discussion, is the knowledge which the user requires before using the program. The difficulty arises directly from the generality of the program. The more general the program, the more information that is required by the program before it can provide the analysis of any particular data set, and the more complex the language required for communication between user and program. The description of input and output information requires the development of a communication system, and this is a difficult skill with a strong educational requirement. Far too little attention has been paid to this aspect of the development of general statistical packages. The need to understand the particular form in which information must be provided for programs is obvious. Input to general programs will usually involve various forms of coded information devised to minimise the actual amount of typed information required to activate the program. To illustrate the variation of forms in which output from a statistical package can appear, we shall consider the analyses of a simple randomised block design with a three-factor treatment structure. The experiment is from an intercropping research programme in which methods of growing crops of maize and cowpea together (in alternate rows) were being investigated. The three treatment factors were maize varieties, cowpea varieties and nitrogen level. There were three blocks with 24 plots in each block. The data presented in Table 5.1 are the cowpea yields (kg/ha) for the 72 plots.

We consider the analysis of cowpea yields using two of the general packages currently available. Any general package necessarily develops its own special terminology and imposes its own restrictions. To the user

Table 5.1. *Cowpea yields* (kg/ha).

Maize variety		1			2			3		
	Block	I	II	III	I	II	III	I	II	III
Cowpea variety	Nitrogen level									
A	n_0	259	645	470	523	540	380	585	455	484
A	n_1	614	470	753	408	321	448	427	305	387
A	n_2	355	570	435	311	457	435	361	586	208
A	n_3	609	837	671	459	483	447	416	357	324
B	n_0	601	707	879	403	308	715	590	490	676
B	n_1	627	470	657	351	469	602	527	321	447
B	n_2	608	590	765	425	262	612	259	263	526
B	n_3	369	499	506	272	421	280	304	295	357

familiar with these characteristics of the package these will appear straightforward and the user will be accustomed to working within the limitations of the package. It is not possible to adopt a totally objective approach to the assessment of packages and I do not pretend to be objective. What I asked of each package was to produce an initial analysis of variance followed by the tables of treatment means allowing all important main effects and interaction comparisons to be interpreted. The consideration of treatment contrasts was not included in this initial analysis. The two packages used will be recognised by many readers and to try to prevent such recognition would be pointless. Where it is felt that the examples do not use the facilities of the package as effectively as is possible then this is at least partly a reflection on the difficulty that I (and my programming experts) experienced in trying to use the package.

The output from the first package is shown in Box 5.1. Some aspects which I believe are particularly commendable are:

    (i) The avoidance of large numbers of superfluous figures (three digits per treatment means is sensible);

    (ii) the information about the percentage subdivision of the SS in the analysis of variance (SS%);

    (iii) the fairly simple form in which the very important standard errors of differences (SED) are provided.

Some aspects which might hinder the unfamiliar user are:

    (i) the use of the term 'stratum' and the repetition of the total variation line of the analysis which arise because the package is more general than is required for this example or many others;

Box 5.1

```
Box 5.1

Source of variation      df      SS      SS%     MS      VR

BLOCK.*UNITS* STRATUM
  BLOCK                   2    73014    4.60   36507    2.802
  MAIZE                   2   409446   25.81  204723   15.712
  COWPEA                  1     6013    0.38    6013    0.462
  NITRO                   3   113064    7.13   37688    2.893
  MAIZE.COWPEA            2     9910    0.62    4955    0.380
  MAIZE.NITRO            6    67563    4.26   11261    0.864
  COWPEA.NITRO           3   172403   10.87   57468    4.411
  MAIZE.COWPEA.NITRO     6   135379    8.54   22563    1.732
  RESIDUAL              46   599358   37.79   13030
TOTAL                  71  1586150  100.00   22340

GRAND TOTAL            71  1586150  100.00

GRAND MEAN                    476
TOTAL NUMBER OF OBSERVATIONS 72

**** TABLES OF MEANS ****

VARIATE: YIELD

  GRAND MEAN     476

        BLOCK      1     2     3
                 444   463   519

        MAIZE      1     2     3
                 582   430   415

        COWPEA     1     2
                 467   485

        NITRO      1     2     3     4
                 539   478   446   439

        COWPEA     1     2
        MAIZE
            1    557   606
            2    434   427
            3    408   421

        NITRO      1     2     3     4
        MAIZE
            1    593   598   554   582
            2    478   433   417   394
            3    547   402   367   342

        NITRO      1     2     3     4
        COWPEA
            1    482   459   413   511
            2    597   497   479   367

        COWPEA     1                 2
        NITRO      1     2     3     4     1     2     3     4
        MAIZE
            1    458   612   453   706   729   585   654   458
            2    481   392   401   463   475   474   433   324
            3    508   373   385   366   585   432   349   319

*****STANDARD ERRORS OF DIFFERENCES OF MEANS *****
```

TABLE	BLOCK	MAIZE	COWPEA	NITRO	MAIZE COWPEA	MAIZE NITRO	COWPEA NITRO
REP	24	24	36	18	12	6	9
SED	33.0	33.0	26.9	38.0	46.6	65.6	53.8

TABLE	MAIZE COWPEA NITRO
REP	3
SED	93.2

```
***** STRATUM STANDARD ERRORS AND COEFFICIENTS OF VARIATION *****
```

STRATUM	DF	SE	CV%
BLOCK.*UNITS*	46	114.1	24.0

(ii) the automatic provision of all sets of treatment means when the analysis of variance suggests that only the effects of maize varieties and the two-way table for cowpea varieties × nitrogen are of any importance. The balance between providing all necessary information, and producing such a forest of numbers that sorting out the important ones is difficult, is a delicate one, and really requires interactive use of a package.

The output from the second package is shown in Box 5.2. Aspects of the output which I believe to be beneficial are:

(i) the simple presentation of the initial analysis of variance with the error mean square clearly emphasised;

(ii) the explanation of methodology for the reader (except for the unexplained terms 'comparisonwise' and 'experimentwise error').

Aspects which make the output unsatisfactory are:

(i) the excessive numbers of digits;

(ii) the emphasis on significance tests both in the analysis of variance and the presentation of tables of means;

(iii) the absence of simple standard errors of difference enabling the reader to interpret the results. None of the tables of means can be produced with such standard errors and the two-way table has no estimate of precision and requires additional calculations by the user before interpretation is possible;

(iv) the use of automatic testing procedures (various alternatives to LSD are available) which encourage incorrect interpretation of results.

The interpretation of treatment comparisons and the reasons for avoiding automatic testing procedures are discussed in detail in Chapter 12. At this stage my concern is to illustrate the very substantial difference in treatment of a standard set of data by two major statistical packages, and the difficulties which the user may encounter in the search for a simple, appropriate analysis for each set of data from a designed experiment.

I believe that there is a major need in the present development of statistical packages for a cooperative attempt to produce an interactive user-friendly, statistically valid and comprehensive general package. The cooperation should include applied statisticians, computer scientists, teachers and marketing experts. Any such package which is both user-friendly and general should include, within its normal mode of operation, an educational component enabling (or requiring) the user to check the adequacy of his or her understanding of the interpretation and assumptions of the analysis, and providing the means to develop an adequate understanding.

Box 5.2

Box 5.2

```
Box 5.2


SOURCE   DF   SUM OF SQUARES    MEAN SQUARE    F VALUE  PR>F    R-SQUARE    CV

MODEL    25   986793.41666      39471.73667     3.03    0.0006  0.622130  23.9973

ERROR    46   599358.58334      13029.53442      ROOT MSE            YIELD MEAN

TOTAL    71   1586152.00000                     114.14699            475.66667

SOURCE               DF          ANOVA SS         F-VALUE      PR > F

BLOCK                 2     73014.08333333         2.80        0.0711
NITRO                 3    113064.44444444         2.89        0.0453
MAIZE                 2    409446.33333333        15.71        0.0001
NITRO*MAIZE           6     67563.22222222         0.86        0.5282
COWPEA                1      6013.38888889         0.46        0.5003
NITRO*COWPEA          3    172402.83333333         4.41        0.0083
MAIZE*COWPEA          2      9910.11111111         0.38        0.6858
NITRO*MAIZE*COWPEA    6    135379.00000000         1.73        0.1350


T TESTS (LSD) FOR VARIABLE: YIELD
NOTE: THIS TEST CONTROLS THE TYPE I COMPARISONWISE ERROR RATE
      NOT THE EXPERIMENTWISE ERROR RATE.

ALPHA=0.05  DF=46  MSE=13029.5
CRITICAL VALUE OF T=2.01290
LEAST SIGNIFICANT DIFFERENCE=66.3277

MEANS WITH THE SAME LETTER ARE NOT SIGNIFICANTLY DIFFERENT.

T          GROUPING      MEAN      N    BLOCK

                 A       519.33    24   3
                 A
         B       A       463.38    24   2
         B
         B               444.29    24   1

LEAST SIGNIFICANT DIFFERENCE=76.5887
MEANS WITH THE SAME LETTER ARE NOT SIGNIFICANTLY DIFFERENT.

T          GROUPING      MEAN      N    NITRO

                 A       539.44    18   1
                 A
         B       A       478.00    18   2
         B
         B               446.00    18   3
         B
         B               439.22    18   4

LEAST SIGNIFICANT DIFFERENCE = 66.3277
MEANS WITH THE SAME LETTER ARE NOT SIGNIFICANTLY DIFFERENT

T          GROUPING      MEAN      N    MAIZE

                 A       581.92    24   1

                 B       430.50    24   2
                 B
                 B       414.58    24   3

LEAST SIGNIFICANT DIFFERENCE = 54.1564
MEANS WITH THE SAME LETTER ARE NOT SIGNIFICANTLY DIFFERENT

T          GROUPING      MEAN      N    COWPEA

                 A       484.81    36   2
                 A
                 A       466.53    36   1


                        MEANS
              NITRO    COWPEA    N      YIELD
                1        1       9    482.333333
                1        2       9    596.555556
                2        1       9    459.222222
                2        2       9    496.777778
                3        1       9    413.111111
                3        2       9    478.888889
                4        1       9    511.444444
                4        2       9    367.000000
```

5.3   Requirements of packages for the analysis of experimental data

I suspect that the requirements for a general package for the analysis of data from designed experiments are not very different from those for data from observational studies. Nevertheless this discussion will be restricted to experimental data and it will be left to others to make comparisons with observational data. The requirements set out below are an attempt to include everything that is important. It is not expected that a program satisfying all the requirements will be available (if such a program emerges, some more requirements can be added!).

1 *Variables* An initial requirement which applies to all analysis programs is that the program can handle many variables for the same design structure, without repetition of information about that structure. This must include the facility to derive new variables as functions or combinations of the original variables.

2 *Structure* The program should accept *any* structure of blocking and treatment factors. Either explicitly, or implicitly, the structure is defined by the set of experimental units and an indexing of each unit within the set of levels for each factor.

3 *Multiple levels of variation* The program must be able to analyse, in a single analysis, multiple error designs; that is, designs in which information is available about treatment effects in different levels of the design structure. The program should be capable of performing all the calculations for each level separately and should present results, with correct information about precision, for separate and combined analyses. Further discussion of the concept of combining information from different levels of a design is included in Chapters 7, 14 and 15.

4 *Fitting order* The program must be able to include terms representing sets of effects in the fitted model in any order as specified by the user, and the definition of the order of fitting terms must be simple. The strong implication is that the program must be usable interactively. The order of fitting is important in the interpretation of results, and the determination of the order or, in many cases, orders, in which terms are fitted, must be directly controllable by the user. Default settings for order are not appropriate and positive action should be required of the user. However, there should be restrictions on the order to avoid the possibility of fitting an interaction term before at least one of the corresponding main effects is fitted.

5 *Treatment contrasts* The program should allow, and generally require, the user to specify contrasts of levels of treatment factors to represent the questions which the experiment is intended to answer (see Chapter 12 for discussion about objectives).

6 *Quantitative variables* The program should allow the inclusion of terms representing regression dependence on quantitative variables in the fitted model. The quantitative variables may be covariates (Chapter 10), or may represent levels of a treatment factor in an alternative form to allow regression of the analysed variable on quantitative treatment factor levels. Within the model quantitative and qualitative terms should have equal status.

7 *Multivariate* The program should allow the joint analysis of several variables, providing a bivariate analysis, or more generally a multivariate analysis of variance.

8 *Subset analysis* The program should allow the analysis of a subset of the experimental units, the subset being definable in terms of the defined structure of the experimental units or through intervention of the user. Correspondingly the program should allow some variables to be defined for only a subset of units.

9 *Non-normal error* The program should permit the definition of non-normal error structures. Essentially the whole set of ideas relating to generalised linear models (discussed in Chapter 11) should be implementable within the program.

10 *User specified additions* The program must present and identify the results from any analysis in such a way that the user can define additional calculations and specify the form of presentation of information within the operation of the program. This will include an extensive provision of graphical facilities, simple arithmetical operations and even matrix manipulation.

11 *Presentation of information* The program must provide the user with all information necessary to interpret the results. This information must be clear and succinct without superfluous detail. The essential requirements must include
   (i) the initial subdivision of information in an analysis of variance including alternative orders of fitting terms, and

(ii) tables of fitted mean values with standard errors for differences between means for all important effects (and not for unimportant effects).

12 *Interactive education* Most important of all, the program must be so written that the facilities provided can be described and presented in such a manner that each user may obtain as much information as is necessary to interpret the results. This will require an initial definition of the general level of statistical knowledge expected of the user, and thereafter the program must allow the possibility of the user requesting information about the methods used. Indeed, the program should provoke the user to assess whether he or she does fully understand the procedures and methods which are being used in the program. Of course, the initial definition could be set at a sufficiently high level of statistical sophistication that no further information could possibly be required. This represents an extreme of user-unfriendliness and should not be considered appropriate in an age when many computer programs are going to be very easily available. Users will use user-friendly programs regardless of whether they are statistically valid.

13 *Data base information* Looking rather beyond the present situation, programs and packages should have facilities for consulting and summarising the results of analyses from similar experiments and experimental material. Consider the analysis of data from the experiment on the effects of two cowpea varieties, three maize varieties and four nitrogen levels, with the variable to be analysed being the yield of cowpea when intercropped with maize. There will have been other experiments on cowpea, possibly involving some of the varieties used in the current experiment, possibly from experiments on intercropping, possibly even including nitrogen levels (though obviously not for an experiment solely on cowpea). The results for these experiments and the consequent analyses should be available in a data bank in a form which can be accessed by the program being used to analyse the current data. The program should be capable of extracting either general information from previous analyses (error mean square, general mean, CV) or specific information about mean yields and standard errors for treatments or factor levels common to both the current and previous data sets.

5.4   The factor philosophy of analysis programs

As mentioned previously the earlier programs for the analysis of experimental data were all based on the pattern of calculations used before computers. The underlying concept was that the units were classified according to several factors, such as blocks, treatments, rows, columns, levels of factor A, levels of factor B and so on. The classes, or levels, of each classification, or factor, are viewed symmetrically as expressed in the model

$$y_{ijk} = \mu + a_i + b_j + c_k + \varepsilon_{ijk}.$$

The analysis of variance provides information about the sets of parameters (a_i), (b_j), (c_k), and an extimate of σ^2, the variance of the ε_{ijk}.

There are two computational problems in the factor-based programs. The first arises from the interrelationships between the different factors; the second from the non-absoluteness of the parameter values.

The majority of simple programs for the analysis of variance handle only designs in which all factors are orthogonal to each other. Most of the classical experimental designs, and most of the designs actually intended in experiments, assume orthogonal relationships between factors. The consequences of orthogonality are that the SS for the different factors are calculated independently. For orthogonal designs there is no need to consider the order of fitting sets of parameters, and the actual calculation of SS can be achieved in various ways, either directly from the raw data, or successively from the residuals from previous stages in the fitting.

When factors are not orthogonal the interpretation of the analysis of variance depends in varying degrees on the pattern of interrelationships between the different factors. The basic theoretical results required for the analysis of non-orthogonal factors is that derived in Section 4.5 on partitioning. For two factors, whose levels are represented by sets of parameters θ_1 and θ_2, respectively, the estimates of θ_2 after adjustment for θ_1 are given by

$$\hat{\theta}_2 = (C^{-1})_{22}(A_2' - C_{21}C_{11}^{-1}A_1')y$$

and the SS for fitting θ_2 allowing for θ_1 is

$$\hat{\theta}_2(C^{-1})_{22}^{-1}\hat{\theta}_2,$$

where the terminology is that used in Chapter 4 for the partitioning of the matrices A, C and C^{-1} for θ_1 and θ_2. The crucial characteristic, computationally, is the pattern of the matrix $(C^{-1})_{22}$. If the off-diagonal terms of this matrix are all equal then the parameters in the set θ_2 are said to be *balanced*. The concept of balance is examined in Chapter 7. More generally any pattern of the matrix $(C^{-1})_{22}$ allows a more or less simple form of analysis, and general analysis programs can be defined to allow different degrees of complexity of this matrix. The degree of complexity permitted within the program defines the class of designs which the program can analyse.

The second problem is that of constraints, discussed in Section 4.4 and arising from the fact that the sets of parameters corresponding to the levels of each factor are not uniquely defined if there are two or more factors. Consider the model

$$E(y_{ij}) = a_i + b_j.$$

It is not possible from any set of observations on (i, j) combinations to deduce absolute values for either set of parameters without imposing an arbitrary restriction on the other set of parameters. Inferences are possible only for differences within a single set of parameters. This restriction may be managed in various ways either by imposing an arbitrary, non-symmetric constraint on the set of parameters (e.g. $a_1 = 0$) or by adopting a symmetric constraint which is suitable in the context of the generality of the $(C^{-1})_{22}$ matrix which the program is designed to handle.

A further difficulty arises from the attempt to retain the symmetry of levels within a factor. If particular treatment contrasts are of interest then these will, almost invariably, be non-symmetrical in nature. To include a subdivision of treatment sums of squares corresponding to particular contrasts will require a different philosophy from the symmetry of the factor approach and will often be difficult or impossible in the context of a general analysis of variance program based on the factor approach.

The advantages of a general analysis of variance program based on a linear model using a factor approach to models are that the computational methods may be simple and efficient within the degree of generality permitted by the program. It is possible to retain the symmetry of factor levels. The disadvantages are that there is a restriction on the designs which can be analysed, and non-symmetric questions about the factor levels have to be considered separately from the main pattern of the program.

There is an alternative approach to analysing data using the factor approach, the 'sweeping' methods mentioned in Section 5.1. The fundamental concept involves estimating the effects of each level of a factor by sweeping through the data, attributing each yield to the relevant level of the factor, calculating a mean effect for each level and subtracting the mean for each level effect from all yields corresponding to that level. The reduction in the SS of the residuals represents the SS due to the factor used in the sweep. For orthogonal designs the set of SS produced immediately is exactly the same as those produced by the more familiar methods. For non-orthogonal designs the sweeping has to be applied in an iterative scheme. This produces the correct SS for sets of factorial effects and enables any form of non-orthogonal design to be handled. Few analysis of variance packages have adopted the sweeping approach, but recently interest in the method has revived and the method is being used as the basis for new packages.

5.5   The regression model for analysis programs

The alternative to a factor-based approach for the construction of analysis programs is to view all models as essentially multiple linear regression models. Instead of a set of parameters (a_i) we now think of the parameters individually, $a_1, a_2, \ldots, a_m$. This approach immediately removes the question of which classes of design can be handled by the program since, for any design which actually contains information about all the required parameters, the appropriate model can be fitted and analysed. The constraint problem must be addressed directly by the elimination of one parameter from each set (or several parameters from sets of parameters for interaction effects). Essentially if a_1 is omitted or set equal to zero then the remaining parameters in the set represent deviations from a_1.

The history of the analysis of data from designed experiments has been dominated by the factor philosophy of analysis because of the easy form of calculations for orthogonal designs. The benefits of the regression approach have been less appreciated because the computational problems of fitting multiple regression models were prohibitive before computers were available. There are three main advantages of the regression approach.

(i) Any data structure can be accommodated, and this includes any experimental design from which some observations are not available. The implications for design will be considered further in the next section.

(ii) The parameters corresponding to treatments can be written in the form most appropriate to the questions which the treatments were chosen to answer. In other words, the model may be expressed directly in terms of treatment contrasts.

(iii) The models may be generalised to allow a much wider class of distributional assumptions.

The advantage of choosing parameters to represent treatment comparisons directly should be considerable. For example, if three equally spaced levels of a quantitative treatment factor are chosen, the pattern of yield response to the factor can be assessed by using a quadratic response relationship. This can be directly represented either by replacing the model

$$y_{ij} = a_i + b_j + \varepsilon_{ij} \qquad i = 1, 2, 3$$

by the contrast model

$$y_{ij} = a_L x_L + a_Q x_Q + b_j + \varepsilon_{ij}$$

where x_L and x_Q have values $(-1, 0, 1)$ and $(1, -2, 1)$ for $i = (1, 2, 3)$, or equivalently by the more obvious (and equivalent) regression model

$$y_{ij} = \beta(x_i - \bar{x}) + \gamma(x_i - \bar{x})^2 + b_j + \varepsilon_{ij}.$$

Models allowing the quantitative response for levels of factor A to vary over levels of factor B can be written similarly.

More generally, any set of treatment contrasts may be represented directly as model parameters. This should have the considerable advantage of encouraging the experimenter to think of the analysis of data directly in relation to the original questions instead of going through two stages: first obtaining a measure of the overall variation among the experimental treatments, and subsequently defining a meaningful set of treatment comparisons.

A recent major development has been that of generalised linear models in which the error structure can be modelled quite separately from the model for the expected value of y. The use of generalised linear models is introduced in Section 11.4 of this book. A full account is given in the monograph by McCullagh and Nelder (1983).

5.6   Implications for design

It is extremely important to realise that the possible scope for the subject of experimental design has been completely changed by the advent of the computer. The chronological development of the theory of experimental design can be seen, with the benefit of hindsight, to consist of a sequence of designs of increasing complexity of analysis. The early orthogonal designs were extended to include non-orthogonal designs as mathematical devices were discovered to allow the simple computational analysis of data. It is certainly possible to argue that the statistical considerations of the efficiency of designs for providing answers to questions became of secondary importance to the consideration of the possibility of the analysis of the data resulting from the design.

Not only was the theoretical development of design theory constricted by the necessity of a practical form of analysis, and incidentally a method of analysis restricted to the algebraically manipulable theory of least squares estimation, but experimental research developed a very limited catalogue of designs deemed suitable for practical experiments. For the vast majority of experiments all the statistical ideas of blocking became identified with the randomised complete block design with the result that many practical experiments have been and still are designed in the randomised complete block format using blocks which are a complete travesty of the principle of blocking.

It is time for a complete reappraisal of the principles of experimental design without the restraints imposed by the limits of computational practice which existed when such admirable books as Cochran and Cox (1957), Cox (1958), or Kempthorne (1952) were written. In the rest of this book I shall attempt to examine the important concepts of the statistical philosophy of experimental design assuming the availability of general packages for the analysis of experimental data which allow the analysis of data from any experimental design. The mathematical theory of general linear models is necessary to permit a comprehensive assessment of the statistical concepts, but these concepts only rarely require mathematical theory for their development; consequently the following chapters employ very little mathematics.

PART II

FIRST SUBJECT

6

Replication

6.0   *Preliminary example*

The student explained that his was a simple problem. He had done an experiment with eight treatments, each treatment replicated ten times. The analysis of variance seemed straightforward, the salient points being as follows:

	df	F
treatments	7	500
error	72	
total	79	

He wanted to be assured simply that his analysis was correct. Apart from a feeling that $F = 500$ suggested that ten replicates constituted overkill, there was nothing apparently suspicious. Requests for more information elicited a description of the units as small discs cut from leaves of a particular plant species, innoculated with spores of a fungus. The treatment consisted of injecting one of two fungicides into either the upper or lower sides of the leaf, using old or young leaves, the total set of treatments being old/young × upper/lower × two fungicides. Suspicions and questions now begin to emerge in the statistician's mind. What was the design? The discs for the old and young leaves must be from different leaves. The analysis implies that all the comparisons are at the same level of variation and hence that all 80 discs came from 80 different leaves. However, many alternative designs are possible. To determine whether the analysis was correct the statistician must identify precisely the design of experiment. How many leaves per plant? How many discs per leaf? How exactly were the leaves chosen?

6.1   The need for replication

In this chapter, we consider various aspects of replication, which is an extremely important concept in experimental design. The first and most vital idea is of replication itself and we commence with the fundamental argument for replication. Consider the situation when farmer A uses variety X in one field and gets a yield of 24.1; while farmer B uses variety Y in one field and gets a yield of 21.4. Farmer A claims that his cultivation method is better than farmer B's; the seed merchant producing variety X claims it is better than variety Y. But, without any further information, the difference could be simply due to different fertility levels in the two fields. Suppose farmer A has also sown a second field also with variety X, and gets a yield of 24.2. The whole situation is now changed. The difference of 21.4 from 24.1 now seems notably large compared with the difference between 24.1 and 24.2. The smallness of the latter difference may lead to an unreasonably optimistic assessment of the first difference. And it is still not possible to deduce whether the important difference is A − B or X − Y. But a second difference is now available against which to measure the first, and this is the crucial benefit of replication.

Suppose further that the two varieties are each grown in several fields, giving yields of 24.1, 24.2, 23.0, 25.4 and 22.8 for X and 21.4, 20.9, 23.1 and 22.6 for Y. The reality of a difference between X and Y (or A and B) is becoming more credible and, provided the allocation of fields to X and Y is not done in any way that favours X, the reality of the difference should by now be fairly convincing; the t statistic for comparing the two sample means is 2.5 on 7 df.

It is important to ensure that the replication is appropriate for the comparison that is to be made. Different varieties are used in different fields, and consequently replication of fields with the same variety is needed to provide a measure of variation against which the difference between fields with different varieties can be measured. Sometimes there is a temptation to take a short cut to achieve replication by measuring sample plots within fields. Suppose that each farmer applied his chosen variety to a single field but, within their field, farmer A marks out five plots and farmer B marks out four plots and the yields from these plots are recorded. Now, if the same yields as those discussed in the previous paragraph are obtained, the same confidence that the apparent difference between X and Y is real cannot be sustained. One field has been selected for X and one for Y and assessment of the variation that would have been found between two fields treated identically is necessary before the difference between the yields in the two fields can be interpreted. The variation between small areas within a field does not, in general, give information about differences between fields. A given field may be very homogeneous or it may be very variable. The small areas within a field represent a different scale of unit from that of complete fields, and information about variation between units for one size of unit may not be a reliable guide to the variation to be expected between units of another size.

6.2 The completely randomised design

When the experiment consists of several replicate observations for each of a number of units, and the set of available units is allocated quite arbitrarily to the required treatments, subject only to the restrictions on the total number of units for each treatment, then the experimental design is called a completely randomised design. A full discussion of the crucial concept of randomisation is deferred until Chapter 9, but it will be assumed here that random allocation of treatments to units is essential in experiments. The completely randomised design was considered briefly when introducing the concept of the analysis of variance in Chapter 2, and is now re-examined to illustrate the application of the formal models and methods of analysis developed in Chapter 4.

We assume, as before, that there are t treatments with replications n_1, $n_2, \ldots, n_t$ and a total number of observations N. The model is written formally as follows:

$$y_{jk} = \mu + t_j + \varepsilon_{jk} \qquad j = 1(1)t \qquad k = 1(1)n_j,$$

where y_{jk} represents the observation for the kth unit to which treatment j is applied. The least squares equations are

$$\hat{\mu} \sum_j n_j + \sum_j \hat{t}_j n_j = Y_{..}$$

$$\hat{\mu} n_j + \hat{t}_j n_j = Y_{j.} \qquad j = 1, 2, \ldots, t.$$

The constraint $\sum_j n_j t_j = 0$ results from the general concept of t_j as deviations from an overall mean and expresses the interdependence of the t_j in a form which allows easy solution of the least squares equations. The resulting parameter estimates are

$$\hat{\mu} = Y_{..}/N$$

$$\hat{t}_j = Y_{j.}/n_j - Y_{..}/N.$$

The fitting SS is

$$y'A\hat{\theta} = \hat{\mu} Y_{..} + \sum_j Y_{j.} \hat{t}_j$$

$$= Y_{..}^2/N + \sum_j (Y_{j.}^2/n_j) - Y_{..}^2/N.$$

The analysis of variance structure is shown in Table 6.1.

Table 6.1.

Source	SS	df	MS
Mean (μ)	$Y_{..}^2/N$	1	
Treatments (t_j)	$\sum_j Y_{j.}^2/n_j - Y_{..}^2/N$	$t-1$	
Residual	S_r	$N-t$	s^2
Total	$\sum_{jk} y_{jk}^2$	N	

We have seen already, in Chapter 2, how a comparison of the treatment mean square with the error mean square provides an assessment of the overall strength of treatment effects. In Chapter 4, the formal justification for the use of the F distribution to assess the statistic

$$F = \text{treatment mean square/error mean square}$$

was developed for the general design model. It is also instructive to examine, for this simple design, the expected value of the treatment mean square.

For the completely randomised design model

$$\text{treatment SS} = \sum_j Y_{j.}^2/n_j - Y_{..}^2/N$$

$$= \sum_j n_j \left\{ \mu + t_j + \left(\sum_k \varepsilon_{jk}/n_j \right) \right\}^2 - N \left(\mu + \sum_{jk} \varepsilon_{jk}/N \right)^2$$

$$E(\text{treatment SS}) = \sum_j (n_j\mu^2 + n_j t_j^2 + \sigma^2) - (N\mu^2 + \sigma^2)$$

(since $\sum_j n_j t_j = 0$)

$$= (t-1)\sigma^2 + \sum_j n_j t_j^2.$$

Hence the expected value of the treatment mean square is

$$\sigma^2 + \sum_j n_j t_j^2/(t-1)$$

which simplifies to

$$\sigma^2 + n \sum_j t_j^2/(t-1)$$

for the case where all treatment have equal replication, $n_.$. Now from the general result in Section 4.2

$$E(\text{residual mean square}) = \sigma^2$$

and therefore that the power of the F test to detect treatment differences depends on the average of the squares of treatment deviations. This suggests that it is perfectly possible for one substantial treatment effect to be lost because other treatment deviations are much smaller and suggests that absolute rules such as 'Only examine detailed treatment differences if the overall F statistic is significant' are foolish.

Treatment means and standard errors

The estimated means for the different treatments are formally $\hat{\mu}+\hat{t}_j$ estimated by $Y_{j.}/n_j$, the sample mean from the n_j observations for treatment j. Equally obviously the variance of a treatment mean is:

$$\sigma^2/n_j,$$

and the variance of a difference between two treatment means, based on n_j and $n_{j'}$ observations, is

$$\sigma^2(1/n_j+1/n_{j'}).$$

To illustrate further the use of the completely randomised design we consider data from an experiment to compare three drugs whose purpose is to relieve pain. The experiment from which the data are derived is a complex one, and the data from the experiment are examined further in later chapters. The basic format of the experiment was such that each patient was given a drug whenever the patient decided that their level of pain required relief. The number of hours relief provided by each drug was recorded (time from administration of drug to the next request for a drug). For each patient the sequence of drugs to be provided was predefined (but unknown to the patient).

The variable considered here is the number of hours relief afforded by the first drug given to each patient. In fact, for each patient the drug provided on the second request was a repeat of the drug provided on the first request. The variation between the relief times for the two successive applications of the same drug provides no proper replication and the total hours of relief per patient from the two administrations is the only relevant measure for comparing the effects of the drugs. The three drugs were labelled D, T_1 and T_3, D being a form of control. The number of patients were 14 for D, 13 for T_1 and 16 for T_3, and the results were:

$$\text{D} \quad 2, 6, 4, 13, 5, 8, 4, 6, 7, 6, 8, 12, 4, 4$$
$$T_1 \quad 2, 0, 3, 3, 0, 0, 8, 1, 4, 2, 2, 1, 3$$
$$T_3 \quad 6, 4, 4, 0, 1, 8, 2, 8, 12, 1, 5, 2, 4, 6, 5$$

The two stages of model fitting are:

  (i) fitting the general mean (residual variation $= 352$);

  (ii) fitting the treatment effects (residual variation $= 237$).

The analysis of variance is shown in Table 6.2, and the treatment means and standard errors are in Table 6.3. Clearly the three drugs have substantially different effects with D providing more hours relief than T_3, and T_3 providing more hours relief than T_1.

Table 6.2.

Source	SS	df	MS
Mean	813	1	813
Treatment	115	2	57.5
Residual	237	40	5.9
Total	1165	43	

Table 6.3.

Treatment means			Standard errors (40 df)		
D	T_1	T_3	$D-T_1$	$D-T_3$	T_1-T_3
6.4	2.2	4.3	0.94	0.89	0.91

6.3   Different levels of variation

We have already referred briefly to the need to ensure that replication is relevant to the treatment comparisons being made. There are many very tempting ways of increasing replication which in many cases do not provide proper replication, and which lead very frequently to misleading results. The repetition of the drug for each patient may be appropriate to assess the real effect of each drug but as mentioned previously the results from the two successive applications do not provide useful replication and to use the variation between the two successive relief times as a measure of precision would be misleading. Some examples of other situations in which inappropriate forms of replication are tempting are:

(i) Different diets are applied to litters of pigs. Because of the shortage of litters, there are only two litters for each treatment. To overcome this lack of replication, it is proposed to record and analyse the weights of individual pigs. Each pig is an identifiable unit, and the variation between pigs is clearly important in assessing the difference between litters. Unfortunately, in most such experiments it is not possible to feed and rear pigs individually, and whole litters are treated together. The variation between pigs within a litter is possibly similar to the variation between pigs in different litters, with the same treatment. But pigs in the same litter might be expected to show less variation because of genetic similarity. Or to show more variation because of competitive feeding. There is no way of telling which pattern is occurring (or whether some other pattern is present) and so the within litter variation cannot be used for treatment comparison.

(ii) In crop experiments it is often possible to record yields for each individual plant within a plot, or for each of several rows within a plot. Again, if all plots were treated identically, we might expect either that the variation between plants within a plot might be less than the variation between plants in different plots because of locally homogeneous soil fertility. Alternatively it might be argued that competition between individual plants within a plot might lead to more variation of plant performance within plots than between plots. The crucial implication of these different possible expected patterns is that it is not valid to use within plot variation to assess the differences between treatments applied to whole plots.

(iii) In many scientific disciplines it is now possible to make very frequent observations automatically through the use of automatic data loggers. Such observations may provide very useful information about the detailed behaviour of the experimental unit which is being monitored. But they do not provide any basis for the comparison of treatments applied to different units. In extreme situations observations could be made so frequently that each observation must be almost identical with the previous one. If enough almost identical (false) replications are observed then any difference between different experimental units would appear relatively large and thereby be assessed as significant.

To demonstrate how misleading the use of multiple observations can be we consider some artificial data. Imagine an educational experiment, in which there are assumed to be effects on the assessment results of the time within a day when the assessment is made, and a longer term learning process over a period of several months which will also change the assessment results systematically. Three different treatments (forms of teaching) are applied to three classes each. The testing of the subject material takes one hour, and, because different tests must involve slightly different material, there is inevitably some variation of the recorded result over and above the diurnal and long term trend effects. Similar effects can be suggested for other fields of applications.

Consider first the average result of a single test for each class shown in

Table 6.4.

Treatment 1			Treatment 2			Treatment 3		
A	B	C	D	E	F	G	H	I
27	43	38	41	30	47	46	34	50

Table 6.4. The analysis of variance for the completely randomised design is:

between treatments SS = 81 on 2 df, MS = 40;

within treatments SS = 447 on 6 df, MS = 74.

Clearly the evidence that the treatments have different effects is weak ($F = 0.54$). There is wide variation between classes for each treatment.

Next, suppose that each class is tested on four consecutive days, the test being at the same time each day to eliminate diurnal variation. The results are shown in Table 6.5.

Table 6.5.

Treatment 1			Treatment 2			Treatment 3		
A	B	C	D	E	F	G	H	I
27	43	38	41	30	47	46	34	50
25	43	36	43	35	42	48	37	44
30	46	37	44	31	46	46	38	52
31	44	41	45	35	48	45	35	49

The analysis of variance if the data are treated as 12 observations, ignoring the possible average differences between days, for each of three sets of data is:

between treatments SS = 288 on 2 df, MS = 144;

within treatments SS = 1394 on 33 df, MS = 42.

The evidence for differential treatment effects now seems much stronger ($F = 3.4$), reflecting the apparent consistency of observations, which essentially is due to the fact that we are using (almost) the same observation four times for each class. With more such replicate observations for each class we would obtain results for the three treatments which were 'even more convincingly different'.

Now suppose that, on each day, only one class can be tested at a time,

Figure 6.1. Diurnal pattern for test performance.

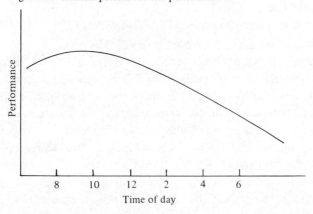

that all nine classes are tested each day, and that there is a diurnal pattern of test performance shown in Figure 6.1. The pattern of the arrangement for testing the nine classes on the four days has been arranged, for this example, so that the set of times for each class is spread evenly over the available time as shown in Figure 6.2. The effect of this is that the estimates of the average treatment differences are not much altered by the diurnal effects.

The test results for the four days might now be as in Table 6.6. Now the

Figure 6.2. Arrangement of treatments within days and times.

		Time								
		1	2	3	4	5	6	7	8	9
Day	1	A	D	G	E	B	H	I	F	C
	2	H	E	I	A	F	C	D	B	G
	3	I	C	F	G	D	A	B	H	E
	4	F	B	E	H	C	I	G	D	A

Table 6.6.

Treatment 1			Treatment 2			Treatment 3		
A	B	C	D	E	F	G	H	I
21	51	28	40	38	41	50	38	49
33	37	40	42	34	50	38	31	48
34	45	36	52	21	50	54	32	46
21	43	49	39	39	42	44	43	53

analysis of variance for the three sets of 12 observations, ignoring the day and time effects, is:

between treatments SS = 322 on 2 df, MS = 162;

within treatments SS = 2464 on 33 df, MS = 75.

The evidence for the existence of different effects of the three treatments is now apparently slightly weaker ($F = 2.2$) because the diurnal effects mask the treatment differences. Essentially, the false replication obtained by repeated observations on the same students has been diluted by the diurnal variation against which the treatments are evenly spaced. This kind of 'evening out' of effects is a form of blocking. It is clearly sensible to try to spread treatments evenly over the different times, but this control of the variation should also be reflected in the analysis of data, the result of which would be to reduce the random, within treatment SS. In this example, of course, the use of replication of observations for each unit is wrong, and modifications of the analysis cannot correct this fundamental defect.

Finally consider the effect of a long term trend as a result of which test results on four days separated by long gaps in time will show systematic increases (see Table 6.7).

Table 6.7.

Treatment 1			Treatment 2			Treatment 3		
A	B	C	D	E	F	G	H	I
21	51	28	40	38	41	50	38	49
39	43	46	48	40	56	44	37	54
46	57	48	64	33	62	66	44	58
39	61	67	57	57	60	62	61	71

The analysis of variance, again simply assuming the 12 values for each treatment are proper replicates, is:

between treatments SS = 322 on 2 df, MS = 162;

within treatments SS = 4492 on 33 df, MS = 136.

The evidence of treatment effects has now virtually disappeared, submerged beneath the learning effect, combined with the diurnal effect. Of course we might be interested in the learning effect *per se*, and this could be examined by considering the change of score over time for each class. This form of repeated measurement experiment will be examined in detail in Chapter 14. In this example, the treatment effects must be measured in terms of the variation between classes since each treatment can only be applied to the experimental unit of a class. As the conditions of the experiment have been changed while preserving the same form of analysis, the apparent strength of evidence for the existence of treatment differences has altered, although the real information available has not changed. The use of more complex analyses of data can also not change the real information available and, though alternative analyses should be preferred because they reflect the structure of observations, there is a danger that using the more appropriate analyses may blind the experimenter to the fact that the wrong form of replication is being used. The only proper measure of the differences between the teaching methods (applied to whole classes) is derived from the class averages using just one measurement per class. The learning effect is measured in terms of the variation over time of observations for the same class. The idea of information at several different levels of unit occurs in many later chapters, most notably in Chapter 14. In this present section, the important principle is that, in an experiment, the replication of observations must be exactly appropriate to the comparison of treatments.

6.4 Identifying and allowing for different levels of variation

Sometimes it is not clear from the description of the form of data whether the replication is at the same level of experimental unit as the application of treatments. Sometimes it is a matter of judgement whether the replication is appropriate. And sometimes it is necessary and desirable to use within unit replication for the comparison of treatments applied to whole units, even though this may not be formally valid.

The lack of clarity frequently stems from an inadequate description of the experiment, and of the form of data. If results are presented properly, with the relevant degrees of freedom quoted for any standard error, then it should be possible to deduce when within unit replication has been used. For the example of the previous section, reference to the use of standard errors based on 33 df combined with the use of a total of nine classes should immediately indicate what has happened. Less directly, if a randomised block design with three blocks of two treatments has been used and effects for many variables are declared to be significant, then the results should be treated with suspicion, particularly if there is any implication that t-values of the order of two are significant since the design gives only 2 df for error, and with 2 df a value of 4.3 is required for significance at the 5% level.

When experiments are not properly described, as in the example at the beginning of this chapter, then the statistician, or the experimenter considering another experimenter's results, is right to be suspicious of unspecified replication, since experience, particularly from reading papers in applied journals of all disciplines, shows that such suspicion is all too frequently justified. Let us return to the initial example of this chapter. The reader with a suspicious mind will recognise that the replication used to calculate the F ratio of 500 might have come from sampling different plants, different leaves from the same plant, or different discs from the same leaf. Also, the different treatments might have been applied to discs from different plants, from different leaves or from the same leaves. In this instance, the truth when it finally emerged was the saddest of the possibilities. The ten-fold replication for each treatment consisted of ten discs from a single leaf. And each different treatment was applied to discs from a leaf from a different plant. Thus, the value of 500 could have been due to the treatments but could equally well be attributable to the fact that the leaves come from different plants. The sad conclusion in this example was that there was no valid replication on which comparisons of the treatments could be based, and that the experiment had been so designed as to be completely useless.

There are some situations where the relevant form of replication is not entirely clear. For example, when comparing different types of soil it is necessary to use several samples of each type of soil as replicates in order to assess the extent to which apparent differences between soil types are meaningful. However, each soil type will occur over a relatively compact area, and therefore the different samples for each soil type will necessarily come from that area. In most investigations it would be proper to be suspicious if replicates of the different 'treatments' were bunched together within small areas. However, such replication provides the only genuine basis for comparison of soil types, since the area is the soil type and differences between soil types can only be meaningfully assessed in terms of the variation within the soil type. It is of course essential to keep each soil sample separate throughout the analysis procedure. The common laboratory procedure of mixing the soil to obtain a fully representative sample and then subsampling from the mixture immediately returns to the situation where the replication is at a different level from that at which comparison is required.

Another situation where the replication must be carefully considered is when records of animal progeny, for example sheep, are obtained from a number of different farms, the progeny being classified by breed and farm. One of the purposes of such investigations is to compare different breeds. However, each farm will usually have sheep from only two or three out of the ten or so breeds for which comparison is required. The question then is whether the records for individual sheep at each farm can be used as replicate observations, or whether only the averages for each breed/farm combination can be used for assessing breed differences. Here, it is a matter of judgement, related to the purpose of the comparison, whether the variation between sheep within a breed/farm combination is relevant to breed comparison. The argument in favour is that breeds are collections of individual sheep, and that therefore breed comparisons should be made relative to variation between sheep within each breed. The opposite argument is that of the previous section, that variation between sheep within a farm may be increased by competition between individuals or decreased by being within a particularly homogeneous environment.

The alternative estimators of replicate variance may be demonstrated by considering the analysis of variance structure (Table 6.8).

Table 6.8.

Source	df
Farms (ignoring breeds)	24
Breeds (adjusting for farms)	9
Residual between farm/breed combinations	$18 (= 51 - 24 - 9)$
Within farm/breed combinations	1965
Total	2017

In this study there were observations (on the birthweight of twin lambs) from 2018 ewes from 25 farms, involving ten breeds but a total of only 52 breed farm combinations (of the 250 possible combinations).

If the interest is in predicting breed differences over a large population of farms then the residual MS between farm/breed combinations forms the basis of our estimation of the standard error. However, if the objective is to estimate the variability within a breed for a single farm then the within farm/breed combinations MS should be used. The ideas of multiple levels of variation involved in these judgements are developed more formally in the next section, and we return to the problem of determining the appropriate variance for comparing treatments in the final chapter of the book.

One extension of the idea of within unit replication involves the effects of treatments on replication. Consider an experiment to compare four hormone treatments for pigs. Sixteen animals are available in four pens of four animals each. An initial plan for the experiment might be to use hormone A on the four pigs in pen 1, hormone B for those in pen 2 and so on, as represented in Figure 6.3(a). However, the replication would be based on variation between animals within a pen, whereas the treatment is applied to a whole pen. A classic case of within unit replication and one

Figure 6.3. Experimental plans for pig-hormone experiment (a) different hormones in different pens, (b) each hormone in each pen.

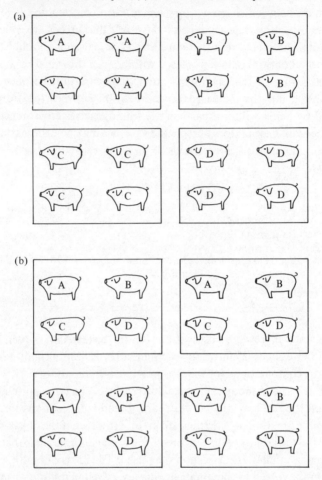

which we have hopefully learnt to avoid. An alternative illustrated in Figure 6.3(b) is to treat the four pigs in pen 1 with hormones A, B, C and D, one pig per hormone, and to repeat the procedure in each pen. This design resembles a randomised block design and may seem appropriate. However, in each pen the animals live and feed together competitively, the only applied differences being the different hormones. If we consider the final difference in, say, growth rate for individual pigs, then the observed differences would be due not only to the direct effects of the hormone, but also to the results of competition arising from the effects of the hormones. Thus, if hormone A makes pigs aggressive, then this will tend to make the hormone A pigs prosper in face of the lesser competition from the more docile hormone B pigs! Therefore, the experiment measures not only the effects of the different hormones, but also the effects of rearing pigs with different hormones together, which latter effect is not practically relevant. The search for appropriate replication has led, in this case, to inappropriate treatment of the individual pig units. The only solution to the replication problem is to use replication of whole pens or to house the pigs individually (which may be inappropriate as a guide to behaviour of the hormone treated pigs in a normal farming environment).

Sometimes it is not possible to achieve sufficient replication at the level appropriate to compare treatments and it is necessary to consider whether within unit replication can be used to provide at least some idea of precision. This question really concerns the sources of the within unit and between unit variation. One subject of research which often leads to this problem is the growth of plants in controlled or artificial environments. Suppose an experimenter wishes to include, in an experiment on the growth of greenhouse plants, the investigation of the effect of heating the greenhouse. Distinct heating levels are only possible if large sections of the greenhouse are used and such sections will usually be few in number. Typically, there might be six greenhouse sections capable of independent heating: if three heating treatments are to be compared, then each may be applied to only two sections. Within each section, it will be possible to have a number of experimental units, probably with different treatments applied to different units. Can the within section variation between units be used as the basis for comparison between heating treatments?

The strictly correct answer, as discussed earlier in the chapter, is 'No'. But since the experimenter has no other way of investigating the effects of heating, we should consider when the answer might be 'Yes, if ...'. To use the within section replication to compare treatments applied to whole sections, it is necessary to assume

(i) that there are no intrinsic differences between the sections which might affect plant growth and production,

(ii) that the variation between units both within and between sections is essentially due to plant-to-plant variation,

(iii) that there is no competitive variation between units within a section, induced by the treatment or the effects of the treatments.

The crucial requirement is that the experimenter can believe that the variation between experimental units is so dominated by plant-to-plant variation that other sources of variation can be ignored. If this is credible, then the experimenter *may* feel justified in using the within section variation to compare the heating treatments. Note, however, that in presenting results for the heating treatments using standard errors based on within section replication, the experimenter no longer has the protection of a properly randomised experiment with genuine replication. The experimenter must demand an act of faith from the reader in accepting the experimenter's assertion that within section variation between units is the same as between section variation between units. The experimenter must make the assertion explicitly and has no statistical basis for arguing against the rejection of the assertion.

6.5   Sampling and components of variation

There is a relationship between the ideas of several levels of variation between experimental units and those of sampling observations in several stages. Consider sampling the output of an industrial process. If there are no systematic trends with time, then we can consider sampling at short intervals, say every ten minutes, then sampling daily and finally sampling monthly. In the absence of trends, a model for the observed values would be

$$y_{ijk} = \mu + m_i + d_{ij} + \varepsilon_{ijk}, \tag{6.1}$$

where the m_i represent monthly variation, the d_{ij} represent daily variation within months, and the ε_{ijk} short term within day variation. In this model, m_i, d_{ij}, and ε_{ijk} are assumed to be independent random variables, all with zero expected values. The variances of m_i, d_{ij}, ε_{ijk} may be represented by σ_m^2, σ_d^2 and σ^2, respectively. Suppose we have n_{ij} observations for day j in month i. The total variation between the observations can be divided into variation between months, between days within months, and within days by the following analysis of variance which corresponds to fitting successive sets of terms in the model (6.1):

$$\sum_{ijk}(y_{ijk}-y_{...})^2 = \sum_{ijk}(y_{i..}-y_{...})^2 + \sum_{ijk}(y_{ij.}-y_{i..})^2 + \sum_{ijk}(y_{ijk}-y_{ij.})^2. \tag{6.2}$$

The component SS on the right hand side of (6.2) are calculated in the form

$$\sum_{ijk}(y_{i..}-y_{...})^2 = \sum_i\left\{\left(\sum_{jk}y_{ijk}\right)^2\bigg/\sum_j(n_{ij})\right\} - \left(\sum_{ijk}y_{ijk}\right)^2\bigg/\sum_{ij}(n_{ij})$$

$$\sum_{ijk}(y_{ij.}-y_{i..})^2 = \sum_{ij}\left\{\left(\sum_k y_{ijk}\right)^2\bigg/n_{ij}\right\} - \sum_i\left\{\left(\sum_{jk}y_{ijk}\right)^2\bigg/\sum_j(n_{ij})\right\}$$

$$\sum_{ijk}(y_{ijk}-y_{ij.})^2 = \sum_{ijk}y_{ijk}^2 - \sum_{ij}\left\{\left(\sum_k y_{ijk}^2\right)^2\bigg/n_{ij}\right\}.$$

For the special case when n_{ij} is constant ($=n$) for each day, the number of days on which observations are made in month i is constant ($=d$) and there are m months, then the expected values of the SS on the right hand side of (6.2) are particularly simple. Using the original model (6.1)

$$y_{ijk} - y_{ij.} = (\mu + m_i + d_{ij} + \varepsilon_{ijk}) - (\mu + m_i + d_{ij} + \varepsilon_{ij.})$$

$$= \varepsilon_{ijk} - \varepsilon_{ij.}$$

$$E\left\{\sum_k (y_{ijk} - y_{ij.})^2\right\} = E\left\{\sum_k (\varepsilon_{ijk} - \varepsilon_{ij.})^2\right\} = (n-1)\sigma^2$$

$$y_{ij.} - y_{i..} = (\mu + m_i + d_{ij} + \varepsilon_{ij.}) - (\mu + m_i + d_{i.} + \varepsilon_{i..})$$

$$= (d_{ij} - d_{i.}) + (\varepsilon_{ij.} - \varepsilon_{i..})$$

$$E\left\{\sum_j (y_{ij.} - y_{i..})^2\right\} = E\left[\left\{\sum_j (d_{ij} - d_{i.}) + \sum_j (\varepsilon_{ij.} - \varepsilon_{i..})\right\}^2\right]$$

$$= E\left\{\sum_j (d_{ij} - d_{i.})^2\right\} + E\left\{\sum_j (\varepsilon_{ij.} - \varepsilon_{i..})^2\right\}$$

$$= (d-1)\sigma_d^2 + (d-1)(\sigma^2/n)$$

$$y_{i..} - y_{...} = (\mu + m_i + d_{i.} + \varepsilon_{i..}) - (\mu + m_. + d_{..} + \varepsilon_{...})$$

$$= (m_i - m_.) + (d_{i.} - d_{..}) + (\varepsilon_{i..} - \varepsilon_{...})$$

$$E\left\{\sum_i (y_{i..} - y_{...})^2\right\} = E\left[\left\{\sum_i (m_i - m_.) + \sum_i (d_{i.} - d_{..}) + \sum_i (\varepsilon_{i..} - \varepsilon_{...})\right\}^2\right]$$

$$= E\{\sum (m_i - m_.)^2\} + E\{\sum (d_{i.} - d_{..})^2\} + E\left\{\sum_e (\varepsilon_{i..} - \varepsilon_{...})^2\right\}$$

$$= (m-1)\sigma_m^2 + (m-1)\sigma_d^2/d + (m-1)\sigma^2/(nd).$$

The expected values of sums of squares and of mean squares can be summarized in the analysis of variance structure (Table 6.9). The expected mean squares here show precisely why the use of an inappropriate level of replication may be dangerous. For the very simple model (6.1) the mean square for variation between days has an expected value greater than that within days by n times the between day variance. If σ_d^2/σ^2 is not negligible then the expected values of the two mean squares will be very different. The pattern of expected values of mean squares in the analysis of variance

Table 6.9.

Source	df	E(SS)	E(MS)
Between months	$m-1$	$nd\{(m-1)\sigma_m^2 + (m-1)\sigma_d^2/d$ $+ (m-1)\sigma^2/(nd)\}$	$\sigma^2 + n\sigma_d^2 + nd\sigma_m^2$
Between days within months	$m(d-1)$	$mn\{(d-1)\sigma_d^2 + (m-1)\sigma^2/n\}$	$\sigma^2 + n\sigma_d^2$
Within days	$md(n-1)$	$md\{(n-1)\sigma^2\}$	σ^2

shows clearly how σ^2, σ_d^2 and σ_m^2 can be estimated either directly from a mean square or from differences between mean squares.

We have discussed this example in terms of a pure sampling procedure. However, if we have different forms of management applied to different days, or different amounts of components of the process used at different times within each day, then the same multi-level structure of units exists in the resulting experiment, and the use of the wrong mean square for calculating standard errors can be misleading and dangerous. We shall discuss the use of these results when we consider, in Chapter 12, treatments which are themselves random samples from populations and, in Chapter 14, designs where different sets of treatments are deliberately applied at different unit sizes.

Finally, note the similarity between the expected mean squares for the hierarchic sampling and completely randomised design model experiment with fixed treatment effects t_j, developed in Section 2:

$$E(\text{between days MS}) \quad \sigma^2 + n\sigma_d^2$$
$$E(\text{within days MS}) \quad \sigma^2$$
$$E(\text{treatment MS}) \quad \sigma^2 + n\sum_j t_j^2/(t-1)$$
$$E(\text{residual MS}) \quad \sigma^2.$$

6.6   How much replication?

The question which is most frequently asked of a statistician when the design of an experiment is being considered is how many replicates should be used. It must be emphasised immediately that there is no simple rule for the amount of replication, which depends on many considerations, including:

(a) the resources available,

(b) the variability of experimental units,

(c) the treatment structure,

(d) the size of effect which is of importance, and

(e) the relative importance of different comparisons.

It may seem superfluous to suggest consideration of the available resources, but it is not unknown for lengthy discussion of the design of an experiment to lead to a decision that, for example, to detect treatment differences of a size which has important practical implications, it is necessary to include six replicate units for each of eight treatments, only to discover that a maximum of 25 units can be made available. The first question, then, for any experimenter is, how many experimental units can be made available? This may sometimes lead to consideration of replication of experiments in time in order to increase the scale of the experiment.

Suppose we have a total of N units available, and it is proposed that t treatments be compared. Then, assuming equal replication of treatments, the replication per treatment is $n = N/t$, and the standard error of a difference between two treatment means is:

$$(2\sigma^2/n)^{1/2}.$$

To assess whether this is sufficiently precise, we must ask first what the likely size of σ^2 is, and secondly what size of treatment difference is important. Suppose that σ is expected to be about 10% of the mean yield. Then, if resources allow $n = 4$, the standard error of a difference between two treatment means will be approximately 7% of the mean yield. This means that an observed difference of 14% between two treatment means will be assessed as significant at the 5% level. However, if the true difference between two treatment effects is 14% of the mean yield, then the probability that the observed difference between the means for those two treatments will be 14% or greater is only 0.5. Hence, the probability of detecting as significant, at a 5% significance level, a true difference of 21% (or three times the standard error) is 0.83; the probability of detecting a true difference of 28% (four times) is about 0.975.

We have assumed that a trustworthy estimate of σ^2 is available. Without any estimate of σ^2 it is, of course, quite impossible to assess whether the replication proposed in an experiment is adequate. Fortunately for any proposed experiment, there are almost always previous experiments of sufficient similarity that a reasonable guess of the value of σ^2 can be made from the analysis of previous data. For any predetermined level of resources in terms of the number of available experimental units, the replication is determined by the number of treatments, and, given an estimate of σ^2, the experimenter can assess the precision and hence the detectable size of any treatment difference. It is then possible to decide whether the experiment is worth proceeding with.

If the resources are not, apparently, limiting then to determine the appropriate level of replication the experimenter needs an estimate of σ^2 and also an estimate of the size of difference which it is important to detect. Returning to the previous discussion of the probability of detecting at a given significance level, α, the observed difference when the true difference is a specified size, d, the necessary replication, n, will be such that

$$d/(2\sigma^2/n)^{1/2} = t_\alpha + \Phi^{-1}(p),$$

where Φ is the normal distribution function and p is the probability of obtaining an observed difference which will be detected as significant. A realistic rule-of-thumb for most situations is to require that the replication r be sufficient to make the standard error of the differences between two treatment means no bigger than $d/3$. This is equivalent to a probability, p, of 0.83 for a significance level, α, of 0.05.

For example, if it is important to detect true differences of $d = 5\%$ of the mean, then still using $\sigma = 10\%$ of the mean, and assuming that for $\alpha = 0.05$ t is approximately 2, we would require

$$\{2(10)^2\}/n \leqslant (5/3)^2$$

or $n \geqslant 72$. If d is 20% of the mean, then a level of replication, n, of 5 should be adequate.

The choice of replication may not be determined by the precision required for a simple comparison between two treatment means. More generally, the crucial precision may be that for a treatment comparison $\sum_j l_j t_j$. The variance for such a comparison is $\sigma^2 \sum_j l_j^2/n$, and the previous argument holds for this general form of variance in place of the previously considered special case with $\sum_j l_j^2 = 2$.

This discussion of the necessary replication has ignored blocking and factorial structure. If blocks are used and the treatments are orthogonal to blocks (see Chapter 7 for discussion of orthogonality) then the replication arguments are unchanged. If the treatments proposed for the experiment have a factorial structure, then the comparisons considered in determining the required precision of the experiment may be main effect comparisons, in which case the hidden replication of the factorial structure should be included in the calculation of replication.

So far, we have assumed that all treatments in the proposed experiment should be equally replicated. This in turn implies that all treatment comparisons are of equal importance, and that there are no restrictions on the possible replication for any of the treatments. Although these are the assumptions that are usually made, they are not necessary and, in some experimental situations, they are not sensible. The classic situation where they are not reasonable is where one treatment is a standard or control. The more important comparisons are of each new treatment with the standard, the comparisons between the new treatments being relatively less important.

For the particular case of t new treatments, a total number of N available units and replication n for the new treatments and n_s for the standard, then

$$tn + n_s = N.$$

The variance for the comparison between the standard and one new treatment is

$$\sigma^2(1/n + 1/n_s).$$

If we write $n_s = kn$ then $(k + t)n = N$, and the variance becomes

$$\sigma^2(k + 1)(k + t)/Nk.$$

It can be shown quite simply that this variance is a minimum when $k = t^{1/2}$. Using this optimal form of unequal replication the variances for (a) comparing a new treatment with the standard, or (b) comparing two new treatments can be substantially different from the variances that would be achieved if equal replication is adopted. The variance formulae for unequal replication are

$$\sigma^2(t^{1/2} + 1)^2/N$$

when comparing a new treatment with the standard, and

$$2\sigma^2 t^{1/2}(t^{1/2} + 1)/N$$

when comparing two new treatments. The formula for equal replication is

$$2\sigma^2(t + 1)/N$$

for both cases. Numerical values, for different values of t, of the variances (divided by σ^2/N) are given in Table 6.10. Thus the ratio of variances for the comparisons between a new treatment and the standard relative to the comparison of two new treatments tends downwards to a limiting value of $1/2$ as t increases. The ratio of variances for the new versus standard comparison using the unequal replication compared with the equal replication also tends to $1/2$, but considerably more slowly.

It is important to emphasise that equal replication for all experimental treatments is not always necessary nor even desirable. If the replication of one or more treatments is limited by resources, as could be the case if, in a plant breeding trial, there are only small numbers of seeds for some

Table 6.10.

	Unequal replication		Equal replication
t	New–standard	New–new	Any comparison
2	5.8	6.8	6.0
4	9.0	12.0	10.0
8	14.7	21.7	18.0
16	25.0	40.0	34.0
32	44.3	75.3	66.0

breeding lines, then to some extent precision can be improved by using more replicates of other lines. In other words, since the variance of a comparison between two treatments with n_1 and n_2 replications is

$$\sigma^2(1/n_1 + 2/n_2)$$

then if $n_1 + n_2 = N$ is limited then of course the variance is minimised when $n_1 = n_2 = N/2$. However, if n_2 is limited, the variance can be reduced by increasing n_1.

One final consideration when determining how much replication is needed is the precision of estimation of σ^2. The estimate of σ^2 is obtained from the residual mean square, and the precision of this estimate is determined by the degrees of freedom for the residual SS. As the residual degrees of freedom are increased, so the precision of the estimate of σ^2 is increased, and this may be most clearly seen by examining the set of values of the t-statistic required for significance at 5%. The values given in Table 6.11 show the substantial benefits of increasing the degrees of freedom up to about 10 df. After 10 df the improvement in the precision of estimating σ^2 tails off, and after 20 df the benefits of additional degrees of freedom are negligible. The conclusions for consideration when determining the amount of replication are that it is desirable that there should always be at least 10 df for the residual SS but that there is really no advantage having more than 20 df.

Table 6.11.

df	5% significance t-value
1	12.71
2	4.30
3	3.18
4	2.78
5	2.57
6	2.45
7	2.36
8	2.31
9	2.26
10	2.23
12	2.18
15	2.13
20	2.09
30	2.04
60	2.00
120	1.98

Exercises 6

(1) Examine critically the information about replication in an applied science journal. Assess the level of replication used in experiments from which results are reported, and judge whether the form of replication used to provide the estimated standard errors is appropriate to the treatment comparisons. (I should be surprised if you do not find examples of inappropriate replication, and also if the necessary information is always provided.)

(2) For a completely randomised design to compare three treatments, two experimenters have different views on the relative importance of the three comparisons:

(i) $t_1 - t_2$, (ii) $t_1 - t_3$, (iii) $t_2 - t_3$

They suggest that the standard errors for these comparisons should be in the ratios

(a) 4:5:6, or

(b) 3:3:4.

Determine the relative replication needed for each suggestion. Assuming a total of 60 observations is available, compare the expected precision of treatment comparisons achieved using the designs appropriate to (a), (b) with the precision for the equal replicate design and present your advice to the experimenters.

7

Blocking

7.0 *Preliminary examples*

(a) An experiment is to be designed to investigate the effects of supplementary heating, supplementary lighting and carbon dioxide on the growth of peppers in glasshouses. The number of possible treatment combinations is eight:

(standard or supplementary heating) × (standard or supplementary lighting) × (control or added CO_2).

Each observation on a treatment combination requires a glasshouse compartment and there are available 12 glasshouse compartments in two sets of six compartments; compartments in different sets are likely to produce rather different yields. The problem is to design an experiment in two blocks each of six units, to compare eight treatments (with a factorial structure). It is intended to have a second experiment in the next year after which the most effective treatment will be recommended for use.

(b) Forty-two treatments are to be compared. They consist of 14 different sources of material treated in three different ways, but the purpose of the experiment is a general comparison of the 42 treatments to select the best six for further work. The available resources allow three replicate units of each treatment, but at most 18 units can be processed in a batch and it is known that there are likely to be substantial batch differences. Seven batches of 18 units will accommodate the three replications of 42 treatments. How shall the treatments be allocated to batches?

7.1 Design and analysis for very simple blocked experiments

The primary idea of blocking is that identifying blocks of homogeneous units allows more precise comparison of treatments through the elimination of the large differences between units in different blocks from the comparison of treatments. It follows that information from blocked experiments is predominantly based on the comparisons that can be made between treatment observations in the same block. If two treatments do not occur together in a block, then it will still be possible to make a valid comparison between the two treatments if each occurs in a block with a common third treatment. Essentially, a comparison between two treatments, A and B, within a block, has variance $2\sigma^2$, where σ^2 is the variance of units within a block. If each of A and B occurs in a block with treatment C then the $A-B$ comparison can be calculated as $(A-C)+(C-B)$, each of the two component comparisons being made within a block, and the variance of the resulting $A-B$ comparison is $2(2\sigma^2)$.

If the indirect comparison of A and B can be made through several different intermediaries, then the precision of the $A-B$ comparison is improved. Suppose we have two blocks of five units each, and wish to compare six treatments, A, B, C, D, E and F, then, provided each treatment appears at least once, and no treatment appears twice in a block, the four treatments (C, D, E and F) will appear in each block and the other two treatments, A and B, in one block each. This simple design is shown in Figure 7.1. To compare any two of C, D, E and F, we obviously use the mean yields from the two observations for each of the two compared treatments. To compare A and B, we use all four other treatments as intermediaries to estimate the difference between the blocks, and compare each of A and B with the mean yields of the other four treatments in the block in which A, or B, appears. The variance of $A-(C+D+E+F)/4$ is $5\sigma^2/4$ and hence the variance of $A-B$ is $5\sigma^2/2$. This may be compared with the variance for the difference between two treatments each replicated once, and occurring together in a block, $2\sigma^2$.

We can also consider the comparison of A, or B, with one of the other four treatments. A first thought might suggest that, since A only occurs in block I and since C also occurs there, the comparison $A-C$ will be based

Figure 7.1. Experimental plan for comparing six treatments in two blocks of five units.

```
        Block

I                 II

A                 B
C                 C
D                 D
E                 E
F                 F
```

only on yields from block I. However, it should be apparent that the C yield in block II also contributes information to the comparison. To use the comparison between A in block I and C in block II, we have to estimate the difference between blocks I and II, and this is done through the treatments (C, D, E and F) occurring in both blocks. However, this means that the estimate of the difference between A and C from A in block I and C in block II involves the differences between C in blocks I and II, and hence involves the $A-C$ difference in block I. Thus, although the principle of the comparison is quite simple, the practical details are more difficult. These details can be approached either through the algebraic solution of the least squares equations, or by fitting the model on a general computer program.

We now consider the least squares equations for this particular simple example in detail, not only to solve the problem but also to demonstrate the benefits from examination of least squares equations. The model is

$$y_{ij} = \mu + b_i + t_j + \varepsilon_{ij}, \quad i = 1, 2; \quad j = A, B, C, D, E, F$$

except $(i, j) = (1, B), (2, A)$. The constraints are

$$b_1 + b_2 = 0 \qquad \sum t_j = 0.$$

The least squares equations are

$$10\hat{\mu} - \hat{t}_A - \hat{t}_B = Y_{..} \tag{7.1}$$

$$5\hat{\mu} + 5\hat{b}_1 - \hat{t}_B = Y_{1.} \tag{7.2}$$

$$5\hat{\mu} + 5\hat{b}_2 - \hat{t}_A = Y_{2.} \tag{7.3}$$

$$\hat{\mu} + \hat{t}_A - \hat{b}_2 = Y_{.A} \tag{7.4}$$

$$\hat{\mu} + \hat{t}_B - \hat{b}_1 = Y_{.B} \tag{7.5}$$

$$2\hat{\mu} + 2\hat{t}_C = Y_{.C}. \tag{7.6}$$

These equations are solved by utilising the symmetry for $(\hat{t}_A, \hat{t}_B)$ and $(\hat{b}_1, \hat{b}_2)$ and considering $(\hat{t}_A - \hat{t}_B)$ and $(\hat{t}_A + \hat{t}_B)$. The steps of the argument are as follows:

(7.2) and (7.3)

$$5(\hat{b}_1 - \hat{b}_2) + (\hat{t}_A - \hat{t}_B) = Y_{1.} - Y_{2.} \tag{7.7}$$

(7.4) and (7.5)

$$(\hat{t}_A - \hat{t}_B) + (\hat{b}_1 - \hat{b}_2) = Y_{.A} - Y_{.B} \tag{7.8}$$

(7.7) and (7.8)

$$4(\hat{t}_A - \hat{t}_B) = 5(Y_{.A} - Y_{.B}) - (Y_{1.} - Y_{2.}) \tag{7.9}$$

(7.4) and (7.5)

$$2\hat{\mu} + (\hat{t}_A + \hat{t}_B) = Y_{.A} + Y_{.B} \tag{7.10}$$

(7.1) and (7.10)

$$6(\hat{t}_A + \hat{t}_B) \qquad = 5(Y_A + Y_B) - Y_{..} \qquad (7.11)$$

(7.9) and (7.11)

$$24\hat{t}_A = 25Y_A - 5Y_B - 3Y_{1.} + 3Y_{2.} - 2Y_{..} \qquad (7.12)$$

(7.1) and (7.10)

$$12\hat{\mu} = Y_{..} + (Y_A + Y_B) \qquad (7.13)$$

(7.6), (7.12) and (7.13)

$$24(\hat{t}_A - \hat{t}_C) = 27Y_A - 3Y_B - 12Y_C - 3Y_{1.} + 3Y_{2.} \qquad (7.14)$$

Hence, from (7.9),

$$\hat{t}_A - \hat{t}_B = 1/4\{5(Y_A - Y_B) - (Y_{1.} - Y_{2.})\}$$

and from (7.14)

$$\hat{t}_A - \hat{t}_C = 1/8\{9Y_A - Y_B - 4Y_C - (Y_{1.} - Y_{2.})\}. \qquad (7.15)$$

In each case, the estimate of the difference clearly involves the correction for the difference between the two blocks. The variances of the two estimated differences can be derived in various ways. The simplest is by considering the estimates as linear combinations of yields:

$$\hat{t}_A - \hat{t}_B = 1/4(4y_{1A} - y_{1C} - y_{1D} - y_{1E} - y_{1F} - 4y_{2B} + y_{2C} + y_{2D} + y_{2E} + y_{2F})$$

$$\hat{t}_A - \hat{t}_C = 1/8(8y_{1A} - 5y_{1C} - y_{1D} - y_{1E} - y_{1F} - 3y_{2C} + y_{2D} + y_{2E} + y_{2F}),$$

$$(7.16)$$

whence

$$\text{Var}(\hat{t}_A - \hat{t}_B) = (\sigma^2/16)(2 \times 16 + 8 \times 1) = 5\sigma^2/2$$

confirming the earlier result, and

$$\text{Var}(\hat{t}_A - \hat{t}_C) = (\sigma^2/64)(64 + 25 + 9 + 6 \times 1) = 13\sigma^2/8.$$

These variances should be compared with those for an unblocked experiment with the same replication, $\sigma^2(1+1) = 2\sigma^2$, and $\sigma^2(1 + 1/2) = 3\sigma^2/2$, but we must remember that the effect of blocking should be to reduce σ^2.

A diagrammatic method of obtaining the linear combination of yields for complex least squares estimates is illustrated in Figure 7.2. The method is essentially just writing out each total in (7.15) as the sum of the individual yields, but the diagram provides a structure, reducing the scope for mistakes. The separate figures in each 'unit' derive from the totals in (7.15); the circled figures represent the total coefficient for each unit (as in (7.16)).

Clearly this derivation of least squares estimates and their variances, although straightforward, is too lengthy to be used in each particular case. General results are needed and these are derived in Section 7.3. Moreover

Figure 7.2. Diagrammatic representation of estimate of t_A-t_C.

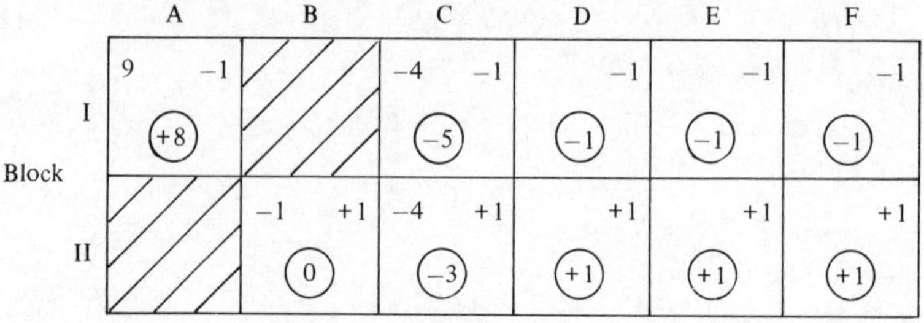

the analysis of experimental data can always be achieved through the use of general linear model computer packages. However, the most important conclusion from this example concerns the simplicity of the design argument. The implications for analysis are not simple but, given the problem of comparing six treatments in two blocks of five units each, the design solution is inevitable, the only choice being which two treatments should appear only once.

7.2   Design principles in blocked experiments

In this section the thesis that the intuitive ideas of block designs are extremely simple is developed through considering a series of examples. In each example, the treatments are assumed to have no structure, and all treatment comparisons are assumed to be equally important.

Example 7.1

Six treatments, A to F, are to be compared using 24 units for which a natural blocking system gives four blocks of six units each. How should the six treatments be allocated to the units?

This is an easy problem. All six treatments can and should appear in each block leading to a randomised complete block design. Any other allocation will involve at least one treatment occurring twice in a block, and consequently some treatment comparisons being not directly estimated in that block, which must lead to less efficient comparison. The treatment allocation to blocks is shown in Figure 7.3 (throughout this chapter treatment allocations to blocks will be shown without randomisation).

Example 7.2

Again, six treatments A to F, this time using 30 units grouped in six natural blocks of five units each. How should the treatments be allocated to units?

Figure 7.3. Experimental plan for comparing six treatments in four blocks of six units.

Block

I	II	III	IV
A	A	A	A
B	B	B	B
C	C	C	C
D	D	D	D
E	E	E	E
F	F	F	F

Since all treatments are assumed equally important, each treatment should occur five times. The number of direct comparisons of different treatments within a block is maximised by avoiding repetition of any treatment within a block; thus, each block will contain five, different, treatments. The problem then reduces to choosing which treatment should be omitted from each block. Since there are six blocks and six treatments, the symmetrical arrangement of omitting a different treatment in each block, as shown in Figure 7.4, has an intuitive appeal. The intuition is correct. The overall average precision of comparisons cannot be improved. The assumption of equal importance for all treatment comparisons provides the justification for the choice of the symmetric design.

Figure 7.4. Experimental plan for comparing six treatments in six blocks of five units.

Block

I	II	III	IV	V	VI
A	A	A	A	A	B
B	B	B	B	C	C
C	C	C	D	D	D
D	D	E	E	E	E
E	F	F	F	F	F

Example 7.3
Again six treatments, A to F, this time using 24 units grouped in six natural blocks of four units each.

The equal importance of treatments leads to the first decision to have four units for each treatment. The primacy of direct comparisons requires that no treatment occurs twice in a block. In each block two treatments must be omitted, and correspondingly each treatment must be omitted from two blocks. This means that symmetry cannot be retained. Treatment A must be omitted twice and, to make direct comparisons with other treatments as even as possible, two different treatments will be omitted with treatment A. Suppose that in block I treatments A and B are omitted, and in block II treatments A and C are omitted. In the remaining four blocks, A occurs in each, B and C in three, and D, E and F in two each. Hence, the direct comparisons of A with B or with C are three, whereas there are two with each of D, E and F. By considering the total possible direct comparisons for A, it can be recognised that no more even arrangement of direct comparisons can be achieved. Treatment A occurs in four blocks, in each of which there are three other units giving three direct comparisons in each block, and a total of 12 direct comparisons with A. Since there are five other treatments, and 12 is not a multiple of five, an uneven division of direct comparisons is inevitable.

An allocation in which each pair of omitted treatments is different seems intuitively attractive. There are two possible systems for the set of omitted pairs:

(AB), (AC), (BC), (DE), (DF), (EF),

or

(AB), (AC), (BD), (CE), (DF), (EF).

The second is based on one cycle (ABDFEC) rather than two (ABC) (DEF), and seems intuitively to offer a more even distribution of comparisons. This is confirmed by the computer-generated variances of estimated treatment differences. For the design with pairs of missing treatments

(AB), (AC), (BC), (DE), (DF), (EF)

there are two variances for treatment differences:

$0.7303 \ \sigma^2$ for (AB), (AC), (BC), (DE), (DF), (EF),

$0.7601 \ \sigma^2$ for the other nine treatment differences.

For the design with pairs of missing treatments

(AB) (BD) (DF) (FE) (EC) (CA)

there are three variances for treatment differences:

$0.7321 \ \sigma^2$ for (AB) (BD) (DF) (FE) (EC) (CA),

$0.7579 \ \sigma^2$ for (AD) (BF) (DE) (FC) (EA) (CB),

$0.7596 \ \sigma^2$ for (AF) (BE) (DC).

The difference is indeed slight with weighted mean variances of $0.7482 \ \sigma^2$ and $0.7479 \ \sigma^2$, though the second design also has the benefit of more homogeneous variances. It would, of course, be ridiculous to argue that the first design is inefficient in any practically meaningful sense.

Example 7.4

Once again, six treatments, again with 30 units, but this time in ten natural blocks of three units each. Again, the equal importance of the treatments implies five occurrences of each treatment. Consideration of the direct comparisons with treatment A suggests that, since A will occur in five blocks, with two other treatments in each block, giving a total of ten direct comparisons to be between five other treatments, there may be a design offering equal frequency of direct comparison for all treatment pairs. Such a design, if it in fact exists, can be constructed logically as follows:

(i) A must occur in five blocks.

(ii) B must occur in two of these blocks and in three others.

(iii) C must occur twice with A, twice with B and a total of five times. Suppose that C occurs in block I, then the remaining four occurrences must include only one with A, and one with B. Since there are only two blocks available in which C can occur without A or B it follows that C must occur in both of these 'empty' blocks, once in a block with A alone, and one in a block with B alone. Therefore, C occurs in block I (with A and B), in block III (with A), in block VI (with B) and in blocks IX and X.

(iv) D must occur twice with A, twice with B, twice with C, and five times in all. Since all blocks now have at least one of A, B or C, it follows that D will occur with two of A, B or C in one block, and with one of them in each of its other four appearances. Without loss of generality we can add D to block II since one of D, E or F, must occur in this block. Therefore, D occurs in block II (with A and B), not in III, in IV (with A), not in V or VI, in VII (with B) and in IX and X (with C).

(v) The allocation of E and F can be considered together. Both must occur in blocks V and VIII. One must occur in III and the other in IV. And one must occur in VI and the other in VII, the one in VI being the same as in IV. Thus, E occurs in III, V, VII, VIII and IX, and F occurs in IV, V, VI, VIII and X.

Figure 7.5. Experimental plan for comparing six treatments in ten blocks of three units.

					Block				
I	II	III	IV	V	VI	VII	VIII	IX	X
A	A	A	A	A	B	B	B	C	C
B	B	C	D	E	C	D	E	D	D
C	D	E	F	F	F	E	F	E	F

This design has been considered in some detail, and it should be clear that there is no real choice in the construction of the design. Any other design fitting the limitations of the design will be identical (except for permutation of the letters) with the one we have constructed, shown in Figure 7.5.

Example 7.5

Still six treatments, this time with 42 units grouped in six blocks of seven units. The symmetry principles advanced in example 7.2 apply again here. The natural design includes six treatments in each block with an additional treatment in each block, the additional treatment being different in each block. The resulting design is shown in Figure 7.6.

Figure 7.6. Experimental plan for comparing six treatments in six blocks of seven units.

Block

I	II	III	IV	V	VI
A	A	A	A	A	A
B	B	B	B	B	B
C	C	C	C	C	C
D	D	D	D	D	D
E	E	E	E	E	E
F	F	F	F	F	F
A	B	C	D	E	F

Example 7.6

For the last time, six treatments, now in 25 units grouped in five blocks of five units each. Again, comparison with example 7.2 is useful. Each block must contain five out of the six treatments, and a different treatment should be omitted from each block. One treatment will inevitably occur in each block and the comparisons involving that treatment will be more

Figure 7.7. Experimental plan for comparing six treatments in five blocks of five units.

Block

I	II	III	IV	V
A	A	A	A	A
B	B	B	B	C
C	C	C	D	D
D	D	E	E	E
E	F	F	F	F

precisely determined than those of the other treatments. The design offering most nearly even precision of comparisons is therefore that shown in Figure 7.7.

Example 7.7

Finally in this section we consider an example discussed initially in Chapter 1, in which seven treatments affecting the growth of rats are to be compared using 35 rats from five litters, or blocks, containing five, six, seven, seven and ten rats, respectively. The two blocks of seven units can be dealt with first, by allocating all seven treatments to each block. The block of ten units must contain all seven treatments, with three of the treatments being duplicated. These three treatments must then be omitted from one or other of the two smaller blocks. The design is shown in Figure 7.8. Clearly, the precision of different comparisons will vary slightly but the design is both obvious and efficient.

Figure 7.8. Experimental plan for comparing seven treatments in five blocks with five, six, seven, seven and ten units.

Block				
I	II	III	IV	V
A	A	A	A	A
B	B	B	B	B
C	C	C	C	C
D	D	D	D	D
E	F	E	E	E
	G	F	F	F
		G	G	G
				E
				F
				G

The examples presented in this section are intended to demonstrate that, when block size is not equal to the number of treatments, then the construction of a sensible design is not difficult. Two particular principles deserve special emphasis. First, if a design such as that for example 7.4, in which a complete, uniform, set of pairwise occurrences of treatments is possible, then simple logical argument will produce that design. If equal pairwise occurrence cannot be achieved then simple logical argument will produce the most nearly equal pattern of occurrences. Second, recognising that the randomised complete design is the most efficient possible blocked design, small modifications of the complete block design, such as omitting a different treatment from each block, will provide designs with high efficiency.

In the previous section we discussed the precision of treatment comparisons when direct and indirect comparisons are combined. The variance for any treatment comparison in an incomplete block design will include a multiplying factor larger than that for the same comparison in a complete block design, assuming the value of σ^2 were the same in both designs. However, the multiplying factors in the variances of treatment comparisons in incomplete block designs are usually not much larger than those for the same comparisons (for the same treatment replication) in complete block designs (or equivalently in unblocked designs). In Table 7.1 the variances for all the different forms of comparison between treatments for the various designs illustrated in this section are listed. For comparison the variances for the same replication pattern in complete blocks, and the variances which would be obtained if only direct comparisons (from joint occurrence of the treatments in a block) were considered, are also given. All variances are written in terms of σ^2, and if the blocks for incomplete block designs are 'better' blocks than those for complete block designs then σ^2 will be smaller for the incomplete block design – sometimes very much smaller. It is clear from Table 7.1 that the increase in the multiplying factor from using incomplete blocks is usually small, the greatest loss in precision occurring for example 7.4, when the smallest block size (half the number of treatments) is used. It should also be noted that the incomplete

Table 7.1. *Comparison of variances for treatment comparisons from incomplete block designs in examples 7.2 to 7.7, with those for complete blocks and for direct comparisons.*

Design	Variance for treatment differences	Complete block variance	Direct block variance
(7.2)	$0.4167\sigma^2$	$0.4000\sigma^2$	$0.5000\sigma^2$
(7.3) AB	$0.5359\sigma^2$	$0.5000\sigma^2$	$0.6667\sigma^2$
AD	$0.5744\sigma^2$	$0.5000\sigma^2$	$1.0000\sigma^2$
AF	$0.5769\sigma^2$		
(7.4)	$0.5000\sigma^2$	$0.4000\sigma^2$	$1.0000\sigma^2$
(7.5)	$0.2917\sigma^2$	$0.2857\sigma^2$	$0.3056\sigma^2$
(7.6) AB	$0.4605\sigma^2$	$0.4500\sigma^2$	$0.5000\sigma^2$
BC	$0.5262\sigma^2$	$0.5000\sigma^2$	$0.6667\sigma^2$
(7.7) AB	$0.4000\sigma^2$	$0.4000\sigma^2$	$0.4000\sigma^2$
AE	$0.4112\sigma^2$	$0.4000\sigma^2$	$0.4688\sigma^2$
AF	$0.4130\sigma^2$	$0.4000\sigma^2$	$0.4688\sigma^2$
EF	$0.4156\sigma^2$	$0.4000\sigma^2$	$0.5000\sigma^2$
FG	$0.4000\sigma^2$	$0.4000\sigma^2$	$0.4000\sigma^2$

block variances are invariably much closer to the complete block variances than to the direct comparison variances. Thus the effective replication of treatment comparisons, though less than the actual replication is much greater than suggested by the direct comparison replication.

One final warning. In discussing incomplete block designs we have assumed equal importance for all treatment comparisons. This will often be true or at least the relative importance of different comparisons will be not clearly identified and it will be reasonable to proceed as if all comparisons are equally important. However, sometimes it will be clear that some treatment comparisons require greater precision than others. In such cases a sensible procedure is to consider first the desired relative precision of different comparisons and to choose the replication for each treatment, n_j, in such a way that the variances of treatment differences for an unblocked experiment, $\sigma^2(n_j^{-1} + n_{j'}^{-1})$, take appropriate relative values. Then the allocation of the treatments, with n_j observations for treatment j, should be chosen so as to achieve joint occurrences of pairs of treatments in numbers proportional to the product of their replications. Wide variation in precision is not a practical option (as was illustrated in exercise 6(2)).

Example 7.8

Suppose it has been decided that the replication of a control treatment should be twice that of other treatments and that the experiment should include eight observations of treatment O and four observations of each of the other seven treatments (A to G). The appropriate blocking system is believed to be six blocks of six units each. Then each block must have at least one control treatment observation. The joint occurrences of pairs of the other treatments will be two or three and a suitable design is that shown in Figure 7.9. The resulting variances of treatment differences are

Figure 7.9. Experimental plan for comparing eight treatments in six blocks of six units.

Block

I	II	III	IV	V	VI
A	A	A	A	B	B
B	B	C	C	C	D
C	E	D	E	D	E
D	F	F	G	F	F
E	G	G	O	O	G
O	O	O	O	O	O

Table 7.2. *Variances of treatment differences for example 7.8.*

	A	B	C	D	E	F	G
B	0.5477						
C	0.5236	0.5475					
D	0.5477	0.5228	0.5236				
E	0.5228	0.5239	0.5475	0.5477			
F	0.5477	0.5228	0.5475	0.5228	0.5467		
G	0.5228	0.5417	0.5475	0.5477	0.5228	0.5239	
O	0.3978	0.3983	0.3866	0.3978	0.3983	0.3983	0.3983

tabulated in Table 7.2, and again it can be seen that the irregular pattern of the design is not reflected in the variances which are very homogeneous.

7.3 The analysis of block–treatment designs

Consider the general design with n_{ij} observations for treatment j in block i. For most designs considered in the previous section, and in the rest of this chapter, n_{ij} will be either 0 or 1. We write the model for the analysis of such designs in the form

$$y_{ijk} = b_i + t_j + \varepsilon_{ijk} \qquad k = 1, \dots, n_{ij},$$

where the b_i's represent block mean yields and the t_j's are deviations of yield for particular treatments from the average treatment yield. The least squares equations are

$$\hat{b}_i N_{i.} + \sum_j \hat{t}_j n_{ij} = Y_{i..} \tag{7.17}$$

$$\sum_i \hat{b}_i n_{ij} + \hat{t}_j N_{.j} = Y_{.j.}, \tag{7.18}$$

where the dot notation for totals is used for both n_{ij} and y_{ijk}. Eliminating $\hat{b}_i$ from (7.18) using (7.17), we obtain

$$\hat{t}_j \left(N_{.j} - \sum_i n_{ij}^2/N_{i.} \right) - \sum_{j' \neq j} t_{j'} \left(\sum_i n_{ij} n_{ij'}/N_{i.} \right) = Y_{.j} - \sum_i n_{ij} Y_{i..}/N_{i.}.$$

These equations are not independent, as can be seen by checking that the sum of the coefficients of any particular t_j over the set of t equations is zero. The constraint $\sum_j t_j = 0$ is therefore needed to solve the equations.

A more important deduction from the set of equations concerns the condition necessary for these equations to be symmetrical for all t treatments. If we consider the coefficient of $t_{j'}$ in the equation for t_j then we can see that all such coefficients will be equal if

$$p_{jj'} = \sum_i n_{ij} n_{ij'}/N_{i.} = \text{constant for all } (j, j').$$

This result was recognised first by Pearce (1963), who called the $p_{jj'}$ the 'sum of weighted concurrences' for treatments j and j'. Pearce's result is that, if all $p_{jj'}$ are equal, then all simple treatment comparisons are equally precise, and the design is said to be *balanced*. This is a wider definition of balance than the classical special case of balanced incomplete block designs, which will be considered later in this chapter.

A simple example, taken from Pearce's paper, demonstrates the general principle of balance and provides a further example of the detailed manipulation of the least squares equations to obtain variances for treatment comparisons.

Example 7.9
Consider a design for three treatments (O, A and B) in two blocks of four units and one of two units:

> block I   O   O   A   B
> block II   O   O   A   B
> block III   A   B.

The sums of weighted concurrences are then

> O and A $\dfrac{2 \times 1}{4} + \dfrac{2 \times 1}{4} = 1$
> O and B
>
> A and B $\dfrac{1 \times 1}{4} + \dfrac{1 \times 1}{4} + \dfrac{1 \times 1}{2} = 1.$

The least squares equations for the three block effects and three treatment effects are

$$
\begin{aligned}
4\hat{b}_1 &\phantom{+2\hat{b}_2} \phantom{+2\hat{b}_3} + 2\hat{t}_O + \hat{t}_A + \hat{t}_B = Y_1 \\
&4\hat{b}_2 \phantom{+2\hat{b}_3} + 2\hat{t}_O + \hat{t}_A + \hat{t}_B = Y_2 \\
&\phantom{4\hat{b}_2} 2\hat{b}_3 \phantom{+2\hat{t}_O} + \hat{t}_A + \hat{t}_B = Y_3 \\
2\hat{b}_1 + 2\hat{b}_2 \phantom{+2\hat{b}_3} + 4\hat{t}_O \phantom{+\hat{t}_A+\hat{t}_B} = Y_O \\
\hat{b}_1 + \hat{b}_2 + \hat{b}_3 \phantom{+4\hat{t}_O} + 3\hat{t}_A \phantom{+\hat{t}_B} = Y_A \\
\hat{b}_1 + \hat{b}_2 + \hat{b}_3 \phantom{+4\hat{t}_O+3\hat{t}_A} + 3\hat{t}_B = Y_B.
\end{aligned}
$$

Eliminating $\hat{b}_1$, $\hat{b}_2$ and $\hat{b}_3$ leads to the set of equations

$$
\begin{bmatrix} 2 & -1 & -1 \\ -1 & 2 & -1 \\ -1 & -1 & 2 \end{bmatrix}
\begin{bmatrix} \hat{t}_O \\ \hat{t}_A \\ \hat{t}_B \end{bmatrix} =
\begin{bmatrix} Y_O - 1/2\,(Y_1 + Y_2) \\ Y_A - 1/4\,(Y_1 + Y_2 + 2Y_3) \\ Y_B - 1/4\,(Y_1 + Y_2 + 2Y_3) \end{bmatrix}.
$$

These can be simply solved with the constraint

$$t_O + t_A + t_B = 0.$$

Alternatively, if we apply the constraint directly to the original equations, we have each block effect expressed in terms of $\hat{t}_O$,

$$4\hat{b}_1 + \hat{t}_O = Y_1$$
$$4\hat{b}_2 + \hat{t}_O = Y_2$$
$$2\hat{b}_3 - \hat{t}_O = Y_3.$$

Substitution in the equation for $\hat{t}_O$ and subsequently in the equations for $\hat{t}_A$ and $\hat{t}_B$ provides direct estimates of treatment effects. Either method leads to the same estimates

$$3\hat{t}_O = Y_O - 1/2(Y_1 + Y_2)$$
$$3\hat{t}_A = Y_A - 1/4(Y_1 + Y_2 + 2Y_3)$$
$$3\hat{t}_B = Y_B - 1/4(Y_1 + Y_2 + 2Y_3),$$

whence

$$3(\hat{t}_O - \hat{t}_A) = Y_O - Y_A - (Y_1 + Y_2)/4 + Y_3/2$$
$$3(\hat{t}_A - \hat{t}_B) = Y_A - Y_B.$$

Hence $\mathrm{Var}(\hat{t}_A - \hat{t}_B) = 2\sigma^2/3$, which, since each of A and B appears in each block, is an obvious result. And, either directly, or using the diagrammatic method of Figure 7.2,

$$12(\hat{t}_O - \hat{t}_A) = 3(y_{101} + y_{102} + y_{201} + y_{202}) - 5(y_{1A} + y_{2A})$$
$$- (y_{1B} + y_{2B}) - 2(y_{3A} + y_{3B})$$

and

$$\mathrm{Var}(\hat{t}_O - \hat{t}_A) = \sigma^2(4 \times 9 + 2 \times 25 + 2 \times 1 + 2 \times 4)/144$$
$$= 2\sigma^2/3.$$

Consider now the fitting SS for the analysis of the general block–treatment model:

$$\text{fitting SS} = \sum_i \hat{b}_i Y_{i..} + \sum_j \hat{t}_j Y_{.j.}.$$

Substituting for $\hat{b}_i$,

$$\text{fitting SS} = \sum_i Y_{i..} \left\{ (Y_{i..}/N_{i.}) - \sum_j (n_{ij}\hat{t}_j/N_{i.}) \right\} + \sum_j \hat{t}_j Y_{.j.}$$
$$= \sum_i Y_{i..}^2/N_{i.} + \sum_j \hat{t}_j \left\{ Y_{.j.} - \sum_i (n_{ij}Y_{i..}/N_{i.}) \right\},$$

and the two terms can be recognised as (a) the SS for blocks ignoring treatment effects, and (b) the SS for treatments allowing for block effects. The second term is a particular example of the general result in Section 4.A5:

$$\theta_2'(A_2' - C_{21}C_{11}^{-1}A_1')y.$$

The practical analysis of data from incomplete block designs is illustrated in two examples.

Example 7.10

An experiment to examine preferences of cabbage root flies for six different substances on which to lay their eggs involved the use of ten cages of flies with three substances available in each cage. The number of eggs laid on the various substances were as shown in Table 7.3.

Table 7.3.

Cage	Substance					
	1	2	3	4	5	6
1	452	69	83			
2	802	143		53		
3	699		32		4	
4	1207			19		32
5	958				8	8
6		328	147			53
7		314		204	223	
8		158			36	5
9			117	14	115	
10			23	16		2

The design may be recognised as that of example 7.4. There are some difficulties in drawing inferences from this data. The eggs laid in each cage will be distributed between the three substances as a result of the 'competition' between the three substances for the flies' attention. But also, the overall rate of egg laying within each cage may be stimulated or depressed by the set of three substances in that cage. Therefore, inferences about differences between egg laying rates for the three substances may be unreliable if extrapolated to external conditions where only a single one of the substances will be available. Nevertheless, as a basis for identifying the relative preferences of flies for the different substances, in the context of the experimental situation, the analysis is informative.

There are large overall differences between cages and, to try to satisfy the assumption that all observations are subject to the same random unit variance, σ^2, a square-root transformation of the egg counts seems appropriate. The philosophy of transformations is discussed in more detail in Chapter 11. The square roots of observed counts are shown in Table 7.4. Cage and substance totals are:

Cage	1	38.7	Substance 1	141.8
	2	47.6	2	68.7
	3	34.1	3	42.5
	4	44.9	4	35.6
	5	36.6	5	36.4
	6	37.5	6	19.4
	7	48.8	Total	344.4
	8	20.8		
	9	25.2		
	10	10.2		

Table 7.4.

Cage	Substance					
	1	2	3	4	5	6
1	21.3	8.3	9.1			
2	28.3	12.0		7.3		
3	26.4		5.7		2.0	
4	34.8			4.4		5.7
5	31.0				2.8	2.8
6		18.1	12.1			7.3
7		17.7		14.2	14.9	
8		12.6			6.0	2.2
9			10.8	3.7	10.7	
10			4.8	4.0		1.4

Fitting cage and treatment effects using a general linear model program gives the following residual SS:

total variation about the mean	= 2431 (29 df)
SS (fitting cage effects only)	= 1973 (20 df)
SS (fitting substance effects only)	= 467 (24 df)
SS (fitting cage and substance effects)	= 114 (15 df).

The analysis of variance can be written in two forms shown in Table 7.5. The first of the analyses in Table 7.5 corresponds to the algebraic form of the fitting SS derived earlier in this section.

The least squares estimates of cage and substance effects from the full model are given in Table 7.6. Allowing for differences between cages, the F

Table 7.5.

	SS	df	MS
Cage effects (ignoring substances)	458	9	51
Substance effects (allowing for cages)	1859	5	372
Error	114	15	7.6
Total	2431	29	
or			
Substance effects (ignoring cages)	1964	5	393
Cage effects (allowing for substances)	353	9	39
Error	114	15	7.6
Total	2431	29	

Table 7.6.

	Effect
Cage 1	−4.64
2	−0.23
3	−4.25
4	1.77
5	−1.37
6	3.72
7	7.94
8	−0.78
9	0.93
10	−3.08
Substance 1	18.62
2	1.05
3	−1.52
4	−5.83
5	−4.70
6	−7.65

Standard error of difference between two substances is 1.95.

ratio to test the barely credible hypothesis of no difference between substances is 49 on 5 and 15 df. The conclusions about the substances are that substance 1 is very much preferred by cabbage root flies, with substances 2 and 3 next in the order of preference. The differences between substances 2 to 6 although clearly not negligible do not show a clear grouping, the order of preference being 2, 3, 5, 4, 6. As mentioned earlier, more detailed numerical interpretations are probably not capable of valid interpretation.

Example 7.11

The experiment on peppers discussed in the preliminary section of this chapter was the first part of a two year programme to find the best treatment. The results (yield − costs) from 24 observations over two years of experiments were as shown in Table 7.7. The philosophy behind the design is simple. In the first year eight treatments can be arranged in two blocks of six by having four treatments in both blocks and the other four treatments in one of the two blocks. The four treatments to be used in both blocks may be chosen arbitrarily or selected to contain each level of each factor exactly twice (a concept we return to in Chapter 13). In the second year, the five most successful treatments are retained. The complete set of 24 observations can be analysed considering the years and blocks as blocking factors and the treatment combinations simply as eight treatments. The resulting residual SS are

$$\text{total variation about the overall mean} = 153.2 \ (23 \ df)$$
$$\text{fitting block effects only} = \ 36.7 \ (20 \ df)$$
$$\text{fitting treatment effects only} = \ 91.3 \ (16 \ df)$$
$$\text{fitting block and treatment effects} = \ \ 6.0 \ (13 \ df).$$

The analysis of variance is shown in Table 7.8, and the least squares

Table 7.7.

Heating	0	0	0	0	1	1	1	1
Lighting	0	0	1	1	0	0	1	1
CO_2	0	1	0	1	0	1	0	1
Year 1 Block 1	11.4	13.2	10.4	—	13.7	—	12.0	12.5
Year 1 Block 2	—	8.4	6.5	6.1	10.8	9.4	—	9.1
Year 2 Block 1	—	13.7	—	—	14.6	16.5	12.8	12.9
						15.4		
Year 2 Block 2	—	10.7	—	—	10.9	10.9	9.0	10.2
							10.1	

Table 7.8.

	SS	df	MS	F
Block effects (ignoring treatments)	116.5	3	38.8	
Treatment effects (allowing for blocks)	30.7	7	4.4	9.6
Error	6.0	13	0.46	
Total	153.2	23		

estimates of the treatment effects (allowing for block effects) are shown in Table 7.9(a). The standard errors for differences between treatment means are tabulated in Table 7.9(b).

The variation in replication of the treatments causes some variation of precision but treatment comparisons for treatments 2, 5, 6, 7 and 8 are similarly precise. The standard error of a treatment difference using four replicates with a σ^2 value of 0.46 (as achieved by blocking) would be 0.48, almost the same as those shown in Table 7.9(b). However, if the variance ignoring block effects ($91.3/16 = 5.7$) is used then the standard error would be 1.69. Thus the non-orthogonality of the block–treatment structure hardly affects the standard errors but the use of blocking reduces the standard errors by a factor greater than three. Note that the analysis has not used the treatment structure. Further analysis would examine the main effect and interaction information in the data. It is, however, immediately obvious that the dominant effect is the main effect of heating.

Table 7.9(a). *Least square estimates*

	H	L	CO_2	Estimate of mean
1	0	0	0	9.23
2	0	0	1	10.72
3	0	1	0	8.25
4	0	1	1	7.88
5	1	0	0	11.72
6	1	0	1	12.01
7	1	1	0	9.86
8	1	1	1	10.39

(b). *Standard errors*

1	2	3	4	5	6	7	8
0.80							
0.86	0.61						
1.06	0.80	0.86					
0.80	0.48	0.61	0.80				
0.84	0.49	0.64	0.81	0.49			
0.81	0.49	0.64	0.84	0.49	0.50		
0.80	0.48	0.61	0.80	0.48	0.49	0.49	

7.4   Balanced incomplete block designs and classes of less balanced designs

There is a vast literature on the theory and indexing of classes of incomplete block designs. The original reason for the investigation of possible incomplete block designs arose from the wish to use practically relevant sets of treatments and practically sensible block sizes. Moreover the interest in such designs began many years before the advent of computers, and so an overriding consideration in developing incomplete block designs was the need to be able to analyse the results using only a manually operated calculator. The main problem in devising designs was therefore to recognise mathematical structures enabling the least squares equations to be solved in a way that involved simple arithmetical calculations. These mathematical structures essentially define classes of restricted designs. Consequently the history of the theory of incomplete block designs is primarily a sequence of increasingly intricate classes of designs, with occasional generalisations to include several previous classes.

Clearly the computational facilities are now quite different, and much of the theory of particular classes of design is now irrelevant. Nevertheless we can learn something about the principles of experimental design by examining some of the major developments of incomplete block design.

The general Pearce criterion of balance defines the condition required for the estimation of treatment differences to be equally precise for each pair of treatments. However, it was recognised many years earlier by Yates that balanced designs could be constructed through symmetry considerations. Essentially if, in a set of blocks of the same size, each pair of treatments occurs together in a block the same number of times then by symmetry all treatment differences will be equally precise. The resulting designs are known as balanced incomplete block (BIB) designs.

First, some terminology. A BIB design has

> t treatments and each block has
> k units. There are
> b blocks and
> r replicates of each treatment. Each pair of treatments occurs
> λ times together in a block.

Trivial examples of BIBs are those which include all possible groups of k out of t treatments. These are known as unreduced designs and their values of b, r and λ may be found by simple combinatorial arguments

$$b_u = \frac{t!}{k!(t-k)!}$$

$$r_u = \frac{(t-1)!}{(k-1)!(t-k)!}$$

$$\lambda_u = \frac{(t-2)!}{(k-2)!(t-k)!}.$$

By considering the total number of experimental units it can be seen that

$$bk = tr.$$

Similarly, by considering the number of units included in the set of blocks containing a particular treatment,

$$r(k-1) = (t-1)\lambda.$$

Hence, for any BIB design,

$$b : r : \lambda = (t-1)t : (t-1)k : (k-1)k,$$

so that, for any pair of values of t and k, the ratios $b : r : \lambda$ are invariant over all possible designs, and in particular $b : r : \lambda$ will be the same as $b_u : r_u : \lambda_u$. It follows that reduced BIB designs can only exist when (b_u, r_u, λ_u) have a common factor greater than 1, and that in this case BIB designs may exist for values

$$b = (m/f)b_u, \quad r = (m/f)r_u, \quad \lambda = (m/f)\lambda_u,$$

where f is the highest common factor of (b, r, λ) and m is any integer less than f. As might perhaps be inferred from the ease with which we constructed example 7.4, designs are easily found whenever existence has been demonstrated to be possible. The set of (t, k) pairs for which designs have been shown to exist with block size not bigger than ten, and total number of units not exceeding 100, are listed in Table 7.10 and the sparseness of the available designs is clearly apparent. It is notable that the number of designs with a small number of replicates ($\leqslant$ five) such as experimenters would expect to use is extremely limited. More detailed design information is given in Cochran and Cox (1957) and Fisher and Yates (1963).

One characteristic of incomplete block designs sometimes considered to be important is whether the blocks can be grouped so that each group of blocks constitutes a complete replicate of the set of treatments. When this division of an experiment into replicate groups of blocks is possible the design is said to be resolvable. None of the incomplete block examples

Table 7.10. *Pairs of values of* (k, t) *for which a BIB design is known to exist for* $k \leqslant 10$ *and* $bk < 100$.

k	t	b	r	λ	k	t	b	r	λ
3	4	4	3	2	4	13	13	4	1
3	5	10	6	3	4	16	20	5	1
3	6	10	5	2	5	6	6	5	4
3	6	20	10	4	5	9	18	10	5
3	6	10	5	2	5	10	18	9	4
3	7	7	3	1	5	11	11	5	2
3	9	12	4	1	6	7	7	6	5
3	10	30	9	2	6	9	12	8	5
3	13	26	6	1	6	10	15	9	5
4	5	5	4	3	6	11	11	6	3
4	6	15	10	6	6	16	16	6	2
4	7	7	4	2	7	8	8	7	6
4	8	14	7	3	8	9	9	8	7
4	9	18	8	3	9	10	10	9	8
4	10	15	6	2					

considered so far in this chapter is resolvable and the concept of resolvable designs is chiefly relevant to designs for large number of treatments, where it is feared that mistakes in allocation may occur more easily in non-resolvable designs.

One important class of BIB designs for large numbers of treatments is the set of lattice designs. A simple lattice design is based on a square array containing the set of treatment identifiers. For a set of 25 treatments the square array would be

$$
\begin{array}{ccccc}
A & B & C & D & E \\
F & G & H & I & J \\
K & L & M & N & O \\
P & Q & R & S & T \\
U & V & W & X & Y.
\end{array}
$$

A lattice design for 25 treatments uses blocks of five units in groups of five blocks. The first group consists of one block for each row of the array. The second group consists of one block for each column of the array. Subsequent groups are defined by superimposing on the 5×5 square, arrays whose elements are the digits 1, 2, 3, 4, 5 arranged so that each digit occurs once in each row and once in each column. Thus the array

1	2	3	4	5
2	3	4	5	1
3	4	5	1	2
4	5	1	2	3
5	1	2	3	4

can be used to generate a group of five blocks with each block defined by the treatments corresponding to a particular digit. For digit 1 we get the block (AJNRV). By discovering four such arrays of digits such that each pair of arrays is orthogonal in the sense that each digit in one array occurs with each digit in the second array we can construct four groups of five blocks which when combined with the two groups for rows and columns give a balanced design. The contents of the 30 blocks are listed in Figure 7.10. Simple lattice designs are clearly resolvable but they are also very clearly restrictive, being limited in the possible numbers of treatments $(t = n^2)$, block size $(t)^{1/2}$ and number of replications $(1 + t^{1/2})$. The last restriction may be eliminated by considering designs omitting one or more

Figure 7.10. Experimental plan for comparing 25 treatments in a lattice design with 30 blocks of five units.

Blocks

I	II	III	IV	V	VI	VII	VIII	IX	X
A	F	K	P	U	A	B	C	D	E
B	G	L	Q	V	F	G	H	I	J
C	H	M	R	W	K	L	M	N	O
D	I	N	S	X	P	Q	R	S	T
E	J	O	T	Y	U	V	W	X	Y

XI	XII	XIII	XIV	XV	XVI	XVII	XVIII	XIX	XX
A	B	C	D	E	A	B	C	D	E
J	F	G	H	I	I	J	F	G	H
N	O	K	L	M	L	M	N	O	K
R	S	T	P	Q	T	P	Q	R	S
V	W	X	Y	U	W	X	Y	U	V

XXI	XXII	XXIII	XXIV	XXV	XXVI	XXVII	XXVIII	XXIX	XXX
A	B	C	D	E	A	B	C	D	E
H	I	J	F	G	G	H	I	J	F
O	K	L	M	N	M	N	O	K	L
Q	R	S	T	P	S	T	P	Q	R
X	Y	U	V	W	Y	U	V	W	X

groups of blocks. The lack of balance in an incomplete lattice design is not extreme and the precision of treatment differences is at two levels depending on whether the two treatments to be compared occur together in a block. Such a design may be described as possessing partial balance with two classes of association, or equivalently precision. The class of partially balanced incomplete block designs includes many other families of designs and has been the source of an enormous literature, now of 'archaeological' interest only.

A final class of designs which deserves brief mention is that of cyclic designs. In the basic form of cyclic designs blocks are generated by cyclic rotation through the treatment letters, starting from an initial block. The initial block must be carefully chosen. Consider the problem of designing an experiment to compare seven treatments in seven blocks of three. We could start with the block (ABC) and by cyclic generation produce the design

(ABC), (BCD), (CDE), (DEF), (EFG), (FGA), (GAB).

This is obviously a poor design with treatment A occurring twice with B and G, once with C and F and not at all with D or E. However, starting with the block (ABD) we obtain

(ABD), (BCE), (CDF), (DEG), (EFA), (FGB), (GAC),

which is a BIB design and the best available design. Cyclic design structure has produced an extensive literature, and may be generalised to factorial structural treatments. It may be recognised as the most sophisticated form of classical mathematical design theory.

7.5   Orthogonality, balance and the practical choice of design

We have introduced two characteristics of block–treatment designs, orthogonality and balance. These two concepts have been extremely important in the development of design theory and we must assess their importance and relevance for designing experiments in the present day.

Consider again the terminology of the general block–treatment design introduced in Section 7.2, with n_{ij} observations of treatment j in block i. An orthogonal design is one in which treatment differences are estimated independently of block differences. A necessary and sufficient condition for this is that

$$n_{ij} = N_{i.} N_{.j}/N_{..} \quad \text{for all } i, j \text{ combinations}$$

when $N_{i.}$, $N_{.j}$ and $N_{..}$ are, respectively, the total numbers of observations in block i, for treatment j, and in the complete experiment. Note that treatments do not have to be equally replicated in each block and blocks may be of different sizes. The crucial characteristic is that the ratios of treatment replications must be constant over blocks.

The principal consequences of orthogonality are that effects of treatments can be interpreted without simultaneously considering inferences about block effects. Consider again the analysis of variance for a randomised block design, rice spacing experiment of Chapter 2 (see Table 7.11).

Table 7.11.

	SS	df	MS	F
Blocks	5.88	3	1.96	4.26
Spacing	23.14	9	2.57	5.59
Error	12.42	27	0.46	
Total	41.44	39		

The SS for blocks and for treatments are independent. Hence, the ratio spacing MS/error MS can be used to test the hypothesis that all the treatment effects, t_j, are zero, independent of the values of the block effects. Correspondingly the ratio block MS/error MS could be used to test the hypothesis that all the block effects, b_i, are zero, independent of the values of the treatment effects. Because the same divisor is used in both test estimates, the two test results are not independent, and it should also be emphasised that there is rarely any interest in drawing formal inferences about block effects. All we are usually interested in is the extent to which blocking has succeeded in reducing variation between units within blocks so that we can decide whether to use similar blocking criteria in future experiments. Nevertheless the orthogonality of treatments and blocks enables the effects of blocks and of treatments to be considered quite separately. The benefits of orthogonality extend to the interpretation of orthogonal treatment contrasts, the philosophy for which was developed in Chapter 4, and the application of which will be considered in more detail in Chapter 12. If the block–treatment structure is orthogonal then inferences about orthogonal treatment contrasts will be independent. All other things being equal, designs should be arranged to be orthogonal and thus have the benefit of allowing independent interpretation of effects. However, non-orthogonality is not a major defect of a design. Rather it produces a (usually) small loss of information. In all the non-orthogonal designs in this chapter (and book) the non-orthogonality of treatments and blocks is very much less than is typically found in a multiple regression study. Consider the two analyses of variance in example 7.10. The variation of SS according to the order of fitting is relatively slight. Certainly the interpretation of the strength of the substance effects, and of the cage effects, is unchanged whether or not we allow for the other set of effects, and this is typical of the non-orthogonality effect in well-designed experiments.

A second consequence of orthogonality, which was extremely important in the era before computers, is the simplicity of the calculations both for the analysis of variance and for the comparison of treatment effects. This benefit of orthogonality is obviously now greatly reduced and whereas 20 or 30 years ago there might have been conflict between the requirements to use the natural blocking system, and to use a blocking system which allowed orthogonal estimation of treatment and block effects, the decision now should be dominated by the choice of the most effective blocking system. This should imply a reduction in the propensity to use the randomised block design since, if natural blocking systems are being properly sought and recognised, it is not very probable that the size of block will be exactly equal to the number of treatments to be compared.

Balance, as defined in Section 7.3, requires that

$$p_{jj'} = \sum_i n_{ij} n_{ij'} / N_{i.}$$

is constant for all (j, j') pairs. When block size, $N_{i.}$, is constant this reduces to requiring that the total number of joint occurrences (in a block) of two treatments should be the same for all treatment pairs and forms the basis of the BIB designs. Because the class of BIB designs which actually exist is extremely limited, it is usual to find that, given a block size of k units and a set of t treatments, the only BIB design which satisfies the requirements uses far more replicates and hence resources, than are available. Since designs which are not balanced but which are sensibly constructed to make the pairwise treatment occurrence as nearly equal as possible achieve very little variation of the precision of treatment differences, balance must be viewed as a pleasing but unimportant luxury. The theoretical developments of incomplete block design theory have led to definition and tabulation of classes of designs with mathematically neat patterns of treatment occurrence in blocks. The philosophy of such tabulation is, presumably, that the prospective experimenter, with a statistical consultant, should attempt to match the conditions for the desired experiment to that design in the available list which requires the smallest modification of the desired conditions.

I believe that this philosophy, whether or not it was appropriate in the past, is not now practically acceptable. It is important to recognise and control potential variation in experimental units by using blocking in the most effective way possible. It is also important to choose the treatments for comparison in the experiment without constraint from considerations of the block structure. The choice of design should consist of choosing the allocation of treatments to blocks in such a way that treatment comparisons are made as efficiently as possible while minimising the degree of non-orthogonality of blocks and treatments. If an orthogonal design is possible it should be used. A small level of non-orthogonality such as occurs when designs are constructed to make pairwise treatment occurrence as nearly equal as possible will rarely cause ambiguity in interpreting results. If the blocking requirements and the number of treatments are compatible with a balanced design then clearly such a design should be used. However, such designs will automatically be obtained, when they are possible, if in each case the design is constructed to satisfy the general principles discussed in this chapter.

Assuming that all comparisons between two treatments are equally important then the relative precision of treatment comparisons is determined primarily by the number of joint occurrences of the treatments within blocks, and secondarily by the number of second order comparisons when the two treatments occur in different blocks but with a number of common, or linking, treatments. For incomplete blocks it follows that (i) in each block there should be no repeated treatments; (ii) that the pairwise occurrence of treatments should be as nearly even as possible; (iii) that if, for example, the joint occurrences of AB and of AC are relatively high then that of BC need not also be relatively high. An implication and extension of this last idea is that the 'chains' of pairs with relatively high joint occurrence should be long, not short (for a very simple example see example 7.3).

A formal expression of these concepts can be based on the matrix of block–treatment occurrences, N, where, as previously,

n_{ij} = number of occurrences of treatment j in block i.

The joint occurrences p_{jk} of treatments j and k are the elements of the matrix $P = N'N$. The second order links, s_{jk}, for treatments j and k are the elements of the matrix $S = P'P$. If p_{jk} is relatively small then s_{jk} should be relatively large.

In practice, the achievement of an 'optimal' design is not usually an absolute objective since in most problems of design there are many designs of almost equal efficiency. It is therefore sufficient to devise methods of design selection which are simple but reasonably efficient. For relatively small numbers of treatments the philosophy evolved in Section 7.2 is adequate. Treatment should be allocated to sets of blocks sequentially, with all the treatment A allocations first, followed by all the treatment B allocations, and so on. Each allocation of the observations for a particular treatment should be arranged to achieve as nearly as possible the mean joint occurrence with the treatments already allocated, and to minimise restrictions on later treatments.

At the other extreme are designs for large numbers of treatments with small numbers of replications and block sizes such that the joint occurrences of treatment pairs are all zero or one. A substantial amount of work on the practical requirements for designs for national trials of large numbers of varieties of a particular crop has been produced by Patterson and co-workers. Details are given in several papers, notably Patterson, Williams and Hunter (1978) and Patterson and Silvey (1980). The designs used in these trials are required to be resolvable. I believe absolute resolvability should not be a critical requirement in designs for which all joint occurrences are zero or one since it may impose unnecessary conditions on the block sizes. In the second example at the beginning of this chapter, the capacity to handle 18 units in a batch is more important than the resolvability concept which would use 14 units in a batch and thereby waste resources by requiring nine batches instead of seven. However, in constructing designs for large numbers of treatments it is sensible to consider each replicate in turn and within each replicate allocate treatments to complete as many blocks as possible, thus achieving near resolvability.

A helpful concept for the construction of these designs for large numbers of treatments is the general idea of lattices. If the blocks of the first replicate are written out as rows, then a second replicate should employ blocks which 'run across' the rows, and the blocks of the third and subsequent replicates should make use of diagonals.

The two philosophies of design construction, first of allocating treatments sequentially when the number of treatments is relatively small, and second of allocating replicates sequentially when the number of treatments is large, inevitably imply an intermediate area and the designer of experiments has to learn how to select the appropriate philosophy for each problem. We conclude this section by discussing three further problems.

Example 7.12

A trial to compare 40 treatments with three replicates of each treatment using blocks of six units each. The first replicate is written down in six

Figure 7.11. Experimental plan for comparing 40 treatments in 20 blocks of six units.

Block

	I	II	III	IV	V	VI	VII
VII	1	7					
VIII	2	8	13	19	25	31	
IX	3	9	14	20	26	32	
X	4	10	15	21	27		37
XI	5	11	16	22		33	38
XII	6		17	23	28	34	39
XIII		12	18	24	29	35	40
XIV					30	36	

XIV	XV	XVI	XVII	XVIII	XIX	XX
	25	26	27	28	29	30
	32	33	34	35	36	31
5	6	1	7	2	3	4
8	18	12	13	9	10	11
14	21	19	20	15	16	17
37	38	39	40	22	23	24

blocks plus the first four units in the seventh block. Each block of the second replicate is constructed by taking one treatment from each of six blocks of the first replicate. If the treatment occurrence in the blocks for the first two replicates is written down as in Figure 7.11 then the construction of the third replicate must be planned to avoid repeating any treatment pair in a block. Since treatments 30 and 36 have occurred in the overflow into block XIV the treatments in each of blocks V and VI must occur one per block in the last six blocks. The remaining 28 treatments are allocated to blocks XIV to XX mainly by using diagonal transects across the two-way array of the blocks for the first two replicates, and the details of these blocks complete the design in Figure 7.11.

Example 7.13

The second example in the problems outlined at the beginning of this chapter required three replicates of 42 treatments arranged in seven blocks of 18. Clearly in an ideal design each pair of treatments should occur together once or twice. However, it rapidly becomes clear that with the rather large blocks some pairs of treatments must occur together three times. The first 18 treatments are allocated to the first block and the next 18 to the second block. The third block contains the remaining six treatments to complete the first replicate, and the remaining spaces in this

Figure 7.12. Experimental plan for comparing 42 treatments in seven blocks of 18 units.

Block

I	II	III	IV	V	VI	VII
1	19	37	7	13	4	10
2	20	38	8	14	5	11
3	21	39	9	15	6	12
4	22	40	10	16	28	19
5	23	41	11	17	29	20
6	24	42	12	18	30	21
7	25	1	25	31	7	16
8	26	2	26	32	8	17
9	27	3	27	33	9	18
10	28	4	28	34	22	40
11	29	5	29	35	23	41
12	30	6	30	36	24	42
13	31	19	37	1	13	31
14	32	20	38	2	14	32
15	33	21	39	3	15	33
16	34	22	40	25	37	34
17	35	23	41	26	38	35
18	36	24	42	27	39	36

block are allocated to six treatments from block I and six from block II. Since there are only four more blocks each treatment of these sets of six must occur again and it follows that some treatments in each of these groups must occur together three times. The occurrence of new pairs is maximised by constructing blocks IV and V to contain the remaining groups of six from blocks I and II and allocating the last six treatments to block IV. The resulting design is shown in Figure 7.12. It would be possible to arrange that some of the treatments occurred with all the treatments with which they had not previously occurred but this increases the number of treatment pairs occurring together three times. In this example and the previous one it may be possible to find better solutions but the improvement in overall precision will be small. We should also note at this point that we have not considered how to allocate treatments to treatment identifiers; this is discussed in Chapter 9.

Example 7.14

Finally we consider a problem for which all possible designs can be evaluated. We can therefore assess the performance of the designs derived through the mathematical classification methods mentioned in the previous section.

Suppose we wish to compare nine treatments (A, B, ..., I) and we have available units which occur naturally in pairs and we can have 18 pairs of units for this experiment. Each treatment must occur four times, and it seems reasonable to assume that a particular treatment should occur with four different treatments in the four blocks in which it appears. There is no BIB design with 18 blocks of two units each for nine treatments. There is a partially balanced design such that the variance of an estimated difference between two treatments takes one of two values depending only on whether the two treatments appear in a block together. The block structure is

(AB) (AC) (AD) (AE) (BC) (BF) (BH) (CG) (CI)
(DE) (DF) (DG) (EH) (EI) (FG) (FH) (GI) (HI).

The pattern in this structure may not be obvious. However, the pattern is identical for all nine treatments as may be seen by rearranging the order in which the same 18 pairs are written, starting with some letter other than A, for example F:

(BF) (FH) (DF) (FG) (BH) (AB) (BC) (EH) (HI)
(DG) (AD) (DE) (CG) (GI) (AC) (AE) (CI) (EI).

The structure is identical, the letters (F, B, H, D, G, A, C, E, I) in the second ordering corresponding to the letters (A, B, C, D, E, F, G, H, I) in the original. The same pattern will be observed for any initial letter, and this symmetry is the mathematical pattern which is the basis of this design. The design gives just two different variances (different by a factor 4/5) for the differences between two estimated treatment effects, and is the only partially balanced design with just two variances of treatment difference for this particular block structure.

There are two possible cyclic designs. The sets of treatments occurring with each treatment are obtained by a cyclic replacement of treatment letters as displayed below for the first design:

treatment A with B, C, H, I
treatment B with C, D, I, A
treatment C with D, E, A, B
treatment D with E, F, A, B
treatment E with F, G, C, D
treatment F with G, H, D, E
treatment G with H, I, E, F
treatment H with I, A, F, G
treatment I with A, B, G, H.

The two cyclic designs are:

(i) (AB), (BC), (CD), (DE), (EF), (FG), (GH), (HI), (IA),
    (AC), (BD), (CE), (DF), (EG), (FH), (GI), (HA), (IB),   and

(ii) (AB), (BC), (CD), (DE), (EF), (FG), (GH), (HI), (IA),
    (AD), (BE), (CF), (DG), (EH), (FI), (GA), (HB), (IC).

If designs outside these two classes are considered it can be shown that there are 13 further distinct designs. A reasonable criterion for comparing these designs is the average variance of a difference between two treatments, the average being over all possible 36 treatment pairs. The average variances for each design, together with the number of different variances and the range of these variances, are listed in Table 7.12 in order of increasing variance. The partial balanced design is (5); the cyclic designs are (2) and (14).

It can be seen from Table 7.12 that, in the sense of average variance, neither the partial balanced design nor the cyclic designs prove the best, though all three provide relatively rather more homogeneous sets of variances. Also, there are designs (numbers (10), (12) and (16)) which provide precise comparisons for particular treatment comparisons when a factorial treatment structure is assumed, inevitably at a price of less precise comparisons for others. Further investigation of the different designs shows that the most important characteristic in determining the average variance is the numbers of second order links for those treatment pairs which do not occur together in a block. The relationship between average variance and mean second order links, s_{jk}, for those (j, k) combinations

Table 7.12.

Design	Average variance	Number of distinct variances	Range of variance
(1)	0.978	9	0.862–1.128
(2)	0.982	4	0.865–1.093
(3)	0.988	6	0.857–1.103
(4)	0.996	9	0.842–1.155
(5)	1.000	2	0.889–1.111
(6)	1.001	21	0.835–1.173
(7)	1.004	19	0.835–1.209
(8)	1.006	11	0.833–1.167
(9)	1.014	21	0.837–1.239
(10)	1.021	8	0.800–1.257
(11)	1.023	20	0.835–1.241
(12)	1.027	17	0.800–1.250
(13)	1.033	6	0.844–1.244
(14)	1.049	4	0.850–1.242
(15)	1.052	10	0.835–1.314
(16)	1.096	9	0.800–1.400

Figure 7.13. Relationship between mean variance and second-order links for the 16 possible designs.

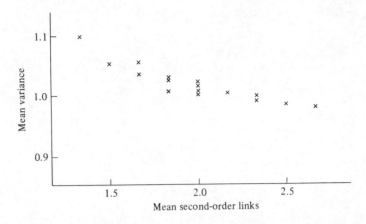

such that $p_{jk}=0$ (using the notation suggested earlier in this section) is shown in Figure 7.13. There will be further consideration of this design situation for the particular case when the nine treatments are a 3×3 factorial set (in Chapter 15).

7.6   The analysis of within block and inter-block information

In Section 7.3 we considered the analysis for general block–treatment designs. However, in that analysis only the information about treatments from comparisons within blocks was considered. The differences between block totals also contain information about treatment differences. Consider the two incomplete block designs shown in Figure 7.14:

(i) $t=6$, $k=5$, $b=6$, $r=5$, $\lambda=4$ (unreduced), and

(ii) $t=10$, $k=4$, $b=15$, $r=6$, $\lambda=2$ (1/14 of unreduced).

The analysis used so far is based on the idea that if, in block I, the observation for treatment A is higher than the other observations in that block then we tend to believe that treatment A is a higher-yielding treatment. We should, however, wish to examine whether treatment A is also relatively high yielding within each of the other blocks in which it occurs before we could rely on the supremacy of A. This is the within block form of comparison which is made more precise through the use of small blocks. Now clearly if in design (i) the total of the observations from block I is markedly larger than the other five block totals we should interpret this as evidence that treatment F gives lower values, though since there is no replication (of blocks with a particular treatment missing) the evidence

Figure 7.14. Incomplete block designs (i) $t=6$, $k=5$, $b=6$, $r=5$, $\lambda=4$, (ii) $t=10$, $k=4$, $b=15$, $r=6$, $\lambda=2$.

Block

(i)	I	II	III	IV	V	VI
	A	A	A	A	A	B
	B	B	B	B	C	C
	C	C	C	D	D	D
	D	D	E	E	E	E
	E	F	F	F	F	F

Block

(ii)	I	II	III	IV	V	VI	VII	VIII
	A	A	A	A	A	A	B	B
	B	B	C	D	E	H	C	D
	C	E	G	F	G	I	I	G
	D	F	H	I	J	J	J	J

	IX	X	XI	XII	XIII	XIV	XV
	B	B	C	C	C	D	D
	E	F	D	E	F	E	F
	H	G	E	F	G	G	H
	I	H	H	J	I	I	J

is slight and no measure of precision is possible. In the second design the potential for evidence is greater. If blocks II, IV, X, XII, XIII and XV gave the six highest block totals then it would be difficult to avoid the suspicion that F yields more than the other treatments. Note that the differences between the individual values making up each block total are irrelevant to this argument. The essential idea (which we shall develop further in Chapter 9) is that the selection of six blocks out of 16 can be made in $(16!)/(6!\,10!)=8008$ ways and that the occurrence of one treatment in all six blocks giving the highest totals is a rather extreme level of coincidence (or significance).

The information from variation between block totals is not limited to arguments about coincidence. By examining the set of 16 block totals and relating the differences between these totals to the sets of treatments occurring in each block we can obtain estimates of differences between treatments. These estimates may not be very precise because blocks are used deliberately to make the variation between units in each block small, and consequently the sampling variation between blocks will tend to be large. Nevertheless, the estimates of treatment differences from inter-block variation do provide information additional to that obtained from the variation within blocks. Somewhat confusingly, in contrast to the inter-block information, the within block information is often referred to as 'intra-block' information, a term which will be familiar to classical scholars but not to other readers. The possibility of obtaining information about treatment effects from two levels of analysis leads to various questions:

(i) How is the inter-block information obtained?

(ii) How should the within block and inter-block information about treatment differences be combined?

(iii) When is it worthwhile to obtain the inter-block information?

We consider these questions using the class of BIB designs as the basis for illustration. However, we start by reconsidering the analysis of the intra-block information. The principles of the analysis are those of Section 7.3 but because of the particular pattern of BIB designs there are certain simplifying features of the analysis.

Within block analysis

The model for the observation for treatment j in block i is

$$y_{ij} = \mu + b_i + t_j + \varepsilon_{ij}. \tag{7.19}$$

The least squares equations are

$$bk\hat{\mu} \qquad\qquad = Y_{..}, \tag{7.20}$$

$$k\hat{\mu} + k\hat{b}_i + \sum_{j(i)} \hat{t}_j = Y_{i.} \tag{7.21}$$

$$r\hat{\mu} + \sum_{i(j)} \hat{b}_i + r\hat{t}_j = Y_{.j} \tag{7.22}$$

when $\sum_{j(i)}$ represents the sum for all treatments (j) which occur in block i, and, similarly, $\sum_{i(j)}$ represents the sum for all blocks (i) in which treatment j occurs. Using (7.21) to eliminate the $\hat{b}_i$ from (7.20) we obtain

$$r\hat{t}_j - r(1/k)\hat{t}_j - \lambda(1/k) \sum_{j' \neq j} \hat{t}_{j'} = Y_{.j} - \sum_{i(j)} (1/k) Y_{i.}$$

since each treatment other than j occurs λ times in the set of blocks, $i(j)$, which include treatment j. Since the treatment effects are, as usual, constrained to sum to zero,

$$t_j = - \sum_{j' \neq j} t_{j'}.$$

Hence (7.21) can be written

$$\hat{t}_j(r - r/k + \lambda/k) = Y_{.j} - 1/k \sum_{i(j)} Y_{i.}$$

whence

$$\hat{t}_j = kQ_j/(\lambda t)$$

using the relationship $\lambda(t-1)=r(k-1)$, where $Q_j=Y_{.j}-1/k\sum_{i(j)} Y_{i.}$ may be thought of as the treatment total effect relative to the block mean yields for those blocks in which the treatment occurs.

The variance of $\hat{t}_j$ and of the estimate of the difference between two treatment effects, $\hat{t}_j-\hat{t}_{j'}$, may be obtained by consideration of the variances and covariances of $Y_{.j}, \sum_{i(j)} Y_{j.}$ and hence Q_j:

$$\text{Var}(Q_j)=r(k-1)\sigma^2/k$$
$$\text{Cov}(Q_j, Q_{j'})= -\lambda\sigma^2/k$$
$$\text{Var}(\hat{t}_j)=\{\delta(t-1)/t\}k\sigma^2/(\lambda t)$$
$$\text{Var}(\hat{t}_j-\hat{t}_{j'})=2k\sigma^2/(\lambda t).$$

In assessing incomplete block designs, the efficiency of the design is measured by comparing the variance of the BIB design with the variance for a complete block design or unblocked design, with the same replication per treatment and assuming, which is most unrealistic, the same σ^2. From the variance for $\hat{t}_j-\hat{t}_{j'}$ the efficiency of the BIB design is as

$$E=(2\sigma^2/r)/(2k\sigma^2/\lambda t)=\lambda t/rk$$

which can be rewritten as $(1-1/k)/(1-1/t)$ whence it is clearly less than 1 since $k<t$.

Note that this efficiency depends critically on k as is shown in numerical values of E given in Table 7.13.

To estimate σ^2, the residual SS (RSS) is derived in a simple algebraic form as follows:

$$\text{RSS}=\sum_{ij} y_{ij}^2-\hat{\mu}Y_{..}-\sum_i \hat{b}_i Y_{i.}-\sum_j \hat{t}_j Y_{.j}.$$

Substituting for $\hat{\mu}$ and $\hat{b}_i$ from (7.20) and (7.21) and simplifying gives

$$\text{RSS}=\sum_{ij} y_{ij}^2-1/k\sum_i Y_{i.}^2-\sum_j \hat{t}_j Q_j.$$

Table 7.13. *Values of E for $k=2, 3, 4, 6, 8$ and $t=4, 6, 8, 12, 26, 24, 32$.*

t	2	3	4	6	8
4	0.67	0.88			
6	0.60	0.80	0.90		
8	0.57	0.76	0.86	0.95	
12	0.55	0.73	0.82	0.91	0.96
16	0.53	0.71	0.80	0.89	0.93
24	0.52	0.70	0.78	0.87	0.91
32	0.52	0.69	0.77	0.86	0.90

This may be recognised as defining the analysis of variance with the last two items being SS for blocks (ignoring treatments) and treatments (allowing for blocks).

Inter-block analysis

The model for block totals may be derived from the original model as

$$Y_{i.} = k\mu + \sum_{j(i)} t_j + kb_i + \sum_{j(i)} \varepsilon_{ij}.$$

Now the b_i represent, with the ε_{ij}, the variation in $Y_{i.}$ additional to that of the treatment and overall mean effects, and we define

$$\eta_i = kb_i + \sum_{j(i)} \varepsilon_{ij}$$

with variance

$$\text{Var}(\eta_i) = k^2 \sigma_b^2 + k\sigma^2 = k\sigma_1^2,$$

where σ_b^2 represents the variance of the population from which the experimental b_i's are a sample.

The least squares equations for μ and t_j are

$$bk\hat{\mu} \qquad\qquad = Y_{..}$$
$$rk\hat{\mu} + r\hat{t}_j + \lambda \sum_{j' \neq j} t_{j'} = \sum_{i(j)} Y_{i..}$$

Hence,

$$\hat{t}_j \quad = \left\{ \sum_{i(j)} Y_{i.} - (r/b) Y_{..} \right\} \bigg/ (r - \lambda)$$

and

$$\hat{t}_j - \hat{t}_{j'} = \left(\sum_{i(j)} Y_{i.} - \sum_{i(j')} Y_{i.} \right) \bigg/ (r - \lambda),$$

whence

$$\text{Var}(\hat{t}_j - \hat{t}_{j'}) = \{1/(r - \lambda)^2\} k\sigma_1^2 (2r - 2\lambda) = 2k\sigma_1^2/(r - \lambda).$$

To estimate σ_1^2 we require the residual mean square of the inter-block analysis which is simply derived:

$$\text{RSS} = \sum_i Y_{i.}^2 - k\hat{\mu} Y_{..} - \sum_j t_j \left(\sum_{i(j)} Y_{i.} \right)$$
$$= \sum Y_{i.}^2 - Y_{..}^2/b - 1/(r - \lambda) \left\{ \left(\sum_{i(j)} Y_{i.} \right)^2 - rk Y_{..}^2/b \right\}.$$

Hence $k\sigma_1^2$ can be estimated from the inter-block residual mean square, by dividing by the residual degrees of freedom, $b - t$. If the total SS for the inter-block analysis is made compatible with the block SS in the intra-block analysis by dividing all the terms in the expression for RSS by k then σ_1^2 is estimated by inter-block RSS/$(b - t)$.

Combination of information

The within block and inter-block analyses provide two independent sets of estimates of treatment effects. There must, inevitably, be a combined estimate, calculated from the two separate estimates of each treatment effect, which will be the best estimate. Since both sets of estimates are unbiassed by the general linear model theory, any linear combination of the estimates will also be unbiassed and the linear combination with the smallest variance will be the best linear unbiassed estimate. This minimum variance combined estimate can be shown to require weighting each estimate inversely by its variance. Thus the best combined estimate is

$$\frac{(kQ_j/\lambda t)/(k\sigma^2/\lambda t) + \left[\left\{\sum_{i(j)} Y_{i.} - (r/b)Y_{..}/(r-\lambda)\right\}\bigg/\{k\sigma_1^2/(r-\lambda)\}\right]}{1/(k\sigma^2/\lambda t) + 1/\{k\sigma_1^2/(r-\lambda)\}}$$

which would be somewhat simplified for calculation. The variance of this combined estimate is

$$[1/(k\sigma^2/\lambda t) + 1/\{k\sigma_1^2/(r-\lambda)\}]^{-1}.$$

There is one further level of complexity which may be added to this combined use of within and inter-block analyses. In the formula for the combined estimate, estimates of σ^2 and σ_1^2 are required to calculate the actual weights. The estimates for σ^2 and σ_1^2 may be obtained from the residual mean squares for the two analyses but the degrees of freedom for the inter-block residual may often be rather few with the result that the second weight is poorly determined and the estimated standard error of the combined estimate may also be poorly estimated. It is possible to obtain a better estimate of σ_1^2 and this is achieved by considering two alternative ways of constructing the analysis of variance. The two forms of analysis of variance are:

(1)	(2)
block SS (ignoring treatments)	treatment SS (ignoring blocks)
treatment SS (allowing for blocks)	block SS (allowing for treatments)
residual SS	residual SS.

The residual SS are, of course, identical, and therefore

blocks (ign. T) + treatments (all. B)
= treatments (ign. B) + blocks (all. T).

By evaluating the expected values of the first three sums of squares it can be shown that

E(block SS/allowing for treatments)
$= (kb - t)\sigma_1^2/k + (t - k)\sigma^2/k.$

Hence, estimating σ^2 from the within block residual MS, an estimate of σ_1^2

Table 7.14.

SS	df
(5) Inter-block treatment SS	$t-1$
(6) Inter-block error SS	$b-t$
(2) Block SS (ignoring treatments)	$b-1$
(3) Treatment SS (allowing for blocks)	$t-1$
(4) Residual	$bk-t-b+1$
(1) Total	$bk-1$

SS	df
(7) Treatment SS (ignoring blocks)	$t-1$
(8) Blocks SS (allowing for treatments)	$b-1$
(4) Residual	$bk-t-b+1$

can be obtained from the block SS (allowing for treatments) based on $(b-1)$ df.

The complete procedure for the possible analysis of variance structure for a BIB design is shown in Table 7.14. The sums of squares are calculated in the order indicated by the numerical labelling (1) to (8).

The logic of the calculation and interpretation is as follows.

(A) The principal form of information about treatment effects is obtained by eliminating the effects of blocks and estimating the treatment differences from the within block comparisons, the precision of these estimates being determined by the residual MS (stages (1), (2), (3) and (4)).

(B) There is additional information on treatments from the differences between block totals and a second set of independent estimates of the treatment effects may be obtained from the inter-block analysis with precision determined by the inter-block error MS (stages (5) and (6)).

(C) If the precision (df) of the standard errors for the inter-block estimates is poor then the second divison of the fitting SS for the initial analysis may be used to get better estimates of the standard errors of inter-block estimates and of the consequent combined estimates (stages (7) and (8)).

Example 7.15

To illustrate the combination of estimates we reconsider the data of example 7.10. First we must obtain the inter-block estimates from the block (cage) totals (see Table 7.15). The two sets of estimates are shown in Table 7.16. Note that the two sets of estimates are similar only in a fairly

Table 7.15.

Cage	Substances present
38.7	1, 2, 3
47.6	1, 2, 4
34.1	1, 3, 5
44.9	1, 4, 6
36.6	1, 5, 6
37.5	2, 3, 6
48.8	2, 4, 5
20.8	2, 5, 6
25.2	3, 4, 5
10.2	3, 4, 6

Table 7.16

Treatment difference	Within block	Inter-block
1–6	26.2	17.3
2–6	8.7	14.5
3–6	6.1	−1.4
4–6	1.8	8.9
5–6	2.9	6.2

broad sense. Also note that the choice of treatment 6 as an origin for the set of differences is arbitrary and does not affect the results.

The standard errors of the two sets of estimates of differences are calculated from the variance formulae, using the estimate of σ_1^2 based on the block SS allowing for treatments, and are

within block, 1.95
inter-block, 9.30.

It is immediately clear that the between block estimates are much less precise than the within block estimates. The combined estimates, weighting by the reciprocals of the variances, are shown in Table 7.17. Clearly in this case, as in many others, the use of inter-block information adds very little.

Is it always necessary or useful to calculate the full analysis of variance with both sets of estimates? In practice often only the within block analysis is calculated and it is proper to consider when we should additionally attempt to use the inter-block information.

The principle of using incomplete blocks stems from a recognition that σ^2 should be reduced by using smaller blocks. As the selection of blocks becomes more successful the ratio σ^2/σ_1^2 will become smaller and consequently the contribution of the inter-block information will become less. Thus the more successful we are in making the within block analysis efficient the less benefit will accrue from using the inter-block information.

Table 7.17.

Treatment difference	Combined estimate
1–6	25.8
2–6	8.9
3–6	5.8
4–6	2.1
5–6	3.0

Standard error = 1.91

Another way of looking at this situation is to consider the use of an incomplete block design as splitting information about treatments into two components. By the allocation of treatments to blocks we design to maximise the proportion of the information in the within block analysis. By choosing the blocks efficiently the precision of that within block proportion of information will be made as much greater than the precision of the inter-block information as is possible.

It therefore follows that we should expect to use the inter-block information when the within block proportion of information is not very large, or the gain from using small blocks is not very substantial. The within block proportion of information is measured by E and is dependent on the ratio k/t. Even when E is not large (say less than 0.8) it will be appropriate to use the inter-block information only when the ratio σ_1^2/σ^2 is not large. Overall the relative information in the two parts of the analysis is in the ratio

within block $k\ \sigma_1^2/(r-\lambda):k\sigma^2/\lambda t$ inter-block

can be rewritten

$$\sigma_1^2 E:\sigma^2(1-E).$$

A further consideration is that the use of the combined estimate involves standard errors based on the combination of two estimates of variances and consequently the inferences about the combined estimates cannot be simply based on the t distribution.

Overall it seems reasonable to recommend that the conclusions from BIB designs be based on the within block analysis only, except where E^* is less than 0.8 and σ_1^2/σ^2 is less than 5, noting that failure of these conditions implies that at least 4% of the total information is contained in the inter-block analysis. In practice this means that the inter-block analysis would be used if small block sizes of two, three or four units are used, with the blocking resulting in considerably less reduction in σ^2 than would be hoped. This suggests that the usual practical procedure of ignoring inter-block information is not unreasonable.

Exercises 7

(1) Devise experimental designs for comparing seven treatments when all comparisons are of equal interest, given
   (a) seven blocks of six units each,
   (b) seven blocks of five units each,
   (c) seven blocks of four units each,
   (d) seven blocks of ten units each,

(e) five blocks of six units each,

(f) four blocks of six units, three blocks of five units and three blocks of four units.

Explain briefly the reasons for your choice.

(2) For an experiment to compare seven treatments 56 units are available. The 56 units occur in seven natural blocks containing eight plots each. Three experimental designs are suggested:

(i) Discard one plot from each block and use a randomised complete block design.

(ii) Split each block into two blocks of four plots each and use a balanced incomplete block design for seven treatments in 14 blocks of four plots.

(iii) Use an extended block design with seven blocks and each treatment duplicated in one block.

Compare the standard errors of a comparison of two treatment effects for each design. State and justify your preference order of the three designs.

(3) An experimenter wishes to compare six treatments in blocks of three or four plots. For (a) three plots per block and (b) four plots per block:

(i) write out the unreduced BIB design;

(ii) find a reduced BIB design if one exists;

(iii) devise sensible unbalanced designs for six blocks and for eight blocks.

(4) You are required to design an experiment to compare four treatments O, A, B, C.

The experimental material available consists of 20 units in two blocks of four units, two blocks of three units and three blocks of two units.

Use a general linear model computer package to obtain the variance–covariance matrix for various trial designs and thus find appropriate designs to satisfy in turn the following criteria:

(i) variances of treatment differences to be as nearly equal as possible;

(ii) variances of comparisons of O with other treatments to be approximately two-thirds those of comparisons between other treatments.

Comment on the implications of your results.

(5) Four treatments are compared in four blocks of variable size, the design being given below. Obtain estimates of the treatment

effects for O, A and B and standard errors of the estimated differences between effects.

block I   O A B C   (four plots)
block II   O A B   (three plots)
block III O A C   (three plots)
block IV   O A C   (three plots).

(6) A horticulturalist wishes to test which of seven new varieties of aubergine plant produces the best crop. He has a maximum of nine greenhouses available in which to conduct trials; each greenhouse is to be allocated aubergine plants of three different varieties. Construct a design for this experiment, assuming that the horticulturist can obtain as many plants of each variety as he needs. What are the advantages and disadvantages of the design? Calculate the efficiencies of the design relative to an orthogonal block design.

Suppose that before the experiment is started another new variety of aubergine plant becomes available. Redesign the experiment.

(7) Construct the most efficient designs you can devise to compare six treatments in blocks of four units each, with no treatment occurring twice in a block:

(*a*) when all treatments comparisons are equally important, using nine blocks;

(*b*) when treatment A is a control for which all comparisons should be as precise as possible, using ten blocks.

Derive variances of the least squares estimates of treatment differences, $A - B$ and $B - C$ in (*b*) and compare these variances with the corresponding values for (*a*).

(8) An experiment was performed to compare five drugs. Three of the drugs, V_1, V_2 and V_3 had an organic base and the remaining two, W_1 and W_2, had an inorganic base. It was found possible to test any subject with three of the drugs and the following design used (with replication) was

subject	1	2	3	4	5
drugs	V_1	V_1	V_1	V_2	V_3
	V_2	V_2	W_1	W_1	W_1
	V_3	V_3	W_2	W_2	W_2

Write down the least squares equations under the usual additive model and show how to estimate the effects of the drugs.

Consider the variances of the estimates of the differences between any two drugs. Show that these are in the ratio

9 : 8 : 11

for

(a) comparison between two organic-based drugs,

(b) comparison between two inorganic-based drugs, and

(c) comparison between an organic- and inorganic-based drug, respectively.

8

Multiple blocking systems and cross-over designs

8.0   *Preliminary examples*

(a) An experiment to examine the pattern of variation over time of a particular chemical constituent of blood involved sampling the blood of nine chickens on 25 weekly occasions. The principal interest is in the variation of the chemical over the 25 times, the nine chickens being included to provide replication. The chemical analysis is complex and long and a set of at most ten blood samples can be analysed concurrently. It is known that there may be substantial differences in the results of the chemical analysis between different sets of samples. How should the 225 samples (25 times for nine chickens) be allocated to sets of ten (or fewer) so that comparisons between the 25 times are made as precise as possible?

(b) In an experiment to compare diets for cows the experimenter has five diet treatments he wishes to compare but, because of the restrictions implied by the blocking pattern of the available units, he has convinced himself that he will have to omit one treatment. The diets have to be fed to fistulated cows (surgically prepared) so that the performance of the cows can be monitored throughout the application of each diet. Nine such cows are available for sufficient time that four periods of observation may be used for each cow, providing a $9 \times 4 = 36$ set of observations. By the time he came to see the statistician, the experimenter had provisionally decided to use two 4×4 Latin squares for the four most important treatments, discarding one cow. How can the statistician produce a better design?

8.1 Latin square designs and Latin rectangles

We have already mentioned in Chapter 7 the possibility that, in considering the units available for the proposed experiment, there may be more than one apparently reasonable blocking structure. In this chapter, we consider the problems of trying to accommodate two or more blocking systems in a single experiment. A simple and well-known example is in the assessment of the wearing performance of car tyres. Different brands of tyre may be fitted in each of four wheel positions for each of several cars. There may be differences in performance between the four positions which are consistent for all cars. There will certainly be overall performance differences between cars. To compare four brands of tyres (A, B, C, D) using four test cars, we would like to allocate tyres to positions for each car so that each brand is tested on each car and also in each position. Can this be achieved?

A solution is shown in Figure 8.1. Construction of such a design by trial and error leads rapidly to a solution, and there are in fact many solutions. A design with t^2 units arranged in a double-blocking classification system with t blocks in each system with t treatments each occurring once in each block of each block system is called a *Latin square design*, the name reflecting the common use of Latin letters, A, B, C, ..., to represent treatments.

The model and corresponding analysis of variance for the Latin square design are simple extensions of those for the randomised block design. The two blocking systems in the design are traditionally referred to as rows and columns, and the model for the yield of the unit in row i and column j is written

$$y_{ij} = \mu + r_i + c_j + t_{k(ij)} + \varepsilon_{ij}. \tag{8.1}$$

There is one unusual feature of this model compared with those used previously. The yields, y_{ij}, are classified by only two of the three classifications, rows, columns and treatments. Each treatment occurs in

Figure 8.1. Experimental plan for four brands of tyre with each brand on each car and in each position.

	Car			
	1	2	3	4
Position 1	A	B	C	D
2	B	D	A	C
3	C	A	D	B
4	D	C	B	A

each row, each treatment occurs in each column and each column 'occurs in' each row. But only two classifications are required to classify, uniquely, each observation. The treatment suffix k is completely defined by the row and column suffixes i and j, and this dependence is represented by the suffix notation $k(ij)$. Any two of the three classifications could be used to define the set of yields; the model (8.1) is that which is traditionally used. All three sets of effects may be estimated orthogonally, using the obvious restrictions that

$$\sum_i r_i = 0, \quad \sum_j c_j = 0, \quad \sum_k t_k = 0.$$

The analysis of variance structure has the form shown in Table 8.1. Methods for the random allocation of treatments in a Latin square design will be discussed in the next chapter.

Although the Latin square design is a neat, mathematical solution to the problem of utilising two blocking factors, it is extremely restrictive. The number of replicates of each treatment must be equal to the number of treatments. Examination of the analysis of variance shows that the degrees of freedom for error are only $(t-1)(t-2)$, providing only 2 df when $t=3$, and 6 df when $t=4$. In neither case would we expect to get an adequate estimate of σ^2. It is usually necessary therefore, when using a Latin square design for three or four treatments, to have more than one square.

If multiple Latin squares are to be used, then it is important to distinguish two different forms of design. The difference is based on whether the row differences (or the column differences) might be expected to be similar for the different squares. Examples of situations where one of the blocking systems should be consistent over squares are:

(i) The experimental unit is a leaf and the blocking systems are (*a*) plants and (*b*) leaf position. If two groups of plants are used for the two squares it

Table 8.1.

Source of variation	SS	df
Rows	$\sum_i Y_{i\cdot}^2/t - Y_{\cdot\cdot}^2/t^2$	$t-1$
Columns	$\sum_j Y_{\cdot j}^2/t - Y_{\cdot\cdot}^2/t^2$	$t-1$
Treatments	$\sum_k Y_k^2/t - Y_{\cdot\cdot}^2/t^2$	$t-1$
Error	by subtraction	$(t-1)(t-2)$
Total	$\sum_{ij} y_{ij}^2 - Y_{\cdot\cdot}^2/t^2$	t^2-1

would seem reasonable to assume that the differences between leaf position should be similar for plants in both groups.

(ii) In the car tyre example, if two groups of four cars each are used to form two squares with the four positions forming the second blocking system for each group, then, if the differences between position are consistent enough to define a blocking system for one group, they should be consistent over the two groups.

The structure of these situations is illustrated in Figure 8.2(a).

The alternative situation, demonstrated in Figure 8.2(b), where the rows (or columns) in the different squares have no relation to each other, might occur for two quite separate Latin squares in an agricultural crop experiment where the geographical arrangement of plots could directly resemble Figure 8.2(b). Another example could be found in psychological experiments where different individuals (one blocking system) are exposed to different treatments on different occasions (the second blocking system). Both the sets of individuals and the sets of occasions will usually be different in the different squares so that neither set of block differences would be consistent between squares.

The model for the latter situation, where row and column effects are particular to the square in which they occur, is

$$y_{hij} = \mu + s_h + r_{hi} + c_{hj} + t_{k(hij)} + \varepsilon_{hij},$$

Figure 8.2. Multiple Latin squares: (a) with common row effects, (b) completely separate squares.

(a)

						Column			
		1	2	3	4	5	6	7	8
Row	1	A	B	C	D	A	B	C	D
	2	B	D	A	C	B	A	D	C
	3	C	A	D	B	C	D	A	B
	4	D	C	B	A	D	C	B	A

(b)

						Column			
		1	2	3	4	5	6	7	8
Row	1	A	B	C	D				
	2	B	A	D	C				
	3	C	D	B	A				
	4	D	C	A	B				
	5					A	B	C	D
	6					B	C	D	A
	7					C	D	A	B
	8					D	A	B	C

where suffixes h, i and j pertain to square, row and column, respectively. The structure of the corresponding analysis of variance for n separate Latin squares, each $t \times t$, for t treatments is given in Table 8.2.

Table 8.2.

Source	SS	df
Squares	$\sum_h Y_{h..}^2/t^2 - Y_{...}^2/(nt^2)$	$n-1$
Rows in squares	$\sum_{hi} Y_{hi.}^2/t - \sum_h Y_{h..}^2/t^2$	$n(t-1)$
Columns in squares	$\sum_{hj} Y_{h.j}^2/t - \sum_h Y_{h..}^2/t^2$	$n(t-1)$
Treatments	$\sum_k T_k^2/(nt) - Y_{...}^2/(nt^2)$	$t-1$
Error	by subtraction	$(nt-n^n-1)(t-1)$
Total	$\sum_{hij} y_{hij}^2 - Y_{...}^2/nt^2$	nt^2-1

When one of the blocking systems is common to the two or more squares, then the appropriate model for the yields is described simply in terms of rows and columns

$$y_{ij} = \mu + r_i + c_j + t_{k(ij)} + \varepsilon_{ij}.$$

Arbitrarily, we assume for this model that row effects are consistent over several sets of columns so that, in the model, i takes values 1 to t and j takes values 1 to nt. The model is formally identical to that written down for a single Latin square. Essentially this form of multiple Latin square design has the same philosophy as a single Latin square design. Because the row effects are assumed to be consistent over columns, a less stringent requirement for treatment occurrences in rows than would be required for separate Latin squares is appropriate, namely that each treatment appears n times in each row. The design may be thought of as a Latin rectangle. A typical Latin rectangle is shown in Figure 8.3. The randomisation procedure will be discussed in the next chapter.

Figure 8.3. Latin rectangle design.

		Column						
	1	2	3	4	5	6	7	8
Row 1	A	B	C	B	D	A	D	C
2	B	D	A	A	B	C	C	D
3	C	C	D	D	A	B	B	A
4	D	A	B	C	C	D	A	B

Table 8.3.

Source	SS	df
Rows	$\sum_{i} Y_{i.}^{2}/(nt) - Y_{..}^{2}/(nt^{2})$	$t-1$
Columns	$\sum_{j} Y_{.j}^{2}/t - Y_{..}^{2}/(nt^{2})$	$nt-1$
Treatments	$\sum_{k} T_{k}^{2}/(nt) - Y_{..}^{2}/(nt^{2})$	$t-1$
Error	by subtraction	$(nt-2)(t-1)$
Total	$\sum_{ij} y_{ij}^{2} - Y_{..}^{2}/(nt^{2})$	$nt^{2}-1$

The analysis for the Latin rectangle follows, as usual, directly from the model (see Table 8.3).

8.2 Multiple orthogonal classifications and sequences of experiments

At various points in the development of the theory of experimental design, the mathematical ideas involved in the construction of useful designs offer a temptation to divert from the path of usefulness. Extensions of the ideas of constructing Latin squares are one such point and, with the excuse that some of the extensions are occasionally useful, we now succumb to temptation and explore, briefly, the problems of superimposing Latin squares.

For the 3×3 Latin square shown in Figure 8.4(a), a second Latin square

Figure 8.4. (a) Latin square, (b) Graeco–Latin square.

(a)

		Column		
		1	2	3
Row	1	A	B	C
	2	B	C	A
	3	C	A	B

(b)

		Column		
		1	2	3
Row	1	Aα	Bβ	Cγ
	2	Bγ	Cα	Aβ
	3	Cβ	Aγ	Bα

using symbols (α, β, γ) can be superimposed as shown in Figure 8.4(b), so that not only does each Greek letter occur in each row and in each column, but also each Greek letter occurs with each Latin letter. There are four orthogonal classifications: rows, columns, Latin letters and Greek letters. The design of Figure 8.4(b) is called a Graeco–Latin square, for obvious reasons. By considering the degrees of freedom in the analysis of variance we can recognise that each of the four orthogonal classifications has 2 df and, since the total degrees of freedom for the nine observations is eight, we can deduce that no further orthogonal classifications may be added. The 3×3 Graeco–Latin square is an example of a completely orthogonal square.

More generally, we might expect to be able to superimpose sets of Latin squares for any size of square, t. It could be argued from considering the degrees of freedom that the total degrees of freedom, $(t^2 - 1)$, could be split into $(t+1)$ sets, each with $(t-1)$ df, corresponding to $(t+1)$ mutually orthogonal classifications: these classifications would be rows, columns and $(t-1)$ superimposed Latin squares. However, this expectation would be only partially correct. For some Latin squares, it is impossible to construct a Graeco–Latin square by superimposing a second Latin square. Two 4×4 Latin squares are shown in Figure 8.5(a) and (b). The first has a second Latin square imposed to achieve a Graeco–Latin square; the second does not permit a Graeco–Latin square to be constructed. The criteria governing the existence of Graeco–Latin or completely orthogonal squares are related to the algebraic theory of Galois fields. It can be shown quite simply that, for any group of elements satisfying the conditions of a Galois field, there is a corresponding completely orthogonal square. From a practical viewpoint, the important consequences are that Graeco–Latin and completely orthogonal squares may be constructed for any Latin square of size 3, 5 or 7 (and of course for many larger sizes). Graeco–Latin squares exist for some Latin squares of size 4 or 8. But there are no Graeco–Latin squares of size 6. There is much literature on the mathematical properties of completely orthogonal squares but this is not appropriate for this book.

Are there any uses of Graeco–Latin squares for the practical experimenter? It is clearly unrealistic in general to expect to use three blocking

Figure 8.5. Two Latin squares, of which only the first leads to a Graeco-Latin square.

(a)				(b)			
Aα	Bβ	Cγ	Dδ	A	B	C	D
Bγ	Aδ	Dα	Cβ	B	C	D	A
Cδ	Dγ	Aβ	Bα	C	D	A	B
Dβ	Cα	Bδ	Aγ	D	A	B	C

criteria represented by rows and columns and Greek letters because of the requirement that the experimental units would have to group into equal size 'blocks' for each of the three blocking criteria, and for each pair of blocking criteria all possible combinations of the three blocking systems would have to be represented exactly once. Such an occurrence in a set of experimental units could happen only by an extraordinary chance (which the disciple of significance would clearly reject!) or by gross premeditation. However, Graeco–Latin squares may be useful when experimental units are used repeatedly for experiments. Consider a set of 64 experimental units used in an 8×8 Latin square design for an experiment, such that at the end of the experiment the units could be used for a further experiment, but with the additional requirement that planning of the succeeding experiment should make allowance for the possible long-term effects of the treatments in the first experiment. Such a situation can arise with experiments on fruit trees or other perennial crop whose useful experimental life may last 20 years, compared with a duration of three or four years for a particular experiment. Other examples may be found in animal experiments.

If the initial Latin square is of the type that allows a subsequent Graeco–Latin square, then the set of treatments of the first experiment may be regarded as a third blocking classification for the second experiment, and all effects of the original two blocking factors and of the treatments for the first experiment will be orthogonal to comparisons of the treatments in the second experiment. An implication of the use of this form of sequence of experiments is that initial experiments must allow for the possibility of later experiments. In particular, the choice of Latin square or of multiple Latin squares could be modified to include only those Latin squares for which Graeco–Latin squares exist. Where it is likely that several experiments will be performed using the same experimental units then, even when the initial design is a randomised block design rather than a Latin square, the choice of initial design may have consequences for the design of subsequent experiments, and it is important for the experimenter and statistician to be clear about such possibilities.

8.3   Non-orthogonal row and column design

Latin square designs are extremely restrictive. The number of replicates per treatment must be t, the number of treatments, or some multiple of t. The requirement for multiple squares may come from the need for a reasonable estimate of σ^2 rather than from a genuine desire for greater replication. And the total number of units is necessarily t^2 or nt^2 which will frequently involve either not using available units or including units which are not entirely suitable. Thus, in animal nutrition experiments, where the two blocking criteria of genetic similarity (litters = blocks) and initial size (weight class = block) provide an ideal Latin square structure, the experimenter may have to omit suitable animals from larger litters, or include undersized and atypical animals from smaller litters.

The ideas of general block design do not apply with quite the same simplicity to double blocking systems as to the single blocking system discussed in Chapter 7. Nevertheless, the range of useful designs is very much wider than is generally realised, and the possibility of matching a suitable design to the natural structure of the experimental units is much greater than the frequent use of the Latin square design would suggest.

Consider first the designs for which the number of units in each block is the same within a block system but different in the two blocking systems, with each combination of the two blocking systems occurring just once. The resultant structure of experimental units can be represented as a rectangular array, and designs imposed on such a structure are described generally as row and column designs. The first group of designs to be considered is for those row and column structures where the number of columns (or rows) is equal to the number of treatments t or some multiple of t.

If the number of columns equals t, then obviously treatments can be arranged orthogonally to rows, with each treatment appearing in each row. Since columns and rows are orthogonal (because of the rectangular form of the row and column array), the design structure is determined by the pattern of allocation of treatments to columns. This is the same problem as the allocation of treatments to blocks for a single blocking factor when the block size is not equal to the number of treatments. The subsequent arrangement of treatments within columns so that each treatment occurs in each row is trivial. The design for treatments within columns may be constructed exactly as in Section 7.2, and the treatments are then reordered within columns to achieve the orthogonality with respect to rows.

This approach is illustrated in Figure 8.6. The balanced design for seven treatments in seven blocks of three units is shown in Figure 8.6(a), the treatments being written down in the natural order of construction. The rearrangement of letters within each column so that each letter appears once in each row is achieved by working through the columns sequentially. There are some arbitrary choices and it is possible to make these in such a way as to fail to achieve a solution. However, with reasonable forethought a solution can always be found. In column 2, A is put in row 2,

Figure 8.6. Experimental plan for seven treatments (a) in seven blocks of three units, (b) in seven columns × three rows.

(a)
Block

	I	II	III	IV	V	VI	VII
	A	A	A	B	B	C	C
	B	D	F	D	E	D	E
	C	E	G	F	G	G	F

(b)
Column

	1	2	3	4	5	6	7
Row 1	A	D	F	B	G	C	E
2	B	A	G	F	E	D	C
3	C	E	A	D	B	G	F

and in column 3, A is put in row 3. Since each of the other letters has occurred only once in the first three columns, their row positions are chosen arbitrarily. In column 4, given the previous arbitrary decisions, B must go in row 1, and we now have a choice between (D in row 2, F in row 3) or (D in row 3, F in row 2). We defer the choice, proceeding to column 5, where there is no choice; B must go in row 3, G consequently in row 1 and E in row 2. Now, in column 6, G must go in row 3, and D consequently in row 2, with C in row 1. This determines the choice in column 4; D must go in row 3 and F in row 2. The disposition in column 7 is now inevitable. The resulting design is shown in Figure 8.6(b).

Now consider again the designs for six treatments developed in Section 7.2. The designs in six columns of five units or six columns of seven units can be adapted to row and column designs as shown in Figure 8.7. These designs are examples of two particularly useful classes of designs, being Latin squares with either a row missing or an additional row. Any Latin square with a row missing or an extra row will have treatments balanced with respect to columns, and such designs are clearly optimal for t treatments in $t \times (t-1)$ or $t \times (t+1)$ row and column designs. Without considering any other designs with treatments balanced with respect to columns, the set of useful designs increases to three times the number of Latin square designs.

Row and column designs in which the number of columns is equal to the number of treatments, and in which the treatments are balanced for their occurrence in columns, are called Youden squares, because they were systematically developed by the statistician W. J. Youden. They are not, of

Figure 8.7. Experimental plans for six treatments (a) in five rows × six columns, (b) in seven rows × six columns.

(a)   Column

		1	2	3	4	5	6
Row	1	A	B	C	D	E	F
	2	B	D	A	F	C	E
	3	C	F	E	B	A	D
	4	D	A	B	E	F	C
	5	E	C	F	A	D	B

(b)   Column

		1	2	3	4	5	6
Row	1	A	B	C	D	E	F
	2	B	F	A	E	D	C
	3	C	D	E	B	F	A
	4	D	A	B	F	C	E
	5	E	C	F	A	B	D
	6	F	E	D	C	A	B
	7	A	B	C	D	E	F

course, squares because the number of rows is not equal to the number of columns, but the name has stuck.

Unbalanced block–treatment designs can also be converted into row and column designs when the number of blocks is equal to the number of treatments. For example, the design from Section 7.2 with six treatments in six columns of four units each, which was argued to be the most efficient available within the restrictions of the experimental units, can be re-arranged into a four row × six column array as shown in Figure 8.8.

The structure of the row and column designs considered so far has always included two pairs of orthogonal sets of effects. Rows and columns have been orthogonal, and treatments and rows have been orthogonal. If we try to adapt other block–treatment designs of the balanced or nearly

Figure 8.8. Experimental plan for six treatments in four rows × six columns.

Column

		1	2	3	4	5	6
Row	1	A	F	E	D	C	B
	2	B	A	F	E	D	C
	3	C	B	A	F	E	D
	4	D	C	B	A	F	E

balanced types in Chapter 7 then it will rarely be possible to arrange treatments and rows to be orthogonal. Consider the design from Section 7.3 for six treatments in ten blocks of three units. If we attempt to transfer this to a three row × ten column design, then treatments can obviously be balanced with respect to columns. But, with three rows of ten units each, it is clear that the nearest we can come to balance is to regard each row as containing two units per treatment, with two treatments omitted (once) in each row, each treatment being omitted once. Again, we can start with the block–treatment design shown in Figure 8.9(a), and construct the row and column design shown in Figure 8.9(b); again the choice is considerable.

Figure 8.9. Experimental plan for six treatments (a) in ten blocks of three units, (b) in three rows × ten columns.

(a)   Block

I	II	III	IV	V	VI	VII	VIII	IX	X
A	A	A	A	A	B	B	B	C	C
B	B	C	D	E	C	D	E	D	D
C	D	E	F	F	F	E	F	E	F

(b)   Column

		1	2	3	4	5	6	7	8	9	10
Row	1	A	D	C	A	F	B	E	E	C	F
	2	B	A	E	D	A	C	B	F	D	C
	3	C	B	A	F	E	F	D	B	E	D

Row and column designs have been classified by various authors, notably Pearce (1963, 1975), by defining the interrelationships between the three classifications, rows, columns and treatments. Thus far only designs for which rows and columns are orthogonal have been considered. We have considered a range of possibilities for the other relationships, and can tabulate these as in Table 8.4.

Table 8.4.

Design	Row and treatment	Column and treatment
Fig. 8.6 7R × 3C	orthogonal	balanced
Fig. 8.7 5R × 6C	orthogonal	balanced
Fig. 8.7 7R × 6C	orthogonal	balanced
Youden squares generally	orthogonal	balanced
Fig. 8.8 4R × 6C	orthogonal	not balanced
Fig. 8.9 3R × 10C	balanced	not balanced

The overall pattern of treatment comparisons in the designs listed in Table 8.4 is always the 'worse' of the two relationships with rows and columns. Thus, except for the last two, all the designs have balanced treatment comparisons. However, it is theoretically possible to have balanced treatment comparisons when neither of the row–treatment comparisons are balanced. The crucial result was presented by Pearce (1963). If there are r rows and c columns and

$\quad\quad n_{1ik}$ observations on treatment k in row i,

and

$\quad\quad n_{2jk}$ observations on treatment k in column j

then, after elimination of row and column effect estimators, the least squares equations for treatment effects are

$$\hat{t}_k\left(N_{.k}-\sum_i n_{1ik}^2/c-\sum_j n_{2jk}^2/r\right)-\sum_{k'\neq k} t_{k'}\left(\sum_i n_{1ik}n_{1ik'}/c+\sum_j n_{2jk}n_{2jk'}/r\right)$$
$$=T_k-\sum_i n_{1ik}Y_{i.}/c-\sum_j n_{2jk}Y_{.j}/r.$$

Hence, if the weighted sum of concurrences for rows and for columns

$$\sum_i n_{1ik}n_{1ik'}/c+\sum_j n_{2jk}n_{2jk'}/r$$

is invariant over all pairs (k, k'), then the treatments are balanced. Clearly, this could be achieved without necessarily having balance with respect to both rows and columns for which the requirements would be the invariance over (k, k') of both

$\quad\quad$ (a) $\displaystyle\sum_i n_{1ik}n_{1ik'}/c$

and

$\quad\quad$ (b) $\displaystyle\sum_j n_{2jk}n_{2jk'}/r.$

As with the designs for a single blocking criterion, the Pearce criterion provides an ideal at which the designer may aim even when the ideal is not achievable.

In constructing row and column designs the allocation of treatments to rows and to columns should be considered separately. For each blocking system the allocation should be such as to achieve orthogonality if that is possible, balance if orthogonality is not possible, and if balance is not possible the joint occurrences in rows (or columns) should be made as nearly equal for all treatment pairs as is possible. Where two treatments have an unusually high (or low) occurrence for one blocking system, the labelling of treatments should be so arranged that those two treatments tend to have an unusually low (or high) occurrence for the other system. When the row and column allocations are complete the joint allocation is

Figure 8.10. Experimental plan for eight treatments in four rows × six columns.

Column

		1	2	3	4	5	6
Row	1	A	B	C	D	E	H
	2	B	A	H	E	F	G
	3	C	E	A	F	G	D
	4	D	F	G	H	C	B

Figure 8.11. Experimental plan for nine treatments in five rows × seven columns.

Column

		1	2	3	4	5	6	7
Row	1	A	B	C	D	E	F	G
	2	B	A	F	H	C	D	E
	3	C	G	D	I	A	H	B
	4	D	H	E	G	F	I	C
	5	E	F	G	A	I	B	H

constructed to be compatible with the separate allocations. This last step is trivial except for incomplete rectangular arrays but may require some trial and error investigations for such arrays. Using these several guidelines we can deal with situations such as (*a*) comparing eight treatments in a 6 × 4 array, (*b*) comparing nine treatments in a 5 × 7 array, or (*c*) comparing six treatments in an incomplete 6 × 8 array, in which only 36 of the possible 48 combinations occur.

Solutions to the first two problems which are at least nearly optimal are shown in Figures 8.10 and 8.11. The intermediate allocations for rows and columns may be deduced from the final design in each case. Problem (*c*) is the most general example we consider, and it is discussed in some detail to illustrate the general approach.

Example 8.1

The structure of the 36 available units is shown in Figure 8.12. We first consider rows and columns separately. The six rows have seven, six, six, six, seven and four units, respectively; the eight columns have six, five, five, five, four, four, four and three units. An obvious allocation of treatments to rows is to have all six treatments (A, B, C, D, E and F) in each of the first five rows; there can be only four treatments in the final row and the two treatments omitted from that row must appear as the extra treatments in rows 1 and 4. If the treatments in row 6 are A, B, C, D, then the set of weighted concurrences for the 15 treatment pairs are shown in Table

Figure 8.12. Experimental row and column structure of 36 units.

Column

	1	2	3	4	5	6	7	8
Row 1	X	X	X	X	X		X	X
2	X	X	X		X	X	X	
3	X		X	X	X	X		X
4	X	X	X	X	X	X		
5	X	X	X	X		X	X	X
6	X	X		X			X	

Table 8.5(a). *Weighted concurrences for rows.*

	B	C	D	E	F
A	1.04	1.04	1.04	0.93	0.93
B		1.04	1.04	0.93	0.93
C			1.04	0.93	0.93
D				0.93	0.93
E					1.07

(b). *Variances of treatment differences allowing for rows only.*

	B	C	D	E	F
A	0.333	0.333	0.333	0.345	0.345
B		0.333	0.333	0.345	0.345
C			0.333	0.345	0.345
D				0.345	0.345
E					0.342

8.5(a). The variances of treatment comparisons that would be achieved if columns were ignored are given in Table 8.5(b).

The allocation to columns is less inevitable. All six treatments (1, 2, 3, 4, 5, 6) will appear in column 1, and columns 2, 3 and 4 will have a different treatment omitted from each; there is then a choice of which three treatments to put in column 8, and then the various ways of completing columns 5 to 7. The choice for column 8 is in terms of how many of the treatments omitted in columns 2 to 4 to include. The easiest pattern is to include all three, or alternatively to exclude all three. The former leads to the most even distribution of weighted concurrences, and produces a column allocation as follows:

> 1st column 1, 2, 3, 4, 5, 6
> 2nd column 1, 2, 3, 4, 5
> 3rd column 1, 2, 3, 4, 6
> 4th column 1, 2, 3, 5, 6
> 5th column 2, 3, 4, 5
> 6th column 1, 2, 5, 6
> 7th column 1, 3, 4, 6
> 8th column 4, 5, 6.

The weighted concurrences and variances of treatment differences, assuming that the columns provide the only blocking systems for the treatments, are shown in Table 8.6(a) and (b).

Note that we have used different sets of symbols for the treatments when considering row and column allocation. The final choice to complete the total design is the identification of (A, B, C, D, E, F) with (1, 2, 3, 4, 5, 6), and this is simply a question of examining the two sets of weighted concurrences and matching them to produce as even a set of totals as possible. This is achieved by equating E with 3 and F with 6 producing the design shown in Figure 8.13. The actual precision of comparison of pairs of treatments can be obtained by using a general design analysis package, and the variances of treatment pairs so obtained are shown in Table 8.7(b), with the sums of the two sets of weighted concurrences in

Table 8.6(a). *Weighted concurrences for columns.*

	2	3	4	5	6
1	1.02	1.02	0.82	0.82	1.07
2		1.02	0.82	1.07	0.82
3			1.07	0.82	0.82
4				0.95	0.95
5					0.95

(b). *Variances of treatment differences allowing for columns only.*

	2	3	4	5	6
1	0.348	0.348	0.365	0.365	0.350
2		0.348	0.365	0.330	0.365
3			0.350	0.365	0.365
4				0.361	0.361
5					0.361

Figure 8.13. Experimental plan for six treatments in 36 units.

Column

		1	2	3	4	5	6	7	8
Row	1	C	E	B	F	A	–	E	D
	2	E	A	D	–	F	C	B	–
	3	F	–	E	A	D	B	–	C
	4	D	C	A	B	E	F	–	–
	5	B	D	F	E	–	A	C	F
	6	A	B	–	C	–	–	D	–

Table 8.7(a). *Overall sums of weighted concurrences.*

	B	C	D	E	E
A	2.06	1.86	1.86	1.95	2.00
B		2.11	1.86	1.95	1.75
C			1.99	1.75	1.88
D				2.00	1.88
E					1.89

(b). *Variances of treatment differences allowing for rows and columns.*

	B	C	D	E	F
A	0.350	0.368	0.368	0.366	0.359
B		0.370	0.367	0.360	0.377
C			0.362	0.380	0.373
D				0.369	0.373
E					0.378

Table 8.7(a). It is clear from these variances that the designs achieve a very even level of comparison with a maximum deviation from the average variance of less than 5%. This design may not be the most efficient that can be achieved, but it will be very close to the optimal design, and this has been achieved by simple application of the principles of good block–treatment design. It should therefore be clear that efficient designs for quite general row and column designs may be constructed by simple methods.

8.4   The practical choice of row and column design

The philosophy of choosing a design for a particular number of treatments and for a particular set of units using two blocking systems is similar to that for a single blocking system. There are some classes of standard designs with particularly desirable properties which should be used if they happen to coincide with the requirements of the particular problem. Where there is no standard design for a problem then the choice lies between modifying the problem slightly to fit a standard design (the Procrustean philosophy) or producing a tailor-made solution to fit the peculiar features of the problem. As with the single blocking system the design of experiments for comparing large numbers of treatments requires special consideration and is discussed later in this section.

A fully orthogonal design requires single or multiple Latin squares. As with the randomised complete block design for a single blocking system, a Latin square design is ideal if appropriate. To a greater extent than for the randomised block design it is rare for the conditions to be exactly appropriate, and many Latin square designs used in practice represent a major compromise of one of the ideal requirements, either through modification of the desired number of treatments or through accepting either more or less replication than is sensible. Although the possibility of single or multiple Latin square designs should always be considered, they should be used only when the circumstances happen to fit the design exactly or so nearly that the amendment to the experimental design problem to persuade it to fit a Latin square solution is slight.

Similarly the conditions for designs such that rows and columns are orthogonal, treatment and rows are orthogonal and treatments are balanced against columns are very restrictive, and the probability of finding an appropriate standard balanced design is small. Therefore to use a double blocking system efficiently it is frequently necessary for the experimenter and statistician to construct a design which is as nearly balanced as possible for each particular situation.

The two examples at the beginning of the chapter illustrate the two possible approaches clearly. In the first there are 25 treatments (= times) for each of nine chickens (= blocks), and the 225 observations must be grouped into sets of ten or less, each set being a block of the second blocking system. This is an example of a problem for which there is a standard design which is very close to the original specifications. Using sets of ten observations obviously cannot provide a suitable design since the 23rd set will contain only five observations. However, if we consider using nine observations per set there is a clear hint of a satisfactory pattern. Using 25 blocks of nine observations with a second blocking factor having nine levels (chickens) provides a design problem in 25 rows and nine columns for 25 treatments. Clearly the 25 treatments can be arranged orthogonally to columns. A treatment occurring nine times must occur in nine rows with a total of 9×8 other observations also occurring in those nine rows which with 24 other treatments would imply a replication of each treatment pair of $\lambda = 72/24 = 3$. The integral λ value is sufficient encouragement for the statistician to consult reference books with a strong expectation of finding a balanced design. With the large numbers involved it is plainly easier to refer to an index of designs (Cochran and Cox, 1957) rather than construct the design of Figure 8.14 directly. The detailed design may alternatively be obtained from programs for computer-aided design, but the initial investigation (finding $\lambda = 3$) is still required to provide evidence that such a program will be able to construct the required design.

The second problem, with five treatments to be compared within a double block structure of four rows × nine columns, offers several possibilities. The experimenter's presumption that it would be necessary to reduce the treatments to four and use two complete Latin squares giving eight replicates of each treatment with a variance of treatment differences of $2\sigma^2/8$ is a reasonable but pessimistic starting point. If it is decided

Figure 8.14. Experimental plan for 25 treatments in 25 rows × nine columns.

Chicken

	A	B	C	D	E	F	G	H	I
Set 1	1	2	3	4	5	6	7	8	9
2	2	4	9	10	24	17	15	22	12
3	3	24	8	23	18	21	13	4	10
4	4	22	25	8	20	12	11	3	19
5	5	15	17	18	8	11	2	13	20
6	6	8	12	13	1	14	24	25	15
7	7	16	5	22	3	10	25	15	13
8	8	10	11	16	6	22	23	1	17
9	9	13	20	5	12	23	1	21	22
10	10	19	14	12	16	2	8	5	21
11	11	18	19	24	10	1	5	9	25
12	12	6	10	25	7	18	20	2	23
13	13	11	4	9	23	25	14	16	2
14	14	3	7	17	11	5	12	23	24
15	15	20	21	1	14	7	10	11	4
16	16	17	18	7	13	19	4	12	1
17	17	14	13	20	19	3	9	10	6
18	18	9	22	14	25	8	21	17	7
19	19	7	23	15	9	20	16	24	8
20	20	5	16	6	4	24	22	14	18
21	21	12	6	11	15	9	3	18	16
22	22	21	24	19	2	13	6	7	11
23	23	1	2	3	22	15	18	19	14
24	24	25	1	2	21	16	17	20	3
25	25	23	15	21	17	4	19	6	5

to accept the reduction to four treatments then hopefully the statistician and experimenter will decide to include the ninth cow giving nine replicates per treatment with each row (= period) having a different one of the four treatments repeated a third time. The variance of treatment differences for this design (treatments balanced over rows, orthogonal to columns), shown in Figure 8.15(a) is $9\sigma^2/40$. If a design is sought for five treatments then with four rows (periods) the advantages of Latin squares with a row omitted should be remembered. Such a design for five treatments arranged for the first five cows and the four periods is highly efficient. The same logical approach for the remaining four cows times four periods would suggest using a Latin square omitting one row and one column. The resulting total design is illustrated in Figure 8.15(b), and the variances of treatment differences are $0.284\sigma^2$ for comparisons including treatment A and $0.313\sigma^2$ for other comparisons. The two designs in Figure 8.15 are optimal for four and for five treatments, and the choice of design reduces to deciding whether the benefit of including the fifth treatment is sufficiently advantageous to outweigh the increase of the variances of treatment difference from $0.225\sigma^2$ to $0.284\sigma^2$ and $0.313\sigma^2$.

For design problems with many treatments it may often be appropriate to use several rectangular arrays, within each of which there are row and column effects. An advantage of using two blocking systems is that precision of comparison between two treatments depends on the joint occurrences of the two treatments either in a row or in a column. Thus,

Figure 8.15. Experimental plans for a four row × nine column structure for (a) four treatments, (b) five treatments.

(a)

	Cow								
	1	2	3	4	5	6	7	8	9
Period 1	A	B	C	A	D	B	C	D	A
2	B	D	B	C	A	D	A	C	B
3	C	A	D	B	C	A	D	B	C
4	D	C	A	D	B	C	B	A	D

(b)

	Cow								
	1	2	3	4	5	6	7	8	9
Period 1	A	B	C	D	E	A	B	C	D
2	B	C	D	E	A	B	C	D	E
3	C	D	E	A	B	C	D	E	A
4	D	E	A	B	C	D	E	A	B

pairs of treatments which do not occur together in any row can usually be arranged to occur together in a column, and vice versa. This is a crucial implication of the Pearce weighted joint occurrence result derived in the previous section. As usual there are standard designs, the major design pattern being the lattice square based on writing the treatments in several square arrays with the rows and columns of each square defining the blocks of each blocking system. Also as usual these designs in their complete, balanced, form are very restrictive; the number of treatments must be the square of an integer, k, and the number of replicates is $(k+1)$ or $(k+1)/2$. The design for $k=7$, 49 treatments with four replicate squares is shown in Figure 8.16.

Often the size of the design requires that the average number of joint occurrences will be less than 1 so that ideally each treatment pair should occur together once or not at all. A simple approach to constructing reasonably efficient designs is to write out rectangular arrays for each replicate using the rows and columns for each array as the blocks of the two blocking systems. If we try this and attempt to construct more than one replicate then we rapidly discover that it is not possible to avoid replication of treatment pairs, unless the row and column arrays are square. For example, consider an eight row × ten column array for 80 treatments. In each row of a second 8×10 array at least two pairs of treatments must occur which also occurred together in some row of the first array, simply because there are only eight rows of the original array. The reason for the advantages of the lattice square should now be clear.

Figure 8.16. Lattice square design for 49 treatments in four arrays each of seven rows × seven columns.

Replicate I

1	2	3	4	5	6	7
8	9	10	11	12	13	14
15	16	17	18	19	20	21
22	23	24	25	26	27	28
29	30	31	32	33	34	35
36	37	38	39	40	41	42
43	44	45	46	47	48	49

Replicate II

1	38	26	14	44	32	20
21	2	39	27	8	45	33
34	15	3	40	28	9	46
47	35	16	4	41	22	10
11	48	29	17	5	42	23
24	12	49	30	18	6	36
37	25	13	43	31	19	7

Replicate III

1	19	30	48	10	28	39
40	2	20	31	49	11	22
23	41	3	21	32	43	12
13	24	42	4	15	33	44
45	14	25	36	5	16	34
35	46	8	26	37	6	17
18	29	47	9	27	38	7

Replicate IV

1	42	27	12	46	31	16
17	2	36	28	13	47	32
33	18	3	37	22	14	48
49	34	19	4	38	23	8
9	43	35	20	5	39	24
25	10	44	29	21	6	40
41	26	11	45	30	15	7

Nevertheless, nearly square arrays of suitably small size can be constructed when many treatments are to be compared.

8.5 Cross-over designs – time as a blocking factor

One particular form of row and column design has time as one blocking factor with the different treatments applied in sequence to each experimental unit, the treatment sequences being different for different units. This practice is particularly common in medical, psychological or agricultural animal experiments, and requires special consideration. The experiment will include a number of patients, subjects or animals who each receive different treatments in each of a number of successive periods. Essentially the experimental unit is redefined to be an observation for an individual subject in a short period of time. There are assumed to be consistent differences between periods in addition to those between subjects. In the simplest case with two treatments, A and B, subjects either receive treatment A in the first period followed by treatment B in the second period, or the reverse ordering. By having equal numbers of subjects for the two orderings, the effects of the order of treatments are eliminated from the comparison between the treatments A and B. The resulting design will be a multiple 2×2 Latin square, known as a cross-over design and illustrated in Figure 8.17.

Before considering some of the technical problems of design and analysis involved in the cross-over design, the general advantages and disadvantages should be examined. The general reason for wishing to use cross-over designs is the anticipated high level of variability between patients, subjects or animals. In many experiments, it is found that the variance of observations for different patients or animals may be more than ten times greater than the variation between observations at different times for the same patient or animal. This suggests that enormous gains in precision can be achieved by observing more than one treatment on each patient or animal. A further advantage is that fewer patients or animals are required. This improvement in statistical precision might seem to provide an overwhelming argument for the use of cross-over designs. However, there are other statistical considerations than precision in the design of

Figure 8.17. Multiple Latin square, or cross-over design for two treatments.

							Subject					
	1	2	3	4	5	6	7	8	9	10	11	12
Period 1	A	B	B	A	A	B	A	B	B	B	A	A
2	B	A	A	B	B	A	B	A	A	A	B	B

experiments, and in particular the questions of validity and of the population for which the experimental results are relevant must be considered.

The difficulty with the cross-over design is that the conclusions are appropriate to units similar to those in the experiment; that is, to subjects for a short time period in the context of a sequence of different treatments. We have to ask if the observed difference between two treatments would be expected to be the same if a treatment is applied consistently to each subject to which it is allocated. This is a problem of interpretation of results from experiment to subsequent use, and it is a problem which must be considered in all experiments. It is particularly acute in cross-over designs because the experiment is so different from subsequent use. After all, no farmer is going to continually swap the diets for his cattle!

The decision whether to use cross-over designs is therefore a balance between a hoped for gain in precision, which may be considerable, and a potential loss of relevance. If it is decided that the benefits of the cross-over design outweigh the deficiencies, then the range of available designs can be considered.

Only rarely would more than four periods be used in a cross-over design, and two- or three-period designs predominate. It is useful to discuss cross-over designs in two contexts. First, we consider briefly designs in which the separate periods are considered simply as a second set of blocks. For these designs we assume that the ordering of time is not relevant, and that the effect of a treatment applied to a subject in one period has no residual effect on the response of the subject in the following period. The second context for considering cross-over designs is when there are assumed to be residual, or carry-over, effects from previous periods. The residual effect complicates the models for the experimental data and the consequent estimation of treatment effects, and we consider such models in the next section.

For the first situation, with no residual effects, the model for the result from the ith subject in the jth time period, during which the subject receives treatment k, is

$$y_{ij} = \mu + s_i + p_j + t_{k(ij)} + \varepsilon_{ij}.$$

This is exactly the same form of model that was required for the Latin square and Latin rectangle designs, and for the incomplete row and column designs of Section 8.3. Essentially, we assume that there may be consistent differences between periods but assume also that, after allowing for these differences, the results are as they would be if the periods occurred in a quite different order.

If these simple model assumptions are felt to be appropriate, then the design problem does not differ from the general row and column design situation considered earlier in this chapter. The number of periods will not usually be large and the complete Latin square will not often provide solutions to practical problems, except that 4×4 Latin squares are used quite frequently and effectively for the comparison of four diets or treatment regimes. The most common situation in dietary trials is probably that with between six and eight treatments using three or four periods, and a number of subjects sufficient to give about four replications per treatment. The general principles for design developed during Section 8.4 will enable any particular experimental situation to be solved. In seeking to achieve near-balance, it is usually advisable to first consider the allocation of treatments to subjects and then to order the treatments within each subject, to achieve improved balance from the distribution of treatments between periods.

Example 8.2
At this point we consider again the example on pain relief discussed initially in Chapter 6. In this experiment patients were given drugs on request. Three drugs were compared in the experiment and each patient received two different drugs. The allocation of drugs to patients at the first request was random within overall restrictions of approximate equality of drug replication and the allocation of the drug at the second request, the second drug being not the same as the first drug, was similarly random. The full results of the trial are shown in Table 8.8. In Chapter 6 the results from the first period only were analysed.

The analysis of variance for a model including patient, period and drug effects, is as shown in Table 8.9. The treatment effect estimates are

$$
\begin{array}{llll}
T_1 - T_2 & +3.42 & \text{SE } 1.05 \\
T_1 - T_3 & +2.04 & \text{SE } 0.99 \\
T_2 - T_3 & -1.38 & \text{SE } 0.96.
\end{array}
$$

The advantage of drug 1 is clear, though it is interesting to note that the results both in terms of the size of the effects and the precision of estimates are little changed from the previous analysis of the first administration data only.

In the same way that we analysed the inter-block information in Chapter 7 we can obtain information about treatments from comparisons between row totals or between column totals. For many row and column designs there are insufficient rows (or columns) to provide useful information about treatment differences. However, in this example there are 43 patients.

Table 8.8. *Hours of relief from pain for each patient after each drug application.*

		Patients							
Period	Drug	1	2	3	4	5	6	7	8
1	T_1	2	6	4	13	5	8	4	
2	T_2	10	8	4	0	5	12	4	
1	T_2	2	0	3	3	0			
2	T_1	8	8	14	11	6			
1	T_1	6	7	6	8	12	4	4	
2	T_3	6	3	0	11	13	13	14	
1	T_3	6	4	4	0	1	8	2	8
2	T_1	14	4	13	9	6	12	6	12
1	T_3	12	1	5	2	1	4	6	5
2	T_2	11	7	12	3	7	5	6	3
1	T_2	0	8	1	4	2	2	1	3
2	T_3	8	7	10	3	12	0	12	5

Table 8.9.

	SS	df	MS
Patients (ignoring drugs)	607	42	
Periods (ignoring drugs)	262	1	
Drugs (allowing for patients and periods)	116	2	58.0
Residual	432	40	10.8
Total	1417	85	

The inter-patient analysis gives the results shown in Table 8.10. The inter-patient estimates of differences of drug effects are

$T_1 - T_2$	$+3.90$	SE 1.33
$T_1 - T_3$	$+1.17$	SE 1.41
$T_2 - T_3$	-2.73	SE 1.43.

The pattern is the same as for the results from patient × period analysis. The standard errors are not very much larger than those for the within patient analysis reflecting the perhaps surprising fact that the between patient variation is not much larger than within patient residual variation. If the results of the two analyses are combined using reciprocal weighting by variances, as described in Section 7.5, the combined estimates and their

Table 8.10.

	SS	df	MS
Between drugs	61	2	30.5
Residual	546	40	13.6
Total	607	42	

standard errors are

$T_1 - T_2$	$+3.60$	SE 0.82
$T_1 - T_3$	$+1.75$	SE 0.81
$T_2 - T_3$	-1.80	SE 0.79.

Note that because the relative sizes of the SEs vary for the three differences the three combined estimates of differences are not exactly consistent. This could be corrected by a least squares analysis.

8.6 Cross-over designs for residual or interaction effects

We first consider designs in which all possible combinations and orders are included. Such a design will be appropriate if either:

(i) the number of treatments is so small and the replication is so large that there really is no case for not acquiring the more complete information provided by using all possible combinations, or

(ii) the possible interactions between treatments and orders, and between treatments in different periods, are sufficiently important, or in doubt, that it is necessary to examine all combinations in order to understand the biological situation.

In medical experiments, there has been considerable investigation of the two-treatment, two-period cross-over, notably by Hills and Armitage. If more than two treatments are to be compared then the subexperiment for each pair of treatments can be analysed separately. The form of analysis given here is based on that described by Hills and Armitage (1979).

Consider the two-treatment, two-period cross-over design with n_1 patients receiving treatment 1 in period 1 followed by treatment 2 in period 2, and n_2 patients receiving treatment 2 in period 1 followed by treatment 1 in period 2. Instead of approaching an analysis through the analysis of variance, the sums and differences of observations for each patient are used to obtain a more simply interpretable analysis. Consider the difference between the observations for the first and second periods for each patient. If it is assumed that there are no residual effects, so that the treatment effects are the same in both periods, then for patients receiving treatment 1 first,

$$d_{1i} = (t_1 - t_2) + (p_1 - p_2) + \varepsilon_{1i},$$

where t_1, t_2 are the treatment effects, p_1, p_2 the period effects, and ε_{1i} represent the error variation. Similarly,

$$d_{2i} = (t_2 - t_1) + (p_1 - p_2) + \varepsilon_{2i}.$$

The two sample means $\bar{d}_1$, $\bar{d}_2$ provide estimates of $(t_1 - t_2) + (p_1 - p_2)$ and $(t_2 - t_1) + (p_1 - p_2)$, respectively. Variances of the sample means are obtained from the sample variances of differences. The difference $\bar{d}_1 - \bar{d}_2$ with variance $s_1^2/n_1 + s_2^2/n_2$ provides an estimate of $2(t_1 - t_2)$, the period effect being eliminated in the subtraction.

Now suppose that there are residual effects of the first period treatment on the second period observation, or equivalently that the difference between treatment effects is modified in the second period. The analysis of differences cannot provide the required information about the simple treatment effect, $t_1 - t_2$. Suppose that in the second period the treatment effects are altered from t_1, t_2 to t_1', t_2'. Then

$$d_{1i} = (t_1 - t_2') + (p_1 - p_2) + \varepsilon_{1i}$$
$$d_{2i} = (t_2 - t_1') + (p_1 - p_2) + \varepsilon_{2i}$$

and

$$\bar{d}_1 - \bar{d}_2 = (t_1 - t_2') - (t_2 - t_1') + \bar{\varepsilon}_1 - \bar{\varepsilon}_2$$
$$= (t_1 - t_2) + (t_1' - t_2') + \bar{\varepsilon}_1 - \bar{\varepsilon}_2.$$

Thus the difference between the means of the within patient differences estimates the sum of the first period treatment effect and the second period treatment effect. The analysis of differences cannot yield any information about the difference between $(t_1 - t_2)$ and $(t_1' - t_2')$. The only source of information about this interaction difference is contained in comparisons between the sums of observations for the different patients.

The model for the sum of observations for each patient includes the sum of the two period effects, and the sum of the treatment effects applied in the two periods:

$$s_{1i} = 2\mu + (t_1 + t_2') + (p_1 + p_2) + \eta_{1i}$$
$$s_{2i} = 2\mu + (t_2 + t_1') + (p_1 + p_2) + \eta_{2i},$$

where the η error terms represent the between patient variation and, if the reasons for using cross-over designs to improve precision are valid, Var(η) will be much larger than Var(ε).

The difference between the sample means of the sums provides information about the difference between $(t_1 - t_2)$ and $(t'_1 - t'_2)$:

$$\bar{s}_1 - \bar{s}_2 = (t_1 + t'_2) - (t_2 + t'_1) + \bar{\eta}_1 - \bar{\eta}_2$$
$$= (t_1 - t_2) - (t'_1 - t'_2) + \bar{\eta}_1 - \bar{\eta}_2.$$

Hence, calculating $\bar{s}_1 - \bar{s}_2$, and the variance of $\bar{s}_1 - \bar{s}_2$ obtainable directly from the sample variances for s_1 and for s_2, we may test the difference between $(t_1 - t_2)$ and $(t'_1 - t'_2)$,

$$(\bar{s}_1 - \bar{s}_2)/\{\mathrm{Var}(\bar{s}_1 - \bar{s}_2)\}^{1/2}.$$

If the test shows negligible evidence of a difference then the estimate of $(t_1 - t_2)$ from $\bar{d}_1 - \bar{d}_2$ can be used.

The problems with this apparently simple approach are twofold. Whereas the estimation of $(t_1 - t_2)$, assuming no interaction or residual effects, is in terms of within patient variation (differences) and is likely to be relatively precise, the test of the existence of interaction or residual effects is in terms of between patient variation (sums), and is likely to be relatively imprecise. But the interaction must be tested before the estimate of the treatment difference can be interpreted. It is quite possible that the lack of precision for the test of interaction would lead to a result that a difference between $(t_1 - t_2)$ and $(t'_1 - t'_2)$ bigger than the average of the two values was not significantly different from zero.

The second problem arises if it is decided that the interaction or residual effects are non-negligible. We then have a choice of estimating treatment differences from within patient differences which are precise but which provide an estimate of $(t_1 - t_2) + (t'_1 - t'_2)$, or of using only the information from the first period, which will be imprecise (because it relies on between patient information), but which provides an estimate of $(t_1 - t_2)$. As always in statistics, then, we must choose between greater precision with a possibly less relevant estimate, or lesser precision with a more relevant estimate. And of course we have returned to the philosophical problem which is discussed at the start of Section 8.5.

Example 8.1 (continued)

The methods for the two-treatment, two-period cross-over are applied to each pair of drugs in turn.

Drugs 1 and 2

	T_1 before T_2, $n=7$	difference (period 1 − period 2)	sum
	mean	−0.14	12.14
	variance	42.14	17.48
	T_2 before T_1, $n=5$		
	mean	−7.80	11.00
	variance	4.20	20.00

To test whether the differences between the drugs are consistent between the two periods we use the difference between the two sum means

$$(12.14 - 11.00)/(17.48/7 + 20.00/5)^{1/2} = 0.45.$$

There is clearly no evidence of interaction. We therefore estimate the difference between the effects of the drugs from the mean difference,

$$\hat{t}_1 - \hat{t}_2 = \tfrac{1}{2}\{-0.14 - (-7.80)\} = 3.83,$$

and the standard error of this estimate is

$$0.5(42.14/7 + 4.20/5)^{1/2} = 1.31.$$

Drugs 1 and 3

	difference	sum
T_1 before T_3, $n=7$		
mean	-3.00	15.29
variance	30.00	40.57
T_3 before T_1, $n=8$		
mean	-5.38	13.62
variance	9.70	37.41

To test interaction

$$(15.29 - 13.62)/(40.57/7 + 37.41/8)^{1/2} = 0.52.$$

Again there is negligible evidence for interaction.

$$\hat{t}_1 - \hat{t}_3 = \tfrac{1}{2}\{-3.00 - (-5.38)\} = 1.19,$$

with standard error

$$0.5(30.00/7 + 9.70/8)^{1/2} = 1.17.$$

Drugs 2 and 3

	difference	sum
T_2 before T_3		
mean	-4.50	9.75
variance	30.57	18.79
T_3 before T_2		
mean	-2.25	11.25
variance	12.50	35.36

To test interaction

$$(11.25 - 9.75)/(18.79/8 + 35.36/8)^{1/2} = 0.58.$$

The evidence for interaction is again negligible.

$$\hat{t}_2 - \hat{t}_3 = \tfrac{1}{2}\{-4.50 - (-2.25)\} = -1.12,$$

with standard error

$$0.5(30.57/8 + 12.50/8)^{1/2} = 1.16.$$

The three sets of data provide a negligible level of evidence of interaction with remarkable consistency. They also provide three reasonably consistent estimates of differences between the drug effects

$$\hat{t}_1 - \hat{t}_2 = 3.83 \quad \text{SE } 1.31$$
$$\hat{t}_1 - \hat{t}_3 = 1.19 \quad \text{SE } 1.17$$
$$\hat{t}_2 - \hat{t}_3 = -1.12 \quad \text{SE } 1.16.$$

Since these estimates are independent they are inevitably not exactly consistent, the difference between t_1 and t_2 being greater than the sum of the differences for $(t_1 - t_3)$ and $(t_3 - t_2)$. A set of consistent estimates can be calculated using simple least squares theory. If the variance estimates are pooled to estimate an assumed common variance, then the estimates obtained from the combined estimation procedure will in fact be those obtained from the within patient analysis in the previous section:

$$\hat{t}_1 - \hat{t}_2 = 3.42$$
$$\hat{t}_1 - \hat{t}_3 = 2.04$$
$$\hat{t}_2 - \hat{t}_3 = -1.38.$$

The example, which has been discussed at some length, shows a relatively small level of between patient variation and the benefits, and the dilemmas of the use of cross-over designs do not materially affect the conclusions.

With only two periods it is not possible to escape from the problem that the treatment difference may differ between periods. If more than two periods are available then we can obtain information about the treatment effects even if these effects vary between periods, provided that we assume a particular form for that interaction variation. Instead of assuming simply that the treatment effects differ between periods we assume that each observation is influenced, not only by the treatment applied in the current period, but also by the treatment applied to the same subject in the previous period; that is, we assume residual treatment effects. With only two periods the model assuming residual effects is the same as that assuming that the treatment effects change. With more than two periods the residual effect model is simpler than assuming different treatment effects in each period.

To allow separation of treatment and residual effects using row and column designs, with periods as rows the choice of appropriate row and column designs must be made with care. We remarked earlier that, if residual effects were of no importance, then any of the row and column designs considered previously were appropriate. This is not true if residual effects are important. Consider the two Latin square designs shown in Figure 8.18 for the comparison of four diets in four periods using four subjects. In (a), treatment A is always preceded by treatment C, B by D, C by B and D by A, so that the estimates of the treatment effect of A and of

Figure 8.18. Latin square designs (a) not suitable, (b) suitable for residual effects.

(a)   Subject

		1	2	3	4
Period	1	A	B	C	D
	2	D	C	A	B
	3	B	A	D	C
	4	C	D	B	A

(b)   Subject

		1	2	3	4
Period	1	A	B	C	D
	2	D	C	A	B
	3	C	D	B	A
	4	B	A	D	C

the residual effect of C will be indistinguishable and the other treatment and residual effects will be similarly linked. In contrast, design (b) has each treatment preceded by each other treatment exactly once. Obviously, design (b) will provide information about residual treatment effects in addition to direct treatment effects, while design (a) will not. However, the investigation of residual effects requires a more complex analysis than the row and column designs considered earlier because the observations in the first period are not subject to residual effects of the treatment in the previous periods, and consequently residual effects are not orthogonal to subject effects or treatment effects.

Consider the model and analysis for the design in Figure 8.18:

$$y_{ij} = \mu + s_i + p_j + t_{k(ij)} + rt_{l(ij)} + \varepsilon_{ij},$$

where s_i, p_j, t_k, rt_l represent subject effects, period effects, direct treatment effects and residual treatment effects, respectively. The $rt_{l(ij)}$ term is included in the model only for $j \neq 1$, that is for periods 2 to 4. The least squares equations for the four sets of effects are not fully orthogonal, as is demonstrated below (using the usual form of constraints):

$$4\hat{\mu} + 4\hat{s}_1 - \hat{rt}_B = Y_{1.}$$
$$4\hat{\mu} + 4\hat{s}_2 - \hat{rt}_A = Y_{2.}$$
$$4\hat{\mu} + 4\hat{s}_3 - \hat{rt}_D = Y_{3.}$$
$$4\hat{\mu} + 4\hat{s}_4 - \hat{rt}_C = Y_{4.}$$
$$4\hat{\mu} + 4\hat{p}_j = Y_{.j}, \quad j = 1, 2, 3, 4$$
$$4\hat{\mu} + 4\hat{t}_A - \hat{rt}_A = T_A$$

$$4\hat{\mu} + 4\hat{t}_B - r\hat{t}_B = T_B$$
$$4\hat{\mu} + 4\hat{t}_C - r\hat{t}_C = T_C$$
$$4\hat{\mu} + 4\hat{t}_C - r\hat{t}_C = T_C$$
$$3\hat{\mu} + 3r\hat{t}_A - \hat{s}_2 - \hat{t}_A = R_A$$
$$3\hat{\mu} + 3r\hat{t}_B - \hat{s}_1 - \hat{t}_B = R_B$$
$$3\hat{\mu} + 3r\hat{t}_C - \hat{s}_4 - \hat{t}_C = R_C$$
$$3\hat{\mu} + 3r\hat{t}_D - \hat{s}_3 - \hat{t}_D = R_D.$$

The non-orthogonality is not complex, the least squares equations being linked in groups of three, for example $(\hat{s}_1, \hat{t}_B, r\hat{t}_B)$, but it does mean that the testing of significance for direct treatment effects, residual treatment effects, and even subject effects (if it is considered useful to test them) requires different orders of fitting terms in the analysis of variance. To test the significance of residual effects, the analysis of variance is calculated in the form shown in Table 8.11. To test treatment effects, allowing for residual effects, a second analysis of variance is calculated (Table 8.12). And, should it be important to assess the significance of between subject variation, then a third ordering would be used, with subject effects fitted last.

In practice, the level of non-orthogonality for designs with four or more periods is rarely such as to cause any substantial change in the pattern of

Table 8.11.

Source	df
Periods	3
Subjects (ignoring residuals)	3
Treatments (ignoring residuals)	3
Residuals (eliminating S and T)	3
Error	3
Total	15

Table 8.12.

Source	df
Periods	3
Subjects (ignoring residuals)	3
Residuals (eliminating S, ignoring T)	3
Treatments (eliminating R)	3
Error	3
Total	15

sums of squares in the different orders of fitting, and the interpretation is not usually difficult. However, the different relevant orders of fitting should be examined to provide properly stringent tests.

Designs allowing investigation of, and adjustment for, residual effects are subject to the usual considerations for row and column designs with additional restrictions of the type illustrated in the earlier discussion of the two 4×4 Latin square designs. If the number of treatments is equal to the number of periods, then obviously a Latin square design is appropriate, and all that is usually necessary is the avoidance of Latin squares which have each treatment always preceded by the same other treatment. The same philosophy applies when the number of treatments is one greater than the number of levels, when a Latin square omitting the last row should be used with the same restrictions. If the number of treatments is greater than the number of periods by more than one, then the usual approach to selecting row and column designs will tend to produce columns such that the treatments occurring in a column with a particular treatment will vary from column to column, and it should not then be difficult to impose the additional requirement that treatments preceding a particular treatment should be as evenly spaced as possible over the set of treatments.

Exercises 8

(1) Construct designs for comparing seven treatments in row and column designs with the following structure:

   (i) 49 units in seven rows × seven columns;

   (ii) 35 units in seven rows × five columns;

   (iii) 70 units in seven rows × ten columns;

   (iv) 48 units in six rows × eight columns;

   (v) 42 units in an incomplete seven rows × eight columns as shown in Figure 8.19.

Figure 8.19. Occurrence of a set of 42 units in seven rows and eight columns.

Column

		1	2	3	4	5	6	7	8
Row	1	X	X	X	X	X	X	X	X
	2	X	X		X	X	X	X	
	3	X	X	X			X	X	X
	4	X	X	X	X	X		X	X
	5	X		X	X	X	X		
	6	X	X		X		X	X	X
	7	X					X	X	X

(2) The 16 students on an MSc course in statistics discover that they form an exceptional body. Four are English, four are Scottish, four are Irish and four are Welsh. From their undergraduate careers, there are four graduates each from the four universities Cambridge, Oxford, Reading and Edinburgh and no two of the same nationality went to the same university. In their choice of option courses in the MSc, four took three pure options (PPP), four took three applied options (AAA), four took two pure and one applied (PPA) and four took one pure and two applied (PAA), and no two with the same choice of options are graduates of the same university, or are of the same nationality. Four students intend to enter the Civil Service, four to remain in academic life, four to train to be actuaries and four to become industrial statisticians, and no two of the same intended profession took the same options or graduated at the same university or are of the same nationality.

Three of the intending academic statisticians are the English Edinburgh graduate who took AAA, the Scottish Oxford graduate who took PPP and the Irish Cambridge graduate who took PAA. The Scottish Edinburgh graduate intends entering the Civil Service, the English Cambridge graduate plans to be an industrial statistician, and the Scottish Reading graduate took AAA. What can you say about the Welshman who went to Oxford?

(3) A design is required to compare four dietary treatments for cows, using seven cows for three periods of observation each in a three row × seven column structure. It is believed that residual effects are not important. One of the treatments is a control. An initial proposal is to modify the Youden square design given below by allocating the control (C) to A, treatment T_1 to B and D, treatment T_2 to C and F and treatment T_3 to E and G:

A D F B G C E
B A G F E D C
C E A D B G F.

How much can you improve the design in the sense of achieving more even precision and smaller variances, for treatment comparisons

    (i) using three replications for C and six replications for T_1, T_2 and T_3,

    (ii) using six replications for C and five replications for T_1, T_2 and T_3?

Figure 8.20. Cross-over design for four milkmaids milking four cows on four days.

Cow

		1	2	3	4
Day	1	A	B	C	D
	2	B	D	A	C
	3	C	A	D	B
	4	D	C	B	A

(4) An experiment to investigate variation between milkmaids was designed as a cross-over design as in Figure 8.20. The model assumed is

$$y_{ij} = \mu + d_i + c_j + m_k + r_l + e_{ij},$$

where d_i = day effect, c_j = cow effect, m_k = milkmaid effect, and r_l = residual effect of previous maid. Using a least squares estimation, show that

$$40(\hat{m}_A - \hat{m}_B) = 11(M_A - M_B) + 4(R_A - R_B) + (C_4 - C_3),$$

where capital letters denote totals, i.e. M_A = total for maid A, and derive the variance of

$$(\hat{m}_A - \hat{m}_B).$$

(5) An experiment to compare five diets for dairy cattle was designed as a change-over design using two 5×5 Latin squares with five periods for each of ten animals, allowing residual effects of diets to be estimated. The five diets were a control diet (C) and a 2×2 factorial set of two different additives each on two levels, $a_1 b_1$ (D), $a_1 b_2$ (E), $a_2 b_1$ (F), $a_2 b_2$ (G). The design is displayed in Figure 8.21.

Show that the least squares estimate of the direct effect of diet C (d_C) is given by

$$240\hat{\mu} + 180\hat{d}_C = 19D_C + 5R_C + A_1 + A_8$$

Figure 8.21. Cross-over design for five diets applied to ten animals over five periods.

Animal

		1	2	3	4	5	6	7	8	9	10
Period	1	F	C	E	G	D	C	G	E	F	D
	2	G	D	F	C	E	E	D	G	C	F
	3	D	F	C	E	G	D	C	F	G	E
	4	E	G	D	F	C	G	F	D	E	C
	5	C	E	G	D	F	F	E	C	D	G

Table 8.13(a). *Total yields.*

	1	2	3	4	5	6	7	8	9	
Period	1	2	3	4	5					total = 30 104
	5985	3153	6212	5932	5822					
Animal	1	2	3	4	5	6	7	8	9	
	2791	3093	3240	2889	3032	3049	3117	2688	2988	
Diet	C	D	E	F	G					
	5483	6097	6124	6217	6183					
Observation following diet (residual total)	C	D	E	F	G					
	5035	4808	4828	4830	4617					

(b). *Residual SS.*

Fitting	Residual SS	df
General mean	118 298	49
+ period effects	108 025	45
+ animal effects	52 461	36
+ residual diet effects	48 569	32
+ direct diet effects	13 186	28
omit residual diet effects	15 472	32

Table 8.14.

Group I

Weeks Patient	Drug A 2	Drug A 4	Drug B 6	Drug B 8
1	553	569	568	568
2	650	632	617	603
3	639	641		657
4	605	594	591	582
5	552	532	545	
6	663	627	616	618
7	702	722	714	727
8	655	644	637	644
9		567	578	580
10	622	631	618	653
11	616	607	614	600
12	681	653	646	
13	555	553	570	547
14	595	600	590	598

Group II

Weeks Patient	Drug B 2	Drug B 4	Drug A 6	Drug A 8
15	652	649	640	613
16	552	551	533	521
17	731	711	676	705
18	598	621	605	605
19	566	574	588	
20	600	600	605	603
21	725		705	753
22	582	589		
23	540	539	540	557
24	676	679		689
25	682		682	

Gaps in the table occur where patients were not available for observation at the relevant time.

where D_C, R_C, A_i are totals for observations for diet C, for observations following diet C and for observations for animal i, respectively. Total yields (lb milk over four weeks) are listed in Table 8.13(a) together with the residual SS obtained from fitting a sequence of models in (b). Construct the analysis of variance and write a full report on the results, quoting numerical values and their standard errors for the important effects.

(6) A clinical trial was conducted to compare the effects of two drugs A and B in alleviating the painful effects of rheumatism. Patients who consented to participate in the trial were allocated at random to one of two groups I and II. Before entering the trial, the amount of swelling present in each patient's finger joints was measured by taking the total circumference of the knuckles of all ten digits, called *totalpip*. Group I was then given drug A for four weeks followed by drug B for a further four weeks, while group II received the two drugs in the reverse order. Patients knew that their treatment was being changed after four weeks, but neither they nor the doctors running the trial knew which group they were in. The totalpip measure was taken from each patient at fortnightly intervals and the results were as in Table 8.14. Analyse the data in any way you consider appropriate and write a short report on your findings.

9

Randomisation

9.1   What is the population?

The fundamental results on which most inferences for data from experimental designs are based were described and proved in Chapter 4. All these results are based on the assumption that the experimental units are a sample from a normally distributed population. This assumption is made for many other sampling situations, and is frequently accepted as being a reasonable approximation. Transformations of the observed variable may be necessary to ensure that the assumptions of normality and homogeneity of variance are satisfied. However, for most analysis of sample data from observational studies, it is possible to envisage a population of which the sample may be realistically accepted as representative.

In designed experiments, the very careful control which is exercised often makes it difficult to identify a population for which the sample is relevant. In field crop experiments, small plots are marked out and treated with great care so that the plots may be regarded as homogeneous, and the treatments may be compared as precisely as possible. In some animal nutrition experiments, animals are surgically treated so that more detailed observations can be made. In psychological experiments, subjects are subjected to stresses in controlled situations, and are required to complete specified tasks, often in very limited times. In industrial experiments, machinery is operated under each set of conditions for a short period of time.

Two questions should be asked about the results from small-scale experiments. There is the general question of whether the conclusions from such rigidly controlled experiments have any validity for future practical situations. This is a difficult question, which extends beyond statistics and the scope of this book, though some examples where the relevance of experimentation seems particularly dubious are discussed. The importance of including a wide range of factors in an experiment to provide greater validity for conclusions has already been discussed in Chapter 3, and will recur in Part III.

The second question is a rather more technical one. When the experimental units have been carefully selected and monitored, is there any sense in which they are a sample from a population? And if the assumption of the existence of a population from which the sample is drawn is untenable, how does this affect the inferences from the data? A solution to this difficulty is the device of randomisation, and in this chapter we discuss the full statistical purpose and theory of randomisation, which is a much more important idea than that suggested in Chapter 2 of simply 'giving each treatment the same chance of allocation to each unit'.

Before developing the formal ideas of randomisation, consider possible inferences from a sample set of data. Suppose eight students, Alan, Carol, Deborah, Jocelyn, Marion, Nigel, Sally and Thomas, take a test and get the following marks:

Alan	14
Carol	15
Deborah	15
Jocelyn	13
Marion	16
Nigel	12
Sally	19
Thomas	10.

Someone looking at their marks might notice that all the four female students had scored higher marks than all four male students. A possible inference, without implying any particular further population of students since these were the only students taking the course leading to the test, might be that females are obviously better than males (at least at this test). The underlying argument is that it is surely improbable that, if there is really no male–female difference, the four highest scores should be achieved by females. It is essentially an extension of the idea of 'coincidence' and is, in fact, well founded since the probability of a prespecified group of four getting the four highest scores is 1 in 70. Since the same level of surprise could have been engendered by the four males getting the highest scores, then the probability of the four highest scores being achieved by four individuals of the same gender is 1 in 35. The logic behind the subsequent inference that gender must be related to genius in that test is based only on the eight scores actually achieved (10, 12, 13, 14, 15, 15, 16 and 19) and implies that *a priori* any pairing of these eight scores and the eight students (A, C, D, J, M, N, S and T) was equally likely. Any less extreme pattern such as

A	16	C	15
J	13	D	15
N	12	M	14
T	10	S	19

would be less convincing evidence for a relationship between gender and ability. However, the crucial idea is that arguments about inferences can be based only on the eight results actually obtained, and that the arguments depend only on considering the actual pairings of students and marks in contrast to all the other possible pairings, and then contending that the result for the actual pairing is (surprisingly) extreme.

9.2   Random treatment allocation

The device of randomisation is intended to provide a valid basis for the general form of argument illustrated by the student results just discussed. The weakness of the argument that gender and ability are related is that it is a purely *post hoc* recognition of pattern in the grouping of high and low scores. Alternative groupings could be noticed: four students are older; four are following a different course; four have red hair; or four live in London. And each of these divisions of the eight students into two groups of four could conceivably be used to 'explain' the results. The advantage of random allocation of treatments in a controlled experiment is that for the division of interest the occurrence of any particular allocation of units to different treatments is exactly as probable as any other allocation. Suppose we are doing a psychological experiment and eight students are randomly or arbitrarily split into two groups, one group being given special training and the other no special training, the students being otherwise apparently similar. Andrew, John, Neil and Timothy have had no specialised training, while those randomly selected for special training are Charles, David, Martin and Stephen. Now, when the scores obtained are:

Andrew	14
Charles	15
David	15
John	13
Martin	16
Neil	12
Stephen	19
Timothy	10

the argument that special training is beneficial (C 15, D 15, M 16, S 19) compared with no training (A 14, J 13, N 12, T 10) is much stronger because the alternative to that argument is that the special training has no effect (null hypothesis) and the random allocation of students to the special training group happened to pick out the four students who were going to score the four highest results. The deliberate random allocation for the one potentially important question provides a stronger basis for the use of the coincidence argument.

The procedure of random allocation of treatments to units within a set of restrictions imposed by the design structure is illustrated by the following three designs:

(i) A completely randomised design with two treatments and eight units, each treatment being allocated to four units. Since each treatment is to be applied to the same number of units, each unit is equally likely to be allocated to treatment A or to treatment B. A proper procedure is, for each unit in turn, to allocate treatment A with a probability of $(4-a)/(8-a-b)$, where a is the number of units already allocated to A and b the number of units already allocated to B. At any point in the allocation procedure, if either A or B has been allocated to four units, then all the remaining units are inevitably allocated to the other treatment.

(ii) A randomised block design with three treatments and nine units divided into three blocks of three units each. Here, the randomisation procedure is restricted, since each treatment must appear once in each block. This, in fact, makes the procedure easier, since in each block of three units one of the three treatments, A, B or C, is selected randomly for the first unit in the block, one of the other two treatments is then selected randomly for the second unit in the block, and the remaining treatment is then allocated to the third unit.

(iii) A completely randomised design with two treatments and 20 units available for testing. One treatment is to be applied to four units and the other treatment, which is a control, to the remaining 16 units. The simple procedure for this case is to select the four units for the first treatment. To do this, we randomly select one unit from the 20, then one from the remaining 19, then one from the remaining 18 and finally from the remaining 17. These four units are allocated to the new treatment and the remaining 16 to the control.

In each case randomisation can be achieved either by selecting a treatment for a particular unit, or by selecting a unit for a particular treatment. For each example only one method has been described, not always the simpler one. Whichever method is used, the logic of the method must be checked to ensure that, for each unit, the probability of allocation to each treatment is constant for that treatment. For example, if in (iii) we tried to choose a treatment for the first unit, then one for the second and so on, we must not make the probabilities of allocation to the first units equal for treatments A and B; in that case the last four units would be virtually certain to be allocated to B because only four are to be allocated to A. Instead, the allocation must be such that the first unit has a 4/5 probability of getting B, and 1/5 of A, and for each subsequent unit the probabilities will depend on the numbers already allocated to A and B.

In clinical trials, the random allocation of treatments to patients is particularly important, but is made difficult by further practical considerations. Essentially none of the clinicians involved in the conduct of the trial should know or be able to deduce the identity of the treatment applied to any particular patient. This is important because such knowledge could not be assumed to have no influence on the general treatment of the patient. The double negative is deliberate. It is the lack of assurance that there would be no effect of knowledge on the treatment of the patient that would leave the conclusions of the trial open to doubt.

The requirement that the trial be conducted 'blind' so that those concerned with the routine administration of the trial cannot know the identity of the treatment for any patient has two implications. First, and obviously, the clinician does not have access to the information about treatment allocation. Second, the clinician should not be able to deduce the identity of the treatments applied to particular patients. Ideally the clinician should not even be able to deduce that the treatments for two patients were different. However, it is important that the clinician be involved in the decisions about the principles used to control experimental variability, and he or she will subsequently be able to recognise the characteristics on which the principles are based. Therefore for small numbers of treatment, 'blocks' should contain more than a single replication of each treatment. If the number of treatments is large, each block will contain a subset of treatments, the principle, but not the detail of the subset arrangements, being determined in advance. These design problems will be considered further in Section 9.7.

9.3   Randomisation tests

The device of randomisation creates a population of experiments that could have been performed, though only one experiment has actually occurred. The identification of the population allows assessment of the value of a test statistic against the background of the distribution of values of that test statistic for all other members of the population of possible experiments, assuming that the null hypothesis of no treatment effect is true. The idea of a randomisation test is developed initially through examples for the three cases for which randomisation procedures were defined in the previous section.

Example 9.1

An experiment is conducted in a city to see whether road improvements affect the incidence of accidents. Eight accident black spots, of a similar nature and each capable of improvement, are selected, and of these eight four are chosen at random and are actually improved, while the other four remain unchanged. The number of accidents for (*a*) the year before the date of improvement and (*b*) the year after the improvement are recorded in Table 9.1. Various criteria to measure the effect (after/before) could be

Table 9.1.

	Improved site				Unchanged site			
	A	B	C	D	E	F	G	H
Accidents before	3	5	4	11	7	1	5	8
Accidents after	0	2	0	5	8	3	2	6
Difference	−3	−3	−4	−6	+1	+2	−3	−2

used. The simple difference (after–before) ignores possible scale differences between the sites, but demonstrates the principles of the test clearly. Consider the hypothesis that improvement has no effect on accidents. If the hypothesis is true we should not expect to find a clear-cut difference between the four differences for 'improved' sites and those for the 'unchanged' sites. Further, if improvement has no effect then the accidents observed at sites A to H would have occurred at those sites regardless of which four sites had been, randomly, selected for improvement. If the total of the four differences for each possible set of four sites out of the eight is calculated, it is possible to assess whether the total which was actually observed for the four randomly selected sites is significantly extreme. We can make this assessment rather tediously by calculating for each set of four sites the total of the four observed differences as follows:

$$
\begin{array}{cccccccc}
A & B & C & D & -3 & -3 & -4 & -6 = -16 \\
A & B & C & E & -3 & -3 & -4 & +1 = -9 \\
A & B & C & F & -3 & -3 & -4 & +2 = -8 \\
A & B & C & G & -3 & -3 & -4 & -3 = -13 \\
\vdots & \vdots & \vdots & \vdots & \vdots & \vdots & \vdots & \vdots \\
E & F & G & H & +1 & +2 & -3 & -2 = -2 \, .
\end{array}
$$

An alternative computational approach to assessing the extremeness of the observed result is to count how many of the possible $70\,(=8!/(4!4!))$ possible allocations would have given a result as extreme as that actually observed. By inspection, we can see that no sets of four sites could have given a total of *more* than -16 and that exactly three sets would give a total of -16, namely:

ABCD, ACDG and BCDG.

Consequently, the significance level of the observed effect is 3/70 or 0.04, and the effect would be classified as 'significant at the 5% level'. Note that we have implicitly used a one-tailed test and the interpretation of the significance level must be made in that context.

Example 9.2

Three treatments, A, B and C, are compared in three randomised blocks of three units each (see Table 9.2). A standard analysis of variance is shown in Table 9.3.

For the randomisation test, all possible random allocations of the treatments A, B and C to the units a, b, c, d, e, f, g, h, i, subject to the blocking restriction, are considered. If for each possible allocation the analysis of variance and the F ratio are calculated the significance of the observed value of 3.7 can be assessed. Since there are 3! possible

Table 9.2.

Treatment	Block			Treatment total
	I	II	III	
A	14(a)	11(d)	14(g)	39
B	10(b)	12(e)	10(h)	32
C	15(c)	17(f)	13(i)	45
Block total	39	40	37	

Table 9.3.

	SS	df	MS	F
Blocks	1.6	2		
Treatments	28.2	2	14.1	3.7
Error	15.1	4	3.8	
Total	44.9	8		

allocations within each block, this could give $(3!)^3$, or 216, sets of calculations, which seems rather laborious. However, these 216 split into 36 groups of six identical results, because any permutation of the three treatments, using the same permutation in each block, would give the same set of treatment totals (with different treatment labels), and therefore the same *F* ratio. The calculations may be further reduced by noting that:

(i) the total SS and block SS will remain unaffected under the hypotheses that for each experimental unit the different treatments would produce identical results and the units remain in the same blocks.

(ii) The implication of (i) is that the sum (treatment SS + error SS) is not altered under permutation of treatment allocation and therefore larger treatment SS implies larger *F*.

(iii) To rank the 36 different allocations in order of increasing *F* ratio the allocations need only be ranked in order of increasing treatment SS, or equivalently in order of increasing values of $(T_A^2 + T_B^2 + T_C^2)$.

The 36 different allocations and the corresponding values of

$$T = (T_A^2 + T_B^2 + T_C^2)$$

that would have been observed if the unit yields are unaffected by treatment are listed in Table 9.4. By inspection, there are seven values of T as large or larger than the observed value of 4570, and hence the significance level of the observed results against the null hypothesis of no treatment difference is $7/36 = 0.19$.

Example 9.3

Two treatments are compared, one of which is applied to four randomly selected mice, and the other to the 16 remaining mice. The survival times of the mice in days are as follows:

treatment A 64, 35, 39, 21

treatment B 7, 15, 32, 5, 43, 19, 26, 8, 6, 19, 45, 31, 33, 14, 21, 12.

To assess how strongly these data support the apparent supremacy of treatment A in prolonging survival, consider what proportion of the sets of four values selected from the 20 observed survival times would give at least as strong evidence of the superiority of treatment A. First, the criterion for the randomisation test must be chosen. Because there are only two treatments the simple criterion is the sum of the observed survival times for treatment A, which for the observed experiment is $64 + 35 + 39 + 21 = 159$. Next, consider how many values there are in the randomisation distribution; that is, how many possible ways are there of

Table 9.4.

Units allocated to treatments			Treatment totals	T	Units allocated to treatments			Treatment totals	T
A	B	C			A	B	C		
adg	beh	cfi	39, 32, 45	4570	aeg	bfh	cdi	40, 37, 39	4490
adg	bei	cfh	39, 35, 42	4530	aeg	bfi	cdh	40, 40, 36	4496
adh	beg	cfi	35, 36, 45	4546	aeh	bfg	cdi	36, 41, 39	4504
adh	bei	cfg	35, 35, 46	4566	aeh	bfi	cdg	36, 40, 40	4496
adi	beg	cfh	38, 36, 42	4524	aei	bfg	cdh	39, 41, 36	4504
adi	beh	cfg	38, 32, 46	4584	aei	bfh	cdg	39, 37, 40	4490
adg	bfh	cei	39, 37, 40	4490	afg	bdh	cei	45, 31, 40	4586
adg	bfi	ceh	39, 40, 37	4490	afg	bdi	ceh	45, 34, 37	4550
adh	bfg	cei	35, 41, 40	4506	afh	bdg	cei	41, 35, 40	4506
adh	bfi	ceg	35, 40, 41	4506	afh	bdi	ceg	41, 34, 41	4518
adi	bfi	ceh	38, 41, 37	4494	afi	bdg	ceh	44, 35, 37	4530
adi	bfh	ceg	38, 37, 41	4494	afi	bdh	ceg	44, 31, 41	4578
aeg	bdh	cfi	40, 31, 45	4586	afg	beh	cdi	45, 32, 39	4570
aeg	bdi	cfh	40, 34, 42	4540	afg	bei	cdh	45, 35, 36	4546
aeh	bdg	cfi	36, 35, 45	4566	afh	beg	cdi	41, 36, 39	4505
aeh	bdi	cfg	36, 34, 46	4568	afh	bei	cdg	41, 35, 40	4506
aei	bdg	cfh	39, 35, 42	4530	afi	beg	cdh	44, 35, 36	4528
aei	bdh	cfg	39, 31, 46	4598	afi	beh	cdg	44, 32, 40	4560

selecting four mice from 20. This is

$$\frac{20!}{16!4!} = 4845.$$

Clearly to consider the set of 4845 possible experiments by hand calculation is not practical, though it would be possible by computer. In this, and other practical cases when the number of values in the randomisation distribution is large, there are two possible approaches. The first is to enumerate all those randomisations which give a treatment total greater than, or equal to, 159. The initial stages of such an enumeration are shown in Table 9.5. The significance level found through such an enumeration is 107/4845, or 2.2%. The other approach, which is often appropriate but rarely used, is to estimate the significance level by sampling from the 4845 possible randomisations. This can easily be achieved by computer. Suppose that, for a sample of 500 randomisations, exactly 14 of the sample experiments give a treatment A total of 159 or more, then the estimated significance level of the observed results would be 14/500 or 2.8%. Using the binomial sampling variance, the variance of this estimated significance level is $(0.028)(0.972)/500 = 0.000054$, the corresponding standard error being 0.0075 or 0.75%.

Note that, in this last example, the calculated significance level is for a one-sided test. To determine the significance level for a two-sided test, it is necessary to determine what difference between treatments A and B, with

Table 9.5. *Randomisations giving a treatment total $\geqslant 159$.*

(1)	64	45	43	39	Total 191
(2)	.	.	.	35	187
(3)	.	.	.	33	185
⋮				⋮	⋮
(14)	64	45	43	8	160
(15)	.	.	.	7	159
(16)	64	45	39	35	183
(17)	.	.	.	33	181
⋮	⋮	⋮	⋮	⋮	⋮
(26)	64	45	39	14	162
(27)	..	..	..	12	160
(28)	64	45	35	33	171
(29)	⋮	⋮	⋮	⋮	⋮
⋮					
(104)	64	33	32	31	160
(105)	45	43	39	35	162
(106)	45	43	39	33	160
(107)	45	43	39	32	159

treatment A giving the lower average survival time, is equivalent to the observed difference. There is an element of subjective choice of criterion here, but a simple form of argument is as follows. For the observed ratio the treatment means are

$$159/4 = 39.75 \text{ for A}$$

and

$$336/16 = 21.00 \text{ for B,}$$

giving a difference of 18.75 between the means. The total of the 20 survival times is 495. To obtain a difference of 18.75 between the two treatment means, with the mean for A being the lower of the two means, the total for treatment A, T_A, must satisfy

$$\frac{T_A}{4} + 18.75 = \frac{495 - T_A}{16},$$

whence $T_A = 39$. By enumerating or estimating the proportion of the randomisation distribution giving treatment A totals either greater than or equal to 159, or less than or equal to 39, a two-sided test significance level for this definition of criterion can be obtained.

9.4   Randomisation theory of the analysis of experimental data

To demonstrate the implications of randomisation theory for the theory of the analysis of data from randomised experiments, we consider the example of the randomised complete block design. The idea of randomisation within a fixed set of experimental units requires a slightly different model from that developed in Chapter 2. There are assumed to be t units in each of b blocks, and the model for unit yields if all units are treated identically is

$$y_{ij} = \mu + b_i + \varepsilon_{ij}, \quad i = 1, 2, \ldots, b, \quad j = 1, 2, \ldots, t,$$

where b_i represents the average deviation for block i from the overall mean, and ε_{ij} represents the deviation for unit j within block i from the mean for that block. These assumptions imply

$$\sum_i b_i = 0$$

and, most importantly,

$$\sum_j \varepsilon_{ij} = 0, \quad i = 1, 2, \ldots, b.$$

The yield that would be obtained if treatment k were applied to unit j in block i would be

$$y_{ij} = \mu + b_i + t_k + \varepsilon_{ij}.$$

However, each treatment is allocated to only one unit in each block, and the allocation procedure is defined by a random variable δ_{ij}^k for which the properties are as follows:

$$\delta_{ij}^k \quad = 1 \quad \text{if treatment } k \text{ occurs on unit } j \text{ in block } i$$
$$= 0 \quad \text{otherwise,}$$
$$\text{prob}(\delta_{ij}^k = 1) \quad = 1/t$$

since there are t units in block i and treatment k occurs on just one,

$$\text{prob}(\delta_{ij'}^k = 1 | \delta_{ij}^k = 1) = 0$$

since treatment k can only occur on one unit in block i,

$$\text{prob}(\delta_{ij}^{k'} = 1 | \delta_{ij}^k = 1) = 0$$

since only one treatment can occur on unit j in block i, and

$$\text{prob}(\delta_{ij'}^{k'} = 1 | \delta_{ij}^k = 1) = 1/(t-1)$$

since once treatment k has been allocated to unit j there remain $(t-1)$ units for treatment k'.

The model for the yield of treatment k in block i, y_{ik}, is therefore

$$y_{ik} = \mu + b_i + t_k + \sum_j \delta_{ij}^k \varepsilon_{ij},$$

where the summation is over all units in block i. Using this model, we consider first the estimates of treatment effects and their variances, then the terms in the analysis of variance identity and finally the relationship between the two.

First, the treatment total for treatment k,

$$T_k = b\mu + bt_k + \sum_i \sum_j \delta_{ij}^k \varepsilon_{ij}.$$

Since $E(\delta_{ij}^k) = 1/t$ and $\sum_j \varepsilon_{ij} = 0$,

$$E(T_k) = b\mu + bt_k$$

whence

$$E(T_k - T_{k'}) = b(t_k - t_{k'})$$

Also

$$T_k - E(T_k) = \sum_i \sum_j \delta_{ij}^k \varepsilon_{ij},$$

and so

$$E[\{T_k - E(T_k)\}^2] = E\left\{\left(\sum_i \sum_j \delta_{ij}^k \varepsilon_{ij}\right)^2\right\}.$$

Expanding the double summation in terms of different sets of ε_{ij}'s gives

$$E\left\{\sum_i \sum_j (\delta_{ij}^k)^2 \varepsilon_{ij}^2 + \sum_i \sum_{j' \neq j} \delta_{ij}^k \delta_{ij'}^k \varepsilon_{ij} \varepsilon_{ij'} + \sum_{i' \neq i} \sum_j \sum_{j'} \delta_{ij}^k \delta_{i'j'}^k \varepsilon_{ij} \varepsilon_{i'j'}\right\}.$$

Since $\sum_j \varepsilon_{ij} = 0$ and $\delta_{ij}^k \delta_{i'j'}^k$ are independent (randomisation is independent in different blocks) the third term is zero. Also, from the properties of $\delta_{ij}^k \cdot \delta_{ij}^k$ the second term is zero. Finally since the possible values of δ_{ij}^k are only 0 or 1, the expectation of $(\delta_{ij}^k)^2$ is $1/t$. Hence,

$$\text{Var}(T_k) = E[\{T_k - E(T_k)\}^2] = (1/t) \sum_{ij} \varepsilon_{ij}^2$$

and

$$\text{Var}(y_{.k}) = \sum_i \sum_j \varepsilon_{ij}^2 / (b^2 t). \tag{9.1}$$

Similarly,

$$E[\{T_k - E(T_k)\}\{T_{k'} - E(T_{k'})\}] = E\left\{\left(\sum_{ij} \delta_{ij}^k \varepsilon_{ij}\right)\left(\sum_{ij} \delta_{ij}^{k'} \varepsilon_{ij}\right)\right\}.$$

Again expansion in terms of different combinations of ε_{ij}'s gives

$$E\left(\sum_i \sum_j \delta_{ij}^k \delta_{ij}^{k'} \varepsilon_{ij}^2 + \sum_i \sum_{j' \neq j} \delta_{ij}^k \delta_{ij'}^{k'} \varepsilon_{ij} \varepsilon_{ij'} + \sum_{i' \neq i} \sum_j \sum_{j'} \delta_{ij}^k \delta_{i'j'}^{k'} \varepsilon_{ij} \varepsilon_{i'j'}\right)$$

Again, the last term is zero because of the independence of δ_{ij}^k and $\delta_{i'j'}^{k'}$ and because $\sum_j \varepsilon_{ij} = 0$. The first term is zero because of the joint properties of δ_{ij}^k and $\delta_{ij}^{k'}$. In the second term, the expected value of $\delta_{ij}^k \delta_{ij'}^{k'}$ is $(1/t)\{1/(t-1)\}$ and

$$\sum_i \sum_{j \neq j'} \varepsilon_{ij} \varepsilon_{ij'} = \sum_i \sum_j \varepsilon_{ij}(-\varepsilon_{ij}) = -\sum_i \sum_j \varepsilon_{ij}^2.$$

Hence

$$E[\{T_k - E(T_k)\}][\{T_{k'} - E(T_{k'})\}] = [-1/\{t(t-1)\}] \sum_i \sum_j \varepsilon_{ij}^2$$

and

$$E[\{(T_k - T_{k'}) - E(T_k - T_{k'})\}^2] = (2/t) \sum_i \sum_j \varepsilon_{ij}^2 + [2/\{t(t-1)\}] \sum_i \sum_j \varepsilon_{ij}^2$$

$$= \{2/(t-1)\} \sum_i \sum_j \varepsilon_{ij}^2$$

so

$$\text{Var}(y_{.k} - y_{.k'}) = [2/\{b^2(t-1)\}] \sum_i \sum_j \varepsilon_{ij}^2. \tag{9.2}$$

Notice at this point from (9.1) and (9.2) that the variance of $(y_{.k} - y_{.k'})$ is not twice the variance of $y_{.k}$ as it would be in the usual linear model analysis of Chapter 4.

Now consider the analysis of variance identity developed in Chapter 2,

$$\sum_{ik} (y_{ik} - y_{..})^2 = t \sum_i (y_{i.} - y_{..})^2 + b \sum_k (y_{.k} - y_{..})^2 + \sum_i \sum_k (y_{ik} - y_{i.} - y_{.k} + y_{..})^2.$$

The first term on the right hand side of the identity involves only the block mean yields, $y_{i.}$, and the overall mean yield, $y_{..}$. These mean yields are unaffected by randomisation since each block total yield includes each treatment effect once, and each unit effect in that block once, and both these sets of effects sum to zero.

The second term on the right hand side of the identity is the treatment SS which can be written

$$\sum_k (y_{.k} - y_{..})^2 = \sum_k (T_k^2/b^2) - Y_{..}^2/b^2 t.$$

Again, $Y_{..}$ is unaffected by randomisation. Also from the earlier results for $E(T_k)$ and $Var(T_k)$,

$$E(T_k^2) = \{b(\mu + t_k)\}^2 + (1/t) \sum_i \sum_j \varepsilon_{ij}^2$$

$$= b^2\mu^2 + b^2 t_k^2 + 2b\mu t_k + (1/t) \sum_i \sum_j \varepsilon_{ij}^2$$

since $\sum_k t_k = 0$. Hence,

$$E\left\{\sum_k (y_{.k} - y_{..})^2\right\} = t\mu^2 + \sum_k t_k^2 + (1/b)^2 \sum_i \sum_j \varepsilon_{ij}^2 - t\mu^2$$

$$= \sum_k t_k^2 + (1/b)^2 \sum_i \sum_j \varepsilon_{ij}^2.$$

Finally, consider again y_{ik},

$$y_{ik} = \mu + b_i + t_k + \sum_j \delta_{ij}^k \varepsilon_{ij}$$

$$E(y_{ik}^2) = \mu^2 + b_i^2 + t_k^2 + 2\mu b_i + 2\mu t_k + 2b_i t_k + (1/t) \sum_j \varepsilon_{ij}^2$$

since all cross products involving ε_{ij} are zero since $\sum_j \varepsilon_{ij} = 0$. Further, since $\sum_i b_i = 0$ and $\sum_k t_k = 0$,

$$E\left(\sum_{ik} y_{ik}^2\right) = bt\mu^2 + t \sum_i b_i^2 + b \sum_k t_k^2 + \sum_i \sum_j \varepsilon_{ij}^2.$$

Hence the total sum of squares, the left hand side of the analysis of variance identity, has expected value

$$E\left\{\sum_{ik} (y_{ik} - y_{..})^2\right\} = bt\mu^2 + t \sum_i b_i^2 + b \sum_k t_k^2 + \sum_i \sum_j \varepsilon_{ij}^2 - bt\mu^2$$

$$= t \sum_i b_i^2 + b \sum_k t_k^2 + \sum_i \sum_j \varepsilon_{ij}^2.$$

The expected values for the analysis of variance structure are therefore as in Table 9.6. If we now write σ^2 for the quantity $\sum_i \sum_j \varepsilon_{ij}^2/\{b(t-1)\}$, which is the expected value of the residual mean square, then the variances for differences between treatment means and for treatments means can be written

$$Var(y_{.k} - y_{.k'}) = 2\sigma^2/b$$

and

$$Var(y_{.k}) \quad = (t-1)\sigma^2/(bt).$$

Table 9.6.

Source	df	E(SS)	E(MS)
Blocks	$b-1$	$t\sum_i b_i^2$	
Treatments	$t-1$	$\dfrac{1}{b}\sum_i\sum_j \varepsilon_{ij}^2 + b\sum_k t_k^2$	$1/\{b(t-1)\}\sum_i\sum_j \varepsilon_{ij}^2 + \{b/(t-1)\}\sum_k t_k^2$
Residual (by subtraction)	$(b-1)(t-1)$	$\{(b-1)/b\}\sum_i\sum_j \varepsilon_{ij}^2$	$1/\{b(t-1)\}\sum_i\sum_j \varepsilon_{ij}^2$
Total	$bt-1$	$t\sum_i b_i^2 + b\sum_k t_k^2 + \sum_i\sum_j \varepsilon_{ij}^2$	

Hence the pattern of mean squares in the analysis of variance and the relation between the residual mean square and the variance of a difference between two treatment means is exactly the same as in the normal distribution theory model. The variance for a single treatment mean is, as noted earlier, smaller by a factor $(t-1)/t$ than would be the case for the normal distribution theory model.

9.5 Practical implications of the two theories for the analysis of experimental data

The practical analysis of experimental data can usually proceed without detailed consideration of which of the two alternative theories of analysis is assumed. However, it is useful to summarise and compare the assumptions and implications of the two theories.

A: Infinite model

(1) This model, for which the theory was developed in Chapter 4, assumes that the experimental units are a sample from a population which, by the implication of the distributional assumptions for the unit errors, must be of infinite size.

(2) The unit errors are assumed to be normally and independently distributed with a common variance σ^2.

(3) The experimental units are assumed to be subject to fixed environmental (block) and treatment effects.

(4) The inferences from the analysis of the data apply to the infinite population from which the experimental units are a sample.

B: Finite model

(1) This model, for which the theory has been developed in the previous section, assumes a finite set of experimental units, all of which are used in the experiment; it further assumes that, within restrictions imposed by the blocking structure of the experiment, the treatments are allocated randomly to experiments.

(2) The unit errors are assumed to be fixed and the set of unit errors in a block is assumed to have a zero sum.

(3) The inferences from the analysis of the data apply strictly to the finite set of units used in the experiment.

Least squares estimates

For both models, the least squares estimates of block and treatment effects are appropriate. The analysis of variance corresponds to the subdivision of variation between the various sets of fitted estimates for the two models, and the relationship between the residual mean square and the variance of a difference between two treatment means is identical for both models. The slightly different result for the variance of a single treatment mean, for the finite model, is a result of the non-independence of treatment means, which is a consequence of the random allocation of treatments to units. Essentially, treatment mean yields are not independent because, if in a particular block treatment A is allocated to unit j, then treatment B must be allocated to one of the remaining units in that block.

Inferences

For the infinite model, with its normal distribution assumptions, the F test for the comparison of mean squares in the analysis of variance, and the t test for the comparison of treatment means, are correct.

For the finite model, it can be shown by considering the moments of the distribution of the ratio

$$\frac{\text{treatment mean square}}{\text{treatment mean square} + \text{error mean square}}$$

that the null hypothesis distribution of the ratio of the treatment and error mean squares is approximately represented by the F distribution on $(t-1)$ and $(b-1)(t-1)$ df, provided the ε_{ij} are homogeneous between blocks. Example 9.2 for data from three blocks of three treatments provides as much support for this general rule as any single example can, the randomisation test giving a significance level of 0.19, which agrees closely with the significance of an F statistic of 3.7 on 2 and 4 df.

For the application of t tests, an analysis based on the finite model has to rely on the general robustness of the t test to departures from an underlying normal distribution.

In addition to the common requirement (3) of fixed environmental and treatment effects, both models assume additive and homogeneous error terms. Departure from these assumptions can frequently be corrected by using a transformation of the observed variable, as discussed in Chapter 11.

Choice between models

It should be clear from the foregoing discussion that the practical analysis, inference and interpretation of experimental data is not materially affected by the choice of model. If the assumptions of the infinite model are felt to be reasonable, then there is no reason not to use that model. However, since those assumptions are often found to be difficult to credit, and the recognition that the assumptions are unrealistic may not come until the experiment is in progress, it is sensible always to assume that treatments should be randomly allocated, unless there are very good practical reasons for using a systematic allocation (some situations where systematic designs may be appropriate will be discussed in Chapters 14 and 17). Not only does the device of randomisation provide an alternative basis for the analysis of experimental data, it also defines a procedure for allocating treatments to units, which would otherwise have to be devised, to avoid a purely subjective allocation.

In the remainder of this book, I shall usually discuss the analysis of data in the context of the infinite model because it is desirable to discuss models for experimental results in terms of a single form of model and the formal model is simpler to write down for the infinite model. The randomisation ideas will never be far away from the development of design principles, however, and it will be assumed throughout that treatments should be allocated randomly unless a non-random allocation is explicitly defined.

9.6   Practical randomisation

In this section we consider the randomisation procedures required for the design structures discussed in earlier chapters, particularly Chapters 7 and 8. The simple practical discussion of randomisation at the end of Chapter 2 must now be modified and extended to allow for the more complex design structures developed.

The first requirement of a randomisation procedure is to define precisely the restrictions within which the randomisation must be applied. Thus, if a completely randomised design is to be used, the numbers of observations for each treatment must be specified. The randomisation scheme used must then ensure that, for each unit, the probabilities that the different treatments are allocated to the unit *must be identical for each unit, regardless of the order of allocation*. The random allocation may be achieved either by allocating units to treatments or by allocating treatments to units. The former requires that the units be individually labelled and that the required number of units for treatment A be selected randomly and sequentially, all units not previously selected being equally likely to be selected at any particular stage of the selection. After all the necessary units for treatment A have been selected, those for B are selected, and thereafter for each other treatment in turn. The procedure of allocating treatments to units is more complex. If treatments A to T are to have $n_A, n_B, \ldots, n_T$ units, respectively, then the allocation probabilities for the first unit are $n_A/\sum n$, $n_B/\sum n \ldots n_T/\sum n$. As units are allocated the probabilities change so that each treatment allocation probability is proportional to the number of units still required for that unit.

If units are grouped in blocks for a single blocking system, then there are three possible components of randomisation. The grouping of treatments into sets for different blocks may involve a random element. The allocation of treatment sets to particular blocks should usually involve a random element. Finally, the actual allocation of treatments within a set to the units within a block should always involve randomisation. In the randomised complete block design all treatments occur in each block so that the first two components of the random allocation procedures have no meaning and only the third component is appropriate. In designs with incomplete blocks all three components should be considered.

Consider, for example, the design shown in Figure 9.1 for comparing six treatments in six blocks of four units each (Example 7.3). This design is not balanced but, in terms of attempting to achieve equality of treatment comparison, it is as close to balance as may be achieved within the restrictions imposed by the blocking system. Nevertheless, the inequality of comparisons means that a particular treatment, for example A, will be compared more precisely with those treatments with which it is jointly omitted from a block than with the other treatments. Thus, comparisons of A with B, A with C, B with D, C with E, D with F and E with F will be more precise than other comparisons. The experimenter must decide whether he wishes to choose which treatment pairs have this very slight advantage of precision, or whether the selection of the six (linked) pairs for omission should be determined randomly. Having determined the six sets of four treatments each, these six sets should be allocated randomly to the six blocks. Finally, the four treatments in the set allocated to a particular block must be randomly allocated to the four units in that block. Thus the design has two components for which random allocation is required, and a third where the selection of the six treatment sets may be deliberately non-random.

The question of whether the allocation of treatments to treatment sets for blocks should be random or deliberate depends on the treatment structure. Consider the problem of Example 7.0(a) (and Example 7.10), in which eight treatments consisting of two heating levels × two lighting levels × two carbon dioxide levels are to be allocated to 12 units in two

Figure 9.1. Design to compare six treatments in six blocks of four units.

```
Block   I     A     B     C     D     (omitting E F)
        II    A     B     C     E     (omitting D F)
       III    A     B     D     F     (omitting C E)
        IV    A     C     E     F     (omitting B D)
         V    B     D     E     F     (omitting A C)
        VI    C     D     E     F     (omitting A B)
```

blocks of six units each. Two sets of six treatments must be selected, one for the first block and the other for the second block. The two sets will inevitably include four common treatments, and each set will also include two of the four remaining treatments. Since the eight treatment combinations are highly structured, it is unlikely that the experimenter will select the four combinations to be included in both treatment sets randomly. If he did so, and then found that the four common treatments were all those with the lower heating level (only a 1 in 70 chance, but with five other similar unfortunate patterns, not a negligible probability), then he would probably wish to abandon that random selection and start again. It would be better to recognise initially that the four common treatments should be representative of both heating levels, both lighting levels and both carbon dioxide levels, and to choose as the common treatments a set such as:

upper level of heating	upper level of lighting	CO_2 1
upper level of heating	lower level of lighting	CO_2 2
lower level of heating	upper level of lighting	CO_2 2
lower level of heating	lower level of lighting	CO_2 1

Such a systematic set is easily constructed from the three stated requirements. Methods of selecting such 'representative' fractional subsets of treatments are discussed in Chapters 13 and 16. Our present purpose is to suggest simply that random selection of treatment sets for incomplete blocks is not always a good idea. The allocation of two more treatments to each block set offers less scope for systematic selection, but the allocation of heating and lighting levels to the two blocks can be made orthogonal to blocks. After selecting the two sets of treatments systematically there should be two stages of randomisation. First the random allocation of sets to blocks, and secondly the random allocation of treatments to units within each block.

The details of the randomisation procedure for any design employing a single blocking system are extremely simple. In all stages of randomisation we have a set of items to be allocated to an equi-numeric set of objects. In a randomised complete block design there are t treatments to be allocated to t units in each block. We may either allocate single units randomly to treatment A, treatment B, ..., treatment T in turn, or we may allocate individual treatments to unit 1, unit 2, ..., unit t in turn; at each stage of either allocation all remaining units (treatments) are equally likely to be selected. In incomplete block designs the allocation of treatment sets to blocks, or of treatments to units within a block, is in exactly the same one-to-one form.

The random allocation of treatments to units in designs with two blocking criteria, such as a Latin square design, is more complex. Unlike the randomised block design, Latin squares cannot be randomised sequentially, for example by randomising treatments in each row subject to the requirement that no treatment appears twice in a column, because such a procedure generally leads to a position from which the column requirements cannot be satisfied. It is therefore necessary to adopt a different approach to randomisation.

If the set of possible Latin squares can be tabulated or indexed in some way, then the random selection of a Latin square can be achieved by simply using the structure of the tabulation. This has been the basis of classical methods for selecting a random Latin square. To understand methods of tabulation, it is important to realise that, for a Latin square, (i) permutations of the letters, or (ii) permutations of the rows, or (iii) permutations of the columns, leaves a design which is still a Latin square. Although we have not used this approach the randomisation procedure for a randomised complete block design can also be recognised as a permutation of treatments within the set of units (or equivalently as a permutation of the units within the set of treatments). For any randomisation test criterion which is symmetric in the treatments, permutation of the letters does not alter the value of the test statistic. We therefore consider only those Latin squares which are distinct in the sense that one cannot be transformed to another simply by relabelling the treatments. The number of distinct squares is small for small t (one for $t=2$, two for $t=3$, 24 for $t=4$) but rapidly becomes too large for tabulation of all distinct squares.

Two methods of indexing the available Latin squares have been devised. The first uses the fact that, for any Latin square, it is possible by reordering first the rows and then columns to produce a Latin square in which the first row and the first column both contain the letters in the alphabetic order $(A, B, C, \ldots)$. A square with the first row and first column 'in order' is called a *standard square*. Latin squares can be divided into sets such that all the squares in one set derive from a single standard square by the reordering of rows and columns. Since each Latin square has a unique associated standard square, a proper method of randomly selecting a Latin square is to select randomly a standard square and then randomly reorder the rows and the columns. However, although the number of standard squares is quite small for small values of t (one for $t=2$ and for $t=3$, four for $t=4$, 56 for $t=5$), for $t=6$ the number is too large for easy tabulation.

The second effort at indexing involves the grouping of standard squares into sets called *transformation sets*, where each standard square in a transformation set can be transformed into any other standard square in the set by first permuting the letters, e.g. (A→B, B→D, D→C), and then reordering the rows and columns to produce a standard square. There are two difficulties associated with transformation sets. The first is that the number of standard squares in a transformation set varies between sets so that, in selecting a random Latin square by first selecting a transformation set, then a standard square and then a Latin square, the initial selection of a transformation set must take account of the different sizes (number of standard squares) of transformation sets. However, the second difficulty is that even the number of transformation sets becomes too large as t increases and this means that the indexing of transformation sets is not a complete solution to the search for a method of randomly selecting a Latin square.

The advice about the practical randomisation of a Latin square design will vary according to the size of square being considered. The basic principle on which the choice should be made is that the set of possible designs considered for selection must be large enough to allow a large number of values in the randomisation distribution.

For $t = 3$ or 4, there really are not enough different Latin squares available (two for $t = 3$ and 24 for $t = 4$) for a randomisation test to be meaningful. Nonetheless, the Latin square should be chosen from all the available squares by randomly selecting a standard square and then randomly permuting the rows and the columns.

For $t = 5$, a standard square should be selected at random from the 56 possible standard squares. This may be done directly or through selection of a transformation set. Subsequently, the rows and columns should be randomly reordered and, if the initial selection was by means of a transformation set, the letters should be randomly permuted.

For $t = 6$, a standard square for any of the larger transformation sets should be selected and the rows, columns and letters should be randomly permuted.

For $t = 7$ or more, it is sufficient to write down any Latin square and randomly permute rows, columns and letters.

For multiple Latin squares the randomisation procedure depends on the form of multiple Latin square design. For the design in which the component Latin squares are entirely separate, the randomisations of each square are independent. If a Latin rectangle design, using Latin squares with common row effects, is required then the reasonable randomisation procedure would be:

  (a) select $n\ t \times t$ standard Latin squares,
  (b) randomise the rows within each square,
  (c) randomise the entire set of nt columns.

The procedure is illustrated for two 5×5 Latin squares in Figure 9.2.

Finally we consider the random allocation of treatments for more general row and column designs. If the rows and columns are orthogonal (that is each row/column combination has the same number of units, usually one) then the randomisation procedure is essentially exactly the same as for Latin squares of size $t \geqslant 7$. Any initial allocation of treatments to the row and column structure, such that the joint occurrence restrictions are satisfied, may be used as the starting point. The rows are randomly reordered; then the columns are randomly reordered; finally, if the treatments are regarded as equally important, the treatment letters are randomly reordered. The procedure is illustrated in Figure 9.3 for a design to compare ten treatments in a 6×5 array.

When rows and columns are not orthogonal the restrictions on possible randomisation are severe. Any set of rows (or columns) which have exactly the same pattern of occurrence in the columns (rows) may be permuted. And of course the treatment letters may be permuted. Other randomisation possibilities may be deduced from the occurrence of arbitrary choices within the design construction procedure.

Figure 9.2. Randomisation of multiple Latin squares (a) two standard squares, (b) rows of each randomised, (c) columns randomised.

(a)

A	B	C	D	E		A	B	C	D	E
B	C	E	A	D		B	E	A	C	D
C	D	A	E	B		C	A	D	E	B
D	E	B	C	A		D	C	E	B	A
E	A	D	B	C		E	D	B	A	C

(b)

A	B	C	D	E		C	A	D	E	B
E	A	D	B	C		E	D	B	A	C
B	C	E	A	D		A	B	C	D	E
D	E	B	C	A		B	E	A	C	D
C	D	A	E	B		D	C	E	B	A

(c)

B	E	A	D	B	E	C	C	A	D
C	A	D	B	A	C	D	E	E	B
E	D	B	C	C	D	E	A	B	A
D	C	E	A	E	A	B	B	D	C
A	B	C	E	D	B	A	D	C	E

Figure 9.3. Randomisation of design for ten treatments in five rows × six columns. (a) Initial, (b) reorder rows, (c) reorder columns, (d) reorder letters.

(a)
A	B	C	D	E	F
I	H	B	G	D	A
C	E	G	A	H	J
D	C	J	F	I	H
J	I	F	E	G	B

(b)
D	C	J	F	I	H
J	I	F	E	G	B
A	B	C	D	E	F
C	E	G	A	H	J
I	H	B	G	D	A

(c)
I	F	J	C	H	D
G	E	F	I	B	J
E	D	C	B	F	A
H	A	G	E	J	C
D	G	B	H	A	I

(d)
E	F	A	J	H	D
G	B	F	E	I	A
B	D	J	I	F	C
H	C	G	B	A	J
D	G	I	H	C	E

9.7 Sequential allocation of treatments in clinical trials

For clinical trials in which the number of treatments is small and for which all the patients are available simultaneously at the beginning of the trial the requirements of randomisation and blindness can be satisfied sufficiently by the use of blocks, each of which contains several replicates of each treatment. Thus, patients can be grouped into appropriate blocks such that each block would contain several replicates of each treatment and treatments can be allocated randomly within blocks. The requirement of blindness may lead to the use of larger blocks than would normally be recommended for other forms of experiment. In practice, however, it is common for patients to be admitted to a clinical trial sequentially, and the allocation of treatments must be random and 'blind' while at the same time achieving, as nearly as possible, balance of the treatments over several different blocking factors. Methods for achieving this sequential balance, while retaining the element of randomisation, are discussed by Pocock (1979). An alternative scheme using traditional blocking methods can also be devised.

The principles of the restricted randomisation methods of Pocock and others are presented here for the two treatment experiment which is an extremely common form of clinical trial. Each patient is classified by each of several blocking factor classifications, and the information on the numbers previously allocated to each treatment from each class of each blocking factor is available. Thus, for example, it may be that patients are classified by age (four classes), sex (two classes) and occupation (four classes). The numbers of patients for each class for each of two treatments, A and B, at a particular stage of the trial are shown in Table 9.7. If a new patient, male, under 30, in occupation class IV becomes available for the trial, then allocation to treatment B would improve the balance of all three blocking factors. A female patient between 50 and 70 in occupation III could not be allocated so as to improve the balance for all three blocking factors. Even in the most clear cut case, however, some element of uncertainty must remain, so that clinicians involved with the trial cannot deduce the identity of the treatment, or even deduce that a patient will be allocated a treatment different from that received by a particular previous patient. Hence, treatment allocation must still include a random component.

The restricted randomisation method defines each allocation as a random event, the probability of allocating a patient to treatment A being P_A. This probability is modified to take account of the degree of imbalance of previous treatment allocations for the class of the new patient for each blocking factor in turn. The simplest form of modification assumes a basic probability of 0.5 and modification by the ratio of previous allocations for each blocking factor, n_B/n_A. For the two examples (male, < 30, IV) and (female, 50–70, III) assuming the previous allocation patterns shown in Table 9.7 the calculations would be as shown in Table 9.8. The probability

Table 9.7.

		Treatment A	Treatment B
Age	<30	10	6
	30–50	12	12
	50–70	4	5
	>70	4	7
Sex	M	17	14
	F	13	16
Occupation	I	5	8
	II	9	13
	III	7	2
	IV	9	7

Table 9.8.

Patient	A : B	Modification to P_A
1: male	17 : 14	14/17
< 30	10 : 6	6/10
IV	9 : 7	7/9
2: female	13 : 16	16/13
50–70	4 : 5	5/4
III	7 : 2	2/7

that patient 1 will be allocated to treatment A would be

$$P_A = 1/2 \times 14/17 \times 6/10 \times 7/9 = 49/255 = 0.19.$$

The probability of allocation to A for patient 2 is

$$P_A = 1/2 \times 16/13 \times 5/4 \times 2/7 = 20/91 = 0.22.$$

There are several other aspects of the restricted randomisation approach which deserve mention. First, more complex methods of modifying the P_A values may be used and the reader is referred to Pocock (1979) for a discussion of these. Second, the restricted randomisation method cannot be used until at least one patient has been allocated to each treatment from each class of each classification (this may not be necessary with modified allocation rules). Third, balance is considered separately for each factor in turn. To allow for all the various different combinations of sex, age and occupation class, balance must be considered for each of the 32 combinations separately. This would usually be too fine a subdivision of the population of patients, giving very small numbers in each combination. However, if the possibility of a very unbalanced allocation for, say, female patients in occupation class I worries the experimenter, then the blocking factors may be redefined to allow balance within this particular combination of classes. Finally, the need for the balancing mechanism to be a probabilistic one is a direct consequence of the need for treatment allocation to be unidentifiable by those directly involved in the trial. This form of secrecy is usually required in medical trials, but not in animal drug trials, and certainly not in agricultural crop experiments or industrial process experimentation. Although the ideas of sequential balancing and restricted randomisation are ingenious, they are, of course, inefficient when the secrecy element is not required, since blocking allocation could then be selected to minimise the treatment comparison variance.

An alternative approach to the problem of simultaneously controlling the variation of patients and achieving blind and random allocation is to define blocks of patient types such that each block may be expected to include, by the end of the sequential arrival of patients, at least three patients per treatment. A random allocation system is then constructed for each block using a block size bigger than that expected. The excess of, say, two replicates per treatment will allow for excess numbers of patients in any block class. To illustrate the procedure suppose that for the trial discussed earlier the expected frequencies for numbers of patients in the combinations of all three classifications are those shown in Table 9.9. Such figures will be approximate, and no effort to make them more exact is useful. If sex is believed to be the most important blocking factor, age the next most important, and occupation least important, then a blocking system as shown in Figure 9.4 might be appropriate, each block being expected to include at least six patients.

Random allocation plans would be prepared for each block by the statistician without the knowledge of the clinician, assuming block sizes four more than those expected. The random allocation for the (male, < 30) block would assume a possible block size of 16 (= 12 + 4) and might be

A   A   A   B   A   B   B   B   A   B   A   B   B   B   A   A.

If now only nine patients in this category occur in the trial then there will be five A's and four B's in the block. If 14 patients occur there will be six A's and eight B's. Since the clinician need not know the block size chosen the allocation is effectively blind.

Although the ideas of sequential randomised allocation are usually related to experiments to compare just two treatments, there is no difficulty in principle in adapting the methods for use in experiments with three or more treatments. Indeed, when more than two treatments are used, the dangers of the clinician penetrating the blindness of either

Table 9.9.

Sex	Age	Occupation			
		I	II	III	IV
M	< 30	4	4	4	0
	30–50	4	4	4	4
	50–70	2	2	4	4
	> 70	0	2	2	4
F	< 30	2	4	6	2
	30–50	2	4	6	4
	50–70	0	4	6	0
	> 70	0	4	4	0

Figure 9.4. Prospective blocking system for trial with three blocking factors.

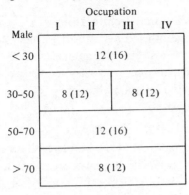

random allocation procedure are much reduced simply because of the multiplicity of treatments.

Absolute blindness in the sense of the clinician being completely incapable of making a better-than-average prediction not only of whether the treatment to be allocated to the current patient is more likely to be A or B, but even of whether the treatment will be different to that applied to the last similar patient, is philosophically impossible. The objective of a trial is to compare two treatments, and any system of allocation must tend to equalise the overall replication of the two treatments: hence, the allocation of treatment A to one patient makes the allocation of treatment B to any other particular patient more likely and hence a limited level of prediction must be feasible. What can and should be attempted in any trial is a degree of randomness within a system of control of variation, retaining near blindness. Either the sequential restricted random allocation or the use of large multi-replicate blocks can achieve this combination of objectives.

Exercises 9

(1) Five pigs are selected randomly from a group of eight and are given a standard diet, the other three receiving a new feed. The weight gains are given below. Use a randomisation test to test whether there is a genuine difference between the effects of the two feeds, defining carefully the criterion you use.

$$Weight \ gains \begin{cases} \text{standard} & 9, \quad 17, \quad 11, \quad 16, \quad 18 \\ \text{new} & 15, \quad 19, \quad 16. \end{cases}$$

(2) A randomisation test is required of the null hypothesis that the five treatment effects are equal for the experimental data for five rice spacing treatments in four blocks given in Table 9.10. By sampling from the full randomisation distribution of $(5!)^3$ values to obtain a sample of 20 values, perform the required randomisation test at the 5% significance level. Think carefully about how you take your random sample of 20 points from the randomisation distribution.

Table 9.10

Treatment	Block			
	I	II	III	IV
1	5.9	5.3	6.5	6.3
2	7.1	6.4	6.6	5.8
3	7.0	6.5	6.3	8.9
4	8.8	7.6	7.0	7.9
5	7.8	6.7	8.2	7.2

(3) Show that there are only 12 distinct 3×3 Latin squares. Show further that, if the ratio of treatment mean square/error mean square for a particular square is R, then the other possible values under randomisation are all either R or $1/R$.

(4) Three treatments, A, B and C, are compared in an experiment for which the design consisted of two quite separate 3×3 Latin squares. Show that under randomisation, and assuming the null hypothesis that the three treatments have identical effects, the distribution of the ratio treatment mean square/error mean square contains 24 distinct values. For the data given in Table 9.11 test the null hypothesis that the treatments have identical effects.

Table 9.11.

A	C	B	B	A	C
4	2	2	4	6	3
B	A	C	A	C	B
3	4	3	4	5	2
C	B	A	C	B	A
3	4	5	4	4	4

(5) The randomisation model for a completely randomised design is

$$y_{ik} = \mu + t_k + \varepsilon_i$$

if treatment k is applied to unit i with

$$y_{.k} = \mu + t_k + \sum_i \delta_i^k \varepsilon_i / n_k,$$

where δ_i^k is the probability that treatment k is allocated to unit i. Identify the probability structure for δ_i^k if treatment k is to be allocated to n_k units and the total number of units is N. Obtain the variance of a difference between two treatment means. Show that the expected value of the residual mean square in the analysis of variance is $\sum \varepsilon_i^2 /(n-1)$ and that the variance of the difference between two treatment means may be estimated by $s^2(1/n_1 + 1/n_2)$, where s^2 is the residual mean square in the analysis of variance.

(6) How many sample points are there in the randomisation distributions for
   (a) a randomised block design for seven treatments in three blocks,
   (b) a BIB design for seven treatments in seven blocks of three units,
   (c) a Youden square for seven treatments in a 7×3 array?

(7) Gordon and Foss (1966) report the results of an experiment on the effect of rocking on the crying of very young babies. On each of 18 days, the babies not crying at a certain instant in a hospital nursery served as subjects. One baby selected at random was rocked for a set period, the remainder serving as controls. The numbers not crying at the end of a specified period were as in Table 9.12. Construct a randomisation test of the hypothesis that rocking has no effect on the propensity of babies to cry.

Table 9.12.

Day	Number of control babies		Number of rocked babies	
	Total	Not crying	Total	Not crying
1	8	3	1	1
2	6	2	1	1
3	5	1	1	1
4	6	1	1	0
5	5	4	1	1
6	9	4	1	1
7	8	5	1	1
8	8	4	1	1
9	5	3	1	1
10	9	8	1	0
11	6	5	1	1
12	9	8	1	1
13	8	5	1	1
14	5	4	1	1
15	6	4	1	1
16	8	7	1	1
17	6	4	1	0
18	8	5	1	1

10

Covariance – extension of linear models

10.0   *Preliminary examples*

(*a*) In a strawberry variety trial, eight varieties were compared in four blocks of eight plots. The blocks were chosen so that each block included eight plots in a row as shown in Figure 10.1. Examination of the yields, displayed in Figure 10.1, revealed a clear trend along the blocks, and on enquiry it was discovered that there had been a hedge at the right hand end of the trial. Clearly, the blocks have not been chosen appropriately and, in analysing the yields from the experiment, it is important to allow for the decline in yield towards the hedge.

(*b*) In the trial of drugs to relieve pain, discussed in Chapters 6 and 8, the data were analysed to estimate differences between treatments, allowing for differences between patients and the effect of the order in which the drugs were administered. However, the timing of the administration of the second drug for each patient was determined by the length of the period of relief afforded by the first drug for the patient. It would be reasonable, in assessing the

Figure 10.1. Experimental plan for eight strawberry varieties in four blocks of eight plots, with hedge.

Block	I	G	V	Rl	F	Re	M	E	P	H
		5.8	6.3	4.9	6.5	4.5	5.2	6.5	3.8	
	II	E	P	M	Re	G	V	F	Rl	E
		6.9	7.6	7.9	5.6	7.0	5.5	4.0	2.7	
										D
	III	V	F	Rl	G	P	E	Re	M	
		7.6	6.4	5.0	6.9	7.4	5.3	5.2	3.2	G
	IV	E	Re	M	P	G	F	V	Rl	
		7.5	7.0	6.1	7.2	6.5	5.6	5.8	1.4	E

effect of the second drug, to try to allow for the variation in the length of the interval from the beginning of the trial before the second drug was administered.

10.1   The use of additional information

Frequently in experiments, the experimenter has additional information about each experimental unit, which might be expected to be related to the yield of the unit. There are various reasons why such additional information may not have been used as the basis for choosing a blocking system for the units. The information may not have been available at the time when the experiment was being designed. The number of blocking factors that could be used may have been too large for all factors to be used in constructing blocks. Or, having thought about the possibility of using the methods described in this chapter, it might have been decided that it was not necessary to use blocking methods.

Examples in agricultural experiments of additional information, which might be expected to be relevant to the yield comparisons being investigated, include:

(i) In animal experiments, with a blocking system based on genetic characteristics of the animals (that is, blocks may be litters), it is clear that the growth or other response of the animals may also be related to the initial weight of each animal.

(ii) In fruit tree experiments, blocking is frequently based on the geographical positions of the trees, but much of the variation in the future yields of individual trees may be predictable from previous yielding records.

(iii) In crop experiments, damage by birds, waterlogging or human error may be recognisably uneven between plots, and plot yield may be closely related to some simple quantitative score of damage.

(iv) In any experiment, where units remain in the same spatial positions, trends in yield variation may become apparent during the experiment, resulting from fertility trends or light trends (particularly in glasshouse experiments).

In medical trials, in which the experimental unit is an individual patient, the amount of information available on each unit is potentially enormous. Patients can be classified by age, sex, weight, occupation, race, location and a variety of historical or health factors. At most two or three of these can be used to construct blocking systems, but it would be foolish to ignore the information available in the other classifications or variables. Similar considerations would apply to psychological experiments.

In chemical experiments, obvious blocking factors include different times, different machines, or different sources of material. In addition, preliminary measurements on the experimental material before treatment may reveal other potential differences.

In all these examples, the general philosophy is the same. The object of the experiment is to compare the experimental treatments as precisely as possible. The precision of comparison is determined primarily by the background variation, represented by the variance, σ^2, of the error term, ε. If some of the error variation can be related to variation in the additional variables, measured on each experimental unit, then the effective background variance σ^2 will be reduced and the treatment comparisons, which may require adjustment to allow for uneven patterns of values of the additional variables, can be made more precise.

For example, suppose that in a simple medical experiment three treatments, A, B and C, are allocated to four patients each. The 12 patients were grouped into four age classes (I, II, III, IV) with three patients per age class so that the experiment is designed as four randomised blocks of three units each, giving results as in Table 10.1. It is discovered after the experiment that the six asterisked patients in Table 10.1 have a different condition from the others. Examination of the observations in some detail suggests that the asterisked observations are rather higher than the general pattern would suggest. For example, the (A-B) differences are 31, 37 and 29 in blocks I, II and IV, when A has the asterisk, and 17 in III when B is asterisked; the (C-A) differences are 9, 11 and 11 when both or neither have asterisks, and 1 in block IV when only A has an asterisk. The (C-B) differences are 40 and 38 when C has an asterisk, 28 when B has an asterisk, and 30 when neither is asterisked.

To assess the differences between the two populations, we might examine the residuals $(y_{ij} - \hat{\mu} - \hat{b}_i - \hat{t}_j)$, retaining the asterisks to indicate the two different groups (see Table 10.2). The mean residual for the asterisked units is $+3.0$, while that for the others is -3.0. The effect of the difference between the two groups could reasonably be estimated as 6. To adjust the comparison between the treatments to allow for the uneven

Table 10.1.

	I	II	III	IV	Total	Mean
A	133*	168*	147	196*	644	161.0
B	102	131	130*	167	530	132.5
C	142*	179*	158	197	676	169.0

Table 10.2.

	I	II	III	IV
A	+0.5*	+1.9*	−4.8	+2.5*
B	−2.0	−6.6	+6.7*	+2.0
C	+1.5*	+4.9*	−1.8	−4.5

allocation of the two populations, the totals for A and B should be corrected to be those which might have been expected if the treatments had each had two units from each group; that is, the total for A is adjusted downwards by 6 from 644 to 638 because A had three units from the asterisk group compared to the average of two asterisked units, and B is adjusted upwards by 6 from 530 to 536.

To see how much improvement in precision has been achieved, consider how the identical yields might be adjusted to an 'average' group value for each unit by reducing the asterisked values by 3.0 and increasing the others by 3.0. The adjusted values and the corresponding residuals are given in Table 10.3. The residual SS is reduced from 179.35 to 43.34, which confirms that the residual variation is substantially accounted for by the difference between the two populations. However, as the adjustments for the effect of the difference between the two groups have themselves been based on the data, the standard error of a difference between two means cannot be calculated from this revised residual SS as if no adjustment had been made. Standard errors based on the reduced residual SS are derived when the full model is developed in the next section.

Table 10.3(a). *Adjusted values*

	I	II	III	IV	Total	Mean
A	130	165	150	193	638	159.5
B	105	134	127	170	536	132.5
C	139	176	161	200	676	169.0

(b). *Residuals.*

	I	II	III	IV
A	0	+1	−1.3	0
B	+0.5	−4.2	+1.2	+2.5
C	−0.5	+2.8	+0.2	−2.5

We have deliberately chosen a very simple example, in which the additional information is in the simplest possible form, to illustrate the principles of detecting the dependence of residual variation on additional variables, and of the adjustment of treatment means to allow for mean effects of this dependence. The difference between the two groups may be represented as a measurement or variate taking one of two values: 1 for the asterisked groups and 0 for the other. Such a variate is called a *covariate*. More generally, a covariate may be either a qualitative or a quantitative measurement. Both forms of covariate are treated in the same way, and several covariates can be used for simultaneous adjustment of treatment comparisons.

10.2   The general theory of covariance analysis

The theory supporting the analysis of covariance is essentially an application of the results for partitioned models (Chapter 4). We first present the theory for a general design with many covariates and then consider the special case of a randomised block design with a single covariate. The general theory is presented using the terminology employed in the appendix to Chapter 4. In discussing the special case the philosophy and the critical assumptions are examined in detail.

The covariance model assumes that the yields of experimental units can be modelled in two stages. First, the usual component of the model representing the experimental design with terms for block effects and treatment effects, both sets of effects being possibly structured, and also the block and treatment effects being combined according to the design. The second component is a set of regression terms expressing the dependence of yield on additional measured or descriptive variables.

If the design model is written

$$y = A\theta + \eta$$

and the regression model is

$$\eta = X\beta + \varepsilon,$$

then the full model is

$$y = A\theta + X\beta + \varepsilon.$$

Assuming the use of necessary constraints, the parameter estimates for the design model are

$$\hat{\theta} = (A'A)^{-1}A'y = C^{-1}A'y$$

and the residual SS is

$$y'y - y'A\hat{\theta} = y'(I - AC^{-1}A')y.$$

The quantities $(I - AC^{-1}A')y$ may be recognised as the residuals after the fitting of the design model. The interpretation of the residual SS and the following results is clarified by the recognition that the matrix

$$\mathbf{R} = \mathbf{I} - \mathbf{A}\mathbf{C}^{-1}\mathbf{A}'$$

is symmetric and idempotent. That is

$$\mathbf{R}' = \mathbf{R}$$

and

$$\mathbf{R}\mathbf{R} = (\mathbf{I} - \mathbf{A}\mathbf{C}^{-1}\mathbf{A}')(\mathbf{I} - \mathbf{A}\mathbf{C}^{-1}\mathbf{A}') = \mathbf{I} - \mathbf{A}\mathbf{C}^{-1}\mathbf{A}'.$$

For the full model, the least squares equations are

$$\begin{pmatrix} \mathbf{A'A} & \mathbf{A'X} \\ \mathbf{X'A} & \mathbf{X'X} \end{pmatrix} \begin{pmatrix} \hat{\theta} \\ \hat{\beta} \end{pmatrix} = \begin{pmatrix} \mathbf{A'y} \\ \mathbf{X'y} \end{pmatrix} \tag{10.1}$$

and eliminating $\hat{\theta}$ we obtain equations for the estimation of the covariance effects

$$(\mathbf{X'X} - \mathbf{X'A}\mathbf{C}^{-1}\mathbf{A'X})\hat{\beta} = \mathbf{X'y} - \mathbf{X'A}\mathbf{C}^{-1}\mathbf{A'y}, \tag{10.2}$$

which we can rewrite in the form

$$\mathbf{X'}(\mathbf{I} - \mathbf{A}\mathbf{C}^{-1}\mathbf{A'})\mathbf{X}\hat{\beta} = \mathbf{X'}(\mathbf{I} - \mathbf{A}\mathbf{C}^{-1}\mathbf{A'})\mathbf{y}$$

or

$$\mathbf{X'R'RX}\hat{\beta} = \mathbf{X'R'Ry}.$$

All the elements in this set of equations are sums of squares and sums of products of the residuals of $\mathbf{y}$ and the residuals of the covariates $\mathbf{x}_i$. Hence we can identify the underlying methodology of covariance analysis as a multiple regression of the residuals of the yield variable, y, on the residuals of the covariates, all residuals being calculated in terms of the basic experimental design represented by the design matrix, $\mathbf{A}$.

The reduction in the residual SS due to fitting β is

$$\hat{\beta}'(\mathbf{X'R'Ry}),$$

where

$$\hat{\beta} = (\mathbf{X'R'RX})^{-1}(\mathbf{X'R'Ry}).$$

The covariance analysis has been developed in terms of fitting $\hat{\theta}$ first and $\hat{\beta}$ subsequently so that the analysis of variance may be written as in Table 10.4. However, the main purpose in introducing the covariates is to

Table 10.4.

Fitting	SS
$\hat{\theta}$ (ignoring $\hat{\beta}$)	$\mathbf{y'A}\mathbf{C}^{-1}\mathbf{A'y}$
$\hat{\beta}$ (allowing for $\hat{\theta}$)	$(\mathbf{X'R'Ry})'(\mathbf{X'R'RX})^{-1}(\mathbf{X'R'Ry})$
Residual	S_r

improve the precision of estimation of treatment parameters. The adjusted treatment effect estimates may be obtained from the full set of least squares equations (10.1);

$$\hat{\boldsymbol{\theta}} = (\mathbf{A}'\mathbf{A})^{-1}(\mathbf{A}'\mathbf{y} - \mathbf{A}'\mathbf{X}\hat{\boldsymbol{\beta}})$$

when $\hat{\boldsymbol{\beta}}$ is given by (10.2), and the variance–covariance matrix of these adjusted effects is

$$\mathbf{V}(\hat{\boldsymbol{\theta}}) = \{\mathbf{A}'\mathbf{A} - \mathbf{A}'\mathbf{X}(\mathbf{X}'\mathbf{X})^{-1}\mathbf{X}'\mathbf{A}\}^{-1}\sigma^2.$$

Whether or not the block and treatment effects were orthogonal in the original design model, it will usually not be true that they are orthogonal after adjustments for the effects of covariance. Hence, to assess the variation due to treatment effects, the reduced model including only the block effects $(\boldsymbol{\theta}_1)$ and the covariance effects $\boldsymbol{\beta}$ is fitted. The fitting SS are:

(i) for $\hat{\boldsymbol{\theta}}_1$ $\mathbf{y}'\mathbf{A}_1\mathbf{C}_{11}^{-1}\mathbf{A}_1'\mathbf{y}$,

(ii) for $\hat{\boldsymbol{\beta}}$ in addition $(\mathbf{X}'\mathbf{R}_1'\mathbf{R}_1\mathbf{y})'(\mathbf{X}'\mathbf{R}_1'\mathbf{R}_1\mathbf{X})^{-1}(\mathbf{X}'\mathbf{R}_1'\mathbf{R}_1\mathbf{y})$,

where $\mathbf{R}_1 = \mathbf{I} - \mathbf{A}_1\mathbf{C}_{11}^{-1}\mathbf{A}_1'$.

The new residual SS, S_r', will of course be greater than the original S_r. The difference $S_r' - S_r$ is the SS for fitting the treatment effects $\hat{\boldsymbol{\theta}}_2$, allowing for block effects and covariance effects.

The calculation of sums of squares for any subset of treatment effects adjusted for the covariance effect is performed in the same manner as for the full treatment SS. That is, a design model is fitted omitting the subset of treatment effects to be tested, and any related effects which derive directly from those effects (for example, interaction effects when main effects of a factor are being considered); the covariance effects are then added to the fitted model, and the difference between the two residual sums of squares after using the full and partial design models represents the variation due to the omitted terms, allowing for the covariance effect.

Degrees of freedom for the various fitted SS are calculated in the usual way for a partitioned model. Effectively, this means that the degrees of freedom for any set of design parameters from the original model, $\boldsymbol{\theta}$, will be unchanged when covariance terms are included in the model.

10.3 Covariance analysis for a randomised block design

To demonstrate the ideas of covariance analysis in more detail, we shall consider the algebraic form of the calculations for a randomised block design with a single covariate. The initial design model is

$$y_{ij} = \mu + b_i + t_j + \varepsilon_{ij}$$

and the full model with a covariate, x, is

$$y_{ij} = \mu + b_i + t_j + \beta(x_{ij} - \bar{x}) + \varepsilon_{ij}.$$

As mentioned in the previous section, covariance analysis takes the form of a regression of the residuals of the principal variate, y, on the residuals of the covariate, x. The residuals represent the variation of y, or x, after allowing for the systematic effects of blocks and treatments. The covariance coefficient is calculated like any simple regression coefficient, as the ratio of the sum of products of residuals of x and y to the sum of squares of the residuals of x.

The sum of squares of the residuals of y is the error SS

$$R_{yy}=\sum_{ij} y_{ij}^2-\sum_i Y_{i.}^2/t-\sum_j Y_{.j}^2/b+Y_{..}^2/(bt)$$

and correspondingly the sum of squares of the residuals of x is

$$R_{xx}=\sum_{ij} x_{ij}^2-\sum_i X_{i.}^2/t-\sum_j X_{.j}^2/b+X_{..}^2/(bt).$$

The sum of products of residuals is $\sum_{ij}(y_{ij}-\hat{y}_{ij})(x_{ij}-\hat{x}_{ij})$, which can be written

$$R_{xy}=\sum_{ij} x_{ij}y_{ij}-\sum_i X_{i.}Y_{i.}/t-\sum_j X_{.j}Y_{.j}/b+X_{..}Y_{..}/(bt).$$

The covariance coefficient is

$$\hat{\beta}=R_{xy}/R_{xx},$$

the covariance SS is

$$(R_{xy})^2/R_{xx},$$

and the residual SS is

$$S_r=R_{yy}-R_{xy}^2/R_{xx}.$$

The treatment effects after adjustment for covariance are estimated by the adjusted treatment means:

$$z_j=y_{.j}-\hat{\beta}(x_{.j}-\bar{x})$$

To examine the practical effect of this adjustment consider the diagram shown in Figure 10.2. If, for three treatments A, B and C it is assumed that the covariance relationship has the same slope, then the three parallel lines in Figure 10.2 represent the covariance relationships for the three treatments. The range of values of x is not the same for the different treatments and the three different ranges are indicated by the horizontal spreads of the three lines. If the x values are ignored and the three mean y values are directly compared then the real differences between the treatments will be misrepresented and the treatment means may occur in wrong order. Covariance analysis standardises the mean y yields so that they are compared at a constant x value. Traditionally the constant x value is usually taken as $\bar{x}$, but since the covariance slope is assumed invariant over treatments the comparison of adjusted means is unaltered by any choice of the standard value of x.

Figure 10.2. Covariance relationships for three treatments with different ranges for the x-variable.

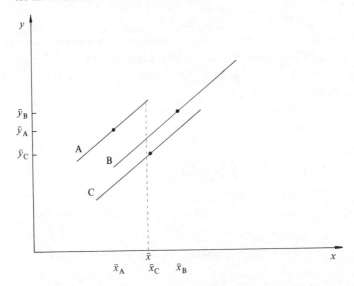

If the covariance slope is positive then yields for treatments with above average x values should be adjusted downwards. In Figure 10.2 treatment B has an unusually high set of x values and would be expected to produce correspondingly higher y values by virtue of the positive covariance whether it is a superior treatment (in terms of y yields) or not.

The difference between two adjusted treatment means is

$$z_j - z_{j'} = (y_{.j} - y_{.j'}) - \hat{\beta}(x_{.j} - x_{.j'}).$$

The variance of this difference is obtained from the variances of its two components. The variance of $(y_{.j} - y_{.j'})$ is $2\sigma^2/b$, and the variance of $\hat{\beta}$ is σ^2/R_{xx}. The covariance between the two components is zero because $\hat{\beta}$ is a linear combination of residuals of y and the linear coefficients (which are residuals of x) sum to zero over blocks for each particular treatment. Hence, the variance of $z_j - z_{j'}$ is

$$\text{Var}(z_j - z_{j'}) = 2\sigma^2/b + \sigma^2(x_{.j} - x_{.j'})^2/R_{xx},$$

where σ^2 is estimated from the residual SS,

$$s^2 = S_r/\{(b-1)(t-1) - 1\}.$$

In practice, this variance, although different for each pair of treatments, will often vary only slightly, and in the interests of simplifying the presentation of results it may be reasonable to present the adjusted treatment means, with a single 'average' variance

$$(2\sigma^2/b)(1 + T_{xx}/R_{xx}),$$

where T_{xx} is the treatment mean square in the analysis of variance for x,

$$T_{xx} = \left\{ \sum_j X^2_{.j}/b - X^2_{..}/(bt) \right\} \Big/ (t-1).$$

To assess the overall significance of the treatment effects allowing for the covariance adjustment, the covariance effect is calculated for the model omitting the treatment effects to obtain the adjusted residual SS omitting treatments, as follows:

$$R'_{yy} = \sum_{ij} y^2_{ij} - \sum_i Y^2_{i.}/t$$

$$R'_{xx} = \sum_{ij} x^2_{ij} - \sum_i X^2_{i.}/t$$

$$R'_{xy} = \sum_{ij} x_{ij} y_{ij} - \sum_i X_{i.}Y_{i.}/t$$

$$S'_r = R'_{yy} - (R'_{xy})^2/R'_{xx}.$$

The adjusted treatment SS, allowing for the effect of covariance, is

$$S'_r - S_r \quad \text{on} \quad (t-1)\,\text{df},$$

and the test of the null hypothesis of no treatment effects is

$$F = \frac{(S'_r - S_r)/(t-1)}{S_r/\{(b-1)(t-1)-1\}}.$$

10.4   Examples of the use of covariance analysis

Example 10.1

For the final time in this book we consider the trial of pain-relieving drugs, previously discussed in Chapters 6 and 8. The data are shown again in Table 10.5.

It seems possible that the level of pain might diminish with time after the operation and that as a result a treatment applied later might be expected

Table 10.5. *Hours of relief from pain for each drug for each patient.*

⎧1st Drug T_1	2	6	4	13	5	8	4	
⎩2nd Drug T_2	10	8	4	0	5	12	4	
⎧1st Drug T_2	2	0	3	3	0			
⎩2nd Drug T_1	8	8	14	11	6			
⎧1st Drug T_1	6	7	6	8	12	4	4	
⎩2nd Drug T_3	6	3	0	11	13	13	14	
⎧1st Drug T_3	6	4	4	0	1	8	2	8
⎩2nd Drug T_1	14	4	13	9	6	12	6	12
⎧1st Drug T_3	12	1	5	2	1	4	6	5
⎩2nd Drug T_2	11	7	12	3	7	5	6	3
⎧1st Drug T_2	0	8	1	4	2	2	1	3
⎩2nd Drug T_3	8	7	10	3	12	0	12	5

to achieve more hours of pain relief. This would mean that the performance of a drug following a less effective drug might appear to be worse than that of the same drug following a more effective drug. A possible model for the results could be

$$y_{ij} = \mu + p_i + o_j + d_{k(ij)} + \beta(x_{ij} - \bar{x}) + \varepsilon_{ij},$$

where p_i, o_j, d_k are patient, order and drug effects and x_{ij} is the hours of relief afforded by the previous drug (x_{ij} will be zero for the first drug for each patient). The residual SS after fitting various models are given in Table 10.6, and the two resulting analyses of variance are given in Table 10.7(a) and (b).

In this instance the use of a covariance term in the model appears to account for most of the apparent differences between drugs. The F statistic

Table 10.6.

Fitting	Residual SS	df
μ	1417	85
μ, p_i	810	43
μ, p_i, o_j	548	42
μ, p_i, o_j, d_k	432	40
$\mu, p_i, o_j, d_k, \beta$	325	39
μ, p_i, o_j, β	351	41

Table 10.7(a).

Source	SS	df	MS
Patients	607	42	
Order	262	1	
Drugs	116	2	
Covariance	107	1	107
Residual	325	39	8.3
Total	1417	85	

(b).

Patients	607	42	
Order	262	1	
Covariance	197	1	
Drugs	26	2	13
Residual	325	39	8.3
Total	1417	85	

to test for differences between the effects of the drugs allowing for the effect of the covariate is 13/8.3 and is clearly unconvincing. However, when the fitted model is examined the fitted value for $\hat{\beta}$ is found to be -0.77, which is clearly not compatible with the assumption made of an increase in the hours of relief when the drug administration is delayed.

The explanation of the results appears to be that because the two drugs which each patient receives are always different the covariance effect is really describing the negative correlation within each pair of drugs. If one

Figure 10.3. Relationship between relief from second drug and relief from first drug for each pair of drugs.

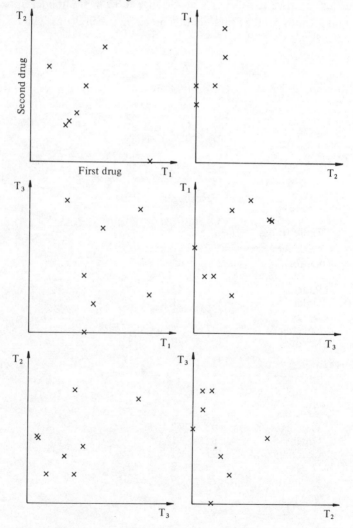

drug is better than average the other will tend to be worse than average. The use of a cross-over trial and the artificial environment which that design imposes on the comparison of drugs will, as mentioned in Section 8.5, produce uninformative results and the negative covariance could be an artifact of the design.

If the relation between the hours relief from the first and second drugs is examined separately for each of the six groups of patients receiving a particular ordered pair of drugs, then the six regression relationships shown in Figure 10.3 are not strong or consistent and there is clearly very little to be gained from any form of covariance. The conclusions from the trial should still be those from the analysis of the first drug administration in Chapter 6, with a small increase in information from the more extensive analysis in Chapter 8.

Example 10.2

An illustration of how covariance analysis can improve the precision of information from an experiment is provided by the strawberry experiment data presented at the beginning of this chapter. The initial analysis of variance of the yield, ignoring the hedge effect, is shown in Table 10.8.

Table 10.8.

	SS	df	MS
Blocks	1.22	3	
Varieties	29.08	7	4.15
Error	42.70	21	2.03
Total	73.00	31	

This suggests that there are large differences between varieties. However, the hedge effect also appears to be substantial, as is shown by the total yields of the four plots at each distance from the hedge:

distance   8   7   6   5   4   3   2   1

total yield 27.8 27.1 23.9 26.2 25.4 21.6 21.5 11.1

The most reasonable simple covariance model would seem to be that the effect of the hedge on yields diminishes as the reciprocal of the distance from the hedge. We therefore try a covariate

$$x = 1/\text{distance}$$

giving eight x values:

0.125, 0.143, 0.167, 0.200, 0.250, 0.333, 0.500, 1.000.

Figure 10.4. Relationship between yield and reciprocal of distance from hedge. Varieties represented by letter, with L for RL and R for Re.

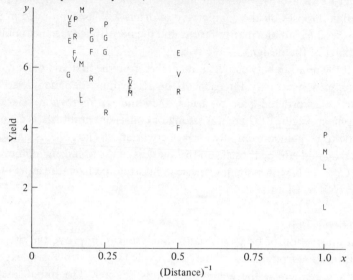

(Distance)$^{-1}$

The graph of individual yields against x shown in Figure 10.4 suggests that the model is a reasonable one.

The block and variety totals for y and x are given in Table 10.9.

Table 10.9.

	y	x
Block 1	43.5	2.718
Block 2	47.2	2.718
Block 3	47.0	2.718
Block 4	47.1	2.718
Variety G	26.2	0.825
Variety V	25.2	1.101
Variety Rl	14.0	2.334
Variety F	22.5	1.176
Variety Re	22.3	1.093
Variety M	22.4	1.667
Variety E	26.2	1.083
Variety P	26.0	1.593
Total	184.8	10.872

To calculate the analysis of covariance, we need

$$\sum_{ij} y_{ij}^2 = 1140.22, \quad \sum_{ij} x_{ij}^2 = 6.1094, \quad \sum_{ij} x_{ij} y_{ij} = 51.974,$$

$$\sum_i Y_{i.}^2/8 = 1068.44, \quad \sum_i X_{i.}^2/8 = 3.6938, \quad \sum_i X_{i.} Y_{i.}/8 = 62.786,$$

$$\sum_j Y_{.j}^2/4 = 1096.30, \quad \sum_j X_{.j}^2/4 = 4.1019, \quad \sum_j X_{.j}Y_{.j}/4 = 60.001,$$

$R_{yy} = 42.70, \quad R_{xx} = 2.0075, \quad R_{xy} = -8.027,$

$\hat{\beta} = -8.027/2.0075 = -4.186,$

covariance SS $= (-8.027)^2/2.0075 = 32.096,$

residual SS $= 10.60.$

The analysis of covariance is shown in Table 10.10, and the adjusted treatment means are calculated as shown in Table 10.11. Note that in Table 10.11 we have adjusted the yields to an x value of 0.125; that is to the level of x at the furthest point from the hedge within the experiment. The predicted yields for an 'average' hedge effect are not appropriate because the hedge has clearly caused a reduction from the 'normal' yield, and the prediction should discount the hedge effect. The complete set of standard errors for comparing the eight varieties is shown in Table 10.12. Clearly the standard errors are very similar, and it is sensible to use the average variance to calculate a common standard error of differences:

$$\sqrt{2(0.53)/4(1 + 0.0583/2.0075)} = 0.52.$$

Table 10.10.

	SS	df	MS
Blocks	1.22	3	
Treatments	29.08	7	
Covariance	32.10	1	32.10
Error	10.60	20	0.53
Total	73.00	31	

Table 10.11.

Variety	$y_{.j}$	$x_{.j}$	$y_{.j} + 4.186(x_{.j} - 0.125)$
G	6.55	0.206	6.89
V	6.30	0.275	6.93
Rl	3.50	0.584	5.42
F	5.62	0.294	6.37
Re	5.58	0.273	6.20
M	5.60	0.417	6.82
E	6.55	0.271	7.16
P	6.50	0.398	7.64
Mean		0.340	

Table 10.12.

	V	Rl	F	Re	M	E	P
G	0.52	0.55	0.52	0.52	0.53	0.52	0.52
V		0.54	0.51	0.51	0.52	0.51	0.52
Re			0.54	0.54	0.52	0.54	0.52
F				0.51	0.52	0.51	0.52
Rl					0.52	0.51	0.52
M						0.52	0.51
E							0.52

This may be compared with the standard error of difference from the analysis without covariance, which is $\sqrt{2(2.03)}/4 = 1.01$. To make an overall test of significance of variety differences, allowing for the hedge covariate effect, we calculate

$$R'_{yy} = 71.78$$
$$R'_{xx} = 2.4156$$
$$R'_{xy} = -10.812$$
$$S'_r = 71.78 - (10.812)^2/2.4156 = 23.39,$$

and the adjusted treatment SS

$$S'_r - S_r = 12.79 \text{ on 7 df,}$$

which gives a second analysis of variance in Table 10.13. The conclusions from the analysis are:

(i) The hedge effect is very substantial, and allowing for it reduces the standard error of estimated differences between variety effects by a factor of $0.522/1.008 = 0.52$.

(ii) The principal difference between varieties is that variety Rl gives the lowest yield, although the hedge adjustment reduces the amount by which Rl yields less than the other varieties. The variation between the remaining varieties can be tested by

Table 10.13.

	SS	df	MS	F
Blocks	1.22	3		
Covariance (ignoring varieties)	48.39	1		
Varieties (allowing for covariance)	12.79	7	1.83	3.45
Error	10.60	20	0.53	
Total	73.00	31		

Table 10.14.

	SS	df	MS	F
Blocks	1.22	3		
Rl v. rest	22.16	1		
Covariance	31.81	1		
Remaining variety variation	7.21	6	1.20	2.26
Error	10.60	20	0.53	
Total	73.00	31		

splitting the variety SS into one component for the comparison between Rl and the rest, and the remaining variation between the other seven. If this is done, and the latter variation assessed, allowing for covariance, the resulting analysis of variance is given in Table 10.14. The analysis suggests that there is still some variation between the other seven varieties, though it is not significant at the 5% level. I believe the best conclusion is that there is a range of yield variation amongst the eight varieties. The most nearly clear-cut difference is the poor yielding of Rl. The best variety is P with an advantage of about one standard error over its nearest rival, E.

(iii) Experimenters should be a lot more careful in choosing blocking structures for their experiments!

10.5 Assumptions and implications of covariance analysis

To examine the additional assumptions required for a covariance analysis, consider the general form of a covariance model:

yield = design model + regression model + error

or, in algebraic notation,

$$y = A\theta + X\beta + \varepsilon.$$

The regression component of the model is clearly assumed to be independent of the design component. In particular, the regression relationship of y on a particular covariate is assumed to have the same slope for all the experimental treatments. This is a very strong assumption, and it is important for the statistician and experimeter to consider the assumption, and to make a positive decision that the assumption is reasonable. Writing the model in the above, simple, form also makes it clear that it would be possible to consider also models in which the regression coefficients were affected by the experimental treatments. Such models, allowing 'interaction' between treatments and covariates, can be fitted with ease using modern statistical computer packages. However, while fitting such models may be useful in checking the assumptions of covariance analysis, the results are of limited value in interpreting comparisons between treatments, which is the original purpose of the experiment.

To appreciate this, consider the situation for three treatments illustrated in Figure 10.5. The three treatments are assumed to have different rates of dependence of yield (y) on the covariate (x) shown by the three different lines. If the treatment means are adjusted to any specified level of the covariate the comparison between the adjusted treatment means depends critically on the level of x which is chosen.

If the parallel line assumption of the covariance model is correct, then comparisons between treatments are independent of the value of the covariate at which the comparison is made. It is, however, important to determine the level of the covariate used for adjustment for each treatment, so that the interpretation of individual adjusted treatment means is practically relevant. Adjustment to the overall mean value of x for the whole experiment is obviously suitable for a covariate such as initial body weight of animals in a growth experiment, where the average initial weight is a typical value that would be relevant in practice. However, as clearly noted for the strawberry experiment, $\bar{x}$ is not always a meaningful level to use in standardising the adjusted yields, and in general the adjustment will be

$$z_j = y_{.j} - \beta(x_{.j} - x_0).$$

for an appropriately chosen x_0.

Figure 10.5. Differing relationships for three treatments.

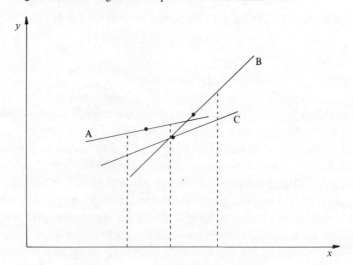

It is sometimes asserted that for a covariance analysis to be valid it is necessary that the covariate be not affected by the experimental treatments. It will often be true that the form of covariate makes it impossible for there to be systematic difference in the value of the covariate *caused* by treatments. However, the model used in covariance analysis does not require that there are no systematic treatment differences for the covariate. There are situations where the covariate is affected by treatments and yet adjustment of treatment comparisons to allow for the variation in the covariate is sensible.

An example occurs in the comparison of onion varieties in a large breeding trial. When the onions are grown from seed, the field germination may fluctuate wildly between different varieties or breeding lines. Although germination may be tested in the laboratory, this does not provide a reliable prediction of field germination, even when the laboratory testing is extensive, which would not be possible for a large variety trial. It is known that onion yields on a per acre basis are substantially affected by crop density. Also, when varieties are selected for large-scale use, their germination rates will be investigated more intensively so that, in future use, any selected variety will be grown at about the optimal density for that variety. In these circumstances, it is essential that varieties in the trial be compared, eliminating the effects of the variation in the achieved field crop density. This comparison may be achieved by using crop density as a covariate in the analysis of the variety trial.

As with any parallel regression model, if the values of the covariates vary too much between treatments, then it is difficult to assess whether the regression slope is consistent for the different treatments. In the context of an experimental comparison of treatments, this means that, if the values of the covariate are very different for different treatments, then it will be difficult to detect treatment differences after allowing for the covariance effect.

10.6 Blocking or covariance

The ideas of covariance analysis appear to offer an alternative to blocking. Since we can adjust comparisons between treatments to allow for variation in a covariate, we can use as many covariates as may be thought appropriate and, using modern computer packages, adjust treatment comparisons simultaneously for all the covariates considered. Why then should we bother with blocking? Is there any reason why we should not simply allocate units to treatments quite arbitrarily, and allow for any disparities between the units in respect of measurable covariates by a multiple covariance adjustment? Are blocking and covariance adjustment simple alternatives?

The arguments of the preceding section about the difficulty of treatment comparison when the values of the covariate differ substantially between treatments suggest that total abandonment of blocking in favour of covariance adjustment could lead to unfortunate patterns of covariate values which could very much reduce the precision of treatment comparisons. Hence, some degree of prior control of treatment allocation would seem desirable, but how much? What should the balance be between blocking and covariance?

In many forms of experiment the primary control of variation is attempted by blocking; adjustment by covariates is introduced to improve precision, usually as an afterthought prompted by some unusual circumstances. There will usually be only one, or occasionally two, covariates, and these will most frequently occur as variables measured after the commencement of the experiment, to record additional information not available or not considered before the experiment. In such experiments, the intention to use covariance adjustments has no implications for the design of the experiment. However, in clinical trials, covariance has come to be regarded as the principal method of controlling variation and this has an important influence on experimental design.

In clinical trials, a great deal of information is available on each individual experimental unit, which of course is an individual patient. Each patient can be classified by age, sex, occupation, various physical characteristics and by factors appropriate to the particular experiment. There are many potential blocking factors, or covariates. Even if the clinical trial is planned with full information about all patients to be included in the trial available at the outset of the trial, then the use of blocking will not be simple. Patients will not occur in equal numbers for

Table 10.15. *Numbers of patients in 16 combinations from four blocking factors.*

		X_1		X_2	
		W_1	W_2	W_1	W_2
Y_1	Z_1	3	5	4	4
	Z_2	6	4	8	1
Y_2	Z_1	3	7	2	5
	Z_2	8	0	3	1

each class of each blocking factor. Typically, the pattern for four blocking factors each with two classes might be as shown in Table 10.15. In most clinical trials, moreover, patients enter the trial sequentially, so that it is impossible to designate a precise blocking structure in advance of the experiment. Methods of sequential allocation intended to avoid extreme forms of non-orthogonality by using restricted, or partially determined, randomisation have been discussed in Section 9.7, but these have not yet been widely accepted. This situation has led some medical statisticians to advocate covariance adjustments as the principal control technique for achieving precision in clinical trials with blocking, or stratification, relegated to a minor, or even non-existent role.

To assess the relative importance of blocking and covariance, we must reconsider the advantages and disadvantages of each.

Blocking is highly effective when the pattern of variation between experimental units is recognisable in terms of a set of distinct groups. If potential variation between units corresponds to qualitative differences between units then blocking is natural. The use of blocking is most advantageous when treatment effects are orthogonal to block effects. The benefits of orthogonality are partly in making the analysis of results easy, though this is relatively unimportant with the statistical computer packages now available, but principally in allowing interpretation of treatment differences to be independent of block effects.

If the blocks are not orthogonal to treatments, then provided the non-orthogonality is not too large, the interpretation of treatment differences is not substantially affected by block differences. Orthogonal designs give independence of interpretation, but impose severe restrictions on block sizes. Mildly non-orthogonal designs allow less restrictive conditions on block sizes, with only a small amount of ambiguity in the interpretation of treatment effects. A further major advantage of blocking or stratification is that it allows the possibility of detecting stratum × treatment interactions when the block size is greater than the number of treatments as could often be the case in clinical trials.

A final advantage of blocking is for the organisation and administration of the experiment. If it is necessary for the experimental units to be assessed on different days or by different assessors the block structure provides an ideal structure for the control of any such management variation.

Covariance allows adjustment for as many covariate factors as are thought appropriate. If the form of the covariance relationship is properly identified then a covariance adjustment for variation between units caused by quantitative differences between units will produce more precise results than the approximation which blocking offers to a continuously varying pattern of yield over the available units. However, if the occurrences of treatments in different classes of a covariate factor are uneven the interpretation of treatment effects will not be independent of the covariate effect. When the unevenness is severe the adjustment of treatment comparisons by the covariate may make the interpretation of treatment comparisons impossible because the precision of comparisons is extremely poor.

Relying solely on covariance adjustment, and allowing totally random allocation of patients to treatments across the whole experiment or even within a few large strata, assumes an avoidance, by chance, of unfortunate allocations. The argument for attempting to avoid unfortunate allocations by design through choosing appropriate blocking factors, with the probability of further adjustment for variation in the blocking factors or other variables by covariance, seems overwhelming.

10.7   Spatial covariance and nearest neighbour analysis

Some agricultural crop experiments, involving large numbers of experimental plots, are performed on large areas of land. Typically, such an experiment may involve a rectangular array of between 60 and 600 plots, the plots being of identical size and shape, the shape being most frequently a long, thin rectangle. Blocking of such an experiment is often viewed as an attempt to insure against unforeseen patterns of variation in the field. There may be some probable directional trend of fertility, which would suggest that blocks should be long in the direction perpendicular to the fertility trend. If blocks are being designed to insure against unforeseen patterns, then two, perpendicular, blocking systems may be used in a row and column design involving incomplete sets of treatments in each row and in each column. The most appropriate designs for this purpose are discussed in Chapter 8.

Because the pattern of spatial yield variation is often largely unknown before the experiment, it might be expected that covariance analysis could be employed to utilise information on yield variation that became available during the experiment. For any specific variable that might be observed during the experiment and thought to be affecting yield, such as pest damage or waterlogging, this can be done in the normal way. However, the information about likely yield patterns may sometimes emerge from the yields themselves. The pattern in the strawberry experiment is an example, though in that case an obvious physical explanation of the pattern was readily available. In many cases, however, although clear yield patterns exist, no direct physical explanation is apparent. The methods described in this section have been developed for agricultural crop experiments. However, there are applications also in laboratory experiments where many samples are tested in moulded plastic sheets containing two-dimensional arrays of containers for sample material.

There are two approaches to using covariance analysis to adjust treatment yields to allow for the underlying but unidentified field yield variation. The first assumes that there is a general two-dimensional response surface of yield which can be represented by a second degree polynomial or two-dimensional Fourier series. The general covariance model discussed in Section 10.2 can allow a response surface model of this kind to be included with the experimental design model.

More recently, there have been developments of models for local spatial variation first suggested over 40 years ago, but largely neglected, possibly because the models require substantial computational facilities. In this form of covariance model, it is assumed that the best information on the level of yield to be expected on a particular plot is contained in the yields from the immediately neighbouring plots. The covariance model (omitting treatment effects) would then take the form

$$y_{ij} = \mu + \beta \sum_{k(ij)} (y_k - \mu) + \varepsilon_{ij}, \tag{10.3}$$

where $k(ij)$ denotes the set of neighbours of y_{ij}.

This form of model cannot be fitted in the usual manner for covariance models because the observed values of the covariate are also values of the principal variate, and all 'dependence' between variables is reciprocal. The model (10.3) is an example of a class of spatial effect models, for which maximum likelihood fitting methods have been developed by Besag (1974) and, for field crop experiments, by Bartlett (1978). If these fitting methods are used when the spatial effect model is the covariance component of an experimental design model, then an iterative procedure is necessary. The covariance effect estimation for the model in Section 10.2 is in terms of the residuals after fitting the design parameters. However, the fitted covariance effects imply patterns of adjacent plot yields and, when these are allowed for, the estimation of treatment effects will change. For example, it may become apparent that an unusually high plot yield for a particular treatment was associated with high yields for the four adjacent plots and, that allowing for those neighbouring yields, the initial yield was much less unusual. This leads to a revised covariance estimate. The general covariance model of Section 10.2 does not give rise to such problems.

This covariance adjustment method is highly effective in some situations (Kempton and Howes, 1981), but it can give ridiculous results if applied inappropriately. These difficulties arise when the experiment contains only two or three replicate plots for each treatment. The iterative method may then be strongly influenced by a small number of pairs of plots, and a form of positive feedback can occur, leading to estimates of the covariance coefficient, and treatment effects which diverge increasingly from sensible values.

Recently there has been a series of important papers developing alternative methods using structured nearest neighbour models for the analysis of experimental data, including Wilkinson *et al.* (1983), Green, Jennison and Seheult (1985) and Besag and Kempton (1986). At the time of writing there is continuing discussion about the merits of different methods and they will not be described in detail here. However, there is a recognisable unity in that all the methods involve the consideration either of first-order differences between adjacent plots, assuming that there is a dominant direction of plot correlation, or of second-order differences. The assumption in all methods is that the pattern of variation of fertility is sufficiently smooth that first- or second-order differences are largely unaffected by the fertility trend effect.

One general implication of these developments is either that blocking is an inappropriate form of control of variation or at least that the block sizes commonly used in field crop experiments may be considerably too large. If conventional blocking designs are to be used then blocks should be small and compact. Nearest neighbour adjustment implies that each pair of adjacent plots should be considered as a block because each neighbour supplies strong evidence of the expected yield level for the neighbouring plot. More specific implications of any of the forms of analysis for nearest neighbour models are that adjacent pairs of treatments should include all possible treatment pairs as nearly equally frequently as is possible. This should improve the precision of comparisons for essentially the same reason that having treatment pairs occur equally frequently together in a block improves precision.

Some of the ideas discussed in Chapter 8, on designs to allow for residual effects, are relevant to the problem of constructing designs so that all pairs of treatments occur adjacently equally frequently. Such designs have been developed for one- or two-dimensional adjacency by various authors, notably by Williams (1952) and Freeman (1979). Examples of neighbour balanced designs using a basic Latin square and an extended Youden square are shown in Figures 10.6 and 10.7.

If only one-dimensional neighbour effects, or trend effects, are deemed important then it is possible to arrange designs in randomised complete blocks with the blocks in sequence so that each treatment has every possible pair of neighbouring treatments. An example given by Dyke and Shelley (1976) which employs nine blocks of four treatments with two end

Figure 10.6. Neighbour-balanced Latin square.

Column

		1	2	3	4	5
Row	1	A	B	C	D	E
	2	B	E	D	A	C
	3	C	D	B	E	A
	4	D	A	E	C	B
	5	E	C	A	B	D

Figure 10.7. Neighbour-balanced Youden square.

Column

		1	2	3	4	5	6	7	8	9	10
Row	1	C	A	B	C	D	A	D	C	B	D
	2	D	B	D	A	C	D	B	A	C	A
	3	B	C	A	D	B	C	A	B	D	C
	4	A	D	C	B	A	B	C	D	A	B

plots is shown in Figure 10.8. This example was originally proposed to investigate the possible transfer effects of treatments from adjacent plots on the yield of a plot. However, the design philosophy is just as appropriate when attempting to allow for smooth trend.

Doubtless much effort and ingenuity will be devoted to identifying the existence of, and devising, designs providing exact neighbour balance for those situations for which such balance is possible. However, in the same way that the construction of incomplete block designs which are as efficient as possible was shown in Chapter 7 to be possible without mathematical theory, so in designs for neighbour effects it is possible to construct designs giving near equality of adjacent occurrence of treatment pairs. Consider the allocation of 12 treatments within a 9×4 array with the intention of using nearest neighbour analysis in the direction in which there are rows of nine plots. Each treatment will occur three times and should have five or six different neighbouring treatments in the occurrences. This is a minimal level of requirement and the range of possible designs is enormous. If we add a further restriction that treatments should

Figure 10.8. Neighbour-balanced design in nine consecutive blocks of four plots.

Block

	I	II	III	IV	V	VI	VII	VIII	IX	
C	BCDA	DBCA	DABC	BACD	CDBA	BDAC	ACBD	BDCA	BADC	B

Figure 10.9. Experimental plan with neighbour effect 'balance' for 12 treatments in four rows × nine rows (a) initial plan, (b) randomised.

(a)
A	B	C	D	E	F	G	H	I
J	K	L	A	H	C	I	E	B
D	G	E	J	I	K	A	F	L
C	J	G	K	B	L	H	D	F

(b)
F	C	J	D	G	H	I	A	L
K	A	F	H	L	E	J	G	B
L	D	B	I	E	A	H	K	C
I	G	E	K	J	F	B	C	D

not be repeated in a row (to be able to allow efficiently for differences between rows) then the design shown in Figure 10.9(a) is quite easily constructed, there still being many points of choice. The randomisation of this design could involve permutation of the four rows, 'rotation' of the nine columns keeping adjacent columns together, and permutation of letters. Such a randomised version is shown in Figure 10.9(b). Further randomisation could be built into the initial construction of the design.

Exercises 10

(1) A trial to compare four varieties of Brussels sprouts was designed as a 4×4 Latin square. Unfortunately, the experimental area was partially waterlogged during the course of the experiment, which produced at least one plot which was clearly not comparable with the other yields. Yields and scores of the extent of waterlogging (recorded for possible future use) are given in Table 10.16. Analyse the data, adjusting yield for the effect of waterlogging by covariance. Present the adjusted mean yields with standard errors.

Table 10.16.

Yields				Waterlogging score			
98(B)	100(D)	127(A)	142(C)	0	0	0	0
141(C)	91(A)	110(D)	124(B)	0	0	0	0
98(D)	102(C)	103(B)	127(A)	0.12	0	0	0
34(A)	71(B)	119(C)	118(D)	0.45	0.09	0	0

(2) A bacteriologist is investigating the effect of various inhibitors on the growth rate of colonies of a species of bacteria. He has four established inhibitors (B, C, D and E) and a new contender, A, which he is convinced on biological grounds should prove more effective. He also knows that temperature has an effect on growth

Table 10.17.

Temp. (°C)	Days										Total Y	Total X
	1		2		3		4		5			
	Y	X	Y	X	Y	X	Y	X	Y	X		
10	B 24	23	D 42	53	C 20	20	A 43	56	E 62	73	191	225
15	A 58	71	B 7	0	E 39	45	C 27	23	D 52	59	183	198
20	D 29	24	E 33	33	A 65	75	B 50	57	C 28	25	205	214
25	E 88	107	C 37	39	B 18	14	D 39	43	A 54	70	236	273
30	C 57	69	A 18	17	D 27	27	E 45	51	B 61	78	208	242
Total	256	298	137	142	169	181	204	230	257	305	1023	1152

rate but does not believe that it affects the action of the inhibitors. On each of five days he prepares five cultures of the bacterium. For each culture the bacteriologist first makes a determination of its potential growth rate (X). Then the culture is treated with one of the inhibitors and left to incubate at a set temperature for eight hours after which a measure of actual growth rate (Y) is made.

Analyse fully the data from this experiment which are given in Table 10.17.

(3) A comparison of herbicide treatments for the control of Ipomea weed in groundnut included 11 forms of herbicide in eight randomised complete blocks. Because of the very variable nature of the weed infestation each plot had, adjacent to the treated plot, a paired control plot which was untreated after the handweeding at the beginning of the experiment. The percentage of Ipomea weed was recorded for all 88 pairs of plots (see Table 10.18), the intention being to analyse the percentage weed cover on the treated plots using the control plot percentage as a covariate. Analyse the data and present the results.

Table 10.18(a). *Treated plots.*

Block	Treatment										
	1	2	3	4	5	6	7	8	9	10	11
I	10.0	5.0	5.0	4.0	0.0	5.5	7.5	12.0	13.0	8.5	16.0
II	12.0	0.0	5.0	5.5	1.0	0.0	9.0	8.0	0.0	19.0	16.0
III	0.5	2.5	2.0	2.0	5.5	4.0	0.0	0.0	0.0	0.0	11.0
IV	5.5	6.0	1.0	1.0	2.5	0.0	10.0	12.0	0.0	1.0	15.5
V	0.0	2.5	0.0	2.5	3.5	2.0	1.5	0.0	0.0	0.0	8.5
VI	1.5	0.0	1.0	1.5	0.0	1.5	0.0	1.0	1.0	1.0	13.0
VII	0.0	0.0	0.0	0.0	0.0	0.0	1.0	7.0	0.5	0.0	0.0
VIII	24.0	0.0	31.0	1.0	0.0	1.0	21.0	0.0	0.5	0.0	0.0

(b). *Paired control plots.*

Block	Treatment										
	1	2	3	4	5	6	7	8	9	10	11
I	25.0	10.5	4.0	8.5	0.0	13.5	25.0	20.5	20.0	22.0	7.0
II	21.5	0.5	11.5	23.5	1.0	2.0	16.0	18.5	0.5	20.0	38.5
III	1.0	4.0	3.0	1.0	20.0	0.0	0.0	1.0	0.0	0.0	17.5
IV	6.0	2.5	16.5	0.0	0.0	2.0	8.5	27.5	0.0	11.0	1.0
V	0.0	10.0	0.0	5.0	15.5	1.5	1.0	3.5	1.0	0.5	5.0
VI	0.0	4.0	3.5	30.0	7.5	7.5	0.0	3.0	6.5	20.5	4.0
VII	8.0	0.0	1.5	0.0	0.0	0.0	0.0	20.0	9.0	2.0	0.0
VIII	18.0	0.0	39.5	10.5	0.0	7.5	17.0	1.0	2.0	4.5	8.0

11

Model assumptions and more general models

11.0   *Preliminary examples*

(*a*) The applied science literature is thickly populated by horrific examples of the thoughtless use of statistics. A very simple example is the following. Four herbicides are compared with a control (untreated), each of the five treatments being applied to four plots, and the number of plants of a particular weed species counted for each plot. The mean counts are presented, with the standard error of a mean 3.8 (12 df).

treatment	A	B	C	D	E(control)
mean count	1.5	3.2	27.0	107.2	153.8

standard error of a mean 3.8 (12 df).

It must be obvious to anyone who thinks about the meaning and basis of calculation of the standard error that these results are not sensible.

(*b*) A more complex situation is represented by data on the number of seeds, from a group of 50, which germinate, for each of 16 treatment conditions. Each treatment had four replicate groups of 50 seeds of *Chenopodium Album*, and the results were as shown in Table 11.1. The first question about the data set in Table 11.1 might be whether there is any need for statistical analysis at all. If statistical analysis is thought useful to add to the qualitative information obviously contained in the data, then should the whole data set be analysed together or only those parts of the data which show more than minimal variation? Should the analysis be of the data as given, or is some transformation needed?

Table 11.1.

Chill	Temperature	Light	Chemical	I	II	III	IV
Unchilled	Constant	Dark	H_2O	0	0	0	2
,,		Dark	KNO_2	2	1	1	0
,,		Light	H_2O	1	0	1	0
,,		Light	KNO_2	5	2	1	1
,,	Alternating	Dark	H_2O	0	0	0	4
,,		Dark	KNO_2	20	25	20	25
,,		Light	H_2O	2	5	6	3
,,		Light	KNO_2	48	50	50	50
Prechilled	Constant	Dark	H_2O	0	0	0	0
,,		Dark	KNO_2	1	3	0	2
,,		Light	H_2O	1	1	2	1
		Light	KNO_2	1	2	1	1
,,	Alternating	Dark	H_2O	2	2	4	1
,,		Dark	KNO_2	13	11	14	12
,,		Light	H_2O	6	3	4	4
,,		Light	KNO_2	45	48	47	47

11.1   The model assumed for general linear model analysis

The simple model we have assumed for the analysis of data from designed experiments is easily written down, but has considerable implications about the form of the data. Thus the randomised block model,

$$y_{ij} = \mu + b_i + t_j + \varepsilon_{ij},$$

with $E(\varepsilon_{ij}) = 0$, $Var(\varepsilon_{ij}) = \sigma^2$, $Cov(\varepsilon_{ij}, \varepsilon_{i'j'}) = 0$, or the randomisation model

$$y_{ik} = \mu + b_i + t_k + \sum_j \delta_{ij}^k \varepsilon_{ij}$$

with $\sum_j \varepsilon_{ij} = 0$ and a probability distribution for δ_{ij}^k each require two major assumptions. First that block effects and treatment effects are additive. Second the variability of observations, represented by ε_{ij} in both models, is not different for different treatments or for different blocks. Neither of these assumptions can be true in all experimental situations, and it would be easy to develop an argument that neither assumption is ever exactly true.

Consider the additivity assumption. This implies that, even if there are substantial differences between blocks, and the purpose of using blocks is to arrange that most of the variation between units does occur between blocks, then the difference between yields for treatments A and B should be about the same in each block. This, in turn, implies that there is no interaction between the treatment factor and those factors which cause the variation between blocks.

The weakness of the variance homogeneity assumption that the variance of observations is consistent over treatments and over blocks is essentially similar. With biological material, it would usually be reasonable that a treatment which produced larger responses from an organism would also lead to considerable variation of that response, whereas a treatment producing a small response might not be expected to produce such large variation. Such simple arguments may not be so immediately applicable in disciplines other than biological sciences, but it must appear reasonable that, where different treatments produce very different mean yields, then it is unlikely that the variance of yields should be similar for the different treatments.

The second stage of the analysis of experimental data often involves the use of t- or F-tests, and the construction of confidence intervals using the t-distribution. These tests or intervals are based on the normal distribution, and so a further assumption is made, that of normality. Even when we appeal to the device of randomisation to avoid the dependence on the normal distribution assumption, t-tests and confidence intervals are still used, with an implicit assumption that the departure from normality is not sufficiently extreme to invalidate the numerical values of the t-distribution.

A further difficulty in analysing experimental data is that some experimental units may not give 'typical' results. This may result in yields which are quantitatively rather different because the chemical constituents of the unit are impure, or the plant material is deformed. In fact, the possible causes of deviant units are very many, including all kinds of abnormal history, and many mechanical errors in measuring or recording information. In extreme cases, there may be major qualitative differences, most often recognised when the cause of the difference is apparent, or there may simply be no recordable yield at all, as when an animal in a drug trial responds not by growing more or less slowly, but by dying! Quantitative deviations are traditionally referred to as outliers; units which give no yield, or for which the yield is qualitatively different, are referred to as missing observations, but there is no clear distinction between the two forms of aberration with many situations providing mixtures of the two.

The first, and most important, point to accept about possible failures of assumptions, outliers and missing values is that there is a substantial subjective element in approaches to these problems. I believe this subjectivity is inevitable and, although there are statistical methods which offer some assistance in making decisions, the methods are not, and cannot be, rules. This situation arises from the nature of statistical analysis, which is based on making assumptions about the provenance of the data, and then comparing statistics calculated from the data with theoretical values calculated on the basis of the assumptions. These assumptions are rarely, if ever, exactly true. When examining evidence for the reasonableness of assumptions, we necessarily make further assumptions to derive test statistics. This may cause the science of statistics to appear impossible to justify but, in practice, the art of statistics can be informative partly because many techniques are not much affected by small perturbations from the assumptions, a characteristic referred to as *robustness*, and partly because an experienced statistician acquires judgement enabling him or her to discern patterns and their strength with confidence.

Enough of this philosophy. In the remainder of the chapter, we examine some methods of examining or testing assumptions; we consider transformations of scale to try to improve the validity of assumptions and a more general approach to fitting models; and we discuss methods to deal with missing observations and how to detect outliers.

11.2 Examining residuals and testing assumptions

There are many methods proposed for testing assumptions. The use of some of these methods is affected by the experimental design, with simpler designs offering more scope for testing assumptions. Thus in a completely randomised design it is possible to calculate the sample variance for each treatment and to examine whether there is a relationship between treatment mean and treatment variance before calculating a complete analysis of variance including pooling the error variance. In contrast, with a randomised block design a simple direct examination of the assumption of homogeneity of variance for all treatments is impossible because each treatment is observed for one unit in each block and the variation of observations for each treatment is partly attributable to block differences. Nonetheless it is often clear from a visual examination of the data that the homogeneity assumption is not satisfied.

It is often asserted that all the information about the fit of the model, or equivalently about the adequacy of the assumptions, is contained in the residuals. This is a truism and it is certainly important to consider the residuals, but it is unfortunately also true that diagnosis, from an examination of the residuals, about the possible failure of assumptions is not a clearly defined procedure. To illustrate the way in which the information from residuals is restricted by their internal relationships we consider the pattern of residuals for a simple randomised block design with five treatments in four blocks. The basic model is

$$y_{ij} = \mu + b_i + t_j + \varepsilon_{ij}$$

and the residuals are

$$r_{ij} = y_{ij} - y_{i.} - y_{.j} + y_{..}.$$

Suppose $\mu = 20$, and block and treatment effects are chosen to be

$$
\begin{aligned}
b_1 &= -2 & t_1 &= +8 \\
b_2 &= +4 & t_2 &= -4 \\
b_3 &= +1 & t_3 &= -5 \\
b_4 &= -3 & t_4 &= -1 \\
& & t_5 &= +2.
\end{aligned}
$$

Then the expected values for the 20 observations, using the usual simple additive model, are given in Table 11.2.

Random effects, ε_{ij}, are obtained using pseudo-random number generation from a normal distribution with $\sigma^2 = 4$, giving a 'typical' set of values (see Table 11.3.). The resulting simulated 'yields' and the corre-

Table 11.2.

Treatment	Block			
	I	II	III	IV
1	26	32	29	25
2	14	20	17	13
3	13	19	16	12
4	17	23	20	16
5	20	26	23	19

Table 11.3.

Random effects	−2.3	+1.0	−0.9	+1.5
	+2.2	+0.3	+3.6	−3.4
	+0.5	+1.4	−1.6	+3.1
	−2.1	−0.4	+1.0	−0.8
	+0.7	−3.5	−2.9	+1.3
Yields	23.7	33.0	28.1	26.5
	16.2	20.3	20.6	9.6
	13.5	20.4	14.4	15.1
	14.9	22.6	21.0	15.2
	20.7	22.5	20.1	20.3
Residuals	−2.0	+1.3	−0.6	+1.3
	+1.6	−0.3	+3.0	−4.5
	−0.2	+0.7	−2.3	+1.9
	−1.4	+0.3	+1.7	−0.6
	+1.9	−2.3	−1.7	+2.0

sponding residuals calculated from these yields are also shown in Table 11.3. This, then, is a typical set of residuals when the assumptions are correct. The pattern induced by the interrelations can be clearly seen. The residuals must sum to zero in each row and each column. Consequently the highest single residual (-4.5) induces other residuals in row 2 and column 4 which are mainly positive with the two largest positive residuals of the whole set occurring in row 2 and in column 4. To assess the information from the residuals the results from this examination of residuals when the assumptions are true may be compared with results from residuals for various sets of data obtained when the assumptions are modified. Six alternative models are considered here.

Three forms of failure of the homogeneity assumption are obtained by
  (i) multiplying the ε's by factors proportional to the expected treatment mean (multiplying factors 1.4, 0.8, 0.75, 0.95, 1.1);
 (ii) as in (i) but with factors proportional to the square of the expected treatment mean (factors 1.96, 0.64, 0.5625, 0.9025, 1.21);
(iii) sorting the 20 ε's into increasing order of absolute size and allocating them in order to the 20 expected values, similarly sorted in order of increasing size (the largest ε, $+3.6$, to the largest expected value, 32).

Three further forms of failure of assumptions.
 (iv) The yields for the 20 combinations are obtained by multiplying block effects (0.9, 1.2, 1.05, 0.85), treatment effects (1.4, 0.8, 0.75, 0.95, 1.15) and random effects $\{(20+\varepsilon)/20\}$. This is a multiplicative model instead of an additive one but for the effects used the results are rather similar, even though treatment effects differ by a factor of almost two.
  (v) The random errors, ε_{ij}, are replaced by $(\varepsilon_{ij}^2-4)/2$. This gives a skew distribution rather different from the normal distribution.
 (vi) The observation for the first block and first treatment is increased by ten. This represents an outlier from the population from which the other observations are derived.

The 'yields' for the six modified models are shown in Table 11.4.

The sets of residuals from the six sets of yields for alternative models are shown in Table 11.5. The strong impression from all the sets of residuals in Table 11.5 and the original set is that they are qualitatively extremely similar. The positions of the high and low residuals change but the underlying pattern is consistent because of the requirements that in each row and each column residuals sum to zero.

A standard method of examining the information in the residuals is to plot the residuals against the fitted values. Such a plot should show no

Table 11.4.

(i)	28.4	33.4	27.7	27.1		(ii)	29.4	34.0	27.2	27.9
	15.8	20.2	19.9	10.3			15.4	20.2	19.3	10.9
	13.4	20.0	14.8	14.3			13.3	19.8	15.1	13.5
	15.0	22.6	20.9	15.2			15.1	22.6	20.9	15.3
	20.8	22.2	19.8	20.4			20.9	21.8	19.5	20.5
(iii)	29.1	35.6	25.5	22.1		(iv)	27.3	35.3	28.1	25.6
	14.7	18.4	18.0	13.5			16.0	19.5	19.8	11.3
	12.6	20.3	15.2	12.3			13.8	19.3	16.6	14.8
	18.0	20.9	21.7	15.1			15.3	22.3	21.0	15.6
	21.5	22.6	25.2	26.4			21.4	22.8	20.7	20.9
(v)	25.4	30.5	27.4	24.1		(vi)	33.7	33.0	28.1	26.5
	14.4	18.0	21.5	16.8			16.2	20.3	20.6	9.6
	11.1	18.0	15.3	14.8			13.5	20.4	14.4	15.1
	17.2	21.1	18.5	14.3			14.9	22.6	21.0	15.2
	18.8	30.1	25.2	17.8			20.7	22.5	20.1	20.3

Table 11.5.

(i)	+0.6	+0.6	−2.0	+0.5		(ii)	+1.1	+0.8	−2.7	+0.8
	+0.6	0.0	+2.8	−3.7			+0.2	+0.1	+2.5	−3.1
	−0.8	+0.8	−1.3	+1.3			−0.8	+0.8	−0.6	+0.6
	−2.0	+0.6	+2.0	−0.6			−2.1	+0.5	+2.1	−0.7
	+1.4	−2.2	−1.5	+2.2			+1.5	−2.5	−1.5	+2.3
(iii)	+1.9	+4.0	−3.6	−2.6		(iv)	−0.2	+2.8	−1.8	−0.7
	−0.5	−1.2	+0.9	+0.8			+0.9	−0.6	+2.3	−2.6
	−1.6	+1.7	−0.9	+0.8			−0.8	−0.3	−0.4	+1.4
	0.0	−1.5	+1.8	−0.4			−1.7	+0.3	+1.6	−0.2
	0.0	−3.3	+1.8	+1.4			+1.5	−2.1	−1.6	+2.2
(v)	+1.2	+0.2	−1.0	−0.3		(vi)	+4.0	−0.6	−2.6	−0.7
	−0.7	−3.2	+2.2	+1.5			+0.1	+0.3	+3.5	−4.0
	−1.1	−0.3	−1.1	+2.4			−1.7	+1.3	−1.8	+2.4
	+2.0	−0.2	−0.9	−1.1			−2.9	+0.9	+2.2	−0.1
	−1.6	+3.6	+0.6	−2.8			+0.4	−1.7	−1.2	+2.5

pattern; in principle any specific pattern implies some form of failure of assumptions and examination of the graph of the residuals should help to diagnose the failure. The graphs for the original data and for data from modified models (ii), (iv) and (v) are shown in Figure 11.1. Patterns which might be perceived include curvilinear trends, systematic trends of vertical spread or non-symmetric vertical distribution. A rapid visual impression suggests to me that the graphs in Figure 11.1(b), (c) and (d) show, if anything, less trend than (a) for which the assumptions are correct. In

Figure 11.1. Plots of residuals against fitted values for (a) data for correct model and (b), (c), (d) models (ii), (iv) and (v).

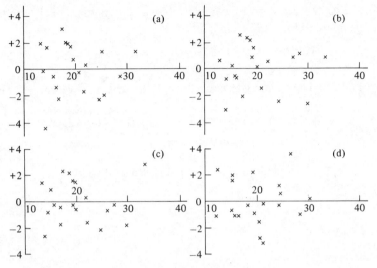

Figure 11.1(a) the single very large negative residual is the source of most apparent patterns. For deviations from assumptions of the degree considered here graphical representation does not help to detect the falsity of the assumptions.

The reader should experiment with more extreme departures from the assumptions to see how far the assumptions have to be wrong before this becomes clearly apparent in the residuals. In general it is difficult to detect failure of assumptions except where the failure is on such a large scale that the invalidity of the assumptions should be immediately apparent from a first look at the data. This might be thought a depressing conclusion but there is a more optimistic view. Since data for which the model assumptions are 'moderately untrue' have the same characteristics as those for which the model is true it would be reasonable to expect conclusions based on the standard methods of analysis to be robust to moderate failure of the assumptions.

What formal methods of testing for failure of specific assumptions are available? For the assumption of additivity a specific test is proposed by Tukey based on the idea that if the additivity assumption fails the most likely pattern of failure is some degree of multiplicative effect – that is 'good' treatments may tend to be even better in high yielding blocks and not quite so good in low yielding blocks. Tukey therefore proposed a single multiplicative term to be added to the model and tested:

$$y_{ij} = \mu + b_i + t_j + \gamma b_i t_j + \varepsilon_{ij}.$$

The test of the hypothesis $\gamma = 0$ provides Tukey's 'one degree of freedom for additivity'.

For the randomised complete block design the Tukey test uses the estimated block and treatment effects

$$\hat{b}_i = y_{i.} - y_{..}$$
$$\hat{t}_j = y_{.j} - y_{..}$$

and considers the regression, through the origin, of y_{ij} on $\hat{b}_i\hat{t}_j$. The SS is

$$\left\{\sum_{ij} y_{ij}(y_{i.} - y_{..})(y_{.j} - y_{..})\right\}^2 \Big/ \left[\left\{\sum_i (y_{i.} - y_{..})^2\right\}\left\{\sum_j (y_{.j} - y_{..})^2\right\}\right].$$

Figure 11.2. Normal probability plots of residuals (r) against expected normal deviate (z) for (a) correct model, (b) non-normal model (v).

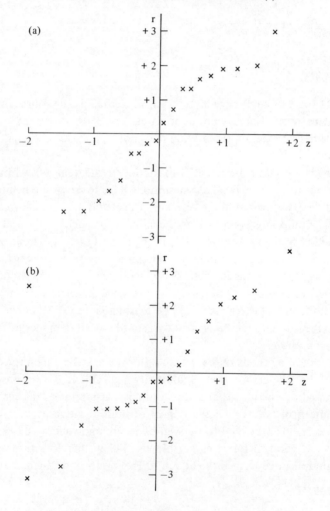

For the original, artificial, set of data for five treatments in four blocks the non-additivity SS is $(4.11)^2/2449 = 0.01$, which is obviously small compared with the error MS of 5.84 and shows no evidence of non-additivity. However, for the data from the multiplicative model (iv) the Tukey SS is 0.63, again small compared with the error MS 3.86.

Another graphical approach to the analysis of residuals, intended to examine the normality assumption, is to plot the set of ordered residuals against expected normal deviates using normal probability paper. The use of normal probability plots for single samples is a standard graphical technique for testing normality. It depends on the ability to distinguish straight lines from curves and therefore is rather subjective, but in the hands (or eyes) of an experienced practitioner graphical plotting is an important statistical technique. The interdependence of the residuals from a block–treatment design reduces the sensitivity of the test but it can still provide useful information. The normal probability plot of the residuals for the original data set is shown in Figure 11.2(a) and shows a nicely linear relationship. A similar plot in Figure 11.2(b) for the non-normal distribution data set (v) does give a hint of non-linearity in the central part of the plot.

11.3 Transformations

Although the formal detection of departures from the assumptions is a difficult area of statistical analysis, it is frequently possible to recognise from consideration of the physical or biological properties of the experimental investigation that the assumptions are quantitatively unreasonable. For data in the form of counts the assumption of a normal distribution should always be regarded with suspicion, not only because counts are discrete whereas the normal distribution assumes a continuous variable, but also because most theoretical model distributions for counts are skewed. Any data for which the mean values for different treatments differ by a factor of three or more must be suspected because it is difficult to believe that substantial changes in mean value can occur without some corresponding change in variability. Similarly if there are substantial differences between blocks then there should be suspicion whether treatment differences can remain constant over blocks.

The indications of situations where the assumption would be treated with suspicion which have been mentioned so far are all data-based. They reflect the experience of statisticians, and such practical experience arising from 'sniffing the data' is an important attribute of the good professional statistician. But more important is a sense, before looking at the data, of those situations where the assumptions are inherently unreasonable. In very many biological investigations the natural inclination of biologists is to think in terms of relative effects rather than additive effects. Most models or descriptions of biological or chemical systems postulate that growth or change is proportional directly, or in a modified form, to present size. Since there will be variation, initially, in the population of experimental units available for an experiment, it is reasonable to expect that, almost always, the variation of measurements observed on the units will be more consistent on a relative scale. It follows that the natural side on which most measurements should be analysed is a log scale.

If we accept that for continuous measurements analysis of log-transformed results is natural then we can restate the question which we have been implying throughout this section. Instead of 'When and how should we transform data to different scales of measurement?' we should ask, 'When is it reasonable to analyse data on anything other than a logged scale?' The answer to the second, more realistic, question is 'Whenever the range of values is such that the pattern of results is almost identical on log-transformed and untransformed scales.' If the range of values is small the log transformation is almost linear and so the analysis will be identical on the original and log-transformed scales. Consider the log transformation function shown in Figure 11.3, and different samples of four values such that the spread of the four values in each sample is similar

Figure 11.3. Log transformation $x = \log_e y$.

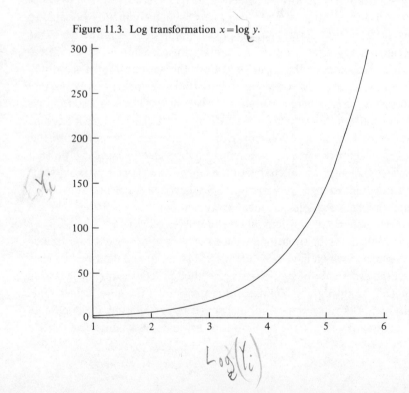

(Y_i)

$Log_e(Y_i)$

Table 11.6.

Sample values on the log scale	Sample values back transformed				Mean	Range
(1) 1.7, 2.0, 2.1, 2.1	5.5	7.3	8.2	8.2	7.3	2.7
(2) 2.1, 2.3, 2.3, 2.6	8.2	10.0	10.0	13.5	10.4	5.3
(3) 2.2, 2.3, 2.6, 2.7	9.0	10.0	13.5	14.9	11.8	5.9
(4) 2.8, 3.1, 3.1, 3.4	16.4	22.2	22.2	30.0	22.8	13.6
(5) 3.7, 4.0, 4.1, 4.2	40.4	54.6	60.3	66.7	55.5	26.3
(6) 3.9, 4.3, 4.3, 4.4	49.4	73.4	73.4	81.5	69.4	32.1
(7) 4.4, 4.4, 4.7, 4.8	81.5	81.5	109.9	121.5	98.6	40.0
(8) 5.1, 5.3, 5.5, 5.7	164.0	200.3	244.7	298.9	227.0	134.9

on the log scale. When the average values of the different samples are close together the spread of values on the unlogged scale is similar as is shown in sets 1, 2 and 3 or in sets 5, 6 and 7 in Table 11.6. However, when sets with markedly different average values are considered, such as sets 2, 4, 6 and 8, then the ranges on the unlogged scales are quite inconsistent.

The statistical moral is that it should be assumed that data for continuous variables should be transformed to a log scale unless there is good reason to believe that this is unnecessary. In that case the normal inclination of the experimenter to wish to analyse the effects of treatments on variables in the form in which they have been recorded or in the form in which the results are most readily interpreted practically should determine the measurement scale for analysis.

We have concentrated so far mainly on the assumption of homogeneity of variance. Does rectification of one assumption imply more or less difficulty with other assumptions? In a simple, and perhaps simplistic, approach to the choice of scale on which to analyse data it is often assumed that there may be a scale of measurement on which all assumptions are simultaneously true. That is, there exists some scale of measurement such that, if observations for a large population of units were measured, the distribution of results for each different treatment would conform to the normal distribution with the same variance for different treatments and consistent treatment differences in different blocks. If this classical statistical philosophy holds then the problem of obtaining a valid statistical analysis is simply that of finding this, unique, scale. On any other scale it follows that all the assumptions will fail.

This unusually optimistic view of statistical life is surprisingly well supported by experience. It is true that for many data sets, for which on the original scale the triple assumptions of homogeneity, normality and additivity all seemed implausible, a simple transformation produced data for which all three assumptions are quite reasonable. Many analyses of data require only a brief consideration of possible transformations, a choice of transformation (usually to logs) and, after a quick visual check that the log-transformed data seems sensible in the context of the observations, a standard analysis and interpretation of data on the transformed scale.

The experimenter who is prepared to expect to use a log transformation for most continuous variables will have few problems. Similarly the regular use of a square root transformation for data in the form of counts will avoid difficulties for the vast majority of data sets. More general families of transformation covering a very wide range of situations are the power transformations discussed by Box and Cox (1964)

$$y^{(\lambda)} = (y^{\lambda} - 1)/\lambda \quad \lambda \neq 0$$
$$= \log(y) \quad \lambda = 0.$$

The major problems of a philosophy of regularly using log or square root transformations for continuous or count measurements are what to do with zero observations and how to present the results. Zeros are the most common manifestation of the more general problem of how to analyse mixtures of quantitative and qualitative data, and this is discussed in Section 6 of this chapter.

Presentation is the biggest obstacle to the acceptance by experimenters of the benefits of the use of transformations. Even when convinced that the assumptions of the analysis of variance and comparisons of treatment means are untrue on the original scale of measurement, the experimenter will wish to discuss results on the scale of original, or practically relevant, measurement. A further requirement of the experimenter is that the values quoted for treatment 'yields' should be the averages of the recorded values on that untransformed scale. Statistically this is clearly wrong. If there is a transformed scale on which all three assumptions are 'correct' then only on that transformed scale are simple comparisons and 'averaging' justified. The appropriate values to represent treatment yields on the original scale are obtained by inverse transformation of treatment mean estimates calculated on the transformed scale. Averaging on the untransformed scale should involve weighting different observations by their (assumed different) precision.

11.4   More general statistical models for analysis of experimental data

Recent developments in statistical computing have allowed the development of a much more powerful and general system of models for analysing experimental data. In the previous section we said that frequently a simple transformation gives transformed data which appear to satisfy all three basic assumptions of the analysis of variance and treatment comparisons. However, not all experimental data sets can be satisfactorily handled by a single transformation. Philosophically there is no reason, apart from innate optimism, to suppose that there is a scale on which all assumptions are simultaneously true. For example, it is not difficult to believe that data could be obtained which are compatible with a model which specifies

(i) that the effects of several treatment factors and blocks can be represented by products of the separate treatment factor and block effects

$$E(y_{ijk}) = \prod_{ijk} a_i b_j c_k,$$

but that

(ii) the population of observations that could be obtained for a particular treatment–block combination might have a normal distribution with variance primarily attributable to characteristics of the experimental material and therefore not related to the treatment or blocking factors,

$$Var(y_{ijk}) = \sigma^2.$$

Equally the reverse alternative of additive treatment and block effects with consistent relative variation provides a credible model. These are simple modifications of the previous form of model. More complex alternative models would allow non-normal distributional assumptions for the error distribution.

A more general model system should therefore permit independent specification of the systematic and random components of the model. Such models have been termed *generalised linear models* and an extended account is given by McCullagh and Nelder (1983). Briefly, it is assumed that the expected value of yield is a function of a linear combination of the parameters representing treatment effects and blocking and covariance effects. The link function relating the expected value of yield, $\mu = E(y)$, to the linear combination of parameters, $\eta = \sum_j a_j \theta_j$, can take many forms. In the general linear model theory previously assumed in this book the link function has been direct equality, $\eta = \mu$. To represent multiplicative effects, the appropriate link function would be $\eta = \log(\mu)$. Other link functions appropriate to particular problems include $\eta = \log\{\mu/(1-\mu)\}$ and $\eta = 1/\mu$.

The random component of the generalised model is defined by the distribution assumed for the yield, y. The general philosophy of using generalised linear models is complicated by the almost inevitable interdependence of the two components of the model, the link between systematic effects and yield parameter, and the distribution of observations. For example, if the error distribution model specifies constant relative variance while the link specifies an additive block–treatment model, then the variance of each observation is a function of the fitted value of the observation and therefore the fitted values must be calculated before the variance. However, the determination of the fitted value requires the assumed variances for the observations. For this example and many others the fitting method involves an iterative scheme. A particularly important extension of the models for the analysis of experimental data is that to allow various distributional assumptions for the yield. For discrete variables it may seem appropriate to use binomial or Poisson distributions. For continuous variables, exponential or gamma distributed errors may be appropriate. The general implementation of generalised linear models should be part of the standard statistical computational facilities available to statisticians and experimenters. The following example illustrates the use of a generalised linear model.

Example 11.1

The data shown in Table 11.7 come from a vaccination study in three different areas, using four batches of vaccine and two methods of vaccination, one of which can be operated with either of two types of needle. The obvious distributional assumption for the observed number of positive responses is that it is binomially distributed. The number tested for each treatment combination varies between 200 and 240. The binomial parameter, p, is also assumed to vary over the different treatment combinations. A suitable model for this variation is constructed by writing the logit of p as a linear combination of the four treatment main effects:

$$\log\{p/(1-p)\} = \mu + a_i + m_j + t_{jk} + v_l.$$

For further discussion of the use of the logit link function see Chapter 4 of McCullagh and Nelder (1983). The set of 12 treatment combinations does not provide orthogonal estimation of all treatment effects. In particular the effects of vaccine 4 and the intradermal method are not at all separable, and the estimation of area effects and the other vaccine effects is not orthogonal. It is natural to fit the needle type effect after the method effect. Because there are only 12 observations it is not realistic to try to include interaction terms. If the terms are fitted in the order in which they occur in the above model the resulting analysis of deviance is given in Table 11.8.

The deviance is a statistic similar conceptually to a residual SS. It is a generalisation of a residual SS and is a multiple of the maximised log likelihood. When different models are compared, the change of deviance may be used to assess the importance of the terms included in one model

Table 11.7.

Area	Method of vaccination	Type of needles	Vaccine batch	Number tested	Number positive	Percentage positive
Staffordshire	Multiple puncture	Fixed	1	228	223	98
	"	Detachable	1	221	210	95
	"	Fixed	2	230	218	95
	Intradermal	—	4	240	238	99
Cardiff	Multiple puncture	Fixed	2	221	181	82
	"	Detachable	2	213	158	74
	"	Fixed	3	200	160	80
	Intradermal	—	4	214	186	87
Sheffield	Multiple puncture	Fixed	1	223	198	85
	"	Detachable	3	228	189	83
	"	Fixed	3	216	177	82
	Intradermal	—	4	224	206	92

Table 11.8.

Fitting order	Deviance (change)	df
Areas	131.6	2
Methods	24.1	1
Needles	3.0	1
Vaccines	0.4	2
Residual	7.5	5

but not the other. However, whereas tests of particular treatment SS require consideration of the SS relative to the residual SS, tests of deviance may be made directly. The asymptotic properties of maximum likelihood estimates enable us to test deviances and changes in deviance by comparing each deviance term with the χ^2 distribution for the appropriate degrees of freedom. These tests are approximate, the quality of the approximation depending on the sample size. With about 200 patients per observation the approximation should be entirely adequate for this example.

The overall fit of the complete model may be tested by comparing the deviance of 7.5 with the χ^2 distribution on 5 df. The conclusion would be that the model fits quite adequately, and the individual contributions to the fitted model may then be examined to assess which of them are important, remembering that if the order of fitting is changed the sets of deviance changes will alter. Clearly the final term in the model, the different vaccines, shows little evidence of any effect. The effect of the needles term is not significant at 5% but is not negligible. If only area and method terms are included the deviance for assessing the fit of the model is 10.9 on 8 df. If areas, methods and needles are included the deviance is 7.9 on 7 df. Either provides evidence of an acceptable fit. There are limited possibilities for considering alternative orders of fitting but the interpretation of results is not materially affected.

The interpretation of the results here is based on the model which includes the needles term in addition to the obviously necessary area and method terms. There are major differences between areas with Staffordshire producing a much higher rate of positive responses than Cardiff or Sheffield. Intradermal vaccination gives a clearly higher rate of positive responses than multiple puncture (this difference could be caused by vaccine 4 being superior to the other three vaccines, but in view of the absence of evidence for variation between vaccines 1, 2 and 3 this seems a less credible interpretation). The fixed needles perform slightly better than

Table 11.9(a). *Fitted effects with standard errors.*

		SE
Staffordshire–Cardiff	1.96	0.21
Staffordshire–Sheffield	1.62	0.20
Intradermal–multiple puncture	0.76	0.17
Fixed–detachable	0.24	0.14

(b). *Fitted logit values.*

	Staffordshire	Cardiff	Sheffield
Multiple puncture			
Fixed	3.32	1.36	1.70
Detachable	3.08	1.12	1.46
Intradermal	4.00	2.04	2.38

(c). *Fitted probabilities.*

	Staffordshire	Cardiff	Sheffield
Multiple puncture			
Fixed	0.965	0.796	0.846
Detachable	0.956	0.754	0.812
Intradermal	0.982	0.885	0.915

the detachable needles. The fitted effects with standard errors are given in Table 11.9, together with the fitted values shown both as logits and as probabilities.

11.5   Missing values and outliers

Many data sets will require modification of the method of analysis to allow for missing observations or observations which are sufficiently discordant to suggest that they provide different information from the remaining observations. Missing observations have, in the past, given rise to much discussion and additional methods of calculation. Much of this is now irrelevant and we shall discuss it only sufficiently to demonstrate the lack of relevance.

The simple modern analytic response to missing observations is to analyse the observations which are available. Even if the original design structure defined an orthogonal analysis structure, the analysis omitting missing observations will be non-orthogonal and consequently it will be necessary to consider the order in which terms are to be fitted and, possibly, the results obtained by different orders of fitting. A nonorthogonal analysis is correct and presents all the available information in the most appropriate manner. There is really no justification for using any other form of analysis, though it is worth considering briefly the philosophy of earlier methods of handling missing values to understand why they should be replaced.

The obvious disadvantage of a missing value is that the pattern of the original design structure is broken. When only a single observation is missing it is algebraically simple to determine an estimate of the missing value such that the residual for the missing/replaced value is zero. The residual SS is expressed as a quadratic function of the missing value x and the numerical observed values. The value of x is then determined which minimises the residual SS. If the analysis of variance is calculated for the original design structure with this substitute x value then the resulting analysis of variance will have the correct residual SS but all other SS will be wrong in the sense that they do not correspond to the SS for effects in any of the possible orders of fitting for the non-orthogonal analysis. This is shown clearly in the example below. The standard errors for comparing treatment means have to be adjusted to take account of the substitution for the missing value. The merit of the missing value replacement method is that it is computationally simple. For a single missing value the distortion of the SS in the analysis of variance is usually not large and the use of the replacement method of analysis will not produce misleading results. It is, however, unnecessary and should not be recommended for general use.

When several values are missing the replacement method may be extended to allow estimation of several values, $x, y, z, \ldots$, to minimise the residual SS. With several replacement values the analysis of variance can become seriously distorted and the potentially ambiguous interpretation should be explicitly recognised through the use of the non-orthogonal analysis of variance considering the possible different orders of fitting terms. The effect of increasing missing values is also illustrated in the following example.

Example 11.2

We consider again a set of data discussed previously in Chapter 10, on yields for four Brussels sprout varieties, from a 4×4 Latin square design in which some plots were affected by waterlogging. The yield data are shown in row and column pattern with the variety letter beside each yield in Figure 11.4. The plot in row 4 column 1 (variety A) was severely waterlogged and the yield is clearly atypical and should be ignored (or adjusted by covariance). The two neighbouring plots in row 3 column 1 and row 4 column 2 were also noted as partially waterlogged and it may be

Figure 11.4. 4 × 4 Latin square design for four Brussels sprout varieties with waterlogging in the bottom left corner.

B	98	D	100	A	127	C	142
C	141	A	91	D	110	B	124
D	97	C	102	B	103	A	127
A	34	B	71	C	119	D	118

appropriate to ignore those also. We shall consider the analyses for (i) the full data, (ii) ignoring the worst affected plot, (iii) ignoring all three affected plots.

The full analysis for all 16 plot yields gives an analysis of variance in Table 11.10. The error mean square, $s^2 = 364$, implies a very high level of random variation, and clearly the full model and analysis is not appropriate.

If the worst affected plot is ignored then the classical method of estimating a replacement value for the observed yield of 34 leads to a replacement value of 103. Using this value in a normal analysis of variance and reducing the error degrees of freedom by one to allow for the estimation of the replacement value gives the analysis of variance as shown in Table 11.11. It is clear that the error variation in this analysis is much more reasonable. However, the sums of squares other than for error are not exactly correct. The non-orthogonal analysis of variance shows how the sums of squares vary according to the order of fitting. Three orders are sufficient to show all the possible sums of squares and the results are shown in Table 11.12. The fluctuations in the sums of squares attributed to treatments, according to the order in which terms are fitted, are not large and the sum of squares calculated for treatments using the replacement

Table 11.10.

Rows	2587	3	862
Columns	3817	3	1272
Treatments	2299	3	766
Error	2183	6	364

Table 11.11.

Rows	581	3	194
Columns	2784	3	928
Treatments	1569	3	523
Error	415	5	83

Table 11.12.

Order	(1)		(2)		(3)	
First	Rows	516	Columns	2780	Treatments	1613
Second	Columns	2787	Treatments	1621	Rows	474
Last	Treatments	1567	Rows	469	Columns	2783

Error = 415.

value method is very close to that obtained when treatments are fitted last. In this case therefore, the use of a single replacement value and the calculation of an approximate sum of squares for treatments is an acceptable procedure in that the conclusions would be very similar to those from the correct analysis. The estimated treatment mean yields are

A	B	C	D
111.9	99.0	126.0	106.2.

The standard error for comparing A with any other variety is 7.44, and the standard error for comparing any other two varieties is 6.44.

When three observations are omitted the effects of non-orthogonality become more noticeable. The estimated replacement values are:

> row 3 column 1 (treatment D) 112
> row 4 column 1 (treatment A) 106
> row 4 column 2 (treatment B) 68.

Note that the replacement value for the third observation rejected as being unreasonably low is in fact less than the rejected value! The analysis of variance calculated using the three replacement values is given in Table 11.13. The non-orthogonal analyses of variance in the three orders is shown in Table 11.14. We can now observe that the sums of squares for rows, columns and treatments vary considerably according to the order of fitting, though the non-orthogonality is not so extreme as to alter the broad pattern of the relative values (the columns SS is always the largest and rows SS the smallest). More important the sums of squares in the approximate analysis based on three replacement values are very different

Table 11.13.

Rows	516	3	172
Columns	2934	3	978
Treatments	1557	3	519
Error	349	3	116

Table 11.14.

Order	(1)		(2)		(3)	
First	Rows	98	Columns	1589	Treatments	710
Second	Column	1561	Treatments	1041	Rows	145
Third	Treatments	1218	Rows	245	Columns	2020

from those from the correct analysis, even if in each case the comparison is made with the sum of squares when each factor is fitted in the last position.

The estimated treatment means using the full non-orthogonal fitted model are

$$\begin{array}{cccc} A & B & C & D \\ 112.8 & 98.2 & 126.0 & 109.9. \end{array}$$

The standard error for comparing A with B or D is 9.6, the standard error for comparing A, B or D with C is 9.0 and the standard error for comparing B with D is 9.5.

The problem of values which are clearly missing is relatively simple. Much more difficult is the problem of dealing with values which appear atypical, and which should possibly be regarded as unrepresentative, or outliers. Methods are available for detecting outliers and the subject is developed in detail in Barnett and Lewis (1978), but there is, inevitably, a strong arbitrary, or subjective element, in all methods. The inevitability follows from the juxtaposition of the reliance of the statistical analysis on distributional assumptions and the admission of the possibility that some of the observations do not follow those assumptions. If some observations do not satisfy the assumption then the conclusions from the analysis will be valid only if those observations are omitted. However, any judgement of whether an individual observation satisfies the assumptions has to be made on the basis of an analysis of the remaining observations, which may still include other deviant observations, the presence of which may mask the first outlier. The assessment of whether a single observation is an outlier, assuming all other observations are valid, depends critically on the extremes of the distribution assumed. Any assumed distribution is likely to be only an approximation and generally the approximation is weakest in the extreme tails of the distribution.

The methods which have been proposed in the literature for detecting outliers are inevitably based on assessment of the largest residual. Tables for testing the significance of a single outlier are given in Barnett and Lewis (1978), who express doubt about the effectiveness of the procedure. A procedure equivalent to testing the largest residual is to examine the residual sum of squares as each observation in turn is omitted from the data.

To demonstrate the difficulties with the procedure of examining the set of residual sums of squares omitting each observation in turn we consider this procedure for two of the simulated data sets used in Section 11.2. For the data from the assumed additive, homogeneous variance model the set of residual SS is shown in Table 11.15. When these are compared with the residual SS of 70.12 for the complete data the observation for block 4 treatment 2 does appear very extreme, although we know from the method of construction of the data set that it is not an outlier.

For data set (vi) in Section 11.2 in which the observation for block I treatment 1 was constructed to be an outlier the residual SS for omitted observations are shown in Table 11.16. In spite of the fact that the ε term for the constructed outlier is twice as large as the non-outlier for block IV treatment 2, in absolute terms the evidence for the existence of an outlier from this second set of residual SS is much less convincing.

The detection and interpretation of outliers is a highly subjective

Table 11.15.

Treatment	Block			
	I	II	III	IV
1	63.52	67.08	69.46	67.44
2	65.53	70.06	54.92	36.67
3	70.05	69.25	60.88	64.45
4	66.90	69.92	65.48	69.46
5	63.88	61.87	65.28	65.49

Table 11.16.

Treatment	Block			
	I	II	III	IV
1	63.52	89.62	78.80	89.44
2	90.28	90.17	69.67	63.92
3	85.42	87.82	84.59	81.16
4	76.40	89.12	82.48	90.29
5	90.01	85.36	87.90	79.95

activity and should be accepted as such. Consequently the statistician should attempt to develop a sound intuition about potential outliers. How can this be done? Undoubtedly the repeated analysis of the same set of data, omitting different values, transforming to different scales, and examining the changes in the pattern of results is a valuable activity. This should now be easily performed on good computer packages. There are still no absolute answers. As observations are omitted or variables transformed to new scales the pattern of results may remain broadly unchanged or there may be sudden changes. Neither invariance nor change provides a recipe for determining whether the original model was right or wrong, whether the omitted observation is an outlier or not. What the multiple analysis provides is information about how much effect a single observation has on the apparent results, and about the dependence of the pattern of results on the scale of measurement. Because the 'right' model is never known we can only examine the alternatives and try to make interpretation with the wealth of information available in the background knowledge of the experimenter. It may sound impossible, but then art usually does.

11.6 The separation of quantitative and qualitative information

One frequent source of difficulty in analysing experimental data is when the variable of interest is quantitative but the observed values include a non-negligible proportion of zeros. Various techniques for modification of zero values, or of 100% values when the observations are proportions, have been devised so that the modified values may be used in a standard form of analysis, usually after some transformation. For example, when a logarithmic transformation is proposed and there are zero values in the data set it is often suggested that

$$\log(x+c)$$

is used, where c is a small number, usually $\leqslant 1$. The choice of c is arbitrary but can have a considerable influence on the pattern of results from the analysis. I believe that this manipulation of data to fit a standard form of analysis is often inappropriate.

The ocurrence of zero values in a quantitative data set frequently indicates not that the observed value is one less than 1, but that no quantitative record is possible. The crucial recognition for such data is that the information is partly quantitative and partly qualitative. In performing any quantitative analysis those treatments giving results showing no response or complete response should be omitted.

Example 11.3

Consider the data set in Example 11.0(*b*). For three of the treatment combinations with a dark environment and the water treatment, all, or almost all, of the replicate sets of 50 seeds failed to produce any germinated seeds. If the replicate variation between these zero values is included in the estimated pooled variation, whatever variance-stabilising transformation or distribution is employed, the resultant variance estimate will be too low and inappropriate for comparing treatment means. Similarly the replicate variation for unchilled seeds at alternating temperatures in the light treated with KNO_2 should not be included in a pooled variance estimate because there is, for three replicates, no quantitative variation. For the seed germination data set we could consider the analysis of a reduced data set (see Table 11.17). The analysis should employ a binomial error structure and a logit link function model including the four main effects and six two-factor interactions.

(i) Fitting main effect terms only, in various orders, produces the changes in deviance shown in Table 11.18 (each on 1 df). The main effects model is not an adequate fit, the model deviance of

Table 11.17.

Chill	Temperature	Light	Chemical	I	II	III	IV
Unchilled	Constant	Dark	KNO_2	2	1	1	0
„	„	Light	KNO_2	5	2	1	1
„	Alternating	Dark	KNO_2	20	25	20	25
„	„	Light	H_2O	2	5	6	3
Chilled	Constant	Dark	KNO_2	1	3	0	2
„	„	Light	H_2O	1	1	2	1
„	„	Light	KNO_2	1	2	1	1
„	Alternating	Dark	H_2O	2	2	4	1
„	„	Dark	KNO_2	13	11	14	12
„	„	Light	H_2O	6	3	4	4
„	„	Light	KNO_2	45	48	47	47

Table 11.18.

Order	(1)		(2)		(3)		(4)	
1	chill	9	chem.	146	light	6	temp.	337
2	temp.	332	temp.	526	chem.	178	chem.	335
3	light	19	light	167	chill	20	light	167
4	chem.	480	chill	0.9	temp.	634	chill	0.9

Deviance for model 106 (39 df)

106 being highly significant. The effects of chemical, temperature and light are clearly large but the evidence for an effect of the chill factor is not convincing.

(ii) Examination of two-factor interaction terms in various orders shows that the interactions between any two of chem., temp. and light give deviance changes ranging between 12 and 34. None of the deviances for two-factor interactions involving the chill factor are significant at 5%.

(iii) The analysis of deviance for the model including main effects and two-factor interactions for temp., chem. and light is shown (in the order of fitting) in Table 11.19.

The overall fit is acceptable, showing not only that the model includes sufficient terms but also that the variation between replicates is compatible with the expected binomial variance. The fitted logit values are shown in Table 11.20. No fitted logit value is quoted for the conditions of constant temperature, H_2O and dark. However, the logit value that would be predicted for the combination is -1.37, corresponding to a proportion of 0.2, which casts some doubt on the validity of the model. The inclusion of a three-factor interaction term in the model might

Table 11.19.

		df
Temperature	337	1
Chemical	335	1
Light	167	1
LC	12	1
TL	20	1
TC	33	1
Residual	43	37

Table 11.20.

Chemical	Light	Constant temperature	Alternating temperature
H_2O	dark	—	-3.06
H_2O	light	-3.66	-2.41
KNO_2	dark	-3.66	-0.62
KNO_2	light	-3.36	$+2.67$

produce a more credible prediction but of course the lack of any useful information about germination under the conditions (constant temperature, dark, H_2O) makes estimation of a three-factor interaction impossible. If the chill effect is assumed to be zero then the six terms in the model are sufficient to ensure that the probabilities corresponding to the fitted logits provide an exact fit to the average observed germination rate for the seven combinations of temp., chem. and light. Note also that the fitted logit values should not be presumed to provide realistic predictions for those treatment combinations which were excluded from the data set.

Have we learnt anything useful from the analysis of this data apart from concluding that germination is virtually impossible for four conditions:

(unchilled, constant temperature, dark H_2O)
(unchilled, constant temperature, light, H_2O)
(unchilled, alternating temperature, dark, H_2O)
(prechilled, constant temperature, dark, H_2O)

and virtually certain for the combination

(unchilled, alternating temperature, light, KNO_2)?

I think that the negative evidence on the effect of pre-chilling seed is of some interest but on balance I believe that the formal statistical analysis, even with the facilities offered by generalised linear models, provides very little useful information. This is an example where statistical quantitative analysis is not really useful because the results are primarily qualitative rather than quantitative. Consequently I would present the results from the experiment in the semi-qualitative form, as in Table 11.21.

Table 11.21.

Light	Chill	Germination percentage			
		Constant temp.		Alternating temp.	
		H_2O	KNO_2	H_2O	KNO_2
Dark	unchilled	Neg.[a]	2–5%	Neg.	45%
Dark	pre-chilled	Neg.	2–5%	2–5%	25%
Light	unchilled	Neg.	2–5%	8%	total
Light	pre-chilled	2–5%	2–5%	8%	93%

[a]Neg. = negligible.

Exercises 11

(1) Two insecticides, A and B, have been developed for controlling boll weevils when growing cotton. It is thought that the insecticides might be most effective when used together. Four powders were prepared, the first containing no insecticide, the second containing A, the third B, and the fourth with equal quantities of A and B.

A field trial of the four powders was carried out using a randomised block design. Each block was divided into eight plots, so that each powder was sprayed on two of the plots within each block, the same amount of powder being sprayed on all 32 plots used in the trial. One-hundred buds were examined from each plot, selected from the centre rows of each plot, and the numbers showing attack by boll weevils are shown in Table 11.22. Consider whether any transformation is required for analysis of the data.

Table 11.22.

Block	Insecticides							
	None		A		B		A and B	
I	9	13	5	16	6	4	3	6
II	16	11	8	14	12	7	4	5
III	33	20	17	15	13	18	13	7
IV	15	13	5	6	10	6	7	4

(2) Groups of 20 chicks of a commercial broiler strain were fed one of four diets:

A based on maize
B based on wheat
C as for A with the addition of 0.1% crystalline copper sulphate
D as for B with the addition of 0.1% crystalline copper sulphate.

The experiment was designed as two Latin squares with two brooders per Latin square, each brooder containing two columns of compartments with four tiers per column. The 32 groups of chicks were allocated randomly to the 32 compartments. Average live weights of the chicks were recorded one and two weeks after birth (see Table 11.23).

The basic analysis of weights or of weight changes (weight at

Table 11.23.

Column	Tier	Diet	Average liveweight (g) for groups of 20 chicks	
			11 Oct. (1 wk)	18 Oct. (2 wks)
1	1	B	93	188
1	2	A	97	182
1	3	C	95	183
1	4	D	106	220
2	1	D	101	179
2	2	B	99	191
2	3	A	93	164
2	4	C	96	185
3	1	A	99	193
3	2	C	93	175
3	3	D	96	194
3	4	B	93	193
4	1	C	92	176
4	2	D	111	228
4	3	B	93	196
4	4	A	98	185
5	1	C	94	176
5	2	A	96	179
5	3	B	97	202
5	4	D	107	227
6	1	A	92	170
6	2	D	102	206
6	3	C	99	172
6	4	B	95	193
7	1	B	93	203
7	2	C	93	176
7	3	D	108	223
7	4	A	103	201
8	1	D	98	204
8	2	B	93	188
8	3	A	95	186
8	4	C	98	188

birth can be taken to be 40 g) is straightforward. However, when the October 18th weights for the two squares are analysed separately and the analyses compared there seems to be a reasonable suspicion that all is not well. Could one of the yields be wrongly recorded?

(3) An experiment to investigate the propensity for water organisms to attach themselves to metal plates of different colours when the plates remain for a long time in the water included plates of seven colours with two replicate plates per colour. The numbers of *Eliminumus modestus* found attached to the plates in each of seven successive months are recorded in Table 11.24. What statistical analysis is suitable?

Table 11.24.

	May	June	July	Aug.	Sept.	Oct.	Nov.
Black A	1	0	6	335	47	0	4
Black B	2	0	49	576	42	1	2
Dark grey A	0	0	4	6	4	0	0
Dark grey B	1	0	1	13	7	0	0
Light grey A	0	0	2	8	0	0	0
Light grey B	0	0	5	20	1	0	0
White A	0	0	0	0	0	0	0
White B	0	0	0	2	0	0	0
Red A	0	0	69	68	300	0	1
Red B	3	0	21	413	44	0	0
Blue A	0	0	52	72	17	0	4
Blue B	0	0	34	223	36	0	1
Yellow A	0	0	0	5	0	0	0
Yellow B	0	0	34	0	0	0	0

PART III

SECOND SUBJECT

12

Experimental objectives, treatments and treatment structures

12.0   *Preliminary examples*

(a) An experiment is proposed to examine the effects of water stress on plant growth and development. It is already determined that the experiment will include plants of three varieties sown at two different sowing dates. The experimental unit is a single plant in a pot. About 50 plants for each variety/sowing date combination are available. The remaining design question concerns how to define the set of treatments for assessing the effects of water stress. Periods when water stress can be applied can be defined in terms of calendar date or of physiological stage of the plants. Several stress periods can be used–should the number be two, three, four or even more? For a given number of stress periods the number of experimental treatments remains to be chosen. Thus, for example, with three stress periods (A, B, C), there are eight possible experimental treatments as in Table 12.1.

From this complete set of possible experimental treatments, sensible choices of the set of treatments to be used in the

Table 12.1.

	A	B	C
(1)	stress	stress	stress
(2)	stress	no stress	no stress
(3)	no stress	stress	no stress
(4)	no stress	no stress	stress
(5)	stress	stress	no stress
(6)	stress	no stress	stress
(7)	no stress	stress	stress
(8)	no stress	no stress	no stress

experiment could be:

(i) (1), (2), (3), (4) and (8), or

(ii) (2), (3), (4) and (8), or

(iii) (1), (5), (6), (7) and (8), or

(iv) all 8, or

(v) (8), (2), (3), (5) and (1)

or possibly other subsets.

(*b*) An experiment to compare different methods of control of an agricultural crop pest has 240 plants in individual pots available as experimental units. Two chemicals, O and E, are to be compared with a standard control treatment, and with an untreated control. O is an oil which requires a surfactant for application and two surfactants, S_1 and S_2, are to be compared in the experiment. Three forms of sprayer (Ed, Ul and Con) are available for comparison, but only Ed and Ul can be used with O and only Ul and Con with E. The experimenter would like to · compare three concentration rates for each of the two new chemicals, O and E, but a single rate is sufficient for the treated control. What is the treatment structure and what replication is appropriate for each treatment combination?

12.1   Different categories of treatment

We have already mentioned, in Chapter 3, that there are many different kinds of experimental objectives, and that the choice of treatments to be used in an experiment requires careful consideration. Although it might seem that the statistician, in attempting to emphasise the importance of relating the choice of experimental treatments to the objective of the experiment, is providing quite superfluous advice, practical experience suggests otherwise. Many experiments are conducted without any precise description of the purpose of the experiment; the treatments are chosen first and the questions that can be answered by comparison of the effects of those treatments are allowed to emerge. As one paper in a reputable journal put it (in 1979), 'The purpose of this experiment is ... to compare the experimental treatments.'

There are many different questions that can be asked before an experiment, some of which are:

(i) Which is the best method (within a defined set of options) of growing a particular crop, or of operating a chemical plant?

(ii) What is the pattern of yield response to a stimulus whose level can be varied?

(iii) Which are the main causes of variation in the yield, or output, from a biological or chemical process?

(iv) What are the advantages, including perhaps the economic advantage, of a new form of medical treatment?

(v) To what extent are the effects of different levels of one treatment factor dependent on the particular level of another treatment factor?

(vi) Which, if any, of a large array of chemicals have any effect on a disease.

Each form of question implies a rather different form of experimental treatments. In addition, there are three major distinctions which should be drawn between types of treatment or of treatment structures which will be discussed further in later sections. These are:

(i) the distinction between treatment factors whose levels are qualitative (different diets, different breeds of animal), and those which are quantitative (drug concentration, diet amount);

(ii) the distinction between treatments which are a representative sample from a population (varieties from a breeding programme), and those which are selected individually because they specifically are of interest. The latter are sometimes referred to as fixed, in contrast to the former, which are referred to as random;

(iii) the distinction between factorial structure for treatments (introduced in Chapter 3) and treatments with a lesser degree of structure or even with no deliberate structure. Sometimes the treatments may be simply a collection of alternatives with no apparent structure. Intermediate structures could involve partial factorial or grouped structures.

Regardless of the precise nature of the treatments, it is important to emphasise that the practical operation of the experiment, including randomisation and the initial form of the analysis, the analysis of variance, is exactly the same for all these distinctions of treatment types. The initial analysis is determined only by the design structure of the experimental units and of the allocation of treatments to those units. The ways in which subsequent analysis will be modified for each of the three distinctions in turn will be discussed in Sections 12.4, 12.5 and 12.6, with consideration in Section 12.7 of the special problems of screening and selection. First, we consider the general problems of interpreting comparisons between treatments, and the presentation of results.

12.2   Comparisons between treatments

The theory of linear contrasts, including the concept of orthogonal contrasts, was developed in Chapter 4. We now consider the use of orthogonal contrasts and other comparisons in interpreting the effects of treatments from experiments.

The classical statistical approach to the interpretation of experimental results advocates the specification, before the experiment is started, of a set of p orthogonal comparisons, L_1 to L_p. The orthogonality enables each comparison to be interpreted independently, even though the tests of significance of the comparisons are not independent because of the use of a common estimate of error variance, s^2. However, although the comparisons are orthogonal, the fact that several significance tests are made simultaneously, and the probabilistic nature of the significance test, leads to consideration of the probability of a false significance, or type I error, over the whole set of comparisons. If we assume that s^2 is approximately equal to σ^2, then for p orthogonal comparisons,

prob($\geq$ one significant result in the situation where all null hypotheses are true) $= 1 - (1 - \alpha)^p \simeq p\alpha$,

where α, the type I error probability of each test, is assumed small (e.g. 0.05 or 0.01). The overall probability of a type I error is then

$$1 - (1 - \alpha)^p,$$

which will be much larger than α if p is substantial.

One approach to this problem has been to define an experimentwise error rate as the overall probability of a type I error if all the null hypotheses are true. If this experimentwise error rate is to be a prescribed level, α, then the type I error probability for each orthogonal comparison test, α', must satisfy

$$1 - (1 - \alpha')^p = \alpha,$$

or

$$\alpha' = 1 - (1 - \alpha)^{1/p}.$$

The classical approach to interpreting the results of experiments has another, practical, difficulty, and this is that the appearance of the results may suggest additional questions beyond the specified p orthogonal comparisons. The classical response to this situation is to suggest a further experiment to test the hypothesis generated by these new questions. This is a theoretical approach to research which the shortage of resources will rarely permit and, in attempting to provide assistance to the justifiable desire of experimenters to interpret their results using information suggested by those results, statisticians have developed a wide range of *multiple comparison methods* or *automatic testing methods*.

Although each of the methods for multiple comparisons was developed for a particular, usually very limited, situation, in practice these methods are used very widely with no apparent thought as to their appropriateness. For many experimenters, and even for editors of journals, they have become automatic in the less desirable sense of being used as a substitute for thought. The scale of the misuse and the narrowness of the situations for which each method is appropriate leads one to agree with Nelder (then the Head of the Statistics Department at Rothamsted), who said in the discussion of a review paper by O'Neill and Wetherill (1971) on multiple comparison methods that 'In my view multiple comparison methods have no place at all in the interpretation of data'. I recommend strongly that multiple comparison methods be avoided unless, after some thought and identifying the situation for which the test you are considering was proposed, you decide that the method is exactly appropriate.

To illustrate why I make this recommendation, I shall consider one particular test as an example. The test is Duncan's test, and it was originally proposed for interpretation of a set of treatments which have no structure or pattern; the treatments can be identified simply as A, B, C, ... , where even the ordering of the letters has no implication of ordering for the corresponding treatments. Duncan's test considers the ordered set of treatment mean yields, and tests each group of two, three, four, or more, successive means to assess whether the spread between the highest and lowest of each group is significantly large in terms of the experimental error mean square, allowing for the number of treatments in the group being tested. The result of using the test is the identification of, possibly overlapping, groups of treatments, such that all the treatments in each group might reasonably be presumed not to differ in their effect on yields.

Examples where Duncan's test is inappropriately applied are many, and occur in almost every volume of every applied science journal in which any statistical analysis is used except where they are replaced by an alternative test of the same general intention. The unthinking use of tests such as Duncan's has been attacked in the very journals where examples of such use occur frequently, by various statisticians including Bryan-Jones and Finney (1983), Maindonald and Cox (1984), Morse and Thompson (1981). However, these tests are embedded in many computer packages and it is important to continue to point out why they are so misleading and inappropriate. A few examples are given here to illustrate some of the forms of misuse.

The simple example shown in Table 12.2 is taken from an experiment to investigate the relationship between potassium level and growth of plants. Any attempt, such as was made by the authors of the paper from which Table 12.2 was taken, to interpret differences between treatments in terms of the Duncan's test information rapidly leads to confusion. For foliage the potassium levels from 1.0 to 9.0 are not significantly different. Level 9.0 is significantly different from 0.5 and lower levels but levels 1.0, 1.5, 3.0 and

Table 12.2. *Results using Duncan's multiple range test.*

Substrate potassium concentration (mM/l)	Foliage dry weight	Roots dry weight
0	4.47d	0.67e
0.25	18.74c	4.51d
0.50	26.40bc	6.30c
1.00	27.34ab	7.48bc
1.50	31.87ab	9.07ab
3.00	32.31ab	8.91ab
6.00	31.68ab	8.46ab
9.00	34.85a	9.39a

6.0 are not significantly different from 0.5. The foliage weight for level 0.25 is also not significantly different from that for level 0.5 but is different from the weights of all higher levels. Fortunately everything is different from level zero. This last statement is true also for root weights but all the other statements have to be modified (slightly).

The propagation of such nonsense may be described as unfortunate or in stronger terms. However, the real wickedness of the use of Duncan's test in this situation, as in many others, is that the true interpretation of the data was submerged in the swamp of significance statements. The true interpretation is obtained from a graph of the results as shown in Figure 12.1. There is a clear asymptotic response pattern for increasing potassium concentration, both for foliage and root dry weights. The superimposed plots also demonstrate the rather more gradual nature of the approach to the asymptote for root dry weight.

An even more impressive piece of misrepresentation is shown in the next example in which a factorial structure for two components of the diet of chicks requires 12 letters to summarise the results of Duncan's test for 24 treatments, shown in Table 12.3.

Table 12.3. *Factorial means with Duncan's multiple range test.*

Methionine	Sodium sulphate				
	0%	0.08%	0.16%	0.24%	0.32%
0%		53a	54a	56c	56a
0.08%	114b	147d	141cd	145d	119bc
0.16%	170e	216fg	217fg	223fg	208f
0.24%	235gh	266ij	266ij	281jk	264i
0.32%	249hi	288jk	300kl	309l	304kl

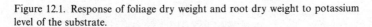

Figure 12.1. Response of foliage dry weight and root dry weight to potassium level of the substrate.

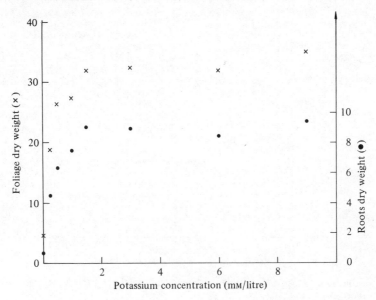

Again graphical representation makes everything clear, which certainly cannot be said for the poor authors. In Figure 12.2 we can see that the effect of methionine is very consistent, for all levels of sodium sulphate. The response pattern is rather weaker for the zero sodium sulphate level and is slightly reduced also for the 0.32% sodium sulphate. There is a slight suggestion that the effect of methionine might continue to increase further at this level of sodium sulphate. Further insight might be gained by fitting response functions. Some might argue that if the obtrusive letters are removed from Table 12.3 the results are clear enough, though I prefer the graphical representation.

Our third example illustrates an alternative form of presentation of the results of a Duncan's test in which the treatments are ordered so that the means increase consistently across the page. The effects of performing this illusion for measurements at four different dates, for the five treatment levels being compared, are shown in Table 12.4. In this case the attempt to unravel the mystery will be left to the reader, though it should be pointed out that for day 25 the author of the results has confused himself into making two incompatible significance statements.

The indictment of multiple comparison methods is on three grounds:

(i) they are regularly used in inappropriate situations,

Figure 12.2. Response of chick growth to methionine for different levels of sodium sulphate 0% (0), 0.08% (1), 0.16% (2), 0.24% (3), 0.32% (4).

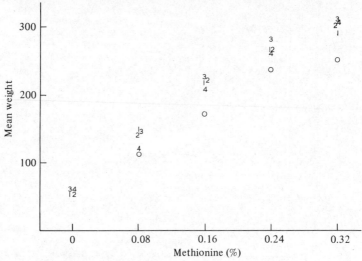

Table 12.4. *Liver/body weight × 100 for five levels on each of four measurement dates.*

Day	Doses and yields						Overall significance of treatments (F test)
11	dose	0.8	0.4	0.0	1.6	3.2	
	yield	2.26	2.29	2.36	2.42	3.19	$p \leqslant 0.01$
18	dose	0.8	0.4	1.6	0.0	3.2	
	yield	2.35	2.87	3.10	3.23	4.03	$p < 0.01$
25	dose	0.8	3.2	0.4	1.6	0.0	
	yield	2.66	2.72	2.76	2.80	2.85	$p < 0.01$
32	dose	0.8	1.6	3.2	0.4	0.0	
	yield	1.85	2.18	2.38	2.50	2.50	$p < 0.05$

Treatment means not significantly different at 5% using Duncan's test are indicated by a common underline.

    (ii) they are used to test hypotheses known to be false, on the evidence of the overall treatment F test, and

    (iii) most importantly, they are used in a routine fashion, and thereby divert many experimenters from a proper analysis of their data.

Even the simple idea of experimentwise error rates is difficult to justify in its simplest form. The experimentwise error rate depends on the idea that it is useful to consider the situation where all null hypotheses are true. However, not only is this situation improbable because experimenters rarely investigate, experimentally, situations where it is credible that all treatments have identical effects, but, in practice, an experimenter only looks at his treatment mean yields if the overall treatment F ratio gives some indication that there are some genuine differences between treatment mean yields. What might be practically useful is some guide to the real significance of apparent effects, given the acceptance that some effects are genuine and that some other effects are negligible. But it would be impossibly complex to devise a general procedure for such situations.

I recommend therefore that multiple comparison methods be avoided; that the idea of experimentwise error rates be retained, but only as a general principle, helpful in the practical interpretation of data which involves the use of significance tests for several orthogonal comparisons.

12.3   Presentation of results

In the later sections of this chapter, we consider some particular approaches to the presentation of results, but it is useful first to consider some general principles of presentation. I believe the first principle should be that the reader of the presented results should be given information in a form which makes it easy for him to make his own interpretation of the results. This means, for example, that the mean yields for all *relevant* treatment comparisons should usually be provided, together with standard errors for the comparison of treatment means and degrees of freedom for the standard errors. The alternative approach of quoting least significance differences or equivalently the smallest difference which, if tested, would be found to be significant at a particular level, reduces the information available to the reader by forcing him to use only the particular significance level chosen by the experimenter, and restricts any attempt to extend the comparisons beyond a simple comparison of two means.

Sometimes, the need for a statistical analysis to provide a concise summary of information will result in the presentation of only a subset of means or possibly of means for groups of treatments. It is still important to present standard errors and their degrees of freedom, but it is also then desirable to provide a summary of the pattern of between treatment variation, and this will usually be most effective in the form of an analysis of variance. Such a summary should at least quote mean squares and degrees of freedom. The practice of quoting only p-values is indefensible.

Presentation in the form of diagrams is often effective, but it is extremely easy to mislead either by design or inadvertently. Figures should not be used when the same information can be conveyed in tabular form, because the tabulated information is usually more precise. The principal benefit of diagrams is undoubtedly for displaying quantitative relationships. In addition to the greater precision of tabulation, it is also much easier to provide information on precision for means in tables than in figures. The use of vertical bars to indicate standard errors in a diagrammatic representation is fraught with possibilities of confusion. In many cases, I believe that the wish to include precision indicators in a diagrammatical representation should suggest that tabular presentation is a superior alternative.

One practice which should be avoided is the expression of results for all treatments as percentages of the result for a single treatment, this being usually some form of control treatment. Although from one viewpoint this is only a change of scale, the fact that the divisor is a variable makes statements of precision much more difficult. In particular, if standard errors are scaled by the same factor that is used for the treatment means, the resulting standard errors will be underestimated so that differences between means will be interpreted as more extremely significant than they should be. A further implication of presenting mean yields relative to a control mean yield is that the yields should be considered on a relative basis which is best achieved by transforming yields to a logarithmic scale.

The last general point about presentation must be a reminder that interpretation must match the objectives of the experiment. There is no such thing as a correct standard form of presentation for experimental data.

12.4 Qualitative or quantitative factors

The distinction between qualitative and quantitative factors is usually quite clear. In using levels of a quantitative factor, we are concerned with the relationship of yield with the varying levels of the factor. The interest in the relationship may be direct – we may wish to establish the form of relationship – or it may be indirect. An example of an indirect use of a relationship is an attempt to identify the optimum level of the factor, for which it is necessary to make some assumption about the form of the relationship, at least in the neighbourhood of the optimum. The fact that there is a numerical component to the definition of a factor level does not necessarily make the factor quantitative. If we are considering various additions to a chemical process, the amount of the additive must be specified but, if the only two alternatives considered for this additive are (*a*) none, or (*b*) some specified amount, then the additive is essentially a qualitative factor.

Most results for qualitative factors will be presented simply as mean yields, with standard errors. When there are only two levels then presentation will often be more effective in the form of differences, particularly where there are other factor levels or treatments. For example, in the experiment on water uptake of amphibians, discussed in Chapter 3, when the analysis of variance had demonstrated which main effects and interactions of the three qualitative factors were important, the interpretation was based primarily on differences between the two levels of the hormone factor, or on differences of these differences.

With more than two qualitative treatments there will usually be a structure for interpretation implied by the choice of treatments – otherwise we would have to deduce that the experimenter did not have any reason for selecting his treatments! Comparisons may be in the orthogonal form discussed in Section 4.7, but this may not always be appropriate, the most frequent example being when a control treatment is used. A more subtle example is shown in the following.

Example 12.1

An experiment to compare the effects of four diets on the growth of young chicks consisted of 32 cages each containing 15 chicks arranged in eight columns (stacks) of four tiers each. The treatments were:

C: a control diet including 0.8% lysine,

L: a lysine enriched diet, being $C + 0.2\%$ of lysine,

W: a wheat diet in which some starch and sugar of diet C is replaced by wheat to produce an extra 0.2% lysine,

A: an amino acid diet in which some other amino acids present in W but not in C or L are added to L.

To understand the interpretation of the experimental results it is necessary to consider the background to the choice of experimental treatments. Previous research had shown that the addition of lysine to the diet of young chicks enhanced their growth and that there was an apparently linear relationship between growth response and the intake of lysine. A diet in which wheat was included to provide the desired additional lysine had been found not to promote growth as effectively as the equivalent lysine-enriched diet. The present experiment was intended to provide information on why the wheat diet was not as effective as expected.

Two possible explanations were to be investigated. First, the wheat diet might include other amino acids which are deleterious to growth. To investigate this explanation the fourth treatment (A) is included. This adds to the lysine diet some of the amino acids present in wheat, which it was thought might cause the relative failure of the wheat diet. The second idea was that the defect of the wheat diet might be that chicks do not eat so much of it as of the lysine diet and so in practice do not get the intended amount of lysine. It is not possible to monitor the amount of each diet eaten by individual birds or by the set of birds in each individual cage, but the total intake for each diet can be recorded. The lysine intake per chick can be estimated for each of the four diets and these intake figures can be used in the assessment of the treatment effects. The design and mean weight data are shown in Table 12.5(a). The results of the initial analysis of variance are shown in Table 12.5(b).

The treatment means are

C: 166

L: 240

W: 190

A: 218   SE of difference = 5.9.

The control diet, C, is not the primary point of reference, being included merely as an origin. Both W and A might be expected to have effects equivalent to L and so basic comparisons, $L-W$, and $L-A$, are important. A further comparison of interest is to examine whether the reduced benefit of the W and A treatments can be explained by variation in the food intake.

Figure 12.3 shows the treatment means plotted against lysine intake

Table 12.5(a).

Tier	Column							
	1	2	3	4	5	6	7	8
1	161C	246L	186W	229A	235L	164C	196A	182W
2	228L	199W	251A	170C	186W	233A	171C	235L
3	224A	169C	241L	183W	159C	191W	251L	175A
4	190W	210A	175C	247L	222A	236L	201W	158C

(b).

Source	SS	df	MS
Brooders	1756	7	251
Tiers	943	3	314
Diets	22473	3	7491
Error	2553	18	142
Total	27725	31	

Figure 12.3. Mean weight of chicks plotted against lysine intake for the four experimental treatments.

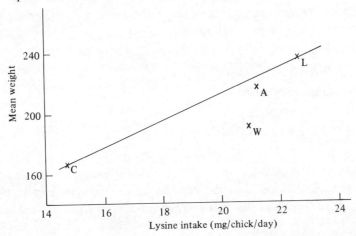

(mg/chick/day) and provides the basis for testing whether the reduced performance of the W diet is attributable to the reduced intake of food (and hence of lysine). Assuming, as previous results show is reasonable, that the increased performance due to lysine is linearly related to actual intake of lysine, the expected mean yield for either the W or A treatments would be given by the value on the straight line joining points C and L in Figure 12.3, at the corresponding food intake value.

The observed intakes are

> C: 14.7
> L: 22.6
> W: 20.9
> A: 21.2.

The additional food intake for W compared with C is 0.78 of the food intake increase from C to L. The comparison of observed and expected performance for W is therefore

$$W - [C + 0.78(L-C)]$$

or

$$W - 0.22C - 0.78L$$

for which the estimated effect is

$$190 - 0.22(166) - 0.78(240) = -33.7.$$

The standard error of this estimate is

$$\{(s^2/8)(1 + 0.22^2 + 0.78^2)\}^{1/2} = 5.4.$$

Since the deficit of the W treatment compared with its 'expected' value based on linear interpolation between C and L is between five and six times its standard error it is clear that the reduction of yield for W is not attributable solely to the reduced intake.

If we perform the same comparison for diet A the comparison is

$$A - \{C + (21.2 - 14.7)/(22.6 - 14.7)(L - C)\}$$

or

$$A - 0.18C - 0.82L.$$

The estimated effect is

$$218 - 0.18(166) - 0.82(240) = -8.7$$

and the standard error of the estimate is

$$[(s^2/8)\{1 + (0.18)^2 + (0.82)^2\}]^{1/2} = 5.5.$$

The deficit for A below its 'expected' value is clearly not significant.

The conclusions from the experiment are therefore clearly that the failure of the wheat diet is not explained either by the reduced intake of chicks on that diet, or by the extra amino acids in the wheat diet which were included in the A diet.

When the treatments are quantitative, the questions being asked by the experimenter should also be quantitative. Such questions are rarely appropriately answered by simple comparisons of pairs of treatment means. The use of three, or more, levels of a quantitative factor implies an interest in the response pattern of yield to increasing amounts of the factor. The experimenter may be interested in estimating the level of the factor which produces maximum yield, or in some other specific characteristic of the response function. Alternatively, he may be interested in a more general description of the forms of response, examining the fit of several different functions to the data. In all cases, a proper first step is to plot the treatment mean yields, y, against the quantitative levels, x.

Example 12.2

In Chapter 2 we examined the analysis of an experiment on rice spacing from which the analysis of variance and the set of treatment means are reproduced in Table 12.6. The yields clearly increase as the density increases, or equivalently as the area per plant decreases. Both forms of relationship are displayed in Figure 12.4. The reader may like to assess which of the two plots appears to give the more linear relationship. The treatment SS can be split into components for regression and residual using the contrast SS method of Chapter 4:

$$SS(\text{regression}) = \sum_j T_j (x_j - \bar{x}) \Big/ \Big\{ 4 \sum_j (x_j - \bar{x})^2 \Big\}.$$

Table 12.6(a).

	SS	df	MS
Blocks	5.88	3	1.96
Treatments	23.14	9	2.57
Error	12.42	27	0.46

(b).

Treatment	Mean yield	Density (plants/m²)	Area per plant (cm²)
30 × 30	6.02	11.1	900
30 × 24	6.48	13.9	720
30 × 20	7.19	16.7	600
30 × 15	6.92	22.2	450
24 × 24	7.60	17.4	576
24 × 20	7.78	20.8	480
24 × 15	7.50	27.8	360
20 × 20	7.71	25.0	400
20 × 15	8.70	33.3	300
15 × 15	8.20	44.4	225

Figure 12.4. Mean yield of rice plotted against plant density (×) and against area per plant (●).

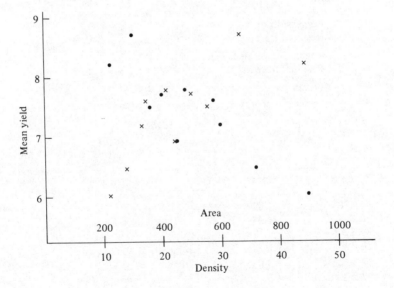

The two alternative explanatory variables give sums of squares summarised in the analysis of variance shown in Table 12.7. The regression SS is very large in either case and there is no benefit from formally examining the ratio of regression mean square to error mean square. The residual mean square is of more interest because it provides information on whether the fit is adequate. Clearly the area regression is considerably better than the density regression. However, the F ratio 0.79/0.46 suggests that there might be further small systematic treatment effects. To examine these we calculate from the fitted yields, using the regression on area, the deviations of observed yield from fitted yield (Table 12.8). The regression equation is

$$y = 9.11 - 0.0034(\text{area}).$$

Examination of the deviations provides some evidence that the fitted regression overestimates yield when the two spacing dimensions are disparate. In other words better yields are obtained when the spacing is more nearly square, a conclusion which would hardly surprise a biologist.

Table 12.7.

Source	SS	df	MS
Regression on area	16.79	1	
Residual for area	6.35	8	0.79
Regression on density	13.40	1	
Residual for density	9.74	8	1.22
Error	12.42	27	0.46

Table 12.8.

Spacing	Observed mean	Fitted yield	Deviation
30 × 30	6.02	6.07	−0.05
30 × 24	6.48	6.66	−0.18
30 × 20	7.19	7.07	+0.12
30 × 15	6.92	7.58	−0.66
24 × 24	7.85	7.35	+0.50
24 × 20	7.78	7.64	+0.14
24 × 15	7.50	7.89	−0.39
20 × 20	7.71	7.75	−0.04
20 × 15	8.70	8.09	+0.61
15 × 15	8.20	8.35	−0.15

Example 12.3
Another experiment on spacing effects, this time on turnips, consisted of
three blocks of 20 treatments, the treatments being all combinations of five
seed rates (0.5, 2, 8, 20, 32 lb/acre) and four row widths (4, 8, 16, 32 inches).
The yields vary greatly between treatments, and substantially between
blocks and have been transferred to a log scale before analysis. The
transformed data are tabulated in Table 12.9. The initial factorial analysis
of variance is given in Table 12.10. The two-way table of mean yields is

Table 12.9. *Log yields (lb per plot) for turnip
experiment.*

Density (lb/acre)	Row spacing (inches)	Block I	II	III
0.5	4	0.00	0.69	1.18
	8	0.81	1.25	1.01
	16	0.69	0.69	1.75
	32	0.92	1.01	0.92
2	4	1.39	1.79	2.22
	8	1.66	2.22	2.40
	16	1.50	1.95	2.63
	32	1.56	1.95	2.22
8	4	2.40	2.60	2.67
	8	2.33	2.63	2.98
	16	2.01	2.37	2.56
	32	1.95	2.17	2.46
20	4	2.56	2.89	2.71
	8	2.56	2.69	3.21
	16	2.56	3.04	2.83
	32	1.87	2.40	2.53
32	4	2.48	2.76	3.14
	8	2.53	2.98	2.89
	16	2.30	2.71	2.80
	32	2.01	2.48	2.56

Table 12.10.

Source	SS	df	MS
Blocking	3.41	2	
Rates	25.61	4	6.40
Row width	0.90	3	0.30
Rates × width	1.00	12	0.083
Error	1.48	38	0.039
Total	32.40	59	

Table 12.11.

Seed rate (lb/acre)	Row width (inches)			
	4	8	16	32
0.5	0.62	1.02	1.04	0.95
2	1.80	2.09	2.03	1.91
8	2.56	2.65	2.31	2.19
20	2.72	2.82	2.81	2.27
32	2.79	2.79	2.60	2.35

shown in Table 12.11. The effect of increasing seed rate dominates the variation in the experiment. However, the effect of row width is clearly significant and possibly modified by seed rate and these effects are investigated in more detail. The choice of row widths increasing by a factor of two implies that the four row widths may be considered as an equally spaced set of four levels (on the log scale) and linear and quadratic terms are fitted on this scale. The mean yields are plotted against row width for each seed rate, using equally spaced levels, in Figure 12.5. The sums of squares corresponding to the linear and quadratic comparisons for each seed rate are tabulated in Table 12.12, together with the residual SS (each sum of squares having a single degree of freedom). Compared with the error mean square of 0.039 the sum of squares for the linear and quadratic terms are generally large and the residual SS small. Three of the linear effect SS are significant at 5%, and two of the quadratic SS are similarly significant. None of the residual SS are anywhere near the 5% significance level. Hence the quadratic regressions provide adequate fits and the quadratic terms are clearly necessary. The fitted equations are

seed rate:

0.05	$y = 0.67 + 0.36x - 0.087x^2$
2	$y = 1.81 + 0.33x - 0.102x^2$
8	$y = 2.59 + 0.01x - 0.052x^2$
20	$y = 2.69 + 0.35x - 0.161x^2$
32	$y = 2.79 + 0.03x - 0.062x^2,$

where $x = \log_2(\text{row width}/4)$. The fitted curves are shown in Figure 12.5 and clearly suggest a shift in the optimal row width towards narrower row widths as the seed rate increases.

Table 12.12.

Row effect	Seed rate (lb/acre)				
	0.5	2	8	20	32
Linear	0.150	0.001	0.304	0.282	0.344
Quadratic	0.183	0.126	0.033	0.310	0.046
Residual	0.010	0.024	0.061	0.027	0.000

Figure 12.5. Fitted quadratic curves of log yield against row spacing for five seed rates, 0.05 (1), 2 (2), 8 (3), 20 (4), 32 (5).

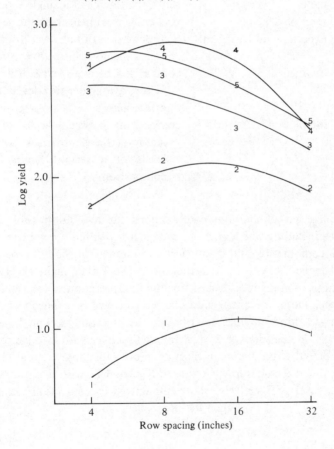

12.5   Treatment structures

Most sets of experimental treatments will include some structure. A major exception is experiments for screening or selecting from large numbers of drugs or new varieties of field crops, and this situation is discussed in Section 12.7. The other common situation where there is no structure is where only two or three treatments are being compared.

The most important concept in treatment structures is that of factorial structures, introduced in Chapter 3, which will be the basis for the ideas in the next four chapters. Here we shall reiterate and extend the rather simplified description of factorial treatments in Chapter 3. The basic components of factorial structure are:

factor	set of treatments of a single type,
level of a factor	particular treatment from the set,
experimental treatment	a combination of one level from each factor included in the experiment,
main effect	comparison between levels of a single factor, averaging over levels of all other factors,
interaction	comparison between levels of one factor of the comparison between levels of a second factor (of a comparison...) averaging over levels of all other factors.

A complete set of factorial combinations includes all possible combinations including one level from each factor. Within a complete set of factorial combinations, all main effect and interaction effect comparisons can be defined so as to be orthogonal. The set of main effects and interactions then provides a basis for the interpretation of the results.

For each factor, the main effect comparisons will be defined as is most appropriate along the lines described in the preceding section for either qualitative or quantitative factors. Interaction effects for two factors are derived directly from the main effect definitions for those factors. Thus, if an experiment includes four factors, A, B, C and D, with two, three, two and four levels, respectively, and treatment effects and mean yields are denoted by t_{ijkl} and $y_{.ijkl}$, then the main effects of A must be

$$A = t_{1...} - t_{0...},$$

estimated by

$$y_{.1...} - y_{.0...}.$$

A suitable pair of main effects for B would be

$$B = t_{.1..} - t_{.0..}$$

and

$$B' = t_{.2..} - 1/2(t_{.1..} + t_{.0..}).$$

Then the interaction effects for factors A and B corresponding to these main effect definitions would be

$$AB = (t_{11..} - t_{10..}) - (t_{01..} - t_{00..})$$

and

$$AB' = \{t_{12..} - 1/2(t_{11..} - t_{10..})\} - \{t_{02..} - 1/2(t_{01..} + t_{00..})\}.$$

Although these interaction effects have been written in terms of a B effect varying over levels of A, they are symmetrical and can equivalently be written as

$$AB = (t_{11..} - t_{01..}) - (t_{10..} - t_{00..})$$

and

$$AB' = (t_{12..} - t_{02..}) - 1/2\{(t_{11..} - t_{01..}) + (t_{10..} - t_{00..})\}.$$

The generalisation to the three-factor interaction effects is straightforward

$$ABC = (t_{111.} - t_{101.} - t_{011.} + t_{001.}) - (t_{110.} - t_{100.} - t_{010.} + t_{000.})$$

and

$$AB'C = \frac{[(t_{121.} - t_{120.}) - 1/2\{(t_{111.} - t_{110.}) + (t_{101.} - t_{100.})\}] -}{[(t_{021.} - t_{020.}) - 1/2\{(t_{011.} - t_{010.}) + (t_{001.} - t_{000.})\}],}$$

where ABC has been written as the contrast of the AB effect between levels of C, and AB'C as the variation of the C effect over AB combinations to illustrate the symmetry of the various possible forms of writing interactions.

Now, since each interaction effect may be interpreted as a modification of a main effect or of a lower level interaction effect the factorial structure provides a sequential structure for interpretation. The main effects should be examined first, followed by two-factor interactions, and then three-factor interactions. It will only very rarely be worthwhile proceeding to four-factor interactions, and sometimes the interpretation of three-factor interactions may be ignored. Also, following from the form of definition of effects, there is a general expectation that main effects will be larger than two-factor interactions, and so on, size being most simply measured by mean squares in the analysis of variance. Where this expectation is fulfilled the interpretation should consist of the consideration first of the important main effects, followed by the discussion of the variation of those effects, as measured by the interactions involving those main effects. If the interactions are of similar size to their related main effects, then the main effects have little importance (being as variable as they are large!). If interactions are negligible, then the interpretation should ignore them.

Factorial treatment structure provides the major statistical contribution to the experimental component of scientific methodology. The classical scientific approach to a comparison of two treatments was to attempt to control all other sources of variation, so that the only difference was the comparison of interest. In contrast, factorial structure provides two complementary advantages. First, the main effect of each factor has been examined over a range of conditions, and it is of greater validity if the interactions are negligible, but more importantly the possibility of inter-action, that is variation of the effect, has been allowed for and examined. Second, and even more important, if interactions are negligible, factorial structure allows the effects of several factors to be assessed independently within the same set of resources. Thus, suppose that three changes to standard conditions are to be assessed, and to examine each change would require 12 observations for the standard, and 12 for the changed condition. Then Figure 12.6(a) shows the minimum number of observations (48) required for the classical scientific approach of changing one factor at a

Figure 12.6. Diagrammatic representations of units required to make three comparisons with specified precision (a) classical scientific approach, (b) factorial approach.

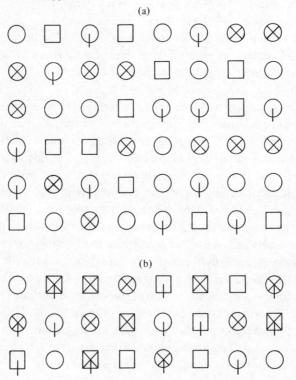

time, controlling all others. The factorial approach, shown in Figure 12.6(b) requires only 24 observations – only half the resources. The secret of the factorial advantage lies in the fact that each observation is included in each comparison. This provides 'hidden' replication in the variation of levels of other factors.

The hidden replication applies even if there are some large interactions, provided there are factors not interacting with the factors involved in these interactions. The advantages of hidden replication also apply if the factorial structure is incomplete (because some combinations of levels are impossible or prone to produce quantitatively different results), though the hidden replication will then be reduced for some main effect comparisons (illustrated in the next section) and the range of validity will also be reduced.

Although factorial structure is the most frequently useful form of structure, it is not the only useful form. Sometimes treatments form natural groups and the relevant treatment comparisons consist of an initial subdivision of the overall variation into between and within groups. Within groups, there may be a further structure; often one of the groups may consist of a control treatment or a small set of control treatments.

Example 12.4

Nine formulations of detergent were compared by washing plates one at a time until they were clean. The experimental procedure required that three basins be used at each time so that the observations were made in threes, and a balanced incomplete block design with 12 blocks was used. The three operators wash at the same rate during the test and the 'yield' is the number of plates washed before the foam disappears. The results are listed in Table 12.13(a), and the within block analysis of variance, (b), was

Table 12.13(a).

Block	Treatments and numbers of plates washed		
1	A 19	B 17	C 11
2	D 6	E 26	F 23
3	G 21	H 19	J 28
4	A 20	D 7	G 20
5	B 17	E 26	H 19
6	C 15	F 23	J 31
7	A 20	E 26	J 31
8	B 16	F 23	G 21
9	C 13	D 7	H 20
10	A 20	F 24	H 19
11	B 17	D 6	J 29
12	C 14	E 24	G 21

Table 12.13(b).

Source	SS	df
Blocks (fitted first)	412.25	11
Detergents (fitted second)	1087.30	8
Error	13.20	16
Total	1512.75	35

obtained using the methods of Section 7.5. The treatments and the adjusted treatment means were:

A: base I + three parts additive = 19.8
B: base I + two parts additive = 17.2
C: base I + one part additive   = 13.2
D: base I   = 6.5
E: base II + three parts additive = 25.3
F: base II + two parts additive = 23.0
G: base II + one part additive   = 21.1
H: base II   = 19.2
J: control   = 29.5.

The comparison of the control with the remaining eight combinations is a natural first step, followed by the subdivision of the sums of squares for the eight combinations into main effect and interaction effect sums of squares.

The subdivision of the treatment SS into components can be calculated directly, because the balance of the design means all component SS are modified by the same factor. Alternatively, it can be calculated by fitting three factors in a general program, the first factor having level one for J and level two for all other treatments, the second having level one for J, two for A, B, C and D, and three for E, F, G and H, and the third having level one for J, two for A and E, three for B and F, four for C and G, and 5 for D and H. The resulting extended analysis of variance is shown in Table 12.14. The very high precision of this experiment makes it inevitable that all treatment effects, including the interaction between base and additive, are highly significant. The pattern of interaction should be examined to assess whether the interaction is large enough to be practically important, as well as significant. Inspection shows that the beneficial effect of the additive in allowing more plates to be washed is much greater for base I than for base II, particularly for the lower levels of additive. Alternatively we can interpret the interaction in terms of the better performance of base II than base I, the difference being large when no additive is used and gradually diminishing as more additive is used.

Table 12.14.

Source	SS	df	MS
Blocks (fitted first)	412.25	11	
Control v. rest	344.29	1	344.29
I v. II	381.60	1	381.60
Additives	311.39	3	103.80
Base × additives	50.02	3	16.67
Error	13.20	16	0.82

12.6 Incomplete structures and varying replication

In developing the ideas of factorial structure during this book we shall predominantly discuss the case of complete factorial structures. There are several reasons for doing so. The complete structure displays most clearly the benefits of factorial structure, that is the capacity to assess interaction and the increased precision due to hidden replication. The mathematical models are neatest when complete factorial structure is used. And, at a purely practical level, the number of possible incomplete structures is enormous and a systematic discussion of incomplete factorial structures would be impossible.

It is, however, important to stress that, although complete factorial structure represents a powerful ideal, the benefits of factorial structure still hold for incomplete factorial structures. It is foolish to use a complete factorial simply to comply with the simple theory. If particular factorial combinations are practically ridiculous or unrealistic then they should not be included. Consider the case of a set of treatments derived from three factors each at two levels, for which the combination of all three lower levels is considered impractical. The set of treatments is

$$a_0b_0c_1 \quad a_1b_0c_0 \quad a_1b_0c_1$$
$$a_0b_1c_0 \quad a_0b_1c_1 \quad a_1b_1c_0 \quad a_1b_1c_1.$$

Suppose an experiment consists of four blocks of these seven treatments. Then both main benefits of factorial structure are retained at a reduced level. Consider first the information on interaction. Considering only observations at level c_1 we clearly have some information on the AB interaction; using level b_1 there is information on the AC interaction; and similarly for BC at level a_1. In each case we have no hidden replication for the interaction as we would have had with the complete structure but there is still information on all two-factor interactions. Where interactions are negligible then we still have most of the hidden replication for main effect comparisons. For the main effect of $A(a_1 - a_0)$ we have estimates based on the three combinations (b_0c_1), (b_1c_0), (b_1c_1), giving an effective replication of 12 for the A main effect. Similarly, in the absence of interactions, the effective replication for the two other main effects is 12.

The analysis of the main effects and interactions is computationally more complex because of the non-orthogonality resulting from the incompleteness of the structure. Further, there will be a small ambiguity in the interpretation of effects but this is of negligible practical importance provided the analysis is correctly produced. Essentially as the full factorial structure is allowed to crumble slowly so the analysis becomes increasingly non-orthogonal and the order of fitting of effects and the resultant interpretation requires more care, though it is unlikely that the non-orthogonality will become so large as to produce a practically serious level of ambiguity, as may often occur in multiple regression.

Another situation where the experimenter should consider an incomplete factorial structure is where the complete factorial structure would include identical treatments. A common example is where two factors represent

(i) alternative forms of a stimulus, and
(ii) different amounts of the stimulus.

If one of the amounts is zero then the zero levels for the alternative forms will be identical. It would be absurd to include a whole set of zero treatments simply for completeness.

The idea of a zero level, which would be identical for alternative levels of another factor, prompts discussion of another completeness pattern which has been implicitly assumed, namely that all treatments, or treatment combinations, should normally be equally replicated. Suppose that we have three alternative formulations of a chemical stimulus and are to use four amounts including a zero (control) with three replications. Then we have $3 \times 3 = 9$ non-zero treatments plus a control making a total of 30 observations. If the alternative formulations do not have different effects then we will wish to interpret comparisons between the four means for the four quantitative levels. The three non-zero means will be based on nine replications while the zero mean is based on three replications. We might therefore consider additional replication of the control treatment. However, if the alternative formulations are very different then, if we have increased the replication of the control, the means for the non-zero levels for a particular alternative formulation will have less replication than the control.

The choice becomes more clear if we restate the problem in the form of specific alternative designs. Suppose we have 60 experimental units available and two alternative designs are considered:

design I
(three forms × three amounts + zero amount) × six replicates = 60 units

design II
three forms × (three amounts + zero) × five replicates = 60 units.

The replication of the means for the control treatment, for a particular combination of form and (non-zero) amount, and for the average over forms for a (non-zero) amount, will be as in Table 12.15. For each design we may calculate the variances of different comparisons, and these are also shown in the table. Thus design I gives greater precision for comparisons not involving the control, whereas design II gives greater precision for comparisons involving the control, the greatest difference between the precision achieved in the two designs being for the comparison of an overall amount mean with the control.

Table 12.15.

	Design I	Design II
Replications		
Control mean	6	15
Combination mean	6	5
Amount mean	18	15
Variances		
Control–combination	$2\sigma^2/6 = 0.33\sigma^2$	$\sigma^2/15 + \sigma^2/5 = 027\sigma^2$
Control–amount	$\sigma^2/6 + \sigma^2/18 = 0.22\sigma^2$	$2\sigma^2/15 = 0.13\sigma^2$
Combination 1–combination 2	$2\sigma^2/6 = 0.33\sigma^2$	$2\sigma^2/5 = 0.40\sigma^2$
Amount 1–amount 2	$2\sigma^2/18 = 0.11\sigma^2$	$2\sigma^2/15 = 0.13\sigma^2$

To illustrate again the way in which decisions about replication for incomplete factorial structures may be considered, we return to the second problem posed in the introduction to this chapter. Two chemicals, O and E, are to be compared with a standard control (S) and an untreated control (C). Chemical O is an oil requiring a surfactant for application and two surfactants, S_1 and S_2, are included. Different sprayers are also to be compared, Ed and Ul for O, and Ul and Con for E. Three concentrations are to be used for the two new chemicals, but not for the controls. The set of treatments is set out in Table 12.16. The total number of distinguishable

Table 12.16.

Chemical	Surfactant added	Sprayer	Concentration		
			1	2	3
O	+S$_1$	Ed	X	X	X
		Ul	X	X	X
	+S$_2$	Ed	X	X	X
		Ul	X	X	X
E		Ul	X	X	X
		Con	X	X	X
		S		X	
		C		X	

treatments is 20 and the number of experimental plants available is 240. It might seem that the obvious design would use 12 replications of each of the 20 treatments in a blocked design to control major patterns of variation within the set of available plants. However, such a design is only one of many reasonable alternatives and it ignores the hidden replication implicit in the partial factorial structure. For example, O will be replicated twice as often as E and therefore will be more precisely assessed (unless one of the surfactants renders chemical O ineffective). Also, the comparison with the controls will be poor because of the relatively low replication of the controls.

Probably the untreated control is included in the experiment merely to assess the general level of pest incidence for this particular experiment, and not for quantitative comparisons with other treatments. This is frequently true for untreated control treatments and it is usually wise not to include such treatments in an analysis of variance with the other treatments, since the error structure for the untreated control is likely to be quite different from that for the other treatments. The replication for the untreated control need not be increased and could even be decreased. But the standard control will be used for comparison with the new chemicals, and will need additional replication. We might therefore consider n_1 replicates of each of the 12 O treatment combinations, n_2 for the six E combinations, and n_3 of the standard control. Precise choice of n_1, n_2 and n_3 will depend on the relative precision required for different comparisons, and must be discussed with the experimenter. At one extreme, $n_2 = 2n_1$, $n_3 = 12n_1$ gives equal overall replication of O, E and S. Since the optimal concentration is not known for O or for E, it is likely that the comparisons of each new chemical with the standard control will involve only one of the three concentrations, and the replication of the control can be reduced accordingly.

The replication pattern employed for this particular problem was to use eight blocks with 30 plants per block with the within block replication structure as shown in Table 12.17. The anticipated treatment contrasts would include

(*a*) between the standard control and the weighted average of the 18 chemical treatments,

(*b*) between the three concentrations,

(*c*) between O and E,

(*d*) main effects and interaction of (S_1, S_2) and (Ed, Ul) for chemical O,

(*e*) interactions of the (*d*) contrasts with the three concentrations,

(*f*) (Ul–Con) for E and interaction with the three concentrations.

The important conclusion to be drawn from these examples, but applicable much more widely, is that choice of replication is the joint responsibility of the experimenter and statistician, and the choice will

Table 12.17.

Chemical	Surfactant added	Sprayer	Concentration		
			1	2	3
O	$+S_1$	Ed	$\times 1$	$\times 1$	$\times 1$
		Ul	$\times 1$	$\times 1$	$\times 1$
	$+S_2$	Ed	$\times 1$	$\times 1$	$\times 1$
		Ul	$\times 1$	$\times 1$	$\times 1$
E		Ul	$\times 2$	$\times 2$	$\times 2$
		Con	$\times 2$	$\times 2$	$\times 2$
		S		$\times 5$	
		C		$\times 1$	

determine the relative precision with which answers are provided to different questions. Invariable use of equal replication is not a proper philosophy nor is 'equal replication' always simply definable as this previous example has shown.

12.7   Treatments as a sample

One of the topics in experimental design which has generated a great deal of often rather heated discussion is the distinction between 'fixed' and 'random' effects. In my opinion, the distinction, though it is clearly necessary, is a very minor one, and I shall discuss it only sufficiently to explain the distinction and the consequent change in the interpretation of the experimental results.

Some experimental treatments are specifically selected at the outset of the planning of the experiment. A new drug to be compared with a control is of interest because it, specifically, has produced some previous good results. The levels of a chemical constituent in a complex chemical process are chosen because it is expected that the range of values will cover the optimum level symmetrically with sufficient range to identify the optimum clearly. In general, the levels of a quantitative factor will be chosen exactly and deliberately to give maximum information in answer to a specific question. Sometimes, however, the treatments are selected to be a representative sample of some population, and the primary purpose of the comparison of the treatments is not to draw conclusions about these particular treatments but rather to assess the variability of the population. This use of representative treatments occurs widely in genetics and breeding experiments, where it is important to assess the levels of variance in different subpopulations, or attributable to different sources. Thus, in a pig breeding experiment, the offspring from several sows for each of several boars may provide data on characteristics measured on individual piglets. These data may then be used to assess the relative levels of variation between (i) different boars, (ii) different sows, and (iii) different piglets within the same litter. The use of representative treatments is described as a random effects model because, in the model for experimental data, the treatment effects, t_j, may be regarded, at the stage of planning the experiment, as being unknown and randomly selected from a definable set of possible treatment effects.

Now the reason why there is no necessity to make a major distinction between fixed, or specifically selected, treatments, and random, or representative, treatments is that the experimental design and the initial analysis of data are identical for the two situations. Essentially, once the particular representative treatments have been chosen, the experimenter considers only those treatments, and the details of the experimental design and the initial analysis of results are conditional on the treatments chosen. The random allocation of treatments to experimental units is the same as for specifically selected treatments. The analysis of variance divides variation into components representing block differences, treatment differences, and residual variation in the usual way. The change which arises because the treatments are a representative sample comes in the interpretation which follows the initial analysis of variance.

When the experimental treatments are a representative sample from a population, comparisons between any two particular treatments are of no interest, because the occurrence of those two particular treatments in the experiment was due to chance. Instead, we wish to estimate the variance of the population from which the treatments are drawn. Obviously, the variance of the potential treatment population, σ_t^2, can be estimated from the treatment mean square. Equally obviously, the treatment mean square is partly attributable to the inherent variation between units. We therefore reconsider the expected value of the treatment mean square (after the manner of Chapter 6) for the case when the t_j are to be regarded as random variables with mean zero and variance σ_t^2. We assume a randomised complete block design or any design which has equal treatment replication and no non-orthogonal block effects:

$$\text{treatment SS} = \sum_j Y_{\cdot j}^2/b - Y_{\cdot\cdot}^2/(bt).$$

Now

$$Y_{\cdot j}^2/b = b\left\{\mu + t_j + (1/b)\sum_i \varepsilon_{ij}\right\}^2$$

$$E(Y_{\cdot j}^2)/b = b\mu^2 + b\sigma_t^2 + \sigma^2$$

and

$$Y_{\cdot\cdot}^2/(bt) = bt\left\{\mu + \sum_j t_j/t + \sum_{ij} \varepsilon_{ij}/(bt)\right\}^2$$

$$E\{Y_{\cdot\cdot}^2/(bt)\} = bt\mu^2 + b\sigma_t^2 + \sigma^2$$

$$E\left\{\sum_j Y_{\cdot j}^2/b - Y_{\cdot\cdot}^2/(bt)\right\} = (bt\mu^2 + bt\sigma_t^2 + t\sigma^2) - (bt\mu^2 + b\sigma_t^2 + \sigma^2)$$

$$= b(t-1)\sigma_t^2 + (t-1)\sigma^2.$$

Hence,

$$E(\text{treatment MS}) = b\sigma_t^2 + \sigma^2,$$

and we can estimate σ_t^2 by

$$(\text{treatment mean square} - \text{error mean square})/b.$$

More generally, if the treatments are unequally replicated, with n_j observations for treatment j, then, assuming a completely randomised design,

$$E(\text{treatment MS}) = \sigma^2 + \left\{ \left(\sum_j n_j \right)^2 - \sum_j (n_j^2) \right\} \sigma_t^2 \bigg/ \left\{ \sum_j n_j(t-1) \right\}.$$

For more complex situations, where the experimental treatments are combinations or hierarchies of 'random' treatments, it is also possible to determine expected values for components of the treatment SS and to estimate the different population variances.

12.8   Screening and selection experiments

There is one substantial area of experimentation which does not fit into the classifications so far considered in this chapter. This is the screening of large numbers of compounds to discover the few which produce a desired effect, or the selection of the best of a large number of alternatives. These two types of investigation often occur as different stages of the same research project, and are met most frequently in breeding programmes or in searches for appropriate chemical compounds for controlling disease.

In both types of investigation, the only structure usually adopted for the experimental treatments is the inclusion of a standard, or control, treatment. Apart from the control treatment, all other treatments are regarded identically, and there is no structure within the non-control set from which the treatments come. The treatments are, of course, a qualitative set.

The essential difference between screening and selection is that, in screening trials, the response variable is qualitative with each compound giving either a positive or a negative result whereas, in selection, the response variable is quantitative. Both forms of investigation are usually multi-stage. Initially, the research scientist has many thousands or hundreds of thousands of candidate drugs or breeding lines. The end point of the investigation is the identification of a small number of these candidates, either for immediate use, or, more commonly, for further investigation of properties other than those of primary interest.

Selection has been discussed from two different philosophical viewpoints. For each it is assumed that there is a true set of expected yields for the candidates available for selection, and that observations in any particular trial can be modelled as

$$y_{ij} = \mu_i + \varepsilon_{ij},$$

where μ_i is the true yield for candidate i, and ε_{ij} represent the response variability inherent in the experimental units. This variability may be subdivided into block variation and random variation, but both components are assumed to be independent of the particular candidate.

One selection philosophy assumes that the objective is to select candidates so as to maximise the probability of including in the selection those candidates which have the highest values of true yield, μ_i. The other philosophy attempts to maximise the average mean value of the μ_i for the set of selected candidates. Obviously the perfect achievement of either objective is the same. The difference comes in the relative value placed on different imperfect achievements. The first objective accepts a small probability of selecting a candidate with a very poor μ_i as necessary to achieve a high probability of including in the selection those candidates with the best μ_i. The second objective accepts a reduced probability of including all the best μ_i candidates to ensure that all selected candidates have μ_i's which are nearly as good as the best.

The choice of philosophy should depend on the particular objective of the research programme. If the selected candidates will inevitably be used, as for example in a breeding programme for cereals, then the interest of the user, the farmer, is to maximise his future yield and this leads to the second philosophy. If, on the other hand, the chance of the selected candidates being ultimately successful is rather small, as in searches for cures for diseases, then maximising the probability of success is important, and the first philosophy would seem appropriate.

One aspect of the design of selection programmes for which intuitive results are available is the division of a selection research programme into stages. If there are to be r stages of selection, and an initial set of n candidates is to be reduced, through selection, to c candidates, then under some rather restrictive assumptions of equal replication within a stage for all candidates included in that stage, the optimal design, using the second philosophy, is to select at the same intensity

$$p = (c/n)^{1/r}$$

at each stage and to allocate equal resources to each stage. Thus if it is required to select four from 100 over three stages of selection the numbers selected after the first and second stages should be 35 and 12, respectively, and the relative replication of observations per candidate in the three stages should be in the ratios 1:3:8. Although this design principle was discovered by considering a very precisely specified situation, it is an appropriate general principle for a much wider range of situations.

Screening problems can be divided into two groups, according to the probability of detection of those individuals which possess the required

characteristic. The extreme case is where the detection of the characteristics is certain. The classic example of this is the screening of blood
samples for a particular constituent which led Dorfman (1943) to propose
a method of group screening whereby many candidates can be assessed
simultaneously using a single test. The principle of group screening
assumes that the cost depends only on the number of tests regardless of
how many candidates are assessed in each test. In a group test which
includes several candidates the test result will be positive if any candidate
included in the group would provide a positive result if tested separately. If
all candidates are negative then the test result will be negative.

The methodology of group screening has been developed principally by
Sobel and Groll (1959), considering only strategies in which, after a
positive result for a group test that group is divided into two subgroups
which are tested independently (unless logically unnecessary). The optimal
design strategy depends on the overall proportion of candidates with the
required characteristic. However, the broad pattern of conclusions from
Sobel and Groll's results is that for a wide range of, relatively small, values
of p a strategy of testing groups of size eight with successive subgroups of
four, two and one is very close to optimal.

If the probability of detection of a characteristic is uncertain, then the
strategies for screening experiments are not so well understood. It is not
clear whether it is useful to pursue experiments involving simultaneous
testing of several candidates on the same experimental unit, with the
implication of further testing for those individuals which appear, as a
group, to have the required characteristic. The ideas of inter-block
information from incomplete block designs considered in Chapter 7 may
have some relevance here but this aspect of design will not be considered
further here.

Exercises 12

(1) In a randomised block experiment on potato scab, eight treatments were replicated four times. Two treatments were identical
controls, the other six consisted of three amounts of sulphur (3, 6
and 12 units) applied either in spring or autumn. The totals for the
eight treatments were:

C_1	C_2	S_3	A_3	S_6	A_6	S_{12}	A_{12}
75	106	38	67	62	73	23	57

Think about the questions that the experimenter should have
been asking to have chosen these treatments. Define a set of

practically sensible orthogonal comparisons and partition the treatment SS into appropriate components. If the error mean square is 30, summarise the results of the experiment.

(2) In a bio-assay experiment, five drugs were compared. Drug one, a control, was applied at three concentrations, 1, 2, 4; drugs two and three were each applied at two concentrations, 0.8 and 1.6, and drugs four and five each at only one concentration 0.6. The experiment was arranged in five randomised blocks and the yield responses for one variable are given in Table 12.18.

Carry out an analysis of variance, splitting the treatment SS into components each with a single degree of freedom, where these are meaningful. Two questions you should attempt to answer are:

(i) Is the relationship between yield and log concentration linear?

(ii) Is the average slope of the relationship between yield and log concentration consistent over drugs?

Thinking about the concepts of main effects and interactions may help clarify your choice of contrasts. Briefly summarise the conclusions to be drawn, including comments about the questions that cannot be answered.

Table 12.18.

Drug	Concentration[a]	Block				
		I	II	III	IV	V
1	1	1.7	1.7	1.8	1.9	1.6
	2	2.0	1.7	2.2	2.1	1.7
	4	2.1	1.9	2.1	2.1	2.1
2	0.8	2.0	1.7	2.1	1.8	1.6
	1.6	2.0	2.0	2.1	1.9	2.0
3	0.8	1.8	1.7	1.7	1.9	1.9
	1.6	2.1	1.8	1.8	2.1	1.9
4	0.6	2.0	1.8	1.8	1.9	1.7
5	0.6	2.2	2.0	2.0	2.0	2.1

[a]No units available.

(3) An incomplete block design was used to investigate the effects on oxygen uptake in microlitres per milligramme of poultry spermatozoa dry tissue per hour of various media, the restriction to

four observations per block being due to the laboratory equipment available. The ten treatments were two controls and the eight combinations of two form of buffering (b_1 = phosphate, b_2 = glutamate) two dilutents (d_1 = sodium chloride, d_2 = mixture of salts) and two nutrients (n_0 = nil, n_1 = glucose):

1	2	3	4	5
C_1	C_2	$b_1 d_1 n_0$	$b_1 d_1 n_1$	$b_1 d_2 n_0$

6	7	8	9	10
$b_1 d_2 n_1$	$b_2 d_1 n_0$	$b_2 d_1 n_1$	$b_2 d_2 n_0$	$b_2 d_2 n_1$

The block design is shown in Table 12.19.

The initial within block analysis of variance and the treatment totals adjusted for blocks ($T_j - \sum_{i(j)} B_i/4$) are given in Tables 12.20 and 12.21. Complete the analysis by splitting the treatment SS into suitable component SS, or by calculating treatment contrasts and their standard errors. Summarise the conclusions to be drawn from the results.

Table 12.19.

Block	Treatments				Block	Treatments			
I	1	2	3	4	IX	2	5	8	10
II	1	2	5	6	X	2	7	8	9
III	1	3	7	8	XI	3	5	9	10
IV	1	4	9	10	XII	3	6	7	10
V	1	5	7	9	XIII	3	4	5	8
VI	1	6	8	10	XIV	4	5	6	7
VII	2	3	6	9	XV	4	6	8	9
VIII	2	4	7	10					

Table 12.20. *Analysis of variance.*

Source	SS
Blocks (ignoring treatment)	124.5
Treatment (eliminating blocks)	292.7
Error	175.2
Total	592.4

Table 12.21. *Adjusted treatment totals.*

1	−12.3
2	−13.0
3	−7.2
4	+9.8
5	−9.6
6	+2.6
7	−14.9
8	+16.4
9	+12.6
10	+15.7

Overall mean = 13.5

(4) In an experiment to examine leakage of propellate from aerosol cans the loss of weight (in grammes) over eight weeks was measured for three different propellates X, Y and Z from cans with seals made by different methods. Four factors involved in the manufacture of seals were examined at the following levels:

depth of operation:	A, B, C;
tool diameter:	1, 2;
weight of sealing coumpound:	0.60, 0.55, 0.50 g;
position of sealing compound:	standard, S, non-standard, T.

The non-standard position of sealing compound could only be used when the weight of sealing compound was 0.50 g. Each observation in Table 12.22 (p. 344) is the mean weight loss for ten cans. First consider what difficulties there might be in the assumptions required for analysis of the data. You may find it helpful initially to regard the different propellates as three 'blocks' within which the different methods for constructing the seals are to be compared. Complete an analysis of the data and present your conclusions.

Table 12.22.

Propellate	Weight of sealing compound (g)	Position of sealing compound	A		B		C	
			1	2	1	2	1	2
X	0.60	S	0.25	0.41	0.40	0.37	0.41	0.49
	0.55	S	0.36	0.29	0.30	0.41	0.29	0.60
	0.50	S	0.38	0.60	0.33	0.41	0.32	0.54
	0.50	T	0.78	0.91	2.10	0.73	0.47	0.32
Y	0.60	S	1.20	1.09	1.13	0.91	0.74	1.00
	0.55	S	1.55	1.33	1.19	1.27	0.75	1.05
	0.50	S	1.80	1.68	1.45	1.27	1.04	1.55
	0.50	T	2.09	2.03	2.08	1.86	1.27	1.35
Z	0.60	S	1.07	1.44	0.66	0.63	0.62	0.28
	0.55	S	1.01	0.62	0.32	0.40	0.81	0.45
	0.50	S	0.64	1.58	0.38	0.36	0.33	0.31
	0.50	T	3.54	3.26	1.37	1.68	1.53	0.83

13

Factorial structure and particular forms of effects

13.0   *Preliminary example*

Suppose eight treatments comprising all combinations of three two-level factors P, Q and R are compared and the resulting treatment means are shown in Table 13.1. What is the best estimate of the difference between the two treatment combinations $p_1q_1r_1$ and $p_1q_1r_0$? Or, in practical terms, if you have decided to recommend the upper levels of factors P and Q, what is the benefit of using r_1 rather than r_0? If you think there is only one possible answer, then you have not yet fully grasped the benefits of factorial structure. With a little effort, you should be able to find at least six different conceivable answers.

Table 13.1.

	r_0	r_1
p_0q_0	17	23
p_0q_1	18	28
p_1q_0	14	22
p_1q^1	19	30

13.1   Factors with two levels only

In Chapters 3 and 12, the general advantages of using experimental treatments with a factorial structure were discussed. In this chapter, we look in more detail at special forms of factorial structure to examine how the general advantages manifest themselves in particular cases, and we develop methods for using further advantages of these special forms.

The first special form of factorial to consider is that in which all factors have two levels. There are many reasons why this is an interesting form to examine in detail. First, it is a sensible practical solution for many scientific situations. We can think of one level of each factor as being the normal mode and the other level as a new, alternative mode. In a classic agricultural example, advisory officers found that two farmers growing mushrooms on a large scale in the same locality were getting very different yields, though much of their management procedure was identical. On careful examination, the advisors distinguished four ways in which the two farmers' procedures differed. The four differences were identified as four factors, and an experiment was planned in which each of the four factors appeared at two levels, the first level being the mode adopted by farmer 1 and the second that of farmer 2. All 16 combinations of levels for the four factors were included in the experiment to determine whether the differences between the two farmers' yields were due to a single factor or to a combination of factors.

Another area where experiments with two-level factors are important is in industrial plants where many different facets of the industrial process might be changed. In attempting to discover which changes are beneficial and which changes produce interdependent effects, an experiment with a number of factors, each at two levels (normal and changed), is often an effective first stage of an investigation.

A third important area for two-level factor experimentation is the investigation of predictions from complex computer simulation models for ecological, chemical or agricultural systems. These models are being extensively developed but the investigation of the implications of such models is often hardly planned at all. Within any such model, there will be many parameter values which could be altered in the practical situation which the model is intended to represent. As an initial stage in the investigation of the properties of the model, consideration of the effects of changing each of, say, six parameters in the model can be most effectively achieved by using a factorial experiment with six factors. For each factor one level is the value originally assumed for the parameter, and the second level is an alternative value.

One other reason for considering two-level factors in some detail is that they are particularly simple to manipulate mathematically. Factorial structures in which all factors have two levels are referred to as 2^n factorials. In a simplified form of the more general notation the levels are defined as 0 and 1, the upper level of a factor, P, is represented by the lower case letter, p, and the lower factor level by the absence of the letter. Thus the treatment combination with all factors at their upper level is represented by pqrst...; the combination with factors P, R and S at their upper levels and all other factors at their lower levels by prs; the combination with only factor R at its upper level by r. The combination with all factors at their lower levels is a special case and conventionally this is represented by (1).

A general factorial treatment combination is written $p_i q_j r_k \ldots$, where each of $i, j, k, \ldots$ may be 0 or 1. Effects are represented by capital letters, P, Q, PQ, PQR, ... It is convenient to use p, pq, pqr, ... to represent both the treatment combinations and the mean yield for that treatment combination, and to use P, PQ, PQR, ... to represent both effects and the least squares estimators of these effects. This might appear to lead to ambiguity and confusion but, in practice, does not, principally because of the simple unbiassed form of least squares estimators.

We consider the detailed properties of two-level factorial structure first for three factors because this enables us to illustrate all the important ideas without employing too large an array of symbols. If the three factors are P, Q and R, then the main effect of P is most simply defined as

$$P = (p + pq + pr + pqr)/4 - \{(1) + q + r + qr\}/4, \qquad (13.1)$$

the simple difference between the mean yield for all observations at the upper level of P and the corresponding mean yield for the lower level of P. This effect can be rewritten as a pseudo-algebraic expression

$$P = \frac{1}{4}(p - 1)(q + 1)(r + 1),$$

the advantage of which will be seen later.

The interaction between two factors, P and Q, is most simply expressed as

$$\{(pq + pqr)/2 - (p + pr)/2\} - [(q + qr)/2 - \{(1) + r\}/2], \qquad (13.2)$$

showing clearly that it is the comparison of the (upper–lower) Q comparison between upper and lower levels of P. This can be rewritten either as

$$\{pq + pqr + (1) + r\}/2 - (p + pr + q + qr)/2$$

or as

$$\frac{1}{2}(p - 1)(q - 1)(r + 1).$$

We can now see that, whereas the use of 2 as the divisor in (13.2) seems logical when thinking about the PQ effect as a difference of difference of mean yields, the resulting estimate is clearly not measured on the same scale as the effect P (13.1). It is usual therefore to define the PQ effect on a scale directly comparable to the P effect,

$$PQ = \frac{1}{4}(p-1)(q-1)(r+1).$$

Similarly, the natural expression of the PQR interaction 'a difference of a difference of a difference' is

$$[\{(pqr-pq)-(pr-p)\}-\{(qr-q)-(r-(1))\}]$$

but, to make the PQR effect directly comparable with those for P and PQ, we define

$$PQR = \frac{1}{4}(p-1)(q-1)(r-1).$$

Two related patterns to note for future reference are:

(i) In the algebraic summary form for an effect each factor included in the effect contributes an $(x-1)$ term and each excluded factor an $(x+1)$ term, the latter representing the averaging of the effect over levels of factors not directly involved in the effect.

(ii) If the effects are rewritten as differences between two groups of treatment combinations, the two groups are, respectively, those with an even number of letters in common with the effect, and those with an odd number of letters in common with the effect. This is a direct result of the definition of interactions as 'differences of differences of differences of...'. The patterns may be confirmed in the further examples below

$$Q = \frac{1}{4}(p+1)(q-1)(r+1)$$

$$= \frac{1}{4}[(pqr+qr+pq+q)-\{pr+r+p+(1)\}]$$

$$QR = \frac{1}{4}(p+1)(q-1)(r-1)$$

$$= \frac{1}{4}[\{pqr+qr+p+(1)\}-(pq+pr+q+r)]$$

$$PR = \frac{1}{4}(p-1)(q+1)(r-1)$$

$$= \frac{1}{4}[(pqr+pr+q+(1))-\{pq+p+qr+r\}].$$

If we define an eighth effect, the overall mean

$$M = \frac{1}{8}\{pqr+pq+pr+qr+p+q+r+(1)\}$$

$$= \frac{1}{8}(p+1)(q+1)(r+1)$$

then we have eight effects and eight treatment combinations, and the relationship between the two can be expressed in matrix form by defining

$$\mathbf{x}' = ((1),\ p,\ q,\ pq,\ r,\ pr,\ qr,\ pqr)$$

$$\mathbf{y}' = (8M,\ 4P,\ 4Q,\ 4PQ,\ 4R,\ 4PR,\ 4QR,\ 4PQR)$$

and writing

$$\mathbf{y} = \mathbf{U}\mathbf{x},$$

where the elements of $\mathbf{U}$ are all ± 1 and the rows of $\mathbf{U}$ are orthogonal:

$$
\mathbf{U} =
\begin{bmatrix}
+1 & +1 & +1 & +1 & +1 & +1 & +1 & +1 \\
-1 & +1 & -1 & +1 & -1 & +1 & -1 & +1 \\
-1 & -1 & +1 & +1 & -1 & -1 & +1 & +1 \\
+1 & -1 & -1 & +1 & +1 & -1 & -1 & +1 \\
-1 & -1 & -1 & -1 & +1 & +1 & +1 & +1 \\
+1 & -1 & +1 & -1 & -1 & +1 & -1 & +1 \\
+1 & +1 & -1 & -1 & -1 & -1 & +1 & +1 \\
-1 & +1 & +1 & -1 & +1 & -1 & -1 & +1
\end{bmatrix}.
$$

Many nice mathematical patterns may be observed in the expression of the relationship between effects $\mathbf{y}$ and treatment combinations $\mathbf{x}$, some of which are statistically useful. For example, the elements of the PQ row of $\mathbf{U}$ are obtained by multiplying the corresponding elements in the rows for P and Q; similarly, the row for PQR is the product of the rows for P and for QR, or of those for Q and for PR.

Also, notice that the relationship between effects, $\mathbf{y}$, and treatment combinations, $\mathbf{x}$, is the same as that between least squares estimators of effects, $\mathbf{y}$, and the mean yields of treatment combinations, $\bar{\mathbf{x}}$. If the experiment has n observations of each treatment combination, then each element of $\bar{\mathbf{x}}$ has variance σ^2/n, and each element of $\mathbf{y}$ has variance $8\sigma^2/n$. Hence the variance of each comparison effect is

$$\mathrm{Var}(P) = \sigma^2/2n.$$

Also, from the general result for the sums of squares for treatment contrasts in Chapter 4

$$\mathrm{SS(contrast)} = \left(\sum_j l_{kj} Y_{.j} \right)^2 \Big/ \left(n \sum_j l_{kj}^2 \right)$$

and so

$$\mathrm{SS(P)} = \frac{P^2}{n\{8(1/4n)^2\}} = 2nP^2$$

with corresponding results for other effects.

All these results can be generalised from three to m factors:

$$\text{effect} = (1/2^{m-1})(p \pm 1)(q \pm 1)(r \pm 1)\ldots(m \pm 1)$$
$$\text{Var(effect)} = \sigma^2/n2^{m-2}$$
$$\text{SS(effect)} = n2^{m-2}(\text{effect})^2.$$

Although modern computational facilities have made the method practically obsolete, no discussion of 2^m factorials could be complete without reference to Yates' algorithm for obtaining estimates of effects. This is defined as follows:

(i) Write the mean yields for treatment combinations in a column in standard order $((1), p, q, pq, r, pr, qr, pqr, s, ps, \ldots)$.

(ii) The first half of a new column consists of the sums of consecutive values from the previous column, $(1) + (2)$, $(3) + (4)$, etc.

(iii) The second half of a new column consists of the differences of consecutive pairs of values from the previous column, $(2) - (1)$, $(4) - (3)$, etc.

(iv) Repeat steps (ii) and (iii) forming a series of new columns from previous columns until m new columns have been completed after the initial column of mean yields.

(v) Then the final column gives the effect vector $\mathbf{y}$ in standard order $(2kM, kP, kQ, kPQ, kR, kPR, kQR, kPQR, kS, kPS, \ldots)$, where $k = 2^{m-1}$.

The transformation from each column to the succeeding column is represented by the matrix:

$$\mathbf{V} = \begin{bmatrix}
+1 & +1 & 0 & 0 & 0 & 0 & 0 & 0 & \cdots \\
0 & 0 & +1 & +1 & 0 & 0 & 0 & 0 & \cdots \\
0 & 0 & 0 & 0 & +1 & +1 & 0 & 0 & \cdots \\
0 & 0 & 0 & 0 & 0 & 0 & +1 & +1 & \cdots \\
\vdots & \vdots & \vdots & \vdots & \vdots & \vdots & \vdots & \vdots & \\
-1 & +1 & 0 & 0 & 0 & 0 & 0 & 0 & \cdots \\
0 & 0 & -1 & +1 & 0 & 0 & 0 & 0 & \cdots \\
0 & 0 & 0 & 0 & -1 & +1 & 0 & 0 & \cdots \\
0 & 0 & 0 & 0 & 0 & 0 & -1 & +1 & \cdots \\
\vdots & \vdots & \vdots & \vdots & \vdots & \vdots & \vdots & \vdots & \cdots
\end{bmatrix}.$$

It is simple to verify for any particular value of m that $\mathbf{V}^m = \mathbf{U}$ but a general proof is outside the scope of this book.

We illustrate the use of Yates' algorithm with the data from Chapter 3 on the water uptake of amphibia. The factors were S = species (toad or frog), M = moisture (wet or dry), H = hormone (control or hormone). The

Table 13.2.

Combination	Species	Moisture	Hormone					Effect
(1)	toad	wet	control	0.36	2.24	33.80	97.10	=4S
s	frog	wet	control	1.88	31.56	63.30	−45.38	=4M
m	toad	dry	control	21.46	24.60	−9.84	43.42	=4SM
sm	frog	dry	control	10.10	38.70	−35.54	−12.58	=4H
h	toad	wet	hormone	21.26	1.52	29.32	29.50	=4SH
sh	frog	wet	hormone	3.34	−10.36	14.10	−25.70	=4MH
mh	toad	dry	hormone	28.16	−17.92	−12.88	−15.22	=4MH
smh	frog	dry	hormone	10.54	−17.62	0.30	13.18	=4SMH

Table 13.3.

Effect	Estimate	SS
S	−11.34	514.38
M	10.86	471.74
SM	−3.14	39.44
H	7.38	217.86
SH	−6.42	164.87
MH	−3.80	57.76
SMH	3.30	43.56

mean results for the eight treatment combinations with the Yates' algorithm calculations are given in Table 13.2. From the final column of Table 13.2, we can calculate estimates for the effects and corresponding sums of squares (see Table 13.3). The standard error of each effect estimate requires the estimate of variance $s^2 = 34.56$ from the analysis of variance in Chapter 3. The standard error for each effect estimate is then $(34.56/4)^{1/2} = 2.94$.

13.2   Improved yield comparisons in terms of effects

Consider again the relationship $y = Ux$ between the vector of effects, y, and the vector of treatment combinations, x. Since the rows of U are orthogonal, U is a scalar multiple of an orthogonal matrix L. For the three-factor structure $U = \sqrt{8}(L)$. Hence,

$$x = U^{-1}y = \frac{1}{\sqrt{8}}L'y = \frac{1}{8}U'y,$$

where L', U' are the transposed forms of L and U. Hence, we can write

$$x = (1/8)\begin{bmatrix} +1 & -1 & -1 & +1 & -1 & +1 & +1 & -1 \\ +1 & +1 & -1 & -1 & -1 & -1 & +1 & +1 \\ +1 & -1 & +1 & -1 & -1 & +1 & -1 & +1 \\ +1 & +1 & +1 & +1 & -1 & -1 & -1 & -1 \\ +1 & -1 & -1 & +1 & +1 & -1 & -1 & +1 \\ +1 & +1 & -1 & -1 & +1 & +1 & -1 & -1 \\ +1 & -1 & +1 & -1 & +1 & -1 & +1 & -1 \\ +1 & +1 & +1 & +1 & +1 & +1 & +1 & +1 \end{bmatrix} y.$$

Hence,

$$x_{ijk} = M + (1/2)\{(-1)^{i-1}P + (-1)^{j-1}Q + (-1)^{i+j-2}PQ + (-1)^{k-1}R$$
$$+ (-1)^{i+k-2}PR + (-1)^{j+k-2}QR + (-1)^{i+j+k-3}PQR\}.$$

$$(13.3)$$

Thus each treatment combination may be expressed in terms of the set of seven treatment effects and, if numerical values are substituted for the effects, estimates of the treatment combination mean yields may be calculated. Of course, if all seven effect estimates are included the estimate of treatment combination means will be simply the original mean yields. For the frog/toad experiment, the predicted yield for toads in dry prior conditions with no hormone is

$$x_{010} = 97.10/8 + (1/2)(11.34 + 10.86 + 3.14 - 7.38 - 6.42$$
$$+ 3.80 + 3.30)$$
$$= 12.14 + (1/2)18.64 = 21.46.$$

However, suppose we had decided that there was clearly no three-factor interaction, then we would obtain a different estimate, since the three-factor interaction effect would be set equal to zero instead of 3.30. The revised estimate of x_{010} would be 19.81. By ignoring other interaction effects, we could obtain other estimates.

Why should we expect any benefits from using estimates different from those actually observed? If we make an assumption, such as $PQR = 0$, which is true, or almost true, then the resulting estimates will be more precise because of the additional information provided by the assumption. This trade-off of assumptions and precision is the basis of any statistical analysis of data. The present form is a particularly blatant example, but is not unusual. To examine the benefit in terms of improved precision, consider again the expression of x_{ijk} in terms of effects (13.3). The eight effects are orthogonal and each element of $\mathbf{y}$ has variance $8\sigma^2/n$ and each comparison effect has variance $\sigma^2/2n$. The estimate of x_{ijk}, setting PQR to zero, includes only seven of the eight terms in the $\mathbf{y}$ vector and therefore the variance of the estimate assuming $PQR = 0$, is

$$(1/8)^2 7(8\sigma^2/n) = (7/8)(\sigma^2/n),$$

a reduction of 1/8 from the original variance of the observed mean, σ^2/n.

Usually, we are most interested in comparisons between two treatment combination means, for example the comparison of hormone and control treatments for frogs in dry pre-experiment conditions. We require an estimate of $x_{011} - x_{010}$, and from the expression (13.3) we obtain

$$x_{011} - x_{010} = R - PR + QR - PQR.$$

To illustrate the effects of making assumptions about interactions, we shall consider five different estimates of $x_{011} - x_{010}$, with the standard errors of those estimates and confidence intervals. The assumptions we shall consider are

(i) no additional assumptions (use observed means),

(ii) assume $PQR = 0$,

(iii) assume PQR = 0, QR = 0,

(iv) assume PQR = 0, PR = 0,

(v) assume PQR = 0, PR = 0, QR = 0.

Since each effect has variance $\sigma^2/2n$, the variance of the alternative estimates of $(x_{111} - x_{110})$ will be

(i) σ^2,

(ii) $3\sigma^2/4$,

(iii) $\sigma^2/2$,

(iv) $\sigma^2/2$,

(v) $\sigma^2/4$.

The results for the frog data are tabulated in Table 13.4 in terms of factors $S (= P)$, $M (= Q)$ and $H (= R)$. The estimates of $x_{011} - x_{010}$ shown in Table 13.4 vary considerably, and so do the 95% confidence intervals. Each point and interval estimate is valid, given the assumptions made, and each could be criticised, not least the estimate (i), which is the value we would use if we had not considered the factorial structure of the treatments. From the analysis of variance for this experiment in Chapter 3, there is little evidence of a three-factor interaction or of an interaction (MH) between the hormone/control factor and the moisture factor. However, the sum of squares for the interaction (SH) between the hormone factor and the species factor is not negligible and this should make us wary about using estimates based on assumptions (iv) or (v). The evidence for an MH interaction is slightly greater than that for the SMH interaction, so the choice of an estimate for $x_{011} - x_{010}$ lies between (ii) and (iii). The best estimate is probably that from (iii), which is the same as was implied in summarising the results of this experiment in Chapter 3.

It may seem that we have spent too much time in discussing a relatively trivial example. However, the central point in this example is the benefit of factorial structure, and in particular the advantages of hidden replication. Using a factorial structure for experimental treatments does provide better information about treatment effects than is implied by the explicit

Table 13.4.

Assumptions	Estimate of $x_{011} - x_{010}$	Standard error of estimate	95% confidence interval
(i) All effects non-zero	6.70	5.88	(−6.88 to 20.28)
(ii) SMH = 0	10.00	5.09	(−1.76 to 21.76)
(iii) SMH = MH = 0	13.80	4.16	(4.19 to 23.41)
(iv) SMH = SH = 0	3.58	4.16	(−6.03 to 13.19)
(v) SMH = MH = SH = 0	7.38	2.94	(0.59 to 14.17)

replication. To use this benefit, it is essential that the estimate of treatment mean yields derived and quoted from the experiment should use the advantages of hidden replication and, to achieve this, we must be prepared to estimate treatment mean yields, ignoring interaction effects for which there is little evidence.

To emphasise this argument, consider the presentation of treatment means as they would be calculated ignoring the factorial structure, but emphasising the species × hormone interaction which the analysis of variance suggests is important (see Table 13.5). Now consider the presentation of the results if we accept the full implications of the analysis of variance that the species × moisture, hormone × moisture and the three-factor interaction may properly be ignored. The means remain unaltered, but the eight treatment combination mean yields are re-estimated using (13.3), ignoring SM, MH and SMH, or equivalently using only the effects S, M, H and SH (see Table 13.6). The differences are not large but the estimates and differences between estimates in Table 13.6 are arguably more appropriate and definitely more precise than those in Table 13.5. The crux of the argument is that we are using a simpler model which we believe is an adequate representation of situation. The more complex model implied by presentation of the observed means is unnecessarily cumbersome.

Table 13.5

Species	Hormone	Wet	Dry	Mean
Toad	control	0.36	21.46	10.91
Frog	control	1.88	10.10	5.99
Toad	hormone	21.26	28.16	24.71
Frog	hormone	3.34	10.54	6.94
Mean		6.71	17.56	12.14

Table 13.6.

Species	Hormone	Wet	Dry	Mean
Toad	control	5.48	16.33	10.91
Frog	control	0.56	11.41	5.99
Toad	hormone	19.28	30.13	24.71
Frog	hormone	1.51	12.36	6.94
Mean		6.71	17.56	12.14

Table 13.7.

	SS	df	MS
Species	515	1	515
Moisture	471	1	471
Hormone	218	1	218
SM	40	1	40
SH	165	1	165
MH	58	1	58
SMH	44	1	44
Error	276	8	34.5
Total	1786	15	

One question which arises when the possibility of ignoring certain interactions is being discussed is which estimate of σ^2 to use. Consider again the analysis of variance for the frog/toad experiment (Table 13.7). If we decide to ignore the SM, MH and SMH interactions, then we might consider omitting the sums of squares of these effects from the analysis of variance as separate components, grouping the sums of squares with the error SS as in Table 13.8.

This procedure is one which has been much discussed by statisticians and research scientists, and it requires careful thought. My belief is that combining insignificant effect SS with the error SS should be avoided unless there are clear benefits to be obtained. There may be advantages from pooling mean squares when there are few degrees of freedom for the error mean square. However, the major philosophical difficulty is that, whereas the original error mean square provides an unbiassed estimate of the variance of an individual observation, the mean square from the combined sum of squares will not be unbiassed. Consider the situation when the effect, whose sum of squares may be pooled, is actually zero. Then that effect SS will be pooled with the error SS, whenever the effect SS

Table 13.8.

	SS	df	MS
Species	51	1	515
Moisture	471	1	471
Hormone	218	1	218
SH	165	1	165
Error	418	11	38
Total	1786	15	

is not significantly larger than the error mean square. This will, on average, reduce the error mean square, since when the error mean square might be most increased we will not pool the sums of squares. In practice, with very large factorial structures, we would be able to examine large numbers of sums of squares for individual interaction effects and, by pooling all the smallest ones, could substantially reduce the error mean square.

Later, we consider the use of interaction sums of squares for the estimation of σ^2. Our arguments for this will be that the interactions concerned are of little interest, being those which involve many factors and for which the practical interpretation is difficult and uninformative. However, we must decide *before* doing the experiment which interaction SS to use to estimate σ^2. Choosing which interaction SS to use in the light of the observed values of the various interaction sums of squares is a dangerous procedure and should be avoided.

13.3 Analysis by considering sums and differences

One other facet of factors with only two levels which may give further insight into the results of 2^m experiments is the consideration of sums and differences of pairs of yields. In considering the definitions of main effects and interactions earlier in this chapter, it is clear that all the effects involving a particular factor, P, include the component $(p-1)$. Sometimes in a 2^m experiment one factor is particularly important. This importance may be recognised before the experiment is performed, or it may become apparent only after the results are obtained. In either case, there are often advantages in considering the analysis of the experimental data in two parts, analysing separately the sums and differences of pairs of yields, each pair differing only in the level of this dominant factor.

The two principal advantages of this approach are, first that the interpretation of the results is likely to be in terms of the pattern of variation of the effect of this important factor over levels of the other factors. This, of course, is essentially a matter of presentation and does not require alteration of the analysis. However, the second advantage is that, if one factor has a very substantial effect, then the variability of sums and differences of pairs of observations for the two levels of that factor might be rather different, contrary to the assumptions of the usual analysis of variance.

The analysis of sums and differences is only practicable if each pair of yields is uniquely defined, and this will be possible only when a randomised block design is used. The frog/toad experiment was not blocked, and therefore it is not possible to establish which of the two yields for frog/wet/hormone should be paired with which of the frog/wet/control yields.

Example 13.1

We therefore use some published data (Rayner, 1967, p. 456) from an experiment on maize in which there were three factors:

(i) two varieties (A and B),

(ii) infestation of plots with witch weed (I) or no infestation (U),

(iii) four fertiliser treatments:

O = control

P = 400 lb superphosphate per morgen,

PM = P + 10 tons farmyard manure per morgen,

PNK = P + 150 lb sulphate of ammonia per morgen
+ 100 lb muriate of potash per morgen.

The yields (lb per 1/200 morgen) from two randomised blocks are shown in Table 13.9.

It is clear immediately that the (U − I) difference is substantial and so we examine the separate analyses of sums (U + I) (Table 13.10) and differences (U − I) (Table 13.11). Note that the 'correction factor' for the mean in the analysis of differences is the sum of squares for the main effect (U − I). The two analyses confirm the suspicion that the very large effect of infestation

Table 13.9.

Variety	Infestation	Treatment	Block I	Block II
A	I	O	10.4	10.0
		P	12.5	10.1
		PM	15.8	13.6
		PNK	11.3	11.4
B	I	O	9.5	9.6
		P	11.8	9.5
		PM	13.5	13.4
		PNK	11.6	9.2
A	U	O	14.8	14.0
		P	19.7	18.0
		PM	24.9	22.0
		PNK	19.9	19.2
B	U	O	12.8	13.0
		P	16.9	16.0
		PM	22.3	20.0
		PNK	17.1	16.6

Table 13.10(a). *Sums* $(U + I)$.

Variety	Treatment	Block		Total
		I	II	
A	O	25.2	24.0	49.2
	P	32.2	28.1	60.3
	PM	40.7	35.6	76.3
	PNK	31.2	30.6	61.8
B	O	22.3	22.6	44.9
	P	28.7	25.5	54.2
	PM	35.8	33.4	69.2
	PNK	28.7	25.8	54.5
	Total	244.8	225.6	470.4

(b). *Analysis of variance.*

	SS	df	MS
Blocks	23.04	1	23.04
Varieties	38.44	1	38.44
Fertilisers	335.49	3	111.83
VF	1.41	3	0.47
Error	11.52	7	1.65
Total	409.90	15	

might distort the assumption of homogeneous variances. The error mean square for differences is markedly lower than that for sums, though the ratio is not significant at 5%. This would suggest that variances of interaction effects involving infestation may be smaller than variances of main effects averaging over infestation levels.

In view of the substantial difference in error variances we may choose to present the results in terms of the separate analyses. The major effect, as we have already observed, is that of infestation. The infestation effect is modified by fertiliser and, to a lesser extent, by variety (difference analysis). There are also main effects of fertiliser and variety (sum analysis). The significant interaction effects are two or three times smaller than the corresponding main effects, and the interpretation should therefore include both main effect means and two-way tables of means. The appropriate means and standard errors are therefore shown in Table 13.12.

The standard errors given in Table 13.12 are for individual quoted values not for differences between these values. Comparison of the values in the difference column with those in the column of means indicates that

Table 13.11(a). *Differences* $(U - I)$.

Variety	Treatment	Block		Total
		I	II	
A	O	4.4	4.0	8.4
	P	7.2	7.9	15.1
	PM	9.1	8.4	17.5
	PNK	8.6	7.8	16.4
B	O	3.3	3.4	6.7
	P	5.1	6.5	11.6
	PM	8.8	6.6	15.4
	PNK	5.5	7.4	12.9
	Total			104.0

(b) *Analysis of variance.*

	SS	df	MS
Varieties	7.29	1	7.29
Fertilisers	44.45	3	14.82
VF	0.66	3	0.22
Error	6.10	8	0.76
Total	58.50	15	
Correction factor (infestation)	676.00	1	676.00

Table 13.12.

	Treatment mean yields		Difference $U - I$	Mean $(U + I)/2$
	Infested	Uninfested		
Mean	11.45	17.95	6.50 SE = 0.22	
O	9.88	13.65	3.78	11.76
P	10.98	17.65	6.67	14.31
PM	14.08	22.30	8.22	18.19
PNK	10.88	18.20	7.32 SE = 0.44	14.54 SE = 0.32
A	11.89	19.06	7.17	15.47
B	11.01	16.84	5.83 SE = 0.31	13.92 SE = 0.23

the infestation effect $(U-I)$ tends to increase with increasing mean value, but, both for fertilisers and for varieties, the differences change relatively more than the means. In practical terms the absolute differences between fertiliser treatments or between varieties are reduced under witch-weed infestation compared with uninfested plots, and the differences are also proportionally smaller for the infested plots.

13.4   Factors with three or more levels

The simplicity of the effects for factors with two levels does not extend to factors with more levels. With three levels, there are two comparisons to be made, and while there are, theoretically at least, many different forms of comparison, there is no pair of comparisons which retains symmetry between all three levels. Consequently, the three levels are treated differently and, particularly as more factors or more levels are included, the effects defined for main effects and interactions become more complex. The range of possible comparisons corresponds to a range of possible questions which may be appropriate to the objectives of the experiment, and it is important that the choice of main effect and interaction comparisons be made in the context of the particular experiment rather than always using a standard set of effects.

In this section we consider some definitions of effects that may be useful. With three levels of a single factor two orthogonal effects may be defined in a multitude of ways, some of which are shown below:

$$(1, \ -1, \quad 0) \quad \text{and} \quad (\quad 1, 1, \ -2)$$
$$(1, \quad 2, \ -3) \quad \text{and} \quad (-5, 4, \quad 1)$$
$$(2, \quad 3, \ -5) \quad \text{and} \quad (-8, 7, \quad 1)$$
$$(1, \quad 3, \ -4) \quad \text{and} \quad (-7, 5, \quad 2)$$

Of these, the first is much the simplest and most frequently useful. The two comparisons split the three levels into two groups, one with two levels and the other with a single level, and the two effects are, respectively, a comparison between the two levels of the first group, and a comparison of the third level with the average of the first two. Thus the crucial choice is which level should be the odd one.

One particular practical situation where the two comparisons $(1, -1, 0)$ and $(1, 1, -2)$ have a simple interpretation is when the three levels of the factor are equally spaced quantitative values. The comparisons

$$t_2 - t_0,$$

and

$$t_2 - 2t_1 + t_0$$

represent the linear and quadratic regression effects of the quantitative factor. This may be recognised directly from the analogy with a regression variable, taking values $-1, 0$ and $+1$. Less directly, by considering the two differences $(t_1 - t_0)$ and $(t_2 - t_1)$ which measure the slopes of the response for the two half intervals, $(t_2 - t_0)$ may be identified as the sum of these two slopes (representing average slope) and $(t_2 - 2t_1 + t_0)$ as the difference (change of slope).

The extension of effects for three level factors to allow for two or more factors tends to increase the complexity of the set of effects, but it is useful to consider in some detail at least the 3^2 factorial structure. The standard set of effects used are

$$P' = (1/3)(p_2 - p_0)(q_2 + q_1 + q_0)$$
$$P'' = (1/6)(p_2 - 2p_1 + p_0)(q_2 + q_1 + q_0)$$
$$Q' = (1/3)(p_2 + p_1 + p_0)(q_2 - q_0)$$
$$Q'' = (1/6)(p_2 + p_1 + p_0)(q_2 - 2q_1 + q_0)$$
$$P'Q' = (1/2)(p_2 - p_0)(q_2 - q_0)$$
$$P'Q'' = (1/4)(p_2 - p_0)(q_2 - 2q_1 + q_0)$$
$$P''Q' = (1/4)(p_2 - 2p_1 + p_0)(q_2 - q_0)$$
$$P''Q'' = (1/8)(p_2 - 2p_1 + p_0)(q_2 - 2q_1 + q_0).$$

Note that the interaction effects are all obtained as direct combinations of main effects and their interpretation follows the same pattern. Thus, $P'Q'$ represents the linear $\times$ linear interaction, the change of the linear effects of P between the extreme levels of Q (by symmetry it is also the change of the linear effect of Q between the extreme levels of P). The $P'Q''$ effect is the linear $\times$ quadratic interaction and has two rather different interpretations. It can be thought of either as the comparison of the linear effects of P between the central and extreme levels of Q; or as the change in the quadratic effect of Q between the extreme levels of P. These interpretations are, of course, all in terms of quantitative factors, but similar interpretations will exist when qualitative factor levels are used, determined by the particular levels used. A final point to note about the effects defined above is the set of multiplicative constants used. In each case, the multiplier is chosen so that the effect is measured as a difference between mean yields per experimental unit.

If we define a ninth effect, the mean,

$$M = (1/9)(p_2 + p_1 + p_0)(q_2 + q_1 + q_0)$$

then, as for two-level factors there is a relationship between the effects, $\mathbf{y}$, and the treatment combinations, $\mathbf{x}$, which in matrix form is

$$\mathbf{y} = \mathbf{U}\mathbf{x}$$

and is written out in full, is

$$
\begin{bmatrix} 9M \\ 3P' \\ 6P'' \\ 3Q' \\ 6Q'' \\ 2P'Q' \\ 4P'Q'' \\ 4P''Q' \\ 8P''Q'' \end{bmatrix} = \begin{bmatrix} 1 & 1 & 1 & 1 & 1 & 1 & 1 & 1 & 1 \\ -1 & -1 & -1 & 0 & 0 & 0 & 1 & 1 & 1 \\ 1 & 1 & 1 & -2 & -2 & -2 & 1 & 1 & 1 \\ -1 & 0 & 1 & -1 & 0 & 1 & -1 & 0 & 1 \\ 1 & -2 & 1 & 1 & -2 & 1 & 1 & -2 & 1 \\ 1 & 0 & -1 & 0 & 0 & 0 & -1 & 0 & 1 \\ -1 & 2 & -1 & 0 & 0 & 0 & 1 & -2 & 1 \\ -1 & 0 & 1 & 2 & 0 & -2 & -1 & 0 & 1 \\ 1 & -2 & 1 & -2 & 4 & -2 & 1 & -2 & 1 \end{bmatrix} \begin{bmatrix} p_0q_0 \\ p_0q_1 \\ p_0q_2 \\ p_1q_0 \\ p_1q_1 \\ p_1q_2 \\ p_2q_0 \\ p_2q_1 \\ p_2q_2 \end{bmatrix}.
$$

The matrix $\mathbf{U}$ can be expressed as the product of two matrices

$$\mathbf{U} = \mathbf{VW},$$

where $\mathbf{W}$ is orthogonal and $\mathbf{V}$ is diagonal with

$$v_{ii}^2 = \sum_j u_{ij}^2$$

giving squared diagonal elements (9, 6, 18, 6, 18, 4, 12, 12, 36) for $\mathbf{V}$. Again, just as for two-level factors, the relationship between effects and combinations can be reversed.

$$
\begin{aligned}
\mathbf{x} &= \mathbf{U}^{-1}\mathbf{y} \\
&= (\mathbf{VW})^{-1}\mathbf{y} \\
&= \mathbf{W}^{-1}\mathbf{V}^{-1}\mathbf{y} \\
&= \mathbf{W}'\mathbf{V}^{-1}\mathbf{y} \\
&= \mathbf{W}'\mathbf{V}\mathbf{V}^{-1}\mathbf{V}^{-1}\mathbf{y} \\
&= \mathbf{U}'\mathbf{V}^{-2}\mathbf{y}.
\end{aligned}
$$

Hence, using the squared diagonal elements of $\mathbf{V}$ and the transpose of $\mathbf{U}$ we can derive comparisons between the treatment combinations in terms of effects;

$$
\begin{aligned}
p_0q_0 &= 9M/9 - 3P'/6 + 6P''/18 - 3Q'/6 + 6Q''/18 + 2P'Q'/4 \\
&\quad - 4P'Q''/12 - 4P''Q'/12 + 8P''Q''/36 \\
&= M - P'/2 + P''/3 - Q'/2 + Q''/3 + P'Q'/2 - P'Q''/3 \\
&\quad - P''Q'/3 + 2P''Q''/9.
\end{aligned}
$$

Similarly

$$p_0q_1 = M - P'/2 + P''/3 - 2Q''/3 + 2P'Q''/3 - 2P''Q''/9,$$

and hence

$$p_0q_0 - p_0q_1 = -Q'/2 - Q'' + P'Q'/2 - P'Q'' - P''Q''/3 + 2P''Q''/3.$$

If all interactions are assumed zero, this comparison reduces to

$$-Q'/2 - Q''.$$

If the linear $\times$ linear interaction is assumed to be non-zero but other interaction effects are assumed zero, then our estimate of $p_0q_0 - p_0q_1$ is

$$-Q'/2 - Q'' + P'Q'/2.$$

Estimates of comparisons between treatment combinations assuming certain interactions to be zero are more precise provided the assumptions are correct. Again the details are more complicated than for two-level factors because the different effects have different variances. Each element of the effect vector, $\mathbf{y}$, has variance $\sigma^2 v_{ii}^2/n$, where n is the replication of each treatment combination. The variances of effects can be calculated as

$$\mathrm{Var}(P') \quad = \mathrm{Var}(Q') \quad = (1/3)^2 6\sigma^2/n \ = 2\sigma^2/3n$$
$$\mathrm{Var}(P'') \quad = \mathrm{Var}(Q'') \quad = (1/6)^2 18\sigma^2/n = \sigma^2/2n$$
$$\mathrm{Var}(P'Q') \qquad\qquad = (1/2)^2 4\sigma^2/n \ = \sigma^2/n$$
$$\mathrm{Var}(P'Q'') = \mathrm{Var}(P''Q') = (1/4)^2 12\sigma^2/n = 3\sigma^2/4n$$
$$\mathrm{Var}(P''Q'') \qquad\qquad = (1/8)^2 36\sigma^2/n = 9\sigma^2/16n.$$

Hence the variance of the three alternative estimates of $(p_0q_0 - p_0q_1)$ considered previously are

(*a*) making no assumptions

$$(1/4)(2\sigma^2/3n) + (\sigma^2/2n) + (1/4)(\sigma^2/n) + (3\sigma^2/4n)$$
$$+ (1/9)(3\sigma^2/4n) + (4/9)(9\sigma^2/16n) = 2\sigma^2/n$$

which is, of course, the variance for comparing two treatment means each based on n observations,

(*b*) assuming all interactions are zero

$$(1/4)(2\sigma^2/3n) + (\sigma^2/2n) = 2\sigma^2/3n,$$

(*c*) assuming all interactions except the linear $\times$ linear are zero

$$(1/4)(2\sigma^2/3n) + (\sigma^2/2n) + (1/4)(\sigma^2/n) = 11\sigma^2/12n.$$

The variance for (*b*) is simply interpreted as the variance for comparing two treatment means each based on $3n$ observations and, since we are assuming zero interactions, the comparison $(p_0q_0 - p_0q_1)$ is estimated by the main effects comparison $(q_0 - q_1)$. The variance for (*c*), which might often be the most reasonable assumption, is much closer to that for (*b*) than to that for (*a*), illustrating again the benefit of factorial structure when some interaction effects are negligible.

The extension of the set of factorial effects from the 2^m to the 3^m case can be continued to allow for more than three levels of a factor and for mixed levels. For three-level factors, although there is theoretically an infinite choice of possible pairs of effects, we have argued that in practice much the most important pair of effects are $(1, -1, 0)$ and $(1, 1, -2)$. With four levels, there are three orthogonal effects to be defined and there is not a single dominant set of three effects. If the levels are equally spaced levels of a quantitative factor, then the three effects may be defined to represent linear, quadratic and cubic effects, as follows:

linear $(-3, -1, 1, \quad 3)$

quadratic $(-1, \quad 1, 1, -1)$

cubic $(\quad 1, -3, 3, -1)$.

Alternatively, the effects can represent successive comparisons of one level compared with the remainder:

$(\quad 3, -1, -1, -1)$

$(\quad 0, \quad 2, -1, -1)$

$(\quad 0, \quad 0, \quad 1, -1)$.

A third alternative involves a hierarchical structure of the levels with two main pairs, each split:

$(\quad 1, \quad 1, -1, -1)$

$(\quad 1, -1, \quad 0, \quad 0)$

$(\quad 0, \quad 0, \quad 1, -1)$.

A small modification of this last set of comparisons leads us back to the 2^2 structure for four levels:

$(\quad 1, \quad 1, -1, -1)$

$(\quad 1, -1, \quad 1, -1)$

$(-1, \quad 1, \quad 1, -1)$.

However, the set of effects are defined the effects vector, $\mathbf{y}$, may be written in terms of the set of treatment combination means, $\mathbf{x}$, through an equation in the form

$$\mathbf{y} = \mathbf{U}\mathbf{x},$$

where $\mathbf{U} = \mathbf{V}\mathbf{W}$ and $\mathbf{W}$ is orthogonal and $\mathbf{V}$ diagonal, exactly as for the 3^m or 2^m structure.

The same structural relationship of effects and treatment combinations holds for mixtures of levels. Thus, for three factors, P, Q and R with three, two and two levels, respectively,

$$\mathbf{P}' = (1/4)(p_2 - p_0)(q_1 + q_0)(r_1 + r_0)$$

$$P'' = (1/8)(p_2 - 2p_1 + p_0)(q_1 + q_0)(r_1 + r_0)$$
$$Q = (1/6)(p_2 + p_1 + p_0)(q_1 - q_0)(r_1 + r_0)$$
$$R = (1/6)(p_2 + p_1 + p_0)(q_1 + q_0)(r_1 - r_0)$$
$$P'Q = (1/4)(p_2 - p_0)(q_1 - q_0)(r_1 + r_0)$$
$$P''Q = (1/8)(p_2 - 2p_1 + p_0)(q_1 - q_0)(r_1 + r_0)$$
$$P'QR = (1/4)(p_2 - p_0)(q_1 - q_0)(r_1 - r_0).$$

The remaining effects may be completed by the reader or read from the matrix $\mathbf{U}$ given below:

$$
\mathbf{y} = \qquad\qquad \mathbf{U} \qquad\qquad \mathbf{x}
$$

$$
\begin{bmatrix}
12M \\
4P' \\
8P'' \\
6Q \\
6R \\
4P'Q \\
8P''Q \\
4P'R \\
8P''R \\
6QR \\
4P'QR \\
8P''QR
\end{bmatrix}
=
\begin{bmatrix}
1 & 1 & 1 & 1 & 1 & 1 & 1 & 1 & 1 & 1 & 1 & 1 \\
-1 & -1 & -1 & -1 & 0 & 0 & 0 & 0 & 1 & 1 & 1 & 1 \\
1 & 1 & 1 & 1 & -2 & -2 & -2 & -2 & 1 & 1 & 1 & 1 \\
-1 & -1 & 1 & 1 & -1 & -1 & 1 & 1 & -1 & -1 & 1 & 1 \\
-1 & 1 & -1 & 1 & -1 & 1 & -1 & 1 & -1 & 1 & -1 & 1 \\
1 & 1 & -1 & -1 & 0 & 0 & 0 & 0 & -1 & -1 & 1 & 1 \\
-1 & -1 & 1 & 1 & 2 & 2 & -2 & -2 & -1 & -1 & 1 & 1 \\
1 & -1 & 1 & -1 & 0 & 0 & 0 & 0 & -1 & 1 & -1 & 1 \\
-1 & 1 & -1 & 1 & 2 & -2 & 2 & -2 & -1 & 1 & -1 & 1 \\
1 & -1 & -1 & 1 & 1 & -1 & -1 & 1 & 1 & -1 & -1 & 1 \\
-1 & 1 & 1 & -1 & 0 & 0 & 0 & 0 & 1 & -1 & -1 & 1 \\
1 & -1 & -1 & 1 & -2 & 2 & 2 & -2 & 1 & -1 & -1 & 1
\end{bmatrix}
\begin{bmatrix}
p_0q_0r_0 \\
p_0q_0r_1 \\
p_0q_1r_0 \\
p_0q_1r_1 \\
p_1q_0r_0 \\
p_1q_0r_1 \\
p_1q_1r_0 \\
p_1q_1r_1 \\
p_2q_0r_0 \\
p_2q_0r_1 \\
p_2q_1r_0 \\
p_2q_1r_1
\end{bmatrix}.
$$

As for the three-level factors $\mathbf{U} = \mathbf{VW}$, where $\mathbf{W}$ is orthogonal and $\mathbf{V}$ is diagonal with

$$v_{ii}^2 = \sum_j u_{ij}^2$$

giving squared diagonal elements of $\mathbf{V}$:

$$(12, 8, 24, 12, 12, 8, 24, 8, 24, 12, 8, 24).$$

And once again we can express the vector of treatment combination yields, $\mathbf{x}$, in terms of the effect vector,

$$\mathbf{x} = \mathbf{U}'\mathbf{V}^{-2}\mathbf{y}.$$

13.5   The use of only a single replicate

The logic of the economy afforded by factorial structure leads naturally to including more and more factors, gradually replacing explicit replication by hidden replication. The extreme situation might seem to occur when we consider only a single replicate of a factorial structure. There are two obvious problems about this possible design. The first is how to utilise relevant blocking structures and this is a major subject which will be developed at length in Chapters 15 and 16. However, the second problem which might seem to make the single-replication design inadmissible is the requirement to estimate σ^2 so that standard errors of treatment effects may be calculated.

Consider, for example, a four-factor experiment with factors P (four levels), Q (three levels), R (two levels) and S (two levels), a total of 48 experimental units in a single replicate. If the factor S were not included, we could have two blocks of 24 units each, with the analysis of variance structure as in Table 13.13, and σ^2 can be estimated from the error mean square. But 23 is a luxuriously large number of degrees of freedom for estimating σ^2 and, by including S, we could obtain additional information not only about S but also about the two-factor interactions of S with P, Q and R. For the four-factor experiment, the full analysis of variance structure will be as shown in Table 13.14 leaving no degrees of freedom for

Table 13.13.

Source	df
Blocks	1
P	3
Q	2
R	1
PQ	6
PR	3
QR	2
PQR	6
Error	23
Total	47

Table 13.14.

Source	df	Source	df
P	3	QS	2
Q	2	RS	1
R	1	PQR	6
S	1	PQS	6
PQ	6	PRS	3
PR	3	QRS	2
PS	3	PQRS	6
QR	2	Total	47

estimating σ^2. Now, in most factorial experiments, the numerical size of effects tends to decline as the number of factors involved in the effect increases. That is, two-factor interaction effects are usually smaller than the corresponding main effects for the two factors; three-factor interaction effects are usually smaller than the corresponding two-factor interaction effects, and so on. There is, of course, no absolute rule that this decline shall occur but statistically such a pattern is often observed. It might be anticipated theoretically by arguing that interactions are essentially modifications of main effects. It is also sometimes argued that the response to changing various quantitative factors could be modelled by a Taylor series expansion, and the terms in that expansion, with more variables included, which represent higher-order interactions, must diminish for convergence of the series.

The general drift of the arguments in the last paragraph is that we may expect to find that three-factor and higher-order interactions are negligibly small. Although only a wide experience can be convincing, most examples in this book do support this expectation as do most examples in a wide range of published books and journals. The conclusion to be drawn is that the three- and four-factor interactions will rarely be of importance, and that the expectations of the mean squares for these effects will often be only slightly larger than σ^2. Hence, we can obtain a sensible estimate of σ^2 by pooling the mean squares for some, or all, three- and four-factor interactions. In the previous four-factor example, if all three- and four-factor interaction sums of squares are used there are 23 degrees of freedom to estimate σ^2. This is ample and we might decide *a priori* that a particular three-factor interaction could be interesting and separate the sum of squares for that interaction effect from the pooled estimate of σ^2. Such decisions should be taken prior to the analysis, or even to examination of the data. The problems inherent in choosing which terms to include in an analysis of variance have been discussed earlier, in Section 13.2, but are so important that the topic needs emphasis. The decision about the form of analysis should be taken at the design stage, rather than when the data is available. Careful selection of estimates of σ^2 may produce very low 'estimates' for σ^2 and some most spectacularly significant results, but the experimenter or statistician who is satisfied by such manipulated results is either a fool or a rogue, and has certainly not understood the purpose of statistics.

Example 13.2

To illustrate the use of a single replicate consider an experiment on the survival of *Salmonella typhimerium*, in which three factors were examined.

Table 13.15.

Sorbic acid (p.p.m.)	pH	Water activity					
		0.78	0.82	0.86	0.90	0.94	0.98
0	5.0	4.20	4.52	5.01	6.14	6.25	8.33
	5.5	4.34	4.31	5.35	5.98	6.70	8.37
	6.0	4.31	4.85	5.06	5.87	6.65	8.19
100	5.0	4.18	4.18	4.29	5.78	6.51	7.59
	5.5	4.39	4.43	4.95	5.28	6.19	7.79
	6.0	4.13	4.29	4.85	5.01	6.52	7.64
200	5.0	4.15	4.37	4.79	5.43	6.43	7.19
	5.5	4.12	4.27	4.40	5.10	6.18	6.92
	6.0	3.93	4.26	4.41	5.20	6.33	7.14

Three levels of sorbic acid, six levels of water activity and three pH levels were combined to give 54 combinations. The data, presented in Table 13.15 are log (density/ml) measured seven days after treatments started. The analysis of variance, including sums of squares for all main effects and two- and three-factor interactions is as in Table 13.16.

The design of the experiment clearly implies the intention of using the three-factor interaction mean square as an estimate of error, and the variance ratios have been calculated using $s^2 = 0.05$ as the divisor. The single replicate clearly provides effective replication for all main effects and the two-factor AS interaction. The major effect is that of water activity, and comparison of the six water activity means is based on an effective nine-fold replication. The sorbic acid means and the pH means have an effective replication of 18, the means for water activity × sorbic acid

Table 13.16.

	SS	df	MS	F
Water activity (A)	81.57	5	16.31	473
Sorbic acid (S)	2.76	2	1.38	40
pH	0.01	2	0.01	0.2
AS	1.32	10	0.13	3.8
ApH	0.45	10	0.04	1.3
SpH	0.23	4	0.06	1.7
ASpH (error)	0.69	20	0.03	
Total	87.03	53		

Table 13.17.

Sorbic acid (p.p.m.)	Water activity						Mean
	0.78	0.82	0.86	0.90	0.94	0.98	
0	4.28	4.56	5.14	6.00	6.53	8.30	5.80
100	4.23	4.30	4.77	5.36	6.41	7.67	5.44
200	4.07	4.30	4.53	5.24	6.31	7.08	5.26
Mean	4.19	4.39	4.79	5.53	6.42	7.68	

SE difference (water activity means) = 0.105
SE difference (sorbic acid means) = 0.075
SE difference (combination means) = 0.183

combinations have effective replication of 3 and the two-way means for pH × sorbic acid have effective replication of 6.

The presentation of results should be as in Table 13.17.

13.6   The use of a fraction of a complete factorial experiment

In the previous section we have seen that it is possible to use a single replicate of a factorial set of treatments to obtain estimates of main effects and two-factor interactions, using high-order interaction mean squares to estimate σ^2 and obtaining sufficient effective replication from the hidden replication of the factorial structure. The three crucial ideas of the use of a single replicate design are:

(i) only a subset of effects are important;

(ii) it is possible to estimate σ^2 from high-order interaction mean squares;

(iii) effective replication of the important treatment effects does not require explicit replication.

Suppose the number of factors which the experimenter wishes to consider is so large that the total number of treatment combinations is greater than can be included in a single experiment. By deliberate choice of the treatment combinations to be included it is possible to obtain all the relevant information from a fraction of the total combinations.

A design using only a fraction of the possible factorial treatment combinations is called a fractional replicate. For example, suppose that we wish to investigate six factors, P, Q, R, S, T and U, each at two levels but that an experiment of 64 observations is too large for the available resources. To assess all the main effects and two-factor interactions requires 6 and 15 degrees of freedom, respectively. Using half of the 64 combinations gives a total of 31 df so that after estimating the main effects and two-factor interactions there are 10 df available to estimate σ^2.

Example 13.3

To illustrate the construction and analysis of a fractional replicate we first consider a trivial example of a half-replicate of a 2^4 factorial structure with factors P, Q, R and S. The combinations to be included in the experiment should satisfy the following criteria:

(i) To estimate a main effect efficiently the two levels of the factor must be equally replicated.

(ii) To estimate main effects independently each pair of main effects should be orthogonal, for which all four combinations of the levels of the two factors should be equally replicated.

These two criteria can be satisfied by several sets of combinations, of which four are shown in Figure 13.1. In fact the number of possibilities is very limited. Either each set of three factors (PQR, PQS, PRS, QRS) has all possible eight combinations present (set I) or there is exactly one set of three factors which does not have all eight combinations present (sets II(PRS), III(PQS), IV(PQR)). All designs satisfying (i) and (ii) are in the form of set I or set II.

If we adopt the arbitrary, but intuitively satisfying, additional requirement that for each set of three factors all combinations shall be equally replicated, then we would select set I as our design; note that for each set there is a complementary design, consisting of the other eight combinations from the total set of 16, which possesses exactly the same properties. The estimation of effects can be represented as in Table 13.18.

Figure 13.1. Subsets of the 2^4 factorial combinations for factors P, Q, R, S satisfying equal replication conditions given in the text.

Set I	Set II	Set III	Set IV
$p_0 q_0 r_0 s_0$	$p_0 q_0 r_0 s_0$	$p_0 q_0 r_0 s_0$	$p_0 q_0 r_0 s_0$
$p_0 q_0 r_1 s_1$	$p_0 q_0 r_1 s_1$	$p_0 q_0 r_1 s_0$	$p_0 q_0 r_0 s_1$
$p_0 q_1 r_0 s_1$	$p_0 q_1 r_0 s_0$	$p_0 q_1 r_0 s_1$	$p_0 q_1 r_1 s_0$
$p_0 q_1 r_1 s_0$	$p_0 q_1 r_1 s_1$	$p_0 q_1 r_1 s_1$	$p_0 q_1 r_1 s_1$
$p_1 q_0 r_0 s_1$	$p_1 q_0 r_0 s_1$	$p_1 q_0 r_0 s_1$	$p_1 q_0 r_1 s_0$
$p_1 q_0 r_1 s_0$	$p_1 q_0 r_1 s_0$	$p_1 q_0 r_1 s_1$	$p_1 q_0 r_1 s_1$
$p_1 q_1 r_0 s_0$	$p_1 q_1 r_0 s_1$	$p_1 q_1 r_0 s_0$	$p_1 q_1 r_0 s_0$
$p_1 q_1 r_1 s_1$	$p_1 q_1 r_1 s_0$	$p_1 q_1 r_1 s_0$	$p_1 q_1 r_0 s_1$

Table 13.18.

Combination	P	Q	R	S	PQ	PR	PS	QR	QS	RS
$p_0q_0r_0s_0$	-1	-1	-1	-1	$+1$	$+1$	$+1$	$+1$	$+1$	$+1$
$p_0q_0r_1s_1$	-1	-1	$+1$	$+1$	$+1$	-1	-1	-1	-1	$+1$
$p_0q_1r_0s_1$	-1	$+1$	-1	$+1$	-1	$+1$	-1	-1	$+1$	-1
$p_0q_1r_1s_0$	-1	$+1$	$+1$	-1	-1	-1	$+1$	$+1$	-1	-1
$p_1q_0r_0s_1$	$+1$	-1	-1	$+1$	-1	-1	$+1$	$+1$	-1	-1
$p_1q_0r_1s_0$	$+1$	-1	$+1$	-1	-1	$+1$	-1	-1	$+1$	-1
$p_1q_1r_0s_0$	$+1$	$+1$	-1	-1	$+1$	-1	-1	-1	-1	$+1$
$p_1q_1r_1s_1$	$+1$	$+1$	$+1$	$+1$	$+1$	$+1$	$+1$	$+1$	$+1$	$+1$

Clearly all main effects are defined to be orthogonal to each other. Equally clearly it is not possible for all ten effects represented to be orthogonal since with eight combinations there can only be seven orthogonal combinations. By inspection there are pairs of two-factor interactions for which the estimates are identical. The estimate of PS is exactly the same as that for QR; $PQ \equiv RS$ and $PR \equiv QS$. In each case the linear contrast of combinations estimates not one parameter of the effects model but two. If the quantity

$$(p_0q_0r_0s_0 - p_0q_0r_1s_1 - p_0q_1r_0s_1 + p_0q_1r_1s_0$$
$$+ p_1q_0r_0s_1 - p_1q_0r_1s_0 - p_1q_1r_0s_0 + p_1q_1r_1s_1)$$

is large then we can only interpret this as evidence that the factors P and S interact if we assume that the effect QR is negligible. This is expressed by saying that PS and QR are aliases.

For an alternative view of the aliasing concept consider the two sets of combinations defined in Table 13.19.

Table 13.19.

Set A	Set B
$y_1 = p_0q_0r_0s_0$	$y_9 = p_0q_0r_0s_1$
$y_2 = p_0q_0r_1s_1$	$y_{10} = p_0q_0r_1s_0$
$y_3 = p_0q_1r_0s_1$	$y_{11} = p_0q_1r_0s_0$
$y_4 = p_0q_1r_1s_0$	$y_{12} = p_0q_1r_1s_1$
$y_5 = p_1q_0r_0s_1$	$y_{13} = p_1q_0r_0s_0$
$y_6 = p_1q_0r_1s_0$	$y_{14} = p_1q_0r_1s_1$
$y_7 = p_1q_1r_0s_0$	$y_{15} = p_1q_1r_0s_1$
$y_8 = p_1q_1r_1s_1$	$y_{16} = p_1q_1r_1s_0$

All 16 treatment combinations are included in one set or the other, and set A is our previous set I. The effect PS is

$$PS = (1/8)(p_1 - p_0)(q_1 + q_0)(r_1 + r_0)(s_1 - s_0),$$

Hence,

$$8PS = (y_1 - y_2 - y_3 + y_4 + y_5 - y_6 - y_7 + y_8)$$
$$- (y_9 - y_{10} - y_{11} + y_{12} + y_{13} - y_{14} - y_{15} + y_{16})$$

while

$$8QR = (y_1 - y_2 - y_3 + y_4 + y_5 - y_6 - y_7 + y_8)$$
$$+ (y_9 - y_{10} - y_{11} + y_{12} + y_{13} - y_{14} - y_{15} + y_{16}).$$

Thus PS is estimated by the difference of two contrasts one from each of the two sets, while QR is estimated by the sum of the same two contrasts. When the complete set of 16 observations is available, PS and QR are orthogonal and may each be estimated separately. When only one set of observations is available, then the estimates of PS and QR will be identical (except possibly in sign).

We may consider whether the main effects also have aliases since, on the argument of the last paragraph, the estimate of P is only 'half' of the normal estimate and might be expected to be 'half' of the normal estimate of some other effect. If we consider the three-factor interaction effects then the estimate of PQR from the half replicate (set A) would be

$$-y_1 + y_2 + y_3 - y_4 + y_5 - y_6 - y_7 + y_8,$$

which is identical with the estimate for S.

Further investigation reveals the complete set of aliases to be

$$P \equiv QRS \qquad PQ \equiv RS$$
$$Q \equiv PRS \qquad P \equiv QS$$
$$R \equiv PQS \qquad PS \equiv QR$$
$$S \equiv PQR$$

The reader will perceive a pattern in these aliases, and mathematical theory provides a basis for deducing such patterns, but we shall defer discussion of this theory until Chapter 16.

The aliasing of P and QRS is not likely to cause problems of interpretation. If the estimate, which could be attributable to either effect, is large then there will rarely be any hesitation in assuming that QRS is negligible and interpreting the estimate as the effect P.

Example 13.4

Consider again the design of an experiment using only 32 observations for a 2^6 factorial structure. For the purposes of estimating all main effects and interactions we require

(i) for each factor the two levels are equally replicated,

(ii) for each pair of factors the four combinations are equally replicated.

To ensure that the estimates are all orthogonal we require

(iii) for any pair of interactions all combinations of levels for all factors involved are equally replicated. This requires that for each set of four factors, all 16 combinations of levels are equally replicated.

These requirements are not sufficient to produce a unique design and, as with the simpler example, we adopt the general requirement that as far as is possible, for each set of factors, all combinations shall be equally replicated. In this case the requirement leads to having all combinations for each set of five factors occurring in the design. The resulting design is shown in Figure 13.2.

Figure 13.2. A half replicate (32 observations) of a 2^5 factorial structure including all combinations of levels for any set of five factors.

$p_0q_0r_0s_0t_0u_0$	$p_0q_1r_0s_0t_0u_1$	$p_1q_0r_0s_0t_0u_1$	$p_1q_1r_0s_0t_0u_0$
$p_0q_0r_0s_0t_1u_1$	$p_0q_1r_0s_0t_1u_0$	$p_1q_0r_0s_0t_1u_0$	$p_1q_1r_0s_0t_1u_1$
$p_0q_0r_0s_1t_0u_1$	$p_0q_1r_0s_1t_0u_0$	$p_1q_0r_0s_1t_0u_0$	$p_1q_1r_0s_1t_0u_1$
$p_0q_0r_0s_1t_1u_0$	$p_0q_1r_0s_1t_1u_1$	$p_1q_0r_0s_1t_1u_1$	$p_1q_1r_0s_1t_1u_0$
$p_0q_0r_1s_0t_0u_1$	$p_0q_1r_1s_0t_0u_0$	$p_1q_0r_1s_0t_0u_0$	$p_1q_1r_1s_0t_0u_1$
$p_0q_0r_1s_0t_1u_0$	$p_0q_1r_1s_0t_1u_1$	$p_1q_0r_1s_0t_1u_1$	$p_1q_1r_1s_0t_1u_0$
$p_0q_0r_1s_1t_0u_0$	$p_0q_1r_1s_1t_0u_1$	$p_1q_0r_1s_1t_0u_1$	$p_1q_1r_1s_1t_0u_0$
$p_0q_0r_1s_1t_1u_1$	$p_0q_1r_1s_1t_1u_0$	$p_1q_0r_1s_1t_1u_0$	$p_1q_1r_1s_1t_1u_1$

The alias structure can be determined to be

$$P \equiv QRSTU \quad PQ \equiv RSTU \quad QR \equiv PSTU \quad RT \equiv PQSU$$
$$Q \equiv PRSTU \quad PR \equiv QSTU \quad QS \equiv PRTU \quad RU \equiv PQST$$
$$R \equiv PQSTU \quad PS \equiv QRTU \quad QT \equiv PRSU \quad ST \equiv PQRU$$
$$S \equiv PQRTU \quad PT \equiv QRSU \quad QU \equiv PRST \quad SU \equiv PQRT$$
$$T \equiv PQRSU \quad PU \equiv QRST \quad RS \equiv PQTU \quad TU \equiv PQRS$$
$$U \equiv PQRST$$

The analysis of variance (see Table 13.20) is in the form discussed earlier.

Example 13.5

The analysis of a fractional replicate for a factorial structure is illustrated for data from an experiment in which four four-level factors were investigated using 64 observations, a one-quarter replicate. The design,

Table 13.20.

	df
Six main effects	6
15 two-factor interactions	15
Error (three-factor interactions)	10
Total	31

which included all possible combinations of levels from three factors, for each possible set of three factors, involved some aliasing of components of different two-factor interactions. The four experimental factors were

MC = moisture content
TE = temperature
TI = duration (time)
PH = pH of solution.

The results for one variable, TSS, are given below. For each combination of MC, TE, TI there is an observation for just one pH level. In the table of results (Table 13.21) the coded pH values and observed TSS values are placed adjacently. The first impression from the data is of a major effect of pH with possibly some modification by MC. There appears to be a

Table 13.21.

MC	TE	TI=0		TI=1		TI=2		TI=3	
		PH	TSS	PH	TSS	PH	TSS	PH	TSS
0	0	0	23.70	2	41.44	3	63.34	1	24.13
0	1	2	41.25	0	22.66	1	24.19	3	65.74
0	2	3	65.78	1	22.35	0	22.99	2	42.84
0	3	1	22.89	3	73.28	2	40.52	0	22.03
1	0	1	32.40	0	20.39	2	48.50	3	75.37
1	1	3	75.24	2	47.87	0	19.10	1	26.87
1	2	2	46.83	3	71.51	1	19.01	0	19.07
1	3	0	19.69	1	19.54	3	69.66	2	29.13
2	0	3	70.62	1	21.26	0	21.58	2	41.68
2	1	1	19.84	3	71.89	2	43.86	0	20.82
2	2	0	21.14	2	44.52	3	61.91	1	20.03
2	3	2	35.90	0	20.29	1	18.96	3	27.48
3	0	2	44.14	3	72.84	1	20.00	0	19.99
3	1	0	21.66	1	19.76	3	69.82	2	58.99
3	2	1	19.14	0	20.57	2	19.28	3	21.93
3	3	3	31.99	2	18.59	0	21.25	1	21.60

Table 13.22.

	SS	df	MS
MC	0.629	3	0.210
TE	0.754	3	0.251
TI	0.071	3	0.024
PH	11.567	3	3.856
Residual	2.805	51	0.055
Total	15.826	63	

possibility that the variability is not homogeneous but the absence of explicit replication makes this difficult to assess formally. It was argued in Chapter 11 that for such situations a log transformation will often provide a valid analysis and logged data are used for this example.

The first stage of the analysis is to calculate the analysis of variance for main effects only (see Table 13.22). Clearly PH provides the dominant contribution to the variation, with MC and TE showing smaller effects.

The aliasing structure involves the pairing of some contrasts from different two-factor interactions. Thus part of the MCTE interaction is aliased with part of the TIPH interaction. Similarly there is some aliasing between MCTI and TEPH and between MCPH and TETI. In view of the patterns of main effects it would be reasonable to expect MCPH and TEPH to be more important than their aliases but the distinction between MCTE or TIPH is not clearly predictable. We therefore calculate two analyses of variance (Tables 13.23a and b). In neither analysis do the interactions appear significant. The TIPH interaction accounts for rather more variation than does MCTE (remember the aliasing is only partial). If the analysis is taken further and the main effects and interactions are split into linear, quadratic and cubic effects then the analysis suggests that the linear × linear component of the MCTE interaction is important. In contrast simple contrasts of the TIPH interaction show no substantial effect. Consequently it was decided to interpret the result in terms of the MCTE interaction rather than the TIPH interaction. The detailed analysis of variance is given in Table 13.24. The important conclusion from the design point of view is that the use of a quarter-replicate enabled the experimenter to examine all four main effects and to obtain some indications about the pattern of interactions. The two-way tables of means are in Table 13.25. (For completeness, the PHTI two-way table, in which no simple pattern has been found, is also shown in Table 13.25.)

Table 13.23(a)

	SS	df	MS
MC	0.629	3	0.210
TE	0.754	3	0.251
TI	0.071	3	0.024
PH	11.567	3	3.856
MCTE	0.705	9	0.078
MCPH	0.446	9	0.050
TEPH	0.436	9	0.048
Residual	1.218	24	0.051

(b)

	SS	df	MS
MC	0.629	3	0.210
TE	0.754	3	0.251
TI	0.071	3	0.023
PH	11.567	3	3.856
MCPH	0.446	9	0.050
TEPH	0.436	9	0.048
TIPH	0.809	9	0.090
Residual	1.114	24	0.046

Table 13.24. *Analysis of variance for the model with fitted linear and quadratic effects.*

Terms	SS	df	MS
MC linear	0.576	1	0.576
other	0.045	2	0.022
TE linear	0.672	1	0.672
other	0.279	2	0.140
TI	0.071	3	0.023
PH linear	10.603	1	10.603
quadratic	0.537	1	0.537
cubic	0.427	1	0.427
MCTE linear × linear	0.285	1	0.285
MCPH linear × linear	0.203	1	0.203
TEPH linear × linear	0.294	1	0.294
Other interaction contrasts	0.805	24	0.033
Residual	1.218	24	0.051

Table 13.25

		PH				TE			
		0	1	2	3	1	2	3	4
MC	1	3.13	3.15	3.75	4.20	3.56	3.55	3.57	3.56
	2	2.97	3.17	3.74	4.29	3.67	3.61	3.50	3.39
	3	3.04	3.00	3.72	3.99	3.53	3.52	3.49	3.21
	4	3.04	3.00	3.44	3.77	3.52	3.60	3.01	3.13
TE	1	3.06	3.18	3.78	4.25				
	2	3.04	3.11	3.86	4.26				
	3	3.04	3.00	3.62	3.92				
	4	3.04	3.03	3.39	3.83				
Mean		3.04	3.08	3.66	4.06				
TI	1	3.07	3.14	3.73	4.06				
	2	3.04	3.03	3.58	4.28				
	3	3.05	3.02	3.58	4.19				
	4	3.02	3.14	3.76	3.73				

Standard error of differences between two means within a two-way table $=0.159$. Standard error of difference between two PH means $=0.080$.

Exercises 13

(1) A 2^4 experiment on yield from a chemical reaction gave treatment totals (from two replicates) shown in Table 13.26. Estimate the effects and calculate the analysis of variance. If the error mean square is 100 (based on 15 df), draw your conclusions from the experiment.

Table 13.26.

	d_0		l_1	
	c_0	c_1	c_0	c_1
a_0b_0	8	31	79	77
b_1	53	12	73	49
a_1b_0	4	9	68	38
b_1	43	36	8	23

If it could be assumed that only main effects and two-factor interactions were non-zero, find estimates of the comparisons:

$(a_1b_0c_0d_0 - a_0b_0c_0d_0)$,

$(a_0b_1c_0d_0 - a_0b_0c_0d_0)$,

$(a_0b_0c_1d_0 - a_0b_0c_0d_0)$

and

$(a_0b_0c_0d_1 - a_0b_0c_0d_0)$;

also find the variance of any one of these estimates.

(2) Write out the matrix relationship between the effects vector, **y**, and the observations vector, **x**, for an experiment with six treatments comprising three levels of factor A with two levels of factor B. The usual form of effects for A should be assumed $((a_2 - a_0)$ and $(a_2 - 2a_1 + a_0))$ with corresponding interaction effects.

Derive the reverse relationship expressing **x** in terms of **y**. Express the variances of the comparisons $(a_1b_1 - a_0b_0)$, $(a_0b_1 - a_0b_0)$, $(a_1b_1 - a_1b_0)$ and $(a_2b_0 - a_0b_0)$ in terms of the variances of the five effects. Examine how these variances are reduced when

  (i) interactions are assumed zero,
  (ii) quadratic main effect and interaction are assumed zero,
  (iii) only the quadratic interaction is assumed zero.

(3) The analysis of variance and treatment means in Tables 13.27(a) and (b) are taken from a fertiliser experiment on potatoes conducted in three randomised blocks of 12 plots. The 12 treatments were all combinations of three nitrogen levels (n_0 = none, n_1 = sulphate of ammonia, n_2 = ammonia bicarbonate), two manure levels (m_0 = none, m_1 = some) and two phosphate levels (p_0 = none, p_1 = some).

Table 13.27(a). *Treatment means (tons/acre).*

	m_0p_0	m_0p_1	m_1p_0	m_1p_1
n_0	4.0	4.2	5.4	6.4
n_1	4.8	5.5	6.2	6.4
n_2	4.4	4.9	5.2	5.9

(b). *Analysis of variance.*

	SS	df	MS
Blocks	18.17	11	
Treatments	23.29	11	2.12
Error	7.04	22	0.32
Total	48.5		

Define a complete set of single degree of freedom contrasts and calculate the corresponding sums of squares.

Obtain estimates of the differences of $n_1 m_0 p_0$, $n_2 m_0 p_0$, $n_0 m_1 p_0$ and $n_0 m_0 p_1$, from $n_0 m_0 p_0$ assuming

(i) all interactions are zero,

(ii) the three factor interaction is zero.

What estimates would you give a farmer who wanted to know what benefits could be expected from adding each one of the fertilisers to the no fertiliser control treatment?

(4) An experiment was concerned with the problem of trying to obtain a new type of coating material for metals used in making aircraft components. The purpose of coating the metal is to increase its strength. The coating powder is shot, using a spray gun, at extremely high velocity through a gas flame onto the metal or substrate. When the particles of coating hit the surface they are squashed and stick to the material. A measure of abrasion loss was then obtained by rubbing the substrate with a rough diamond pin over a fixed period of time.

Table 13.28.

Treatment	Score	Treatment	Score
(1)	1.4	e	1.7
a	1.2	ae	2.0
b	3.6	be	3.1
ab	1.2	abe	1.2
c	1.5	ce	1.9
ac	1.4	ace	1.2
bc	1.5	bce	1.0
abc	1.6	abce	1.8
d	5.0	de	9.5
ad	9.0	ade	5.9
bd	12.0	bde	12.6
abd	5.4	abde	6.3
cd	4.2	cde	8.0
acd	4.4	acde	4.2
bcd	9.3	bcde	7.7
abcd	2.8	abcde	6.0

The effects of five factors were studied, with each factor at two levels. The factors were:

factor A: method of cleaning or preparing the substrate,

factor B: composition of the substrate,

factor C: preheat temperature of the substrate before applying the coating powder,

factor D: type of coating powder,

factor E: size of the powder particles.

The 2^5 experiment was completely randomised using just one replicate. The scores were as shown in Table 13.28.

Analyse these data paying particular attention to the possibilities that some responses may be outliers, that the residual variance about an appropriate linear model may not be homogeneous over the whole experiment, and that a more parsimonious factorial model may be achieved by first transforming the observed responses.

Summarise the conclusions from the experiment.

14

Split unit designs and repeated measurements

14.0 *Preliminary examples*

Both of the following examples come from situations where the experimenter designed an experiment without consulting a statistician. When the experimenter came, with the experimental data, to consult the statistician the first problem was to identify the structure of the design and then to provide a suitable analysis.

(*a*) The first experiment was concerned with the influences on the production of glasshouse tomatoes of differing air and soil temperatures. Eight glasshouse compartments were available, and these were paired in four 'blocks'. Each compartment contained two large troughs in which the tomatoes were grown and in each half of each trough the soil temperature could be heated to a required level or left unheated. In one compartment of each block the minimum air temperature was kept at 55°F and in the other the minimum air temperature was 60°F. In each trough, one half was maintained at the control soil temperature while the other half was at an increased temperature, the increased temperature being 65°F for one trough, and 75°F for the other trough. The design layout for one pair of compartments is shown in Figure 14.1. Yields of tomatoes from each half trough were recorded.

(*b*) The second experiment was also concerned with heating, this time the heating characteristics of different forms of plastic pot situated in cold frames and of different forms of covers for the frames. Six wooden frames were divided into four quarters and four different forms of pot were allocated to the four quarters randomly in each frame. The frames were covered either by glass or by polythene, three frames having glass covers and three having polythene covers. All the heat was derived from solar

Figure 14.1. Design layout for one pair of compartments for the tomato experiment on the effects of air and soil heating.

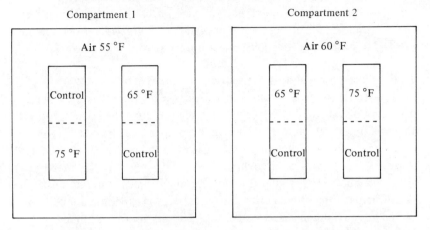

Figure 14.2. Design layout for experiment on effects of different materials on soil heating in pots (a) week 1, (b) week 2. P = polythene, G = glass.

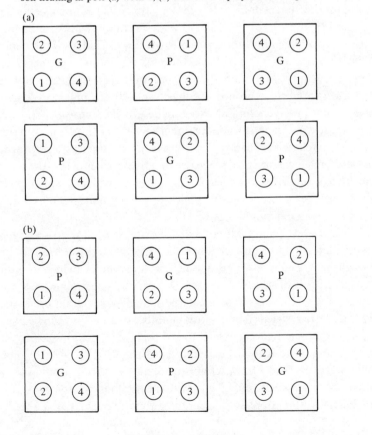

energy. After one week of observation, the covers were exchanged so that each frame that had previously been covered by glass was now covered by polythene and vice versa. The pot positions were unaltered. The design is represented in Figure 14.2. For each pot, in each week, the soil temperature was recorded.

14.1   The practical need for split units

In many investigations in which two or more treatment factors are being investigated, the size of experimental unit which is appropriate for one treatment factor is different from that appropriate to another treatment factor. In the chemical industry, a process may have several stages, starting with a large mixing stage from which the material is subdivided into smaller quantities for the later chemical interactions; if the constituents of the original mixture are varied, the experimental unit must be the large container used for the initial stage, but for treatment factors involving changing the conditions for the subsequent stages of the process, the unit may be a subdivision of the large initial experimental unit.

In experiments on dairy cattle, the experimental treatments may include different treatments of the pasture and variation of milking conditions. Pasture treatments must be applied to groups of animals if they are to be representative of farming practice, but milking method treatments may be applied to individual animals. This division of treatments into those necessarily applied to groups of subjects and those applied to individuals also occurs in psychological experiments. In field crop experiments, or glasshouse experiments, some treatments such as cultivation or irrigation methods or electrical heating conditions must be applied to large areas for the results to have practical relevance, but treatments such as the variety of the crop or involving chemical treatment of individual plants or parts of plants may be advantageously applied to smaller units, such as individual plants or even individual leaves.

When the levels of one treatment factor must be applied to large units and the levels of a second treatment factor may be applied to small units, though they could also be applied to large units (which are simply amalgamations of small units), then there are two possibilities open to the experimenter. The first is to use large units. This will almost inevitably place a severe restriction on the total number of units that can be available for the experiment, and will lead to few replicate units for each treatment combination. The second possibility is to use different sizes of units for different treatments. The latter approach may be thought of in two ways: either as using small units, but with restrictions on the allocation of treatment combinations so that combinations involving the same level of a 'large unit' factor are grouped together on small units forming a large unit; alternatively the large units used for one treatment factor can be envisaged as being split into smaller units for the levels of other factors. Diagrammatically the design for four blocks each containing 15 small units grouped into five large units each split into three subunits is shown in Figure 14.3. The name 'split plot' design was coined for agricultural field crop experiments taking the two-stage procedure for design as the central concept. The more general term 'split unit' design retains that concept.

As with the incomplete block designs discussed in Chapter 7 information about treatments occurs in two different strata of the experiment. Unlike the incomplete block design, where the information in both strata was about the same treatment comparisons, the different strata in split unit designs contain information about different sets of treatment comparisons. The design of Figure 14.3 may be thought of as consisting of two randomised block designs. In the first, involving variation between the large units, the subdivision of the large units into small units is ignored, and the yields are assumed to be influenced by the block differences, and the differences between the treatment levels applied to the large units. The model for a yield from block i and the large unit receiving treatment level p_j in block i is

$$z_{ij} = \mu + b_i + p_j + \eta_{ij}. \tag{14.1}$$

The second level of variation is that between split units within each large or main unit. At this level the large units are viewed as blocks, and all the

Figure 14.3. Experimental plan for a split plot design with four blocks each divided into five large plots, each further divided into three split plots.

Block I						Block II			
Large unit						Large unit			
1	2	3	4	5	1	2	3	4	5
p_2q_1	p_5q_1	p_3q_2	p_1q_3	p_4q_1	p_3q_2	p_4q_3	p_2q_2	p_5q_1	p_1q_2
p_2q_3	p_5q_3	p_3q_1	p_1q_1	p_4q_3	p_3q_3	p_4q_2	p_2q_1	p_5q_3	p_1q_1
p_2q_2	p_5q_2	p_3q_3	p_1q_2	p_4q_2	p_3q_1	p_4q_1	p_2q_3	p_5q_2	p_1q_3

Block III						Block IV			
Large unit						Large unit			
1	2	3	4	5	1	2	3	4	5
p_5q_1	p_1q_3	p_3q_3	p_2q_2	p_4q_3	p_2q_3	p_4q_1	p_1q_2	p_3q_1	p_5q_2
p_5q_2	p_1q_2	p_3q_1	p_2q_3	p_4q_2	p_2q_1	p_4q_3	p_1q_1	p_3q_2	p_5q_3
p_5q_3	p_1q_1	p_3q_2	p_2q_1	p_4q_1	p_2q_2	p_4q_2	p_1q_3	p_3q_3	p_5q_1

influences on the large units are subsumed into 'block' effects, with treatment effects due to the different treatments applied to the different split units within each large unit. The large unit effects are of course themselves attributable partly to blocks and partly to main unit treatments as described in model (14.1), but for consideration of the variation between split units, this is not relevant. However, the treatment effects on split unit yields are not simply the main effects of the split unit treatments but are modified by the interaction between the split unit factor levels and the particular main unit treatment. The model for a yield from the split unit receiving treatment level q_k, within the main unit m_{ij}, is therefore

$$y_{ijk} = \mu + m_{ij} + q_k + (pq)_{jk} + \varepsilon_{ijk}. \tag{14.2}$$

The two models (14.1) and (14.2) can be placed in context by considering an overall model which can be constructed by describing the logical stages of the design of the experiment:

$y_{ijk}\ldots$	the yield in block i for main unit treatment p_j and split unit treatment q_k
$=$	depends on
μ	the general level of average yield
$+b_i$	and the effects of block i
$+p_j$	and the effects of treatment p_j
$+\varepsilon'_{ij}$	these treatments being allocated randomly to main units
$+q_k$	and the effects of treatment q_k
$+(pq)_{jk}$	modified by the treatment p_j
$+\varepsilon_{ijk}$	the q_k treatment levels being allocated randomly to split units within each main unit.

The resulting model

$$y_{ijk} = \mu + b_i + p_j + \varepsilon'_{ij} + q_k + (pq)_{jk} + \varepsilon_{ijk} \tag{14.3}$$

leads directly to the submodels (14.1) and (14.2) if we write

$$y_{ij.} = z_{ij} \text{ with } \eta_{ij} = \varepsilon'_{ij} + \varepsilon_{ij.}$$

and

$$m_{ij} = b_i + p_j + \varepsilon'_{ij}.$$

The least squares estimates and the corresponding analysis of variance may be derived by minimising the sums of squares

$$S_1 = \sum_{ijk} \{y_{ijk} - \mu - m_{ij} - q_k - (pq)_{jk}\}^2 \tag{14.4}$$

and

$$S_2 = \sum_{ij} (y_{ij.} - \mu - b_i - p_j)^2.$$

However, the interrelationship of the two analyses of variance is simpler if instead of S_2 we minimise

$$S'_2 = \sum_{ijk} (y_{ij.} - \mu - b_i - p_j)^2. \tag{14.5}$$

This change has no effect on the parameter estimates because k does not occur within the summation, but it has the advantage that the total sum of squares about the mean derived from (14.5) is the same as the sum of squares for fitting m_{ij} in (14.4) so that the two analyses of variance are linked. The form of the analysis of variance structure is as in Table 14.1. The two residual mean squares E_a and E_b allow estimation of the variances of the two random components in the model (14.3),

$$\sigma^2 = E(\varepsilon_{ijk}^2) \text{ and } \sigma'^2 = E(\varepsilon_{ij.}'^2)$$

The expected values of E_b and E_a are

$$E(E_b) = E(\varepsilon_{ijk}^2) = \sigma^2,$$

and

$$E(E_a) = q E(\eta_{ij}^2) = q(\sigma'^2 + \sigma^2/q)$$
$$= \sigma^2 + q\sigma'^2.$$

Hence σ^2 is estimated by E_b, and σ'^2 is estimated by $(E_a - E_b)/q$. The fact that information about treatment differences exists at two different levels leads to a rather more complex structure of standard errors than the usual fully randomised factorial experiment. The basis for calculating all

Table 14.1.

Source	SS	df	MS
Blocks	$\sum_i (Y_{i..}^2/pq) - (Y_{...}^2/bpq)$	$b-1$	
P main effect	$\sum_j (Y_{.j.}^2/bq) - (Y_{...}^2/bpq)$	$p-1$	
Main unit residual	by subtraction	$(b-1)(p-1)$	E_a
Main unit total variation	$\sum_{ij} (Y_{ij.}^2/q) - (Y_{...}^2/bpq)$	$bp-1$	
Q main effect	$\sum_k (Y_{..k}^2/bp) - (Y_{...}^2/bpq)$	$q-1$	
PQ interaction	$\sum_{jk} (Y_{.jk}^2/b) - \sum_j (Y_{.j.}^2/bq)$	$(p-1)(q-1)$	
	$\quad - \sum_k (Y_{..k}^2/bp) + (Y_{...}^2/bpq)$		
Split unit residual	by subtraction	$(b-1)p(q-1)$	E_b
Total	$\sum_{ijk} y_{ijk}^2 - (Y_{...}^2/bpq)$	$bpq-1$	

standard errors is the complete model (14.3). Consider the table of two-way treatment means shown in Figure 14.4. The arrows show the different comparisons of interest.

(1) $y_{.1.} - y_{.2.}$ is a main effect comparison of main unit treatment levels. From (14.3) the comparison is:

$$(y_{.1.} - y_{.2.}) = (p_1 - p_2) + (\varepsilon'_{.1} - \varepsilon'_{.2}) + (\varepsilon_{.1.} - \varepsilon_{.2.}),$$

all other terms either cancelling or being eliminated by the usual constraints. Hence

$$\mathrm{Var}(y_{.1.} - y_{.2.}) = 2\sigma'^2/b + 2\sigma^2/bq = 2(\sigma^2 + q\sigma'^2)/bq,$$

which is estimated by $2E_a/bq$ which is the usual form of variance for a randomised block design, with the extra divisor q.

(2) $y_{..1} - y_{..2}$ is a main effect comparison of split unit treatment levels. From (14.3):

$$y_{..1} - y_{..2} = (q_1 - q_2) + (\varepsilon_{..1} - \varepsilon_{..2})$$

and the variance is:

$$\mathrm{Var}(y_{..1} - y_{..2}) = 2\sigma^2/bp,$$

which is estimated by $2E_b/bp$, again the usual randomised block form of variance.

(3) $y_{.11} - y_{.12}$ is a comparison of split unit treatment levels for level 1 of the main unit treatments:

$$y_{.11} - y_{.12} = (q_1 - q_2) + \{(pq)_{11} - (pq)_{12}\} + (\varepsilon_{.11} - \varepsilon_{.12})$$

Figure 14.4. Pattern of standard errors for two-way table of means for a split plot design (definitions of standard errors (1), (2), (3), (4), (5) in text).

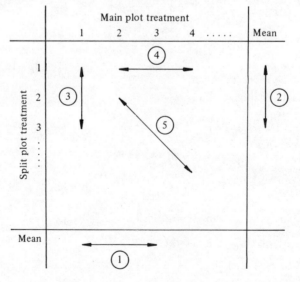

for which the variance is:

$$\text{Var}(y_{.11} - y_{.12}) = 2\sigma^2/b$$

estimated by $2E_b/b$.

(4) $y_{.11} - y_{.21}$ is a comparison of main unit treatment levels for level 1 of the split unit treatments:

$$y_{.11} - y_{.12} = (p_1 - p_2) + (\varepsilon'_{.1} - \varepsilon'_{.2}) + \{(pq)_{11} - (pq)_{21}\} + (\varepsilon_{.11} - \varepsilon_{.12})$$

for which the variance is

$$\text{Var}(y_{.11} - y_{.21}) = 2\sigma'^2/b + 2\sigma^2/b.$$

This is the first case where the estimate of the variance is not simple, being

$$2\{E_a + (q-1)E_b\}/(bq).$$

(5) $y_{.11} - y_{.22}$ involves different main unit treatment levels and different split unit treatment levels:

$$y_{.11} - y_{.22} = (p_1 - p_2) + (\varepsilon'_{.1} - \varepsilon'_{.2}) + (q_1 - q_2)$$
$$+ \{(pq)_{11} - (pq)_{22}\} + (\varepsilon_{.11} - \varepsilon_{.22})$$

for which the variance is

$$\text{Var}(y_{.11} - y_{.22}) = 2\sigma'^2/b + 2\sigma^2/b$$

exactly the same as for the previous comparison (4).

The form of the variance estimate for these last two comparisons causes major problems of statistical inference, because the estimate involves two different error mean squares with different degrees of freedom and different expectations. The linear combination $E_a + (q-1)E_b$ does not have a χ^2 distribution and therefore, strictly, we are not able to calculate confidence intervals or significance levels for comparisons (4) or (5). In practice if we use the smaller of the two error degrees of freedom which will always be that for E_a, $(b-1)(p-1)$, then we will be acting in a cautious fashion, obtaining confidence intervals that are unnecessarily large, and requiring unnecessarily extreme values in determining significance. Such a procedure is termed 'conservative' and is usually regarded as desirable in statistics, though whether it is sensible to emphasise the importance of avoiding type I errors rather than type II errors is often debatable in research work. In any case the complexity of the standard errors and the consequent difficulty with degrees of freedom provide a major drawback to the use of the split unit design.

14.2 Advantages and disadvantages of split unit designs

Although the main reason for using split unit designs, and indeed the whole reason for their existence, is practical necessity, there are some experimenters and statisticians who maintain that there are situations where split unit designs are, in theory, to be preferred to fully randomised factorials. The argument is simple. In a split unit design, information about treatment differences occurs at two levels. The form of the model (4.3) implies that the expected error mean square of main unit comparisons will be larger than that for split unit comparisons. Since the total set of units is the same as would have been used for a fully randomised factorial treatment set, the single error mean square for that experiment would have been the weighted average of the two error mean squares for the split unit experiment. Consequently, in the split unit experiment, comparisons based on the split unit error mean square will be more precise than the same comparisons would have been in the fully randomised factorial experiment, while comparisons based on the main unit error mean square will be less precise than in the fully randomised factorial experiment.

This form of argument is formally correct, but can be misleading, and it is sometimes simplified further to a form in which it contains only the most miniscule grain of truth. Split designs, it is said, are appropriate when one factor main effect is of little interest, and the interest is in the interaction effect and the other factor main effect, which are estimated in terms of split unit comparisons.

Now, the grain of truth is that, indeed, the precision of information on split unit treatment main effects and interactions in the form of F statistics within the analysis of variance should be expected to be better than that of the main unit treatment main effects. However, the disadvantages of the split unit design are many and four disadvantages are discussed here in detail.

(1) The information occurs at two levels and the estimation of both error variances is on fewer degrees of freedom than would be available to estimate the single error variance for the fully randomised factorial.

(2) If the main unit error mean square, E_a, the split unit error mean square, E_b, and the fully randomised factorial error mean square, s^2, are compared, then, since s^2 is a weighted average of E_a and E_b, and the weights are the degrees of freedom, $(b-1)(p-1)$ for E_a, $(b-1)p(q-1)$ for E_b, the difference $(E_a - s^2)$ is usually much greater than the difference $(s^2 - E_b)$. In other words the gain in precision for some effects is much less than the loss for others.

To illustrate the scales of these two disadvantages, the values of the two sets of error degrees of freedom, and the single error degrees of freedom are shown in Table 14.2, together with the ratio of

$$(E_a - s^2)/(s^2 - E_b)$$

for all possible split unit designs with 24, 36 or 48 plots, and with an explicit replication level of four or less. It is plain from these values that the degrees of freedom for the main unit error are too small for effective estimation of variance except when q = 2, that is when there are two split unit treatment levels. In the few cases where there are at least 8 df for the main unit error, the ratio of the loss/gain for the main unit and split unit comparisons is usually very low, and the gain and loss of precision are then similar. However, this almost invariably occurs with only two split plots in

Table 14.2. *Error degrees of freedom.*

b	p	q	Main unit	Split unit	Randomised factorial	$\dfrac{E_a - s^2}{s^2 - E_b}$
2	2	6	1	10	11	10.0
2	6	2	5	6	11	1.2
2	3	4	2	9	11	4.5
2	4	3	3	8	11	2.7
3	2	4	2	12	14	6.0
3	4	2	6	8	14	1.3
4	2	3	3	12	15	4.0
4	3	2	6	9	15	1.5
2	2	9	1	16	17	16.0
2	9	2	8	9	17	1.1
2	3	6	2	15	17	7.5
2	6	3	5	12	17	2.4
3	2	6	2	20	22	10.0
3	6	2	10	12	22	1.2
3	3	4	4	18	22	4.5
3	4	3	6	16	22	2.7
4	3	3	6	18	24	3.0
2	2	12	1	22	23	22.0
2	12	2	11	12	23	1.1
2	3	8	2	21	23	10.5
2	8	3	7	16	23	2.3
3	2	8	2	28	30	14.0
3	8	2	14	16	30	1.1
2	4	6	3	20	23	6.7
2	6	4	5	18	23	3.6
4	2	6	3	30	33	10.0
4	6	2	15	18	33	1.2
3	4	4	6	24	30	4.0
4	3	4	6	27	33	4.5
4	4	3	9	24	33	2.7

each main plot when it might be expected that the gain in precision would be relatively small. The last line of Table 14.2 provides a case when the main plot error degrees of freedom rise to nine, with three split unit treatments. However, the ratio of loss/gain is 2.7, emphasising that the precision of main unit comparisons is very much reduced.

(3) If instead of considering F ratios in the analysis of variance we consider the tables of treatment mean values, which will usually be the main focus of interest for the experimenter, then the advantages of split unit designs for investigating interaction become less clear. Particularly when there are substantial interaction effects the experimenter will interpret the results in terms of comparisons between pairs or groups of treatment combinations. Consider again Figure 14.4 in which the five types of comparison for which standard errors were derived in the previous section are illustrated. In terms of precision, compared with the fully randomised factorial allocation within each block, the main effect comparison (2) and the interaction comparison (3) may be expected to be more precise using a split unit design. Main effect comparisons (1) will be much less precise. Interaction comparisons (4) and (5) will be less precise than in the fully randomised factorial allocation because of the reduction in degrees of freedom for part of the combined standard error. Essentially when considering the comparison of treatment mean values, the practical assessment of interaction effects necessarily involves main effect differences, and hence the benefits of split unit designs for estimating interaction effects largely disappear.

(4) The split unit design, as has been made clear in the previous section, requires four different standard errors in the presentation of results, which causes particular problems in the interpretation of graphical presentation of means. And, of course, the necessity of using standard errors for which exact t-tests and confidence intervals are not possible, and which are based on two estimated variances of which one may have few degrees of freedom, also makes interpretation and presentation more complex.

In conclusion, the argument for using split unit designs, because of their greater precision for interactions and main effects of the split unit factor, is an over-simplification. In many cases split unit designs will lead to no useful estimates of the main effects of the main unit factor, or of comparisons of levels of the main unit factor for particular levels of the split unit factor. In addition, as we shall see in the next chapter, there are much better alternative designs if it is thought desirable to design the experiment to have different comparisons at different levels of precision. The only sound advice I can offer is to avoid split unit designs except when they are essential for practical reasons, as described at the beginning of this chapter.

14.3 Extensions of the split unit idea

The split unit concept offers a wide range of extensions, all deriving from practical requirements of particular experiments. The warning against using split unit designs, for reasons other than that of practical necessity, applies even more strongly to the extensions discussed in this section. Nevertheless designs using many different forms of splitting units are used, some by necessity, others by accident.

Split unit designs have been discussed in the context of an initial randomised block design, but split unit versions of any of the designs discussed in Chapters 7 and 8 can be easily constructed. The main unit part of the analysis of variance is that for the parent design. The split unit analysis involves no block or row and column component, since it is entirely composed of comparisons within main units. Variances of comparisons between split unit factor levels, whether main effect comparisons or at a particular level of the main unit, all involve only the split unit error mean square, and are in the simple form of the variance of a difference between two sample means considered in the previous section. Variances of comparisons between levels of the main unit factor have the same efficiency as those for the parent design, except that, when the comparisons are made at a particular split unit treatment level, there is an additional error component from the split unit error mean square.

The first major extension of the split unit concept involves splitting the split units again for a third treatment factor. The design now consists of three levels. Blocks are divided into main units to which the levels of the first treatment factor are applied; the main units are split into split units to which the levels of the second treatment factor are applied; and the split units are split into split-split units to which the levels of the third treatment factor are applied. The design is illustrated in Figure 14.5. The model is a direct extension of the simple split unit model:

$$y_{ijkl} = \mu + b_i + p_j + \varepsilon''_{ij} + q_k + (pq)_{jk} + \varepsilon'_{ijk} + r_l + (pr)_{jl}$$
$$+ (qr)_{kl} + (pqr)_{jkl} + \varepsilon_{ijkl}.$$

The analysis of variance structure is in three strata, and is summarised in Table 14.3. The expectations of E_a, E_b and E_c are $\sigma^2 + r\sigma'^2 + qr\sigma''^2$, $\sigma^2 + r\sigma'^2$ and σ^2, respectively. Hence, σ^2, σ'^2, σ''^2 can be estimated and standard errors calculated for all comparisons of treatment means.

Figure 14.5. Experimental plan for a split plot design.

Block I

Main plot

1		2		3		4	
Split plot		Split plot		Split plot		Split plot	
1	2	1	2	1	2	1	2
$p_3q_1r_2$	$p_3q_2r_1$	$p_4q_2r_1$	$p_4q_1r_3$	$p_2q_1r_2$	$p_2q_2r_1$	$p_1q_1r_2$	$p_1q_2r_2$
$p_3q_1r_1$	$p_3q_2r_3$	$p_4q_2r_2$	$p_4q_1r_1$	$p_2q_1r_1$	$p_2q_2r_3$	$p_1q_1r_1$	$p_1q_2r_3$
$p_3q_1r_3$	$p_3q_2r_2$	$p_4q_2r_3$	$p_4q_1r_2$	$p_2q_1r_3$	$p_2q_2r_2$	$p_1q_1r_3$	$p_1q_2r_1$

Block II

$p_4q_1r_2$	$p_4q_2r_3$	$p_1q_1r_1$	$p_1q_2r_2$	$p_3q_2r_2$	$p_3q_1r_3$	$p_2q_2r_1$	$p_2q_1r_3$
$p_4q_1r_3$	$p_4q_2r_2$	$p_1q_1r_3$	$p_1q_2r_1$	$p_3q_2r_3$	$p_3q_1r_1$	$p_2q_2r_2$	$p_2q_1r_2$
$p_4q_1r_1$	$p_4q_2r_1$	$p_1q_1r_2$	$p_1q_2r_3$	$p_3q_2r_1$	$p_3q_1r_2$	$p_2q_2r_3$	$p_2q_1r_1$

Block III

$p_1q_2r_3$	$p_1q_1r_2$	$p_3q_1r_3$	$p_3q_2r_3$	$p_4q_2r_1$	$p_4q_1r_3$	$p_2q_1r_2$	$p_2q_2r_1$
$p_1q_2r_2$	$p_1q_1r_1$	$p_3q_1r_1$	$p_3q_2r_2$	$p_4q_2r_3$	$p_4q_1r_2$	$p_2q_1r_1$	$p_2q_2r_3$
$p_1q_2r_1$	$p_1q_1r_3$	$p_3q_1r_2$	$p_3q_2r_1$	$p_4q_2r_2$	$p_4q_1r_1$	$p_2q_1r_3$	$p_2q_2r_2$

Table 14.3.

Source	df	MS
Blocks	$b-1$	
P main effect	$p-1$	
Main unit residual	$(p-1)(b-1)$	E_a
Main unit total	$bp-1$	
Q main effect	$q-1$	
PQ interaction	$(p-1)(q-1)$	
Split unit residual	$(b-1)p(q-1)$	E_b
Split unit total	$bpq-1$	
R main effect	$r-1$	
PR interaction	$(p-1)(r-1)$	
QR interaction	$(q-1)(r-1)$	
PQR interaction	$(p-1)(q-1)(r-1)$	
Split split unit residual	$(b-1)pq(r-1)$	E_c
Split split unit total	$bpqr-1$	

*Called 'split block' and 'sub unit treat.
(Steel + Torrie) in strips'*
*395 (Cochran)
and 'strip plot'
(Moser)*

There is a second type of extension of the split unit idea which is often found, and this is the criss-cross design. A two-factor treatment structure is used and, as in the simple split unit design, the levels of factor P require large units from practical necessity. In addition, the levels of factor Q also require large units. Again, it would be possible to use a large basic unit, but this will usually lead to a very low level of replication for all treatment combinations, because of the limitation on total resources. The alternative approach is to consider a set of small units, arranged in a rectangular array in each block, and to apply the levels of P to whole rows of units, and levels of Q to whole columns. The resulting design is shown in Figure 14.6; the origin of the name should be obvious.

i.e. would need

Figure 14.6. Experimental plan for a criss-cross design.

Block I

P_4q_3	P_4q_4	P_4q_2	P_4q_1
P_1q_3	P_1q_4	P_1q_2	P_1q_1
P_2q_3	P_2q_4	P_2q_2	P_2q_1
P_5q_3	P_5q_4	P_5q_2	P_5q_1
P_3q_3	P_3q_4	P_3q_2	P_3q_1
P_6q_3	P_6q_4	P_6q_2	P_6q_1

Large units (do not have this many available)

$P_4 Q_3$
$P_4 Q_4$
$P_4 Q_2$
$P_6 Q_1$

Block II

P_2q_1	P_2q_3	P_2q_2	P_2q_4
P_6q_1	P_6q_3	P_6q_2	P_6q_4
P_4q_1	P_4q_3	P_4q_2	P_4q_4
P_3q_1	P_3q_3	P_3q_2	P_3q_4
P_1q_1	P_1q_3	P_1q_2	P_1q_4
P_5q_1	P_5q_3	P_5q_2	P_5q_4

Block III

P_1q_2	P_1q_4	P_1q_3	P_1q_1
P_2q_2	P_2q_4	P_2q_3	P_2q_1
P_4q_2	P_4q_4	P_4q_3	P_4q_1
P_5q_2	P_5q_4	P_5q_3	P_5q_1
P_6q_2	P_6q_4	P_6q_3	P_6q_1
P_3q_2	P_3q_4	P_3q_3	P_3q_1

The criss-cross design requires a complex analysis of variance, and uses information in a very inefficient manner. Except when practically essential, it should not be used, but it does occur sufficiently frequently, sometimes without deliberate intent, for the complexity of analysis to be discussed in detail. Some of the ways in which criss-cross and related designs occur by accident in highly structured experiments are discussed in the next section.

The information from a criss-cross design is in three strata. Levels of P are applied to whole rows and therefore main effects of P are estimated from comparisons between rows within blocks. Similarly, main effects of Q are estimated from comparisons between columns within blocks. However, neither comparisons between rows, nor comparisons between columns, provide any information about the interaction effects. To obtain interaction information, we must consider the variation between the split units, which are splits of both row units and column units. The full model is:

$$y_{ijk} = \mu + b_i + p_j + \varepsilon'_{ij} + q_k + \varepsilon''_{ik} + (pq)_{jk} + \varepsilon_{ijk}.$$

One conceptual difficulty with this model is that there is no independent randomisation at the level of the smallest unit; the randomisations of rows and columns completely determine the pattern of treatment occurrence. Nevertheless, it is clear that there are three levels of information, and consequently three sections to the analysis of variance. The analysis of variance structure is shown in Table 14.4. Note that the block SS is a component of both the row main unit variation and the column main unit variation (and must be included in the calculation of the column main unit residual), but can only occur once in the total analysis of variance.

Table 14.4.

Source	df	MS
Blocks	$b-1$	
P main effect	$p-1$	
Row main unit residual	$(p-1)(b-1)$	E_a
Row main unit total	$bp-1$	
Q main effect	$q-1$	
Column main unit residual	$(b-1)(q-1)$	E_b
Column main unit total	$bq-1$	
PQ interaction	$(p-1)(q-1)$	
Split unit residual	$(p-1)(b-1)(q-1)$	E_c
Total	$bpq-1$	

The variances, σ^2, σ'^2 and σ''^2 associated with the three error terms ε_{ijk}, ε'_{ij}, ε''_{ik}, can be estimated from the three residual mean squares. The expected values of E_a, E_b and E_c are $q\sigma'^2 + \sigma^2$, $p\sigma''^2 + \sigma^2$ and σ^2, so that estimates of σ^2, σ'^2 and σ''^2 are:

$$E_c,$$

$$(E_a - E_c)/q,$$

and

$$(E_b - E_c)/r.$$

The spread of information over the three levels leads to a profusion of standard errors for comparisons within the two-way table of treatment means. Consider the two-way structure of treatment means displayed in Figure 14.7. The standard errors of the main effect comparisons ($p_1 - p_4$, $q_2 - q_3$) are as usual derived from the corresponding residual mean squares, E_a and E_b. The other three comparisons in Figure 14.7 require different combinations of estimated error variances. Comparison (3) is estimated by

$$y_{.11} - y_{.21},$$

which can be written, from the general model:

$$y_{.11} - y_{.21} = (p_1 - p_2) + \{(pq)_{11} - (pq)_{21}\} + (\varepsilon'_{.1} - \varepsilon'_{.2}) + (\varepsilon_{.11} - \varepsilon_{.21})$$

from which

$$\mathrm{Var}(y_{.11} - y_{.21}) = 2\sigma'^2/b + 2\sigma^2/b$$

Figure 14.7. Pattern of standard errors for a criss-cross design (details in text).

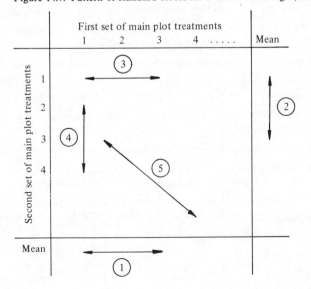

which is estimated by

$$2\{E_a + (q-1)E_c\}/bq.$$

Similarly, the variance for comparison (4) is estimated by

$$2\{E_b + (p-1)E_c\}/bp.$$

Comparison (5) is

$$y_{.23} - y_{.45}$$

for which the error terms involved are

$$(\varepsilon'_{.2} - \varepsilon'_{.4}) + (\varepsilon''_{.3} - \varepsilon''_{.5}) + (\varepsilon_{.23} - \varepsilon_{.45})$$

whence

$$\mathrm{Var}(y_{.23} - y_{.45}) = 2(\sigma'^2 + \sigma''^2 + \sigma^2)/b,$$

which is estimated by

$$2\{pE_a + qE_b + (pq - p - q)E_c\}/bpq.$$

Clearly, assigning exact degrees of freedom to these estimates of standard errors is impossible, and indeed having adequate ($\geqslant 10$) df for each of E_a, E_b and E_c is rarely possible. However, if the need arises because of practical reasons, then the criss-cross design with all its attendant difficulties of interpretation may be useful. Further ramifications of split unit ideas may be produced at will, and may sometimes be necessary practically, but there are rarely sufficient degrees of freedom for the main unit components of analysis to be of much value.

Example 14.1

To demonstrate the practical analysis of split unit and criss-cross designs we consider some data from an experiment investigating the long term effects on pasture composition of different patterns of grazing. Three treatment factors are included

> Period (P): the length of the period when plots were grazed, with three levels (3, 9, 18 days).
>
> Spring grazing cycles (SP): the number of cycles. Either two (two periods grazing with two (long) gaps), or four (four periods grazing with four (shorter) gaps).
>
> Summer grazing cycles (S): two or four cycles, as for spring grazing cycles.

The experimental design is a 3×3 Latin square for the three periods, with each of the nine main plots split twice in a criss-cross design for the two grazing cycle factors. Please note that the author was not consulted about this design: nor was any other statistician! The design is shown in Figure 14.8. The data, which is the percentage area covered by the principal grass, is also given in Figure 14.8.

Figure 14.8. Design and data for 3×3 Latin square with each main plot split in a criss-cross design.

	S			SP			SP	
[4]	12.5	26.2	[4]	59.2	49.9	[4]	55.0	27.3
SP	(18)		S	(9)		S	(3)	
[2]	33.4	44.2	[2]	47.6	15.8	[2]	35.9	18.3
	S			S			S	
[2]	56.2	52.3	[2]	67.7	62.2	[2]	28.0	29.4
SP	(9)		SP	(3)		SP	(18)	
[4]	27.5	25.1	[4]	24.1	27.5	[4]	19.5	29.9
	S			S			SP	
[2]	57.2	69.5	[2]	30.3	26.6	[4]	61.9	26.2
SP	(3)		SP	(18)		S	(9)	
[4]	16.9	19.5	[4]	11.0	17.6	[4]	46.5	15.4

The calculation of the analysis of variance requires, first, the set of main plot totals with the associated totals for rows, columns and periods, which are shown in Table 14.5(a). The other totals required are for the $3 \times 2 \times 2$ treatment combinations, with all subtotals, given in Table 14.5(b). The model on which the analysis is based includes four levels of variation; for main plots, for two sets of half plots for main effect comparisons of the two grazing cycle factors, and for the quarter plots for grazing cycle combinations

$$y_{ijlm} = \mu + r_i + c_j + p_{k(ij)} + \varepsilon'''_{ij} + sp_l + (p, sp)_{kl} + \varepsilon''_{ijl}$$
$$+ s_m + (p, s)_{km} + \varepsilon'_{ijm} + (sp, s)_{lm} + (p, sp, s)_{klm} + \varepsilon_{ijlm}.$$

The analysis of variance provides a summary of the results as shown in Table 14.6. The initial conclusions which should be noted from the analysis of variance are:

(i) The major treatment effects are those of
  spring grazing cycle
  grazing period length
  summer grazing cycle, and
  period $\times$ spring grazing cycle.

All other interaction effects may be ignored.

Table 14.5(a). *Main plot totals.*

	Main Plot Totals (period length)			Total
	116.3(18)	172.5(9)	136.5(3)	425.3
	161.1(9)	181.4(3)	106.8(18)	449.3
	163.1(3)	85.5(18)	150.0(9)	398.6
Total	440.5	439.4	393.3	1273.2

(b). *Treatment totals.*

Grazing	Cycle	Period			Total
Spring	Summer	3	9	18	
2	2	160.8	146.4	91.7	398.9
2	4	186.7	177.3	100.2	464.2
4	2	59.2	56.3	43.0	158.5
4	4	74.3	103.6	73.7	251.6
2	total	347.5	323.7	191.9	863.1
4	total	133.5	159.9	116.7	410.1
Total	2	220.0	202.7	134.7	557.4
Total	4	261.0	280.9	173.9	715.8
Total	total	481.0	483.6	308.6	1273.2

(ii) The pattern of error mean squares follows the usual expectation with remarkable consistency. The whole plot error mean square (E_a) is the largest, the lowest level error mean square (E_d) is the smallest and the two half plot error mean squares are intermediate.

To present the results we require standard errors for comparing whole plot treatment means, both forms of half plot treatment means and the two-way table of means for period × spring grazing cycle. The comparisons and their variance may be derived from the full model for y_{ijlm}.

Periods

$$y_1 - y_2 = (p_1 - p_2) + (\varepsilon_1''' - \varepsilon_2''') + (\varepsilon_{1.}'' - \varepsilon_{2.}'') + (\varepsilon_{1.}' - \varepsilon_{2.}') + (\varepsilon_{2..} - \varepsilon_{2..})$$
$$\mathrm{Var}(y_1 - y_2) = 2\sigma'''^2/3 + 2\sigma''^2/6 + 2\sigma'^2/6 + 2\sigma^2/12,$$

which is estimated by

$$2E_a/12 = 2 \times 107.19/12 = 17.86.$$

Spring grazing cycles

$$y_{..1.} - y_{..2.} = (sp_1 - sp_2) + (\varepsilon_{..1}'' - \varepsilon_{..2}'') + (\varepsilon_{..1.} - \varepsilon_{..2.})$$

Table 14.6.

	SS	df	MS	E(MS)
Rows	107.20	2	53.60	
Columns	120.95	2	60.48	
Periods	1676.49	2	838.24	
Error	214.38	2	$107.19 = E_a$	$\sigma^2 + 2\sigma'^2 + 2\sigma''^2 + 4\sigma'''^2$
Total	2119.02	8		
Spring grazing cycle	5700.25	1	5700.25	
PSP	823.20	2	411.60	
Error	478.33	6	$79.72 = E_b$	$\sigma^2 + 2\sigma''^2$
Spring half plots	9120.80	17		
Summer grazing cycle	696.96	1	696.96	
PS	80.78	2	40.39	
Error	366.80	6	$61.33 = E_c$	$\sigma^2 + 2\sigma'^2$
Summer half plots	3263.56	26		
SPS	21.47	1	21.47	
PSPS	51.74	2	51.75	
Error	177.35	6	$29.56 = E_d$	σ^2
Total	10515.90	35		

$$\text{Var}(y_{..1.} - y_{..2.}) = 2\sigma''^2/9 + 2\sigma^2/18,$$

which is estimated by $2E_b/18 = 2 \times 79.72/18 = 8.86$.

Summer grazing cycles

The pattern is exactly the same as for spring grazing cycles and the variance is estimated by

$$2E_c/18 = 2 \times 61.33/18 = 6.81.$$

Periods × spring grazing

$$y_{11} - y_{12} = (sp_1 - sp_2) + \{(psp)_{11} - (psp)_{12}\} + (\varepsilon''_{11.} - \varepsilon''_{12.})$$
$$+ (\varepsilon_{11.} - \varepsilon_{12.})$$
$$\text{Var}(y_{11} - y_{12}) = 2\sigma''^2/3 + 2\sigma^2/6)$$

which is estimated by

$$2E_b/6 = 2 \times 79.72/6 = 26.57.$$
$$y_{11} - y_{21} = (p_1 - p_2) + \{(psp)_{11} - (psp)_{21}\} + (\varepsilon'''_1 - \varepsilon'''_2)$$
$$+ (\varepsilon''_{11} - \varepsilon''_{21}) + \varepsilon_{11.} - \varepsilon_{21.})$$
$$\text{Var}(y_{11} - y_{21}) = 2\sigma'''^2/3 + 2\sigma''^2/3 + 2\sigma^2/6,$$

Table 14.7(a). *Spring grazing.*

	Period			Mean
	3	9	18	
Two cycles	57.9	54.0	32.0	48.0
Four cycles	22.2	26.6	19.4	22.8
Mean	40.1	40.3	25.7	

SE for the difference between two period means = 4.2 (2 df)
SE for the difference between two cycle means = 3.0 (6 df)
SE for within table vertical comparisons   = 5.2 (6 df)
SE for within table other comparisons   = 5.1 (? df)

(b). *Summer grazing.*

Two cycles 31.0
Four cycles 39.8
SE for difference between two cycle means = 2.6 (6 df)

which is estimated by

$$2(E_a + E_b - E_c + E_d)/12 = 2 \times 155.14/12 = 25.86.$$

The presentation of results is therefore as in Table 14.7.

The major effect on the percentage of the principal grass is caused by the number of cycles during the spring grazing period. The use of four cycles reduces the percentage to less than half that observed for two cycles of grazing. This effect is most pronounced for the shorter grazing periods, being largest for three day grazing periods and much smaller for 18 day grazing periods. The percentages of the principal grass are generally lower for the 18 day grazing periods. Finally it should be noted that the number of cycles during the summer grazing period also has a significant effect but in the opposite direction to that of the spring grazing cycle effect. Using four grazing cycles for the summer grazing tends to increase the percentage of the principal grass, but the increase of 9% is less than half the decrease of 25% apparently due to the four grazing cycles in the spring grazing period. Definitely a complex situation!

14.4   Identification of multiple strata designs

Many experiments whose design is built up gradually, rather than being totally planned at the outset, involve multiple strata comparisons without deliberate intent on the part of the experimenter, and it is often necessary to sort out the actual structure of the experiment, and hence of the analysis of variance, from a rather confused description. In Chapter 6, we considered an example where small discs were cut from leaves and treated in one of eight different ways. The experimenter, consulting a statistician about the analysis, stated simply that each of the eight treatments was replicated ten times, and offered an analysis of variance in the form

treatments   7 df $F = 500$
error   72 df.

After much questioning, it was revealed that the design of the experiment involved eight leaves, one from each of eight plants, the ten discs for a single treatment being taken from a single leaf, each treatment using a different leaf. Two levels of variation, between discs within a leaf, and between leaves, are involved, and the design and analysis of variance can be regarded as a split unit design of a non-standard (and non-useful) form, giving the two stage analysis of variance as shown in Table 14.8. The analysis of variance structure identifies clearly the absence of any information about the precision of treatment comparisons. The treatments must be compared in terms of the main unit error variable, about which there is no information; there is abundant information about the split unit error variance but there are no treatment comparisons for which that variance is relevant.

The experimenter's original description could have implied many different designs. For example, the design could have been defined in three stages:

(i) Ten plants, each providing two pairs of leaves, one pair of old leaves and one pair of new leaves (old/new is one factor of the 2^3 treatment set).

(ii) Each leaf provides two discs, one inocculated with chemical C_1 and one with chemical C_2.

(iii) The two leaves of a pair, each providing two discs, are allocated one for inocculation into the upper side of the leaf and the other for inocculation in the lower side of the leaf.

Table 14.8.

	df
Treatments	7
Main unit (leaf) error	0
Main unit (leaf) total	7
Split unit (disc) error	72
Total	79

Table 14.9.

	df
Plants (blocks)	9
New and old (age)	1
Leaf pairs error	9
Leaf pairs total	19
Upper × lower (side)	1
Age × side	1
Leaves error	18
Leaves total	39
Chemicals	1
Age × chemicals	1
Side × chemicals	1
ASC	1
Discs error	36
Total	79

The analysis of variance for such a design would have had four strata: plants, leaf pairs, leaves, and discs, only three of which provide information about the treatments. This should be recognisable as a split-split unit design, with plants as blocks and leaf pairs as main units, leading to the analysis of variance structure as shown in Table 14.9.

Before the analysis of variance and subsequent interpretation of treatment means can be calculated the statistician will often have to diagnose the design actually used. As we have already emphasised in Chapter 6, it is essential to identify the different levels of replication and thus of variation in an experiment, and to use the appropriate levels of variation for each treatment comparison. This recognition will frequently lead to analyses of the split unit family.

In many experiments, experimenters go to considerable lengths to balance all possible patterns, and this often leads to criss-cross structures with multiple strate in two-way blocking structures. The second example at the beginning of this chapter is such a design, with frames as blocks, positions within each frame as one form of main plot, and times within each frame as the other form of main plot. Another example from a psychology experiment illustrates the process of recognising this type of structure in more detail.

Example 14.2

The experiment involved 16 groups with four individuals per group. The 16 groups were blocked into four sets of four groups each. The principal treatments were four group stress environments (S); each group was subjected to each stress treatment in turn, the order of the four treatments being arranged according to a 4×4 Latin square within each set of four groups. The individuals within each group were two males and two females, and two drug levels (high and low) were allocated so that the four sex and drug combinations all occur in each group. Each individual remained on the same drug level throughout the four stress treatments and, of course, remained the same sex! To assess the performance of individuals under stress and drug treatment, four test papers were used, all the individuals in a set of four groups having the same test at a particular time, the test papers being changed between sets and times according to a Latin square design.

This is a very complex design, built up gradually by introducing ideas of control and of treatment comparisons sequentially. The three major stages of the design are represented in Figure 14.9. To unravel the design and the corresponding analysis structure, we consider the different treatment comparisons and the stratum from which information is available.

Figure 14.9. Diagrammatic representation of the design structure for the experiment of Example 14.2.

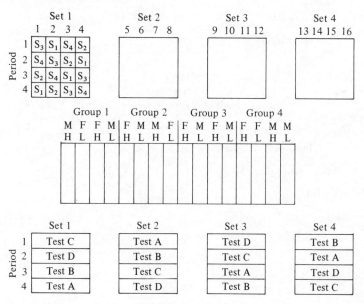

(i) Test papers – this may not be of interest, but if it is, then information comes from the 16 combinations of four sets × four times.

(ii) Stress environments – these are applicable to groups at different times and therefore information comes from the 64 combinations of 16 groups × four times.

(iii) Test × stress – from the same stratum as stress because stress is applied to groups within sets.

(iv) Sex, drug, sex × drug – these are applied to individuals consistently over time so that time has no relevance to these comparisons, and the relevant stratum is individuals within groups.

Table 14.10.

Stratum	Sources	df
Sets × times	Sets	3
	Times	3
	Tests	3
	Sets × times error	6
	Sets × times total	15
Groups × times	Groups within sets	12
	Stresses	3
	Tests × stresses	9
	Groups × times error	24
	Group × times total	63
Individuals	Sets	3
	Groups within sets	12
	Sex	1
	Drug	1
	Sex × drug	1
	Individuals error	45
	Individuals total	63
Individual × times	Stress × sex	7
	Stress × drug	3
	Stress × sex × drug	3
	Test × sex	3
	Test × drug	3
	Test × sex × drug	3
	Stress × test × sex	9
	Stress × test × drug	9
	Stress × test × sex × drug	9
	Error	99
	Total	255

(v) All other interaction effects can only be assessed by considering the stratum of individual × time combinations which consist of the smallest possible units.

The structure of the resulting analysis of variance is set out in Table 14.10. Essentially, the individuals and times classifications provide a criss-cross structure, while individuals within groups within sets provide a split-split unit structure.

This example is an unusually complex one, but hopefully it demonstrates the Latin square and criss-cross structures which arise naturally when an experimenter attempts to control systematically the patterns of treatment allocation.

14.5 Time as a split unit factor and repeated measurements

In many agricultural and other biological experiments with both animals and plants, it is common to take measurements on each individual unit at several different times, and to regard time as a split unit factor. The reason for this form of experiment is that the experimenter is interested in how the effects of treatments vary with time. This is the classic expression of interaction and therefore the investigation of treatment × time interaction within a split unit stratum of analysis seems an obvious approach. There are, however, some major difficulties in this use of the split unit concept.

First there is an important philosophical difficulty that multiple measurements are different variables and not different experimental units. The number of units in the experiment cannot be increased by taking more measurements. By taking more observations on each unit more information is obtained about the set of units originally included in the experiment. Even if we are prepared to confuse the fundamental concepts of experimental units and observed variables there are several technical difficulties about the practice of regarding time as a split unit factor.

Two rather different situations in which time is used as split unit factor must be distinguished. In some experiments, usually with plants, the sample of experimental material at each time is different. Measurement of the variables of principal interest for plants usually involves destruction of the plants. Thus, weight measurements require the plant to be removed from its growth environment, and measurement of chemical constituents require further destruction. Plant material is usually relatively cheap so that the use of repeated 'harvests' is economically viable. Rather than using different experimental plots for each treatment × time combination it is often practically convenient to use more substantial plots and to subsample at the various times. The positions of the subplots to be used at the different times can be selected from the plots either randomly or systematically. Random sampling makes a split unit form of analysis more reasonable than does systematic sampling, but the objections discussed below still require consideration.

Alternatively the repeated measurements at the different times may be taken on exactly the same experimental units. This will typically be the case in animal experiments, and occurs sufficiently frequently for such experiments to be distinguished as a special form, called 'repeated measurements experiments'. With the rapid development of technology producing machinery capable of taking measurements increasingly frequently for both animal and plant experimentation, the repeated measurements experiment is becoming extremely important, and there is a flourishing large literature on the subject.

The objections to the use of time as a split unit factor come under three headings.

(1) *Assumptions*

When variables are measured at different times, the changes induced by time on the plant or animal will usually be much greater than the changes effected by administered treatments. The physiological state of the plant or animal may be quite different at the end of the measurement period from the condition at the beginning. The consequence of this likely large variation in the condition of the experimental material is that the assumptions required for the analysis of variance are unlikely to be even approximately true. In particular, the variance over the experimental units is likely to be extremely heterogeneous over time. While transformations can be sought and applied it will be considerably more difficult to find variance-stabilising transformations than for data where there are no time changes. Although the additivity and normality assumptions are less likely than the variance homogeneity assumption to cause difficulties, they too must be regarded with suspicion when large changes of measurements are taking place. In summary, the variation over time may well mean that the same variable measured at different times cannot be regarded as sufficiently similar for a single quantitative analysis to be meaningful.

(2) *Non-randomness of time*

A second class of objections centres on the fact that time is not randomly ordered. In a split unit analysis, it is assumed that the correlation pattern of observations for different split unit treatments is independent of the particular treatments being considered. Time is ordered, and observations on a single unit taken at adjacent times will be inherently more correlated than will observations separated by a long time gap on a single unit, after allowance is made for the average difference between the values of variables at the two times. This objection is particularly critical for the repeated measurement situation, where there is no element of randomisation in the taking of measurements. However, even in plot experiments, where the subplots may be selected randomly at the different times, the development over time of the whole plot may still produce greater correlation of measurements at adjacent times than at a wider interval.

(3) Very many measurements

The possibility of making very many successive measurements on the same experimental unit raises questions of how much additional information can be obtained from such measurements, and of how to interpret the very large sets of data thus obtained. It is important to realise that more data does not inevitably imply more information. In the limiting cases two observations taken almost simultaneously can provide no more information than a single observation. There is clearly no advantage in repeating observations except at sufficient time intervals that the experimenter might reasonably expect that a major detectable change has taken place. However, this immediately invokes the first objection.

There are three alternatives to treating time as a split unit factor. In each, the analysis is based on the original structure of the experiment, regarding the repeated measurements as different variables.

(1) Derived variables

In the first alternative, analyses can be performed for each separate time variable and for additional variables representing averages or changes over time. Thus, if a measurement is repeated at five different times $(t_1, t_2, t_3, t_4, t_5)$, analyses may usefully be performed for each measurement time $(y_1, y_2, y_3, y_4, y_5)$, for the average

$$(y_1 + y_2 + y_3 + y_4 + y_5)/5,$$

for simple differences

$$(y_2 - y_1, y_3 - y_2, y_4 - y_3, y_5 - y_4),$$

or for more complex comparisons such as

$$y_5 + y_4 - y_2 - y_1,$$
$$2y_5 + y_4 - y_2 - 2y_1,$$

or

$$2y_5 - y_4 - 2y_3 - y_2 + 2y_1.$$

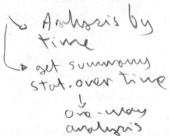

Obviously, not all these variables will be sensible in any particular situation, but each of them, and some others, may be useful. The last two represent the linear rate of increase of yield from t_1 to t_5, and the quadratic change of yield over the five times. Often, these two and the average will contain much of the information.

One difficulty, which must be recognised in this approach, is that the number of analyses performed can easily be allowed to become large, and it is important to remember that the different variables analysed may not be independent. The analyses and the interpretation of results from the analyses would correspondingly not be independent. The purpose of conducting many analyses is to try to discover the important patterns of treatment variation or *lack of variation* so that a sensible overall interpretation can be achieved.

Example 14.3

To illustrate the use of contrasts to interpret repeated measurements we consider the following data on the growth of rats. Three treatments were compared with different numbers of rats per treatment. The rats were weighed before the onset of the treatments and after one, two, three and four weeks. The results are shown in Table 14.11. We consider the analyses not only for the four sets of measurements taken after the treatments were applied, but also for the four simple differences $(y_1 - y_0, y_2 - y_1, y_3 - y_2, y_4 - y_3)$ and for the overall average, $(y_1 + y_2 + y_3 + y_4)/4$, the linear effect $(-3y_1 - y_2 + y_3 + 3y_4)/10$ and the quadratic effect $(y_1 - y_2 - y_3 + y_4)/2$ calculated from the last four weeks' observations. The observed values for these variables are also shown in Table 14.11.

The basic structure of the design is a completely randomised design with unequal numbers, and the analysis of variance structure is:

Source	df
Treatments	2
Rats within treatments	24
Total	26

The analyses for all 11 variables, not including the pre-treatment measurement variable are summarised in Table 14.12 and the table of means and standard errors of differences between two means are given in Table 14.13.

To determine the appropriate interpretation of the results we consider each analysis and set of treatment means in turn. It is immediately apparent that the differences between the results for the control and Thyroxin treatments are small and are not significant at 5% for any

Table 14.11.

y_0	y_1	y_2	y_3	y_4	$y_1 - y_0$	$y_2 - y_1$	$y_3 - y_2$	$y_4 - y_3$	Mean	Linear	Quadratic
Control											
57	86	114	139	172	29	28	25	33	127.75	27.3	2.5
60	93	123	146	177	33	30	23	31	134.75	27.5	0.5
52	77	111	144	185	25	34	33	41	129.25	35.7	3.5
49	67	100	129	164	18	33	29	35	115.00	32.0	1.0
56	81	104	121	151	25	23	17	30	114.25	22.7	3.5
46	70	102	131	153	24	32	29	22	114.00	27.0	-5.0
51	71	94	110	141	20	23	16	31	104.00	22.6	4.0
63	91	112	130	154	28	21	18	24	121.75	20.7	1.5
49	67	90	112	140	18	23	22	28	102.25	24.1	2.5
57	82	110	139	169	25	28	29	30	125.00	29.0	1.0
Thyroxin											
59	85	121	156	191	26	36	35	35	138.25	35.3	-0.5
54	71	90	110	138	17	19	20	28	102.25	22.1	4.5
56	75	108	151	189	19	33	43	38	130.75	38.5	2.5
59	85	116	148	177	26	31	32	29	131.50	30.8	-1.0
57	72	97	120	144	15	25	23	24	108.25	23.9	-0.5
52	73	97	116	140	21	24	19	24	106.50	22.0	0
52	70	105	138	171	18	35	33	33	121.00	33.6	-1.0
Thiouracil											
61	86	109	120	129	25	23	11	9	111.00	14.0	-7.0
59	80	101	111	126	21	21	10	11	103.50	13.6	-5.0
53	79	100	106	133	26	21	6	27	104.50	16.8	3.0
59	88	100	111	122	29	12	11	11	105.25	11.3	-0.5
51	75	101	123	140	24	26	22	17	109.75	21.7	-4.5
51	75	92	100	119	24	17	8	19	96.50	14.0	1.0
51	78	95	103	108	22	17	8	5	96.00	9.8	-6.0
56	69	93	114	138	11	24	21	24	103.50	22.8	0
46	61	78	90	107	15	17	12	17	84.00	15.0	0
53	72	89	104	122	19	17	15	18	96.75	16.5	0.5

Table 14.12.

	Variables	SS	df	MS	F
	y_1				
Between treatments		36	2	18	0.3
Within treatments		1650	24	69	
	y_2				
Between treatments		601	2	300	2.5
Within treatments		2876	24	120	
	y_3				
Between treatments		3564	2	1782	9.8
Within treatments		4356	24	182	
	y_4				
Between treatments		9204	2	4602	17.1
Within treatments		6442	24	268	
	$y_1 - y_0$				
Between treatments		82	2	41	1.7
Within treatments		582	24	24	
	$y_2 - y_1$				
Between treatments		477	2	238	9.4
Within treatments		609	24	25	
	$y_3 - y_2$				
Between treatments		1316	2	658	15.1
Within treatments		1047	24	44	
	$y_4 - y_3$				
Between treatments		1334	2	667	18.8
Within treatments		853	24	36	
	mean				
Between treatments		2085	2	1042	8.5
Within treatments		2932	24	122	
	linear				
Between treatments		1022	2	511	19.2
Within treatments		637	24	26	
	quadratic				
Between treatments		62.8	2	31.4	4.0
Within treatments		189.4	24	7.9	

variable. We therefore examine how the results for Thiouracil differ from the other two treatments.

First consider the results for the weights on the four separate occasions. There is no difference at time 1; at time 2 there is a small reduction in weight for the Thiouracil treated rats, which, in itself would not be convincing; at times 3 and 4 there are increasingly large reductions. The last three growth increments all show clear reductions due to Thiouracil, the reductions in incremental growth being similar for each of the latter two intervals.

Table 14.13.

	Control (1)	Thyroxin (2)	Thiouracil (3)	SE of difference	
				(1 − 3)	(1 − 2) or (2 − 3)
y_1	78	76	76	3.7	4.1
y_2	106	105	96	4.9	5.4
y_3	130	134	108	6.0	6.6
y_4	161	164	124	7.3	8.1
$y_1 - y_0$	24.5	20.3	21.6	2.2	2.4
$y_2 - y_1$	27.5	29.0	19.5	2.2	2.5
$y_3 - y_2$	24.1	29.3	12.4	3.0	3.3
$y_4 - y_3$	30.5	30.1	15.8	2.7	3.0
Mean	119	120	101	4.9	5.4
Linear	27.3	29.5	15.6	2.3	2.5
Quadratic	1.5	0.6	−2.0	1.3	1.4

The mean weight reflects the results for the individual weights showing a reduction due to Thiouracil but with a less extreme significance level. The linear effect provides the most clear-cut difference between Thiouracil and the other two treatments with Thiouracil giving a growth rate only a little over half the growth rate for the other two.

The quadratic effect analysis is curious. None of the three treatments shows a quadratic effect that is significantly different from zero (5% significance level), but the range of mean quadratic effects suggests a difference between the treatments with the control showing a positive quadratic effect and the Thiouracil showing a negative effect, suggesting that weight might be approaching an upper limit for the Thiouracil treatment. Thyroxin seems to have no quadratic effect.

Overall we might base our interpretation on the results for y_1 and the linear effect. There appear to be no differential effects of treatments up to the first measurement occasion. Thereafter the rats treated with Thiouracil grow at roughly half the rate for the control with a suggestion that the difference in growth rates is tending to expand.

(2) Multivariate analysis

The second alternative to treating time as a split unit factor, which we shall not discuss in detail, is to treat the multiple observations as a set of multivariate measurements, and to calculate a multivariate analysis of variance. In many ways, the difficulty with this approach is the reverse of the difficulty with the use of contrasts. The separate analysis of many different variables, including contrast variables, leads to problems of interpreting together the many separate, and interdependent, sets of results. With multivariate analysis, all the variables are included in a single overall analysis, with inevitably significant overall variation between treatments and the problem is to separate out the important systematic patterns of variation.

(3) *Response functions*

The third approach to the analysis of repeated measurements is really an extension of the ideas of contrasts. Instead of looking at particular contrasts representing some aspects of the pattern of yield response over time, the pattern of yield response is examined directly by fitting response functions, for each individual experimental unit. The response functions used should be practically realistic and should be tested for statistical fit. The parameters, or functions of the parameters, can then be analysed using the original structure of the experimental units. The analysis of linear and quadratic regression coefficients, which were the last two contrasts considered when the analysis of contrasts was discussed, is equivalent to the analysis of fitted linear and quadratic parameters if a response function

$$y = a + bt + ct^2$$

was fitted to the data for each rat. If some other response function, such as

$$y = A(1 - e^{-k(t + t_0)})$$

were fitted, then the parameters A, k and t_0 could be analysed, though most interest would probably be concentrated on A, the asymptotic limit to size, and k, the rate of approach to that asymptote. For the rat data, the quadratic response is probably adequate, but for data in which the response is considered over a longer time span the more general response function approach may be useful.

14.6   Systematic treatment variation within main units

We have discussed the use of repeated measurements from an experimental unit. When these measurements are based on separate samples from within an experimental field plot we have to consider how to select the samples from within the plot. The systematic selection of samples for different harvests within a plot will tend to give more precise comparisons than random sampling because the samples can be taken in identical fashion. Starting from the same end of each plot will provide a similar environment for the sequence of samples for different treatments. Systematic selection of samples is usually recommended for multiple harvests but randomised selection should be used if a split plot analysis is to be applied to the data.

An analogous idea to systematic sampling from a plot is the application of systematically varying treatments to successive sections of each plot. The motivation for the use of a systematic arrangement of treatment levels stems from the desire to use resources efficiently. In a conventional randomised design to compare three or four different spacing, or nutrient treatments, it is necessary to have substantial 'guard' or 'discard' areas round each differently treated plot, so that the harvested area in the middle of the plot shall be representative of a whole field receiving the particular treatment of that plot. The more different the conditions produced by the different treatments, the greater the need for large guard areas. In many experiments, the harvested areas from each plot may be no more than 30 or 40% of the total plot area.

Systematic split plot treatments have been suggested and used to increase the harvested area proportion of the experiment for quantitative treatment factors of various types. Nelder (1962) and Bleasdale (1967) suggested various designs for spacing experiments, and many experiments have subsequently used systematic spacing factors. Fertiliser factor levels have been applied systematically by Cleaver, Greenwood and Wood (1970). Trials of insecticides and herbicides have used systematic placement of successive concentrations. Similar ideas have been used in some medical and animal trials, where treatments are applied in order of increasing concentration, because the effect of a subsequent treatment at a lower concentration may be masked by the residual effects of the previous treatment. In field crop experiments, systematic arrangements may be used for two treatment factors, the systematic change of level occurring in perpendicular directions for the two factors. In all cases the use of systematic treatment allocation must be considered cautiously, but particularly in early stages of investigations experiments using systematic treatment arrangements can be extremely informative, provided some proper replication is also included.

The principle behind the use of systematic designs for field crop experiments is that if treatment levels are placed in increasing order, then the two treatment levels on either side of a particular level will be similar to that level, and guard areas can be eliminated, provided that the adjacent treatment levels are sufficiently similar, so that the environment of the particular level may be regarded as representative of application of that level in a large area. Traditionally, it has been suggested that changes of crop density or of fertiliser level of 10% should satisfy the requirement of similarity of adjacent levels, but this is a 'rule of thumb' which may be too lenient or too stringent in particular circumstances. Particular forms of systematic design for spacing arrangement experiments are discussed in Chapter 17.

As for repeated measurement data, it will usually not be appropriate to analyse data from experiments with systematic arrangements of split plot factor levels by calculating the analysis of variance for a split plot design as if the systematic treatment levels had been randomised within each main plot. Although such an analysis might be argued to be valid because the dominant contribution to σ^2 was believed to be the plant variation, rather than soil variation, doubts about the validity of assumptions would be difficult to satisfy. In any case the objectives of the experiment will usually be more appropriately satisfied by first fitting a yield response function over the levels of the systematically applied treatment factor for each main plot, and then analysing the fitted response functions.

The implications of this philosophy of analysis, which is exactly the same as the third approach for the analysis of repeated measurements, are that, when levels of a quantitative factor are applied systematically, there should be other treatment factors applied to main plots, and also some explicit replication of main plots, unless the number of combinations of main plot factors is very large. Essentially, the analysis of variation of the fitted response functions uses the structure of the main plots only, so that

Figure 14.10. Experimental design layout using systematic treatment variation within a 2×3 main plot structure.

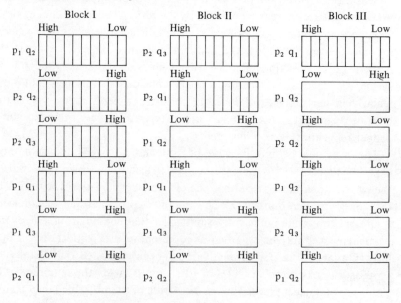

the experimental design of the main plots must constitute a valid design, in particular providing adequate estimation of the main plot error variance. The layout for a typical 'systematic' design is shown in Figure 14.10. The design has ten levels of the systematic split plot factor level within main plots for a 2×3 factorial structure in three randomised blocks of six main plots each. Not only is the allocation of the six main plot treatments (p_1q_1, p_1q_2, p_1q_3, p_2q_1, p_2q_2, p_2q_3) to main plots randomised in each block, but the direction in which the split plot treatments change systematically from low to high is also randomised for each main plot.

Exercises 14

(1) In an experiment to investigate the benefits of growing maize and cowpea together, a criss-cross design was used. Four cowpea varieties were allocated to the four rows of each block and five maize varieties to the columns of each block. Three blocks of this design were used, the orders of the cowpea varieties and of the maize varieties being randomised in each block.

Calculate the analysis of variance for the sets of total yields and sums of squares in Table 14.14, and briefly summarise the conclusions to be drawn from the results.

Table 14.14.

Cowpea variety			Maize variety			Total
	1	2	3	4	5	
A	80.1	50.6	41.5	46.5	57.8	276.4
B	59.1	34.2	41.2	45.3	53.7	233.5
C	100.5	85.7	71.7	90.9	68.9	417.7
D	36.6	49.6	74.8	54.3	56.9	272.2
Total	276.3	220.1	229.2	236.9	237.3	1199.8

$\sum y^2 \qquad = 27\ 122$ $\qquad \sum (\text{Cowpea–maize total})^2 \qquad 78\ 665$

$\sum (\text{Block total})^2 = 487\ 581$ $\qquad \sum (\text{Block–cowpea total})^2 = 129\ 092$

$\sum (\text{Cowpea total})^2 = 379\ 485$ $\qquad \sum (\text{Block–maize total})^2 \quad = 99\ 199$

$\sum (\text{Maize total})^2 = 289\ 751$ $\qquad \sum (\text{Grand total})^2 \qquad = 1\ 439\ 520$

(2) An experiment on competition between grass and legumes was designed as a 2^4 in two blocks with eight main plots per block and two split plots per main plot. The main plot treatments were two cutting regimes (C), two grass species (G) and two legume species

Table 14.15.

	p_0	p_1
$c_0 g_0 l_0$	20	34
$c_0 g_0 l_1$	14	29
$c_0 g_1 l_0$	24	37
$c_0 g_1 l_1$	18	32
$c_1 g_0 l_0$	27	31
$c_1 g_0 l_1$	24	25
$c_1 g_1 l_0$	33	37
$c_1 g_1 l_1$	26	29

Table 14.16.

	SS
Blocks	4.5
Main plot treatments	119.0
Main plot total	132.5
P	144.5
P × main plot treatments	62.5
Total	342.0

(L); the split plot treatments were two proportions, grass:legume (P).

The treatment totals for yield per plot (in a simplified form) are given in Table 14.15. Complete the analysis of variance, started in Table 14.16, separating treatment sums of squares into single degree of freedom components. Summarise the conclusions to be drawn from the experiment, presenting appropriate tables of means and all necessary standard errors of differences between means.

If it is assumed that all three-factor and higher-order interactions are zero, obtain an estimate of the difference $(c_0 g_1 l_0 p_0 - c_1 g_1 l_0 p_0)$ and the standard error of this estimate.

(3) An experiment on air and soil heating effects on glasshouse tomatoes was designed in a split-split plot form. The main plots were whole glasshouse compartments in which the air was maintained at either 60°F or 55°F. Each compartment contained two large troughs to be regarded as the split plots and each trough was divided into two halves to make two split-split plots.

In each trough, one half was heated and the other was unheated (control). In each compartment, the soil in the heated half of one trough was kept at a minimum temperature of 65°F, and the soil in the heated half of the other trough had a minimum temperature of 75°F. There were four blocks (pairs of compartments) and a diagrammatic representation of the design is shown in Figure 14.11.

The model for the yield variation can be written:

$$y_{ijk} = \mu + b_i + \varepsilon''_{ij} + \varepsilon'_{ijk} + \varepsilon_{ijkl} + \text{(treatment effect)},$$

where the suffices (i, j, k, l) correspond to blocks, compartments within blocks, troughs within compartments and half troughs

Figure 14.11. Full design layout for tomato air and soil heating experiment (exercise question 3).

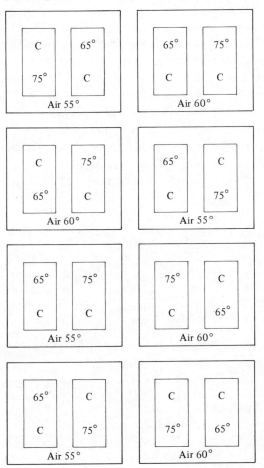

Table 14.17.

	Air 55	Air 60	Mean
Control	$y_{.1.1}$	$y_{.2.1}$	$y_{...1}$
Soil 65	$y_{.112}$	$y_{.212}$	$y_{..12}$
75	$y_{.122}$	$y_{.222}$	$y_{..22}$
Mean	$y_{.1..}$	$y_{.2..}$	

within troughs. There are only six distinguishable treatments for which the mean yields may be written as in Table 14.17.

For each (practically) interesting comparison, of which I reckon there are eight, derive the estimated variance, using the error mean squares of 3.51 for main plots, 0.67 for split plots, and 0.62 for split-split plots.

(4) An experiment on the heating characteristics of four different types of plastic pot, and two different forms of cover, was conducted in six cold frames. In the first stage of the experiment, each frame was divided into four quarters, and each different type of pot allocated to one quarter in each frame. Three of the frames were covered with glass, and three with polythene. The design is represented diagrammatically in Figure 14.12. After a week, the covers on each frame are changed (G→P, P→G), and a further set

Figure 14.12. Design for experiment on the effects of different materials on soil heating.

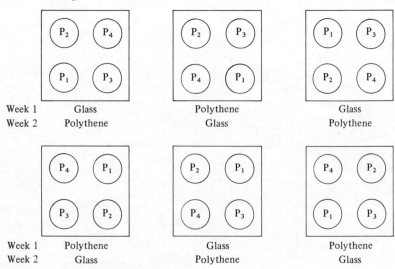

Week 1	Glass	Polythene	Glass
Week 2	Polythene	Glass	Polythene

Week 1	Polythene	Glass	Polythene
Week 2	Glass	Polythene	Glass

of 24 observations obtained, the pots remaining in the same positions.

What is the correct structure for the analysis of variance of the complete set of 48 observations?

(5) In a randomised block design levels of factor A are applied to plots running across the block in a N–S direction and levels of factor B to strips running across the block in an E–W direction. Each level of B is split for levels of C, each level of C running across all A levels, the randomisation of levels of C being independent for each level of B. Write down the model for this design and outline the analysis of variance. What is the estimated variance of the comparison between two levels of A for given level of C, averaged over levels of B?

15

Incomplete block size for factorial experiments

15.0 *Preliminary examples*

(a) An experiment on the effect of different spacings on the productivity of cassava and cowpea crops grown in alternate rows, has been designed as a 3×3 factorial in four randomised blocks of nine plots each. The first factor is the spacing between rows, and the second is the within row density. The experimenter would also like to include in the experiment two varieties of cassava and two varieties of cowpea. The problem is how to allocate the four variety combinations while retaining the blocking structure.

(b) In an animal feeding experiment, six dietary treatments are to be compared. The diets are all possible combinations of three different levels of molasses at two energy levels. Twenty-four sheep are available and each sheep can be fed a different diet in each of two time periods. It is expected that there will be large differences in nutritional performances between sheep and some systematic differences between the results for the first and second periods. The structure of the experimental units therefore has two blocking classifications, giving a 24×2 row and column structure. The treatments have a 3×2 structure and main effect comparisons, particularly between the three levels of molasses are the main area of interest.

15.1 **Small blocks and many factorial combinations**

There is an apparent incompatibility between two fundamental principles discussed in earlier chapters. The idea of blocking is to group together experimental units which are likely to give similar yields if treated with the same treatment, and to base comparisons between treatments on the differences between observations on units within a group. The general implication of the principle of blocking is that blocks will contain small numbers of experimental units, rarely more than 15 and frequently eight or fewer. The principle of factorial structure, on the other hand, leads to experiments with many different experimental treatment combinations. When discussing blocking in Chapter 7, we developed a philosophy for designs using blocks with fewer units than the total number of treatments, but only the case where all treatment comparisons were of equal importance was considered. With factorial structures, there is an implied ordering of importance of effects with, in general, main effects being more important than interactions, two-factor interactions being more important than higher level interactions and so on. In the context of a priority ordering of effects, and with block sizes less than the total number of treatment combinations, how do we allocate treatment combinations to blocks?

An alternative situation producing the same problem is when additional treatments are to be added to an existing experiment, either when the experiment is a continuing one and the additional treatments are for the second stage, or when the additional treatments are conceived after the initial block–treatment pattern is determined. This second situation is simpler than the more general case, and we consider an initial example of this type to illustrate the principles of confounding treatment comparisons.

Example 15.1
An experiment to compare two nitrogen levels, three spatial arrangements and two management systems for a maize crop is planned in four randomised blocks of 12 plots each. The experimenter wishes to include also two different genotypes of maize, but does not wish to have more than 12 plots per block. The initial (unrandomised) design of the experiment is shown in Figure 15.1, and the analysis of variance structure is given in Table 15.1. It is worth noting that 33 df are amply sufficient for the estimation of σ^2 and that the resources which are represented in these 33 df could be more effectively used used to estimate further treatment effects.

To introduce two different genotypes, while (a) retaining all the information on the first three treatment factors, (b) keeping a block size of 12 plots, and (c) obtaining information on the genotype main effects and the interactions of genotypes with each other factor, certain simple restrictions on the design can be specified. First, all 24 treatment combinations should be replicated equally so that all the main effects and interactions of the four factors are mutually orthogonal. Second, each

Figure 15.1. Experimental plan for a $2 \times 3 \times 2$ structure in four blocks of 12 units.

Block

I	II	III	IV
$n_1 s_1 m_1$	$n_1 s_1 m_1$	$n_1 s_1 m_1$	$n_1 s_1 m_1$
$n_1 s_1 m_2$	$n_1 s_1 m_2$	$n_1 s_1 m_2$	$n_1 s_1 m_2$
$n_1 s_2 m_1$	$n_1 s_2 m_1$	$n_1 s_2 m_1$	$n_1 s_2 m_1$
$n_1 s_2 m_2$	$n_1 s_2 m_2$	$n_1 s_2 m_2$	$n_1 s_2 m_2$
$n_1 s_3 m_1$	$n_1 s_3 m_1$	$n_1 s_3 m_1$	$n_1 s_3 m_1$
$n_1 s_3 m_2$	$n_1 s_3 m_2$	$n_1 s_3 m_2$	$n_1 s_3 m_2$
$n_2 s_1 m_1$	$n_2 s_1 m_1$	$n_2 s_1 m_1$	$n_2 s_1 m_1$
$n_2 s_1 m_2$	$n_2 s_1 m_2$	$n_2 s_1 m_2$	$n_2 s_1 m_2$
$n_2 s_2 m_1$	$n_2 s_2 m_1$	$n_2 s_2 m_1$	$n_2 s_2 m_1$
$n_2 s_2 m_2$	$n_2 s_2 m_2$	$n_2 s_2 m_2$	$n_2 s_2 m_2$
$n_2 s_3 m_1$	$n_2 s_3 m_1$	$n_2 s_3 m_1$	$n_2 s_3 m_1$
$n_2 s_3 m_2$	$n_2 s_3 m_2$	$n_2 s_3 m_2$	$n_2 s_3 m_2$

Table 15.1.

Effect	df
Blocks	3
Nitrogen (N)	1
Spacings (S)	2
Management (M)	1
NS interaction	2
NM interaction	1
SM interaction	2
NSM interaction	2
Error	33
Total	47

block of 12 units already includes all 12 combinations of the three original factors so that all estimates of effects for these three factors are unaffected by block differences.

Finally, the two genotypes (g_1 and g_2) should be allocated in such a way that each effect involving the factor G is not affected by block differences. A sufficient requirement for a treatment effect to be estimated independently of block effects is that each level or combination of levels included in the definition of the effect shall occur equally frequently in each block. We can now consider the allocation restrictions for g_1 and g_2 in terms of the main effects of G and the interactions involving G.

(i) To be able to estimate the main effect of G

each genotype must occur six times in each block and twice with each of the original 12 treatment combinations.

(ii) To be able to estimate the GN interaction

each of the four (g, n) combinations must occur thrice in each block and twice with each of the six (s, m) combinations.

(iii) To be able to estimate the GS interaction

each of the six (g, s) combinations must occur twice in each block and twice with each of the four (n, m) combinations.

(iv) To be able to estimate the GNS interaction

each of the 12 (g, n, s) combinations must occur once in each block, and twice with each level of m.

(v) (Also GMS)

(Swap n and m.)

(vi) To be able to estimate the GNM interaction

each of the eight (g, n, m) combinations must occur equally frequently in each block.

(vii) To be able to estimate the GNSM interaction

each of the 24 (g, n, s, m) combinations must occur equally frequently in each block.

Clearly the requirements for (vi) and (vii) are impossible in blocks of 12 units. The other requirements are all individually possible, and the design problem is simply a question of whether they are simultaneously possible. The best way to convince oneself that, in fact, the requirements (i) to (v) can be achieved simultaneously is to try adding the genotypes in two blocks of the original design given in Figure 15.1 (with the treatments written in a systematic order instead of a randomised order to clarify the logic of the allocation). It will be found that choice is very limited, and that the design 'constructs itself'. Two essentially different designs are possible, and these are shown in Figure 15.2 (a) and (b). The difference between the two designs is in the pattern of occurrence of the eight (g, n, m) combinations. In Figure 15.2(a), four of these combinations occur thrice in the first block, and the other four occur thrice in the second block. In Figure 15.2(b), all eight occur in each block, four occurring twice in block I, and the other four twice in block II. In either design the comparison of the eight combinations must involve the differences between blocks. The choice between the designs is considered further in Section 15.3.

Since the complete set of treatment effects is defined to be orthogonal, all effects which are orthogonal to blocks can be estimated independently of block effects and of each other. The calculation of effect estimates and sums of squares in an analysis of variance is exactly the same as for any other factorial or blocked experiment, except for the GNM and GNSM interaction effects. The structure of the analysis of variance for the four block design which consists of two replicates of either of the plans shown in Figure 15.2 can be specified by the degrees of freedom for the fitted terms; see Table 15.2. The GNM interaction, discussed earlier, is said to be confounded with blocks and is necessarily omitted from the analysis. The four-factor interaction GNSM is also omitted but would not usually be included in the analysis. The experiment we have designed is called a confounded experiment. It is important, however, not to concentrate on

Table 15.2.

Effect	df
Blocks	3
Nitrogen (N)	1
Spacing (S)	2
Management (M)	1
Genotype (G)	1
NS interaction	2
NM interaction	1
SM interaction	2
GN interaction	1
GS interaction	2
GM interaction	1
NSM interaction	2
NSG interaction	2
SMG interaction	2
Error	24
Total	47

Figure 15.2. Two alternative designs for a $2 \times 3 \times 2 \times 2$ structure in two blocks of 12 units.

(a)   Block

I	II
$n_1 s_1 m_1 g_1$	$n_1 s_1 m_1 g_2$
$n_1 s_1 m_2 g_2$	$n_1 s_1 m_2 g_1$
$n_1 s_2 m_1 g_1$	$n_1 s_2 m_1 g_2$
$n_1 s_2 m_2 g_2$	$n_1 s_2 m_2 g_1$
$n_1 s_3 m_1 g_1$	$n_1 s_3 m_1 g_2$
$n_1 s_3 m_2 g_2$	$n_1 s_3 m_2 g_1$
$n_2 s_1 m_1 g_2$	$n_2 s_1 m_1 g_1$
$n_2 s_1 m_2 g_1$	$n_2 s_1 m_2 g_2$
$n_2 s_2 m_1 g_2$	$n_2 s_2 m_1 g_1$
$n_2 s_2 m_2 g_1$	$n_2 s_2 m_2 g_2$
$n_2 s_3 m_1 g_2$	$n_2 s_3 m_1 g_1$
$n_2 s_3 m_2 g_1$	$n_2 s_3 m_2 g_2$

(b)   Block

I	II
$n_1 s_1 m_1 g_1$	$n_1 s_1 m_1 g_2$
$n_1 s_1 m_2 g_2$	$n_1 s_1 m_2 g_1$
$n_1 s_2 m_1 g_2$	$n_1 s_2 m_1 g_1$
$n_1 s_2 m_2 g_1$	$n_1 s_2 m_2 g_2$
$n_1 s_3 m_1 g_1$	$n_1 s_3 m_1 g_2$
$n_1 s_3 m_2 g_2$	$n_1 s_3 m_2 g_1$
$n_2 s_1 m_1 g_2$	$n_2 s_1 m_1 g_1$
$n_2 s_1 m_2 g_1$	$n_2 s_1 m_2 g_2$
$n_2 s_2 m_1 g_1$	$n_2 s_2 m_1 g_2$
$n_2 s_2 m_2 g_2$	$n_2 s_2 m_2 g_1$
$n_2 s_3 m_1 g_2$	$n_2 s_3 m_1 g_1$
$n_2 s_3 m_2 g_1$	$n_2 s_3 m_2 g_2$

the information which has been confounded in a confounded experiment, but rather to consider what has been gained. All the main effects and two-factor interactions for all four factors can be estimated, and all the three-factor interactions involving factor S. Compared with the original design involving only three factors, we have additional information on nine treatment comparisons, and we still have a very adequate estimate of σ^2, based on 24 df. Indeed, we could still be accused of not using the available resources efficiently because the 24 df for estimating σ^2 is more than sufficient.

We have discussed the design construction in the context of adding an extra factor to an existing experiment, but if the full design problem of a $2 \times 2 \times 2 \times 3$ factorial structure in four blocks of 12 units is considered directly, we should again arrive at the designs of Figure 15.2.

We have deliberately started the discussion of confounded experiments by considering an experimental problem with many factors, and with different numbers of levels per factor because the principle of avoiding confounding is a simple one which applies equally to large or small, complex or simple, experiments. We continue the discussion by considering two further, practical, large experiments, and then discuss some smaller factorial structures in which all factors have the same number of levels, before going on to consider extensions of the basic principle.

Example 15.2

Four factors, two (P, Q) with three levels and two (R, S) with two levels are to be compared, again using blocks of 12 units, but this time having a total of only 36 observations so that each treatment combination occurs just once.

The effects for which it might be hoped that confounding could be avoided are those for which the number of combinations of levels is a divisor of 12, so that the combinations of levels can be equally replicated in a block of 12. These effects, and the necessary restrictions, are:

(i)   main effects of P and Q, each having three levels, to appear four times in each block;

(ii) main effects of R and S, each with two levels, each level to occur six times per block;

(iii) two-factor interaction, RS, four combinations of levels, each three times per block;

(iv) two-factor interactions, PR, PS, QR, QS, each with six combinations of levels, each combination twice per block;

(v)   three-factor interactions, PRS, QRS, 12 combinations, all 12 to occur in each block.

Figure 15.3. Two alternative designs for a $3 \times 3 \times 2 \times 2$ structure in three blocks of 12 units.

(a)

	Block	
I	II	III
$p_1 q_1 r_1 s_1$	$p_2 q_1 r_1 s_1$	$p_3 q_1 r_1 s_1$
$p_2 q_1 r_1 s_2$	$p_3 q_1 r_1 s_2$	$p_1 q_1 r_1 s_2$
$p_3 q_1 r_2 s_1$	$p_1 q_1 r_2 s_1$	$p_2 q_1 r_2 s_1$
$p_1 q_1 r_2 s_2$	$p_2 q_1 r_2 s_2$	$p_3 q_1 r_2 s_2$
$p_2 q_2 r_1 s_1$	$p_3 q_2 r_1 s_1$	$p_1 q_2 r_1 s_1$
$p_3 q_2 r_1 s_2$	$p_1 q_2 r_1 s_2$	$p_2 q_2 r_1 s_2$
$p_1 q_2 r_2 s_1$	$p_2 q_2 r_2 s_1$	$p_3 q_2 r_2 s_1$
$p_2 q_2 r_2 s_2$	$p_3 q_2 r_2 s_2$	$p_1 q_2 r_2 s_2$
$p_3 q_3 r_1 s_1$	$p_1 q_3 r_1 s_1$	$p_2 q_3 r_1 s_1$
$p_1 q_3 r_1 s_2$	$p_2 q_3 r_1 s_2$	$p_3 q_3 r_1 s_2$
$p_2 q_3 r_2 s_1$	$p_3 q_3 r_2 s_1$	$p_1 q_3 r_2 s_1$
$p_3 q_3 r_2 s_2$	$p_1 q_3 r_2 s_2$	$p_2 q_3 r_2 s_2$

(b)

	Block	
I	II	III
$p_1 q_1 r_1 s_1$	$p_2 q_1 r_1 s_1$	$p_3 q_1 r_1 s_1$
$p_1 q_1 r_1 s_2$	$p_2 q_1 r_1 s_2$	$p_3 q_1 r_1 s_2$
$p_1 q_1 r_2 s_1$	$p_2 q_1 r_2 s_1$	$p_3 q_1 r_2 s_1$
$p_1 q_1 r_2 s_2$	$p_2 q_1 r_2 s_2$	$p_3 q_1 r_2 s_2$
$p_2 q_2 r_1 s_1$	$p_3 q_2 r_1 s_1$	$p_1 q_2 r_1 s_1$
$p_2 q_2 r_1 s_2$	$p_3 q_2 r_1 s_2$	$p_1 q_2 r_1 s_2$
$p_2 q_2 r_2 s_1$	$p_3 q_2 r_2 s_1$	$p_1 q_2 r_2 s_1$
$p_2 q_2 r_2 s_2$	$p_3 q_2 r_2 s_2$	$p_1 q_2 r_2 s_2$
$p_3 q_3 r_1 s_1$	$p_1 q_3 r_1 s_1$	$p_2 q_3 r_1 s_1$
$p_3 q_3 r_1 s_2$	$p_1 q_3 r_1 s_2$	$p_2 q_3 r_1 s_2$
$p_3 q_3 r_2 s_1$	$p_1 q_3 r_2 s_1$	$p_2 q_3 r_2 s_1$
$p_3 q_3 r_2 s_2$	$p_1 q_3 r_2 s_2$	$p_2 q_3 r_2 s_2$

Table 15.3.

Effect	df
Blocks	2
P	2
Q	2
R	1
S	1
RS interaction	1
PR interaction	2
PS interaction	2
QR interaction	2
QS interaction	2
PRS interaction	2
QRS interaction	2
Error	14
Total	35

Effects which will not allow equal occurrence of all combinations in each block are PQ, PQR, PQS, PQRS. Two possible designs are given in Figure 15.3, the design in (*a*) having all combinations of (p, q) appearing in each block, three duplicated in each block, while in (*b*) different groups of three (p, q) combinations appear in each block. The construction of the design may be followed more easily by first recognising that all QRS combinations occur in each block and that the levels of P are added to these combinations so as to satisfy the requirements.

For either design, and for the other designs possible under the restrictions (i) to (v), which are intermediate between (*a*) and (*b*), the analysis of variance structure is straightforward (see Table 15.3), including all effects, for which the design has been constructed to provide estimates, and of course the block effects.

Note that the error degrees of freedom for the estimation of σ^2 are still adequate at 14, but are nearing the reasonable minimum, so that the experiment provides an efficient use of resources.

Example 15.3

A very large experiment encountered in experimentation on intercropping in which there were five factors, three (P, Q, R) with three levels and two (S, T) with two levels. It was proposed to use 18 plots per block, which was probably too large. The factors were:

P: three densities of maize,
Q: three densities of beans,

R: three genotypes of beans,
S: two genotypes of maize,
T: two row arrangements,

and the design in Figure 15.4 allows all main effects and two-factor interactions, except ST, to remain non-confounded, and also allows the estimation of three-factor interactions PQS, PQT, PRS, PRT, QRS. When the levels of T are added to complete the treatment combinations for each block a choice emerges. Either each block contains only two (S, T) combinations and all QRT combinations occur in each block, or all four (S, T) combinations occur in each block, although unevenly, but not all QRT combinations can be included in each block. The design in Figure 15.4 is the latter form based on the practical desire to obtain some information about the ST interaction.

Figure 15.4. Design for a $3 \times 3 \times 2 \times 3 \times 2$ structure in six blocks of 18 units.

PQS combinations in all blocks	\multicolumn{6}{c}{Block}					
	I	II	III	IV	V	VI
$p_1q_1s_1$	r_1t_1	r_1t_2	r_2t_1	r_2t_2	r_3t_1	r_3t_2
$p_1q_1s_2$	r_2t_2	r_2t_1	r_3t_2	r_3t_1	r_1t_2	r_1t_1
$p_1q_2s_1$	r_2t_1	r_2t_2	r_3t_1	r_3t_2	r_1t_1	r_1t_2
$p_1q_2s_2$	r_3t_2	r_3t_1	r_1t_2	r_1t_1	r_2t_2	r_2t_1
$p_1q_3s_1$	r_3t_1	r_3t_2	r_1t_1	r_1t_2	r_2t_1	r_2t_2
$p_1q_3s_2$	r_1t_2	r_1t_1	r_2t_2	r_2t_1	r_3t_2	r_3t_1
$p_2q_1s_1$	r_2t_2	r_2t_1	r_3t_2	r_3t_1	r_1t_2	r_1t_1
$p_2q_1s_2$	r_3t_1	r_3t_2	r_1t_1	r_1t_2	r_2t_1	r_2t_2
$p_2q_2s_1$	r_3t_2	r_3t_1	r_1t_2	r_1t_1	r_2t_2	r_2t_1
$p_2q_2s_2$	r_1t_1	r_1t_2	r_2t_1	r_2t_2	r_3t_1	r_3t_2
$p_2q_3s_1$	r_1t_2	r_1t_1	r_2t_2	r_2t_1	r_3t_2	r_3t_1
$p_2q_3s_2$	r_2t_1	r_2t_2	r_3t_1	r_3t_2	r_1t_1	r_1t_2
$p_3q_1s_1$	r_3t_1	r_3t_2	r_1t_1	r_1t_2	r_2t_1	r_2t_2
$p_3q_1s_2$	r_1t_2	r_1t_1	r_2t_2	r_2t_1	r_3t_2	r_3t_1
$p_3q_2s_1$	r_1t_1	r_1t_2	r_2t_1	r_2t_2	r_3t_1	r_3t_2
$p_3q_2s_2$	r_2t_2	r_2t_1	r_3t_2	r_3t_1	r_1t_2	r_1t_1
$p_3q_3s_1$	r_2t_1	r_2t_2	r_3t_1	r_3t_2	r_1t_1	r_1t_2
$p_3q_3s_2$	r_3t_2	r_3t_1	r_1t_2	r_1t_1	r_2t_2	r_2t_1

15.2   Factors with a common number of levels

In the experiments considered in the previous section, the number of levels per factor varies within each experiment. This is likely to be true of many, possibly of most, practical experimental situations. However, when all factors have the same number of levels, the principles of design are simpler because of the symmetry of all factors.

First, we consider some designs for factors with two levels.

Example 15.4

Three factors in blocks of four. Main effects and two-factor interactions can be estimated independently of blocks by allocating combinations to blocks as shown in Figure 15.5(a). All four combinations of levels for factors P and Q are first allocated to each block. The levels of the third factor, R, are completely determined by the choice in the first combination in the block and the requirements not to confound main effects or two-factor interactions.

Example 15.5

Four factors in blocks of eight. The combinations of levels can be allocated to blocks so that all effects up to three-factor interactions are independent of blocks. To construct the design first allocate the eight combinations of levels from the first three factors to each block. As in Example 15.4, after the arbitrary choice for the first combination in the block the levels of the last factor are completely determined by the requirements of avoiding confounding of interactions up to three factors. The resulting design is shown in Figure 15.5(b).

Example 15.6

Four factors in blocks of four. Main effects and most interactions between two factors can be kept clear of confounding as shown in Figure 15.5(c). All four combinations of levels from the first two factors occur in each block. The choice of levels for the third factor is determined as in Example 15.4. The choice of levels for the fourth factor produces the first failure of our optimistic assumption that, if the number of combinations required for an interaction effect is divisible into the block size, then it will be possible to arrange for that interaction to be not confounded. The allocation of levels of the fourth factor must coincide with the pattern of levels of one of the first three factors. In Figure 15.5(c), the two-factor interaction, RS, is confounded with blocks.

Figure 15.5. (a) $2 \times 2 \times 2$ in two blocks of four, (b) $2 \times 2 \times 2 \times 2$ in two blocks of eight, (c) $2 \times 2 \times 2 \times 2$ in four blocks of four, (d) $2 \times 2 \times 2 \times 2 \times 2$ in four blocks of eight.

(a)

Block

I	II
$p_0 q_0 r_0$	$p_0 q_0 r_1$
$p_0 q_1 r_1$	$p_0 q_1 r_0$
$p_1 q_0 r_1$	$p_1 q_0 r_0$
$p_1 q_1 r_0$	$p_1 q_1 r_1$

(b)

Block

I	II
$p_0 q_0 r_0 s_0$	$p_0 q_0 r_0 s_1$
$p_0 q_0 r_1 s_1$	$p_0 q_0 r_1 s_0$
$p_0 q_1 r_0 s_1$	$p_0 q_1 r_0 s_0$
$p_0 q_1 r_1 s_0$	$p_0 q_1 r_1 s_1$
$p_1 q_0 r_0 s_1$	$p_1 q_0 r_0 s_0$
$p_1 q_0 r_1 s_0$	$p_1 q_0 r_1 s_1$
$p_1 q_1 r_0 s_0$	$p_1 q_1 r_0 s_1$
$p_1 q_1 r_1 s_1$	$p_1 q_1 r_1 s_0$

(c)

Block

I	II	III	IV
$p_0 q_0 r_0 s_0$	$p_0 q_0 r_1 s_1$	$p_0 q_0 r_0 s_1$	$p_0 q_0 r_1 s_0$
$p_0 q_1 r_1 s_1$	$p_0 q_1 r_0 s_0$	$p_0 q_1 r_1 s_0$	$p_0 q_1 r_0 s_1$
$p_1 q_0 r_1 s_1$	$p_1 q_0 r_0 s_0$	$p_1 q_0 r_1 s_0$	$p_1 q_0 r_0 s_1$
$p_1 q_1 r_0 s_0$	$p_1 q_1 r_1 s_1$	$p_1 q_1 r_0 s_1$	$p_1 q_1 r_1 s_0$

(d)

Block

I	II	III	IV
$p_0 q_0 r_0 s_0 t_0$	$p_0 q_0 r_0 s_1 t_0$	$p_0 q_0 r_0 s_0 t_1$	$p_0 q_0 r_0 s_1 t_1$
$p_0 q_0 r_1 s_1 t_1$	$p_0 q_0 r_1 s_0 t_1$	$p_0 q_0 r_1 s_1 t_0$	$p_0 q_0 r_1 s_0 t_0$
$p_0 q_1 r_0 s_1 t_0$	$p_0 q_1 r_0 s_0 t_0$	$p_0 q_1 r_0 s_1 t_1$	$p_0 q_1 r_0 s_0 t_1$
$p_0 q_1 r_1 s_0 t_1$	$p_0 q_1 r_1 s_1 t_1$	$p_0 q_1 r_1 s_0 t_0$	$p_0 q_1 r_1 s_1 t_0$
$p_1 q_0 r_0 s_1 t_1$	$p_1 q_0 r_0 s_0 t_1$	$p_1 q_0 r_0 s_1 t_0$	$p_1 q_0 r_0 s_0 t_0$
$p_1 q_0 r_1 s_0 t_0$	$p_1 q_0 r_1 s_1 t_0$	$p_1 q_0 r_1 s_0 t_1$	$p_1 q_0 r_1 s_1 t_1$
$p_1 q_1 r_0 s_0 t_1$	$p_1 q_1 r_0 s_1 t_1$	$p_1 q_1 r_0 s_0 t_0$	$p_1 q_1 r_0 s_1 t_0$
$p_1 q_1 r_1 s_1 t_0$	$p_1 q_1 r_1 s_0 t_0$	$p_1 q_1 r_1 s_1 t_1$	$p_1 q_1 r_1 s_0 t_1$

Example 15.7

Five factors in blocks of eight. All main effects and two-factor interactions can be estimated from the design shown in Figure 15.5(d). Obviously, with eight units per block, many three-factor interactions can be arranged to be orthogonal to blocks. Examination of the design shows that the only three-factor interactions which have been confounded are PRT and QST.

   While the simple logical methods of design construction employed in the previous section are still effective for 2^n structures, the increasing number of desirable restrictions as n increases makes it necessary to develop a method based less on trial and error. In fact, for 2^n, a very simple method is available, and this is described in Section 16.2 of the next chapter which is concerned with mathematical theory for the construction of confounded designs. This theory demonstrates that, for example, we could not have improved on the four designs constructed in Figure 15.5.
   The second investigation of designs where all factors have the same number of levels is, naturally, for three-level factors. The two simplest cases for two and three factors demonstrate clearly the principles for confounding with three-level factors.

Example 15.8

Two three-level factors. To arrange main effects orthogonal to blocks it is necessary to have all three levels of each factor occurring equally frequently in each block, which requires blocks of three units or of six units. Since a design with three units per block implies a complementary design with six units per block, in which the six combinations in a block are simply those left out of the corresponding block in the three-unit block design, only the three-unit block design is examined. The problem then is to arrange the nine treatment combinations in three blocks of three so that each level of each factor occurs once in each block. There are two possible solutions shown in Figure 15.6. The problem is exactly equivalent to the 3×3 Latin square design, the rows, columns and treatments of the Latin square corresponding to levels of factor P, levels of factor Q, and blocks for the confounded design. The two solutions of Figure 15.6 correspond to the two distinct forms of the 3×3 Latin square.

Example 15.9

Three three-level factors in three blocks of nine units each. Again, the ideas of 3×3 Latin squares help to clarify the possible choices. To avoid confounding the two-factor interactions, each block must contain all nine combinations of levels for each pair of factors. The first requirement is that

Figure 15.6. Two alternative designs for a 3×3 structure in three blocks of three units.

(a)   Block

	I	II	III
	$p_0 q_0$	$p_0 q_1$	$p_0 q_2$
	$p_1 q_1$	$p_1 q_2$	$p_1 q_0$
	$p_2 q_2$	$p_2 q_0$	$p_2 q_1$

(b)   Block

	I	II	III
	$p_0 q_0$	$p_0 q_1$	$p_0 q_2$
	$p_1 q_2$	$p_1 q_0$	$p_1 q_1$
	$p_2 q_1$	$p_2 q_2$	$p_2 q_0$

each block contains all nine (p, q) combinations. The levels of R must be allocated so that each (p, r) combination and each (q, r) combination occurs once in a block. If the nine (p, q) combinations are written in a 3×3 square with levels of P corresponding to rows and levels of Q corresponding to columns, then each R level must occur once in each row and once in each column. There are four essentially different patterns. The first is shown in 15.7(a), the other acceptable patterns are obtained by swapping either the p_2 and p_3 columns or the q_1 and q_2 rows, or both. Each of the four patterns produces a distinct block I and the other two blocks are totally determined by the first block. The four designs are shown in Figure 15.7(b), in which only the levels of r are shown, the nine combinations of (p, q) being in the same standard order for each design.

For the analysis of results from the experiments discussed in this section, it would be sufficient from a practical viewpoint to remark that all non-confounded effects are estimated, and the corresponding sums of squares calculated, in the usual way. The error SS, providing an estimate of σ^2, is calculated by subtracting those treatment SS included in the analysis from the total SS. However, further insight into the properties of confounded designs may be obtained through considering unit contrasts and the corresponding sums of squares.

For the 2^3 in blocks of four, shown in Figure 15.5(a), there are seven orthogonal comparisons between units within a single replicate. We know that these should include the seven treatment comparisons, and the block

Figure 15.7. (a) Allocation pattern for levels of R to PQ combinations; (b) four alternative designs for a $3 \times 3 \times 3$ structure in three blocks of nine units.

(a)

	p_0	p_1	p_2
q_0	r_0	r_1	r_2
q_1	r_1	r_2	r_0
q_2	r_2	r_0	r_1

(b)

	Block				Block		
	I	II	III		I	II	III
p_0q_0	r_0	r_1	r_2		r_0	r_1	r_2
p_0q_1	r_1	r_2	r_0		r_1	r_2	r_0
p_0q_2	r_2	r_0	r_1		r_2	r_0	r_1
p_1q_0	r_1	r_2	r_0		r_2	r_0	r_1
p_1q_1	r_2	r_0	r_1		r_0	r_1	r_2
p_1q_2	r_0	r_1	r_2		r_1	r_2	r_0
p_1q_0	r_2	r_0	r_1		r_1	r_2	r_0
p_2q_1	r_0	r_1	r_2		r_2	r_0	r_1
p_2q_2	r_1	r_2	r_0		r_0	r_1	r_2

	I	II	III		I	II	III
	r_0	r_1	r_2		r_0	r_1	r_2
	r_2	r_0	r_1		r_2	r_0	r_1
	r_1	r_2	r_0		r_1	r_2	r_0
	r_1	r_2	r_0		r_2	r_0	r_1
	r_0	r_1	r_2		r_1	r_2	r_0
	r_2	r_0	r_1		r_0	r_1	r_2
	r_2	r_0	r_1		r_1	r_2	r_0
	r_1	r_2	r_0		r_0	r_1	r_2
	r_0	r_1	r_2		r_2	r_0	r_1

comparison, with three main effect comparisons and three two-factor interaction comparisons independent of the block comparisons. It follows that the three-factor interaction comparison and the block comparison must be identical, since each is orthogonal to all six other orthogonal treatment contrasts. This is confirmed by the set of contrasts displayed in Table 15.4, where it can be seen that column seven is the only contrast which can be constructed to be orthogonal to the first six columns, and that it represents both the three-factor interaction $(p-1)(q-1)(r-1)$ and also the difference between block I and block II.

Table 15.4.

			Contrasts			
1(P)	2(Q)	3(R)	4(PQ)	5(PR)	6(QR)	7(PQR ≡ block)
-1	-1	-1	$+1$	$+1$	$+1$	-1
-1	-1	$+1$	$+1$	-1	-1	$+1$
-1	$+1$	-1	-1	$+1$	-1	$+1$
-1	$+1$	$+1$	-1	-1	$+1$	-1
$+1$	-1	-1	-1	-1	$+1$	$+1$
$+1$	-1	$+1$	-1	$+1$	-1	-1
$+1$	$+1$	-1	$+1$	-1	-1	-1
$+1$	$+1$	$+1$	$+1$	$+1$	$+1$	$+1$

The row labels from top to bottom are: $p_0q_0r_0$, $p_0q_0r_1$, $p_0q_1r_0$, $p_0q_1r_1$, $p_1q_0r_0$, $p_1q_0r_1$, $p_1q_1r_0$, $p_1q_1r_1$.

Similar identification of contrasts can be written down for the other 2^n designs considered earlier and shown in Figure 15.5(b), (c) and (d). The reader is invited to confirm that the block contrasts in these designs are identical with treatment contrasts as follows:

(b) 2^4 in two blocks of eight units; the block difference is identical to the four-factor interaction PQRS.

(c) 2^4 in four blocks of four units; there are three block contrasts:

(i) $(b_1 + b_2 - b_3 - b_4)$,

which is identical with the three-factor interaction PQR;

(ii) $(b_1 - b_2 + b_3 - b_4)$,

which is identical with the three-factor interaction PQS; and

(iii) $(b_1 - b_2 - b_3 + b_4)$,

which is identical with the two-factor interaction RS. Note that the four-factor interaction is not confounded. The reason for this unsought and essentially useless 'benefit' will become clear in the next chapter.

(d) 2^5 in four blocks of eight units; the pattern is similar to that for (c), the three block contrasts being:

(i) $(b_1 + b_2 - b_3 - b_4)$,

which is identical to the four-factor interaction PQRS;

   (ii) $(b_1 - b_2 + b_3 - b_4)$,

which is identical to the three-factor interaction PQS; and

   (iii) $(b_1 - b_2 - b_3 + b_4)$,

which is identical to the three-factor interaction RST. And again the highest-order interaction PQRST is not confounded.

If we consider individual contrasts for the 3^2 design in three blocks of three units each, shown in Figure 15.6(a), then the eight orthogonal contrasts must include two main effect contrasts for each factor, and two block contrasts. Also, there are four interaction contrasts which are orthogonal to the four main effect contrasts. The set of eight orthogonal contrasts, shown in the first eight columns of Table 15.5, includes the four main effect contrasts (contrasts 1 to 4), and the two block contrasts (5 and 6) together with two contrasts (7 and 8) orthogonal to the previous six. Although there is an element of choice of contrasts for three-level factors as mentioned in Section 13.4, in most practical circumstances the contrasts $(1, -1, 0)$ and $(1, 1, -2)$ are the most relevant contrasts, and with this form of contrasts the choice for the contrasts in Table 15.5, including columns 7 and 8, is minimal. The last four columns of Table 15.5 comprise the normal form of the four interaction contrasts for PQ, and it can be seen that they bear no resemblance either to the block contrasts in columns 5 and 6 or to the two 'surplus' contrasts in columns 7 and 8.

Unlike the 2^n experiments, we are not able to identify in terms of practically interpretable effects either those effects which have been confounded or those effects which in addition to the main effects, remain unconfounded. The contrasts in columns 5 to 8 must represent interaction comparisons, since they are orthogonal to the main effect contrasts, but

Table 15.5.

	1	2	3	4	5	6	7	8	9	10	11	12
p_0q_0	-1	+1	-1	+1	-1	+1	-1	+1	+1	-1	-1	+1
p_0q_1	-1	+1	0	-2	+1	+1	+1	+1	0	0	+2	-2
p_0q_2	-1	+1	+1	+1	0	-2	0	-2	-1	+1	-1	-1
p_1q_0	0	-2	-1	+1	0	-2	+1	+1	0	+2	0	-2
p_1q_1	0	-2	0	-2	-1	+1	0	-2	0	0	0	+4
p_1q_2	0	-2	+1	+1	+1	+1	-1	+1	0	-2	0	-2
p_2q_0	+1	+1	-1	+1	+1	+1	0	-2	-1	-1	+1	+1
p_2q_1	+1	+1	0	-2	0	-2	-1	+1	0	0	-2	-2
p_2q_2	+1	+1	+1	+1	-1	+1	+1	+1	+1	+2	+1	+1

they are not apparently related to the form of interaction effect in which we have previously been interested. We return to this question of interpretation of the confounded effects in Section 15.4.

15.3 Incompletely confounded effects

In discussing the analysis of confounded experiments, we have considered only those effects which are either completely unaffected by block differences, or identical with block differences. Classical design theory has concentrated on designs for which all effects are either completely confounded or entirely unconfounded. However, with the computational difficulties previously associated with incompletely confounded designs now eliminated by modern computational practice, it is important also to consider a wider range of designs to identify which designs are practically appropriate. This leads to the consideration of incompletely confounded effects, which are not orthogonal to block differences nor identical with block differences.

Several designs with incompletely confounded effects have already been introduced. In the $3 \times 2 \times 2 \times 2$ example in the first section of this chapter, the $2 \times 2 \times 2$ interaction could not be arranged to be independent of blocks when blocks of 12 units were used. Two alternative designs were constructed in Figure 15.2. In the first (15.2(a)), the $2 \times 2 \times 2$ interaction was exactly identified with the differences between the two blocks and was completely confounded. However, in Figure 15.2(b), reproduced in Figure 15.8, all eight combinations of the three factors occurred, albeit unequally, in each block. If we consider the GNM interaction effect

$$\text{GNM} \propto (g_2 - g_1)(n_2 - n_1)(m_2 - m_1)(s_1 + s_2 + s_3)$$
$$= \{(g_2 n_2 m_2 + g_2 n_1 m_1 + g_1 n_2 m_1 + g_1 n_1 m_2)(s_1 + s_2 + s_3)\}$$
$$- \{(g_2 n_2 m_1 + g_2 n_1 m_2 + g_1 n_2 m_2 + g_1 n_1 m_1)(s_1 + s_2 + s_3)\}$$

the pattern of $+$ and $-$ terms in the two blocks shown in Figure 15.8 is easily identified.

Clearly, the estimation of the GNM interaction effect is equivalent to the estimation of the difference between two treatments such that, in two blocks of 12, one treatment occurs eight times in one block and four times in the other, while the other treatment occurs four times in the first block and eight times in the other. Least squares equations for this situation are

$$12\hat{\mu} + 12\hat{b}_1 + 4\hat{t}_1 = B_1$$
$$12\hat{\mu} + 12\hat{b}_2 + 4\hat{t}_2 = B_2$$
$$12\hat{\mu} + 4\hat{b}_1 + 12\hat{t}_1 = T_1$$
$$12\hat{\mu} + 4\hat{b}_2 + 12\hat{t}_2 = T_2,$$

Figure 15.8. Design for a $2 \times 3 \times 2 \times 2$ structure in two blocks of 12 units (Figure 15.2(b)) showing $+$ and $-$ terms for the NMG ($2 \times 2 \times 2$) interaction.

Block

I		II	
$n_1 s_1 m_1 g_1$	$(-)$	$n_1 s_1 m_1 g_2$	$(+)$
$n_1 s_1 m_2 g_2$	$(-)$	$n_1 s_1 m_2 g_1$	$(+)$
$n_1 s_2 m_1 g_2$	$(+)$	$n_1 s_2 m_1 g_1$	$(-)$
$n_1 s_2 m_2 g_1$	$(+)$	$n_1 s_2 m_2 g_2$	$(-)$
$n_1 s_3 m_1 g_1$	$(-)$	$n_1 s_3 m_1 g_2$	$(+)$
$n_1 s_3 m_2 g_2$	$(-)$	$n_1 s_3 m_2 g_1$	$(+)$
$n_2 s_1 m_1 g_2$	$(-)$	$n_2 s_1 m_1 g_1$	$(+)$
$n_2 s_1 m_2 g_1$	$(-)$	$n_2 s_1 m_2 g_2$	$(+)$
$n_2 s_2 m_1 g_1$	$(+)$	$n_2 s_2 m_1 g_2$	$(-)$
$n_2 s_2 m_2 g_2$	$(+)$	$n_2 s_2 m_2 g_1$	$(-)$
$n_2 s_3 m_1 g_2$	$(-)$	$n_2 s_3 m_1 g_1$	$(+)$
$n_2 s_3 m_2 g_1$	$(-)$	$n_2 s_3 m_2 g_2$	$(+)$

and the variance of the estimate of $\hat{t}_2 - \hat{t}_1$ is $3\sigma^2/16$, compared with the usual $2\sigma^2/12$, giving an efficiency of $\frac{8}{9}$ for the effect GNM. In the design of Figure 15.2(b) the eight GNM combinations are allocated to blocks as evenly as possible within the constraints imposed by the requirement to completely avoid confounding all main effects and two-factor interactions. And the cost of using blocks of 12 units instead of 24 is to lose $\frac{1}{9}$ of the information about the GNM three-factor interaction. In the analysis of variance we can add a term for the GNM interaction with a consequent reduction in error df;

GNM (allowing for blocks)   1 df
error   23 df.

Another example from Section 15.1 illustrates a slightly more complex case of incomplete confounding. The $3 \times 3 \times 2 \times 2$ design confounded in three blocks of 12 units in Figure 15.3(a) inevitably fails to avoid some confounding of the PQ (3×3) interaction. The levels of PQ for the design of Figure 15.3(a) are shown in Figure 15.9 without the levels of the other two factors. Note first that, if the problem has been presented as that of comparing nine treatments in three blocks of 12 assuming all treatment comparisons to be of equal importance, the design of Figure 15.9 would have been the obvious design, using the general philosophy of Chapter 7.

Figure 15.9. Levels of PQ for the design of Figure 15.3(a).

Block

I	II	III
$P_1 q_1$	$P_1 q_1$	$P_1 q_1$
$P_2 q_1$	$P_2 q_1$	$P_2 q_1$
$P_3 q_1$	$P_3 q_1$	$P_3 q_1$
$P_1 q_2$	$P_1 q_2$	$P_1 q_2$
$P_2 q_2$	$P_2 q_2$	$P_2 q_2$
$P_3 q_2$	$P_3 q_2$	$P_3 q_2$
$P_1 q_3$	$P_1 q_3$	$P_1 q_3$
$P_2 q_3$	$P_2 q_3$	$P_2 q_3$
$P_3 q_3$	$P_3 q_3$	$P_3 q_3$
$P_1 q_1$	$P_2 q_1$	$P_3 q_1$
$P_2 q_2$	$P_3 q_2$	$P_1 q_2$
$P_3 q_3$	$P_1 q_3$	$P_2 q_3$

To assess the loss of efficiency of information on the PQ interaction, consider the nine combinations of PQ levels as nine treatments and the eight orthogonal contrasts between these nine treatments. The variance of the estimated difference between any two treatments which are duplicated in the same block is $2\sigma^2/4$ (efficiency of 1.0), and between two treatments which are duplicated in different blocks, the variance of the estimated difference is $23\sigma^2/45$ (efficiency of 0.98). The overall average variance of treatment comparisons for the design is therefore calculated to be

$$\{9(2\sigma^2/4) + 27(23\sigma^2/45)\}/36 = 61\sigma^2/120.$$

The main effects of P and Q are orthogonal to blocks, and therefore are estimated fully efficiently.

Since the eight orthogonal treatment comparisons between the nine PQ combinations consist of four main effects and four interactions, the average efficiency of the interaction comparisons is

$$(2\sigma^2/4)/(62\sigma^2/120) = 60/62 = 0.97.$$

The detailed arithmetic of efficiency is not important. The important information concerns the high efficiency of all treatment comparisons for the nine combinations which ensures that the efficiency of all effects estimates is high.

At this point it is useful to recognise that the general principles of block–treatment designs developed in Chapter 7 can be used to minimise the

Table 15.6.

Treatment structure	Block size	Non-orthogonal interaction	
$2 \times 2 \times 3$	6	2×2	combinations duplicated in pairs such that each pair includes each level of each factor
$2 \times 3 \times 3$	6	3×3	combinations omitted in threes, including each level of each factor
$2 \times 3 \times 4$	12	2×4	combinations duplicated in fours balanced over levels of each factor
$2 \times 2 \times 3 \times 3$	6	2×2 3×3	duplicated in pairs omitted in threes
$2 \times 3 \times 3 \times 3$	6	3×3	omitted in threes
$2 \times 3 \times 3 \times 4$	12	2×4 3×3	duplicated in fours duplicated in threes
$2 \times 3 \times 4 \times 4$	12	2×4 4×4	duplicated in fours omitted in fours

effect of confounding on the estimation of particular treatment interaction effects when it is impossible to arrange for those effects to be orthogonal to blocks. If the combinations for these interaction effects are treated as a set of equally important treatments, the principles of Chapter 7 will suggest a simple form of allocation of combinations to blocks. The allocation must of course take into account restrictions required for those effects which can be made orthogonal to blocks. In general the efficiency of estimation of effects which are necessarily non-orthogonal to blocks, but which are allocated as evenly as possible, will be high.

For factorial structures where all factors have two, three or four levels and for small block sizes which allow main effects to be orthogonal to blocks the allocation problem for all non-orthogonal two-factor interactions will always be simple. Either a simple fraction of the combinations have to be duplicated in each block, or a simple fraction of the combinations have to be omitted from each block. The possible situations are listed in Table 15.6 for some of the more common treatment structures.

15.4   Partial confounding

In some of the examples of Section 15.2, when the set of 2^n treatment combinations was divided into more than two blocks, it was not possible to avoid confounding a relatively low-level effect. In the 2^4 in four blocks of four units a two-factor interaction was inevitably confounded. In the 2^5 in four blocks of eight units two three-factor interactions were necessarily confounded. In practice, we would be most unlikely to use only a single replicate of a 2^4 and usually two or three replicates would be used. There is then a choice of either allowing the same two-factor interaction to be confounded between blocks in each replicate, or arranging that different two-factor interactions are confounded in different replicates. The former choice provides no information on one particular two-factor interaction. The latter allows each two-factor interaction effect to be estimated from information for those blocks in which the effect was not confounded. We would then have some information on each effect of interest but, for each *partially confounded* effect, we would have only partial information.

Consider the 2^4 design in four blocks of four units each. In the design in Figure 15.5(c) a two-factor interaction, RS, is confounded with blocks. A two replicate design with a total of 32 experimental units would usually be practically relevant. Assume that the design from Figure 15.5(c) has been used for the first replicate, shown again in the first four blocks of Figure 15.10. The three effects confounded between blocks are PQR, PQS and RS.

Figure 15.10. Two replicates of a $2 \times 2 \times 2 \times 2$ structure each in four blocks of four units, with different two-factor interactions confounded in the two replicates.

Replicate 1

Block

I	II	III	IV
$p_0 q_0 r_0 s_0$	$p_0 q_0 r_1 s_1$	$p_0 q_0 r_0 s_1$	$p_0 q_0 r_1 s_0$
$p_0 q_1 r_1 s_1$	$p_0 q_1 r_0 s_0$	$p_0 q_1 r_1 s_0$	$p_0 q_1 r_0 s_1$
$p_1 q_0 r_1 s_1$	$p_1 q_0 r_0 s_0$	$p_1 q_0 r_1 s_0$	$p_1 q_0 r_0 s_1$
$p_1 q_1 r_0 s_0$	$p_1 q_1 r_1 s_1$	$p_1 q_1 r_0 s_1$	$p_1 q_0 r_1 s_0$

Replicate 2

Block

V	VI	VII	VIII
$p_0 q_0 r_0 s_0$	$p_0 q_0 r_1 s_0$	$p_0 q_0 r_0 s_1$	$p_0 q_0 r_1 s_1$
$p_0 q_1 r_1 s_1$	$p_0 q_1 r_0 s_1$	$p_0 q_1 r_1 s_0$	$p_0 q_1 r_0 s_0$
$p_1 q_0 r_0 s_1$	$p_1 q_0 r_1 s_1$	$p_1 q_0 r_0 s_0$	$p_1 q_0 r_1 s_0$
$p_1 q_1 r_1 s_0$	$p_1 q_1 r_0 s_0$	$p_1 q_1 r_1 s_1$	$p_1 q_1 r_0 s_1$

In the second replicate, we must ensure that the two-factor interaction which is confounded should not be RS. We may choose which of the five two-factor interactions, other than RS, should remain unconfounded in both replicates. Suppose all interactions involving P are thought to be important, and that QS is also thought likely to be important. We can then choose that QR should be confounded in the second replicate. This is achieved by allocating the levels of R last, after arranging that each block includes all eight (P, Q, S) combinations. Alternatively we may recognise that confounding QR means that in two blocks the levels of R are identical to those of Q, and in the other two blocks the levels of R are the opposite of those of Q. The second replicate in Figure 15.10 is constructed so that all main effects and two-factor interactions other than QR are orthogonal to blocks. By inspection the two other effects confounded with blocks in the second replicate are seen to be PQS (again) and PRS.

What form does the analysis of the two-replicate design of Figure 15.10 take? First consider the analysis from the intuitive approach. The four main effects and the four two-factor interactions PQ, PR, PS, QS are all orthogonal to blocks in both replicates, and the estimates of these effects and the sum of squares for the analysis of variance can be calculated in the usual way. So also, if it is of any interest, can the estimate and sum of squares for the three-factor interaction QRS. For RS and QR, in each replicate one block contrast is identical with the estimate of one of these effects. The estimate of each effect is therefore calculated using only the data from the replicate in which that effect is not confounded. Similarly for the three-factor interactions PQR and PRS, each confounded in one replicate, estimates and sums of squares could be calculated from the other replicate only.

For effects confounded in one or other replicate, the sum of squares in the analysis of variance is most simply calculated directly from the estimate of that effect in the replicate in which it is not confounded. If the calculation of sums of squares for partially confounded effects is attempted by using the sum of squares for all combinations of levels of the factors involved and subtracting the sum of squares for main effects and, if appropriate, interaction effects with fewer factors, then all sums of squares involved in the calculation must be based only on the replicates in which the effect of interest is not confounded. This would mean that the sum of squares for main effects to be deducted in calculating the interaction sum of squares will not be the same as those calculated for the main effect itself. The estimates of effects are defined in Table 15.7, and the analysis of variance structure is in Table 15.8. The inclusion of separate sums of squares for the three- and four-factor interactions is debatable. It will

Table 15.7. *Estimate of effects.*

Effect	Estimate	Var(estimate)
P		
Q		
R		
S	from data of both replicates	$2\sigma^2/16$
PQ		
PR		
PS		
QS		
QR	from replicate 1 data only	$2\sigma^2/8$
RS	from replicate 2 data only	$2\sigma^2/8$
QRS	from data of both replicates	$2\sigma^2/16$
PQR	from replicate 2 data only	$2\sigma^2/8$
PRS	from replicate 1 data only	$2\sigma^2/8$
PQS	none	–
PQRS	from data of both replicates	$2\sigma^2/16$

Table 15.8. *Basis of calculation for effect SS.*

Source	SS	df
Blocks	using block totals as usual	7
P	using both replicates	1
Q	using both replicates	1
R	using both replicates	1
S	using both replicates	1
PQ	using both replicates	1
PR	using both replicates	1
PS	using both replicates	1
QS	using both replicates	1
QR	using replicate 1 only	1
RS	using replicate 2 only	1
QRS	using both replicates	1
PQR	using replicate 2 only	1
PRS	using replicate 1 only	1
PQRS	using both replicates	1
Error	by subtraction as usual	10

usually be sensible to pool them with the error to give 14 df for estimating σ^2.

Partial confounding can be used in any multi-replicate 2^n experiment. It is most common when:

(i) the number of factors is three, four, or five since for more factors it is unlikely that an experiment will have more than one replicate;

(ii) the number of confounded effects in a single replicate is three, or possibly seven since, if only one effect is confounded in a single replicate, it will almost always be the highest possible order interaction which will be of little interest, and which can therefore be confounded in all replicates, obviating any need for partial confounding.

If the number of replicates is more than two, then partial confounding will involve a set of effects in each replicate, and there will be a choice between confounding different effects in each replicate or having some effects confounded more than once. The important consideration is the relative information (precision) for each of the effects to be estimated. Thus, suppose it was decided to use four replicates of a 2^4 in blocks of four, then different strategies of design could be employed:

(a) different confounding in each replicate
   (i) PQR, PQS, RS
   (ii) PRS, QRS, PQ
   (iii) PQS, QRS, PR
   (iv) PQR, PRS, QS;

(b) same confounding in all replicates,
   PQR, PQS, RS;

(c) two confounding patterns
   (i), (iii) PQR, PQS, RS
   (ii), (iv) PQS, PRS, QR;

(d) three confounding patterns avoiding confounding any two factor interaction involving P
   (i), (iv) PQR, PQS, RS
   (ii) PQS, PRS, QR
   (iii) PQR, PRS, QS.

The information available on each interaction effect for each design strategy, relative to the precision for any effect unaffected by confounding, is summarised in Table 15.9. It can be seen that making the choice of the required relative information for each two-factor interaction has the apparent effect of producing levels of information for three-factor interactions in precisely the opposite pattern to that which would seem

Table 15.9.

Effect	Strategy			
	A	B	C	D
PQ	75%	100%	100%	100%
PR	75%	100%	100%	100%
PS	100%	100%	100%	100%
QR	100%	100%	50%	75%
QS	75%	100%	100%	75%
RS	75%	0	50%	50%
PQR	50%	0	50%	25%
PQS	50%	0	0	25%
PRS	50%	100%	50%	50%
QRS	50%	100%	100%	100%

desirable from the selected pattern of information for two-factor inter-actions. Thus the reason for using design D should be that the principal interest is in factor P so that two-factor interactions involving P are estimated as precisely as possible. However, the consequence for three-factor interactions is that those involving P have the worst precision. This undesirable pay-off between the patterns of relative precision for two-factor interactions and for three-factor interactions is inevitable for reasons which will become apparent in the next chapter. The precision for the two-factor interactions should usually be the dominant consideration in choosing patterns of partial confounding.

Although partial confounding is important for two-level factors with varying patterns of numbers of factors and block sizes, probably the best known partially confounded designs are those for two designs involving three-level factors. These two design problems were considered in Section 15.2 but the concept of partial confounding is necessary to obtain the maximum advantage from these designs.

First, the 3^2 design using blocks of three. The two designs which allow main effects of each factor to be estimated with full precision are shown again in Figure 15.11. The consideration of confounded and non-confounded contrasts in Table 15.5 demonstrated that, in the design 15.11(a), the block differences are equivalent to comparisons between three sets of treatment combination

$$I_1 = p_0 q_0 + p_1 q_1 + p_2 q_2,$$
$$I_2 = p_0 q_1 + p_1 q_2 + p_2 q_0,$$

and

$$I_3 = p_0 q_2 + p_1 q_0 + p_2 q_1,$$

Figure 15.11. The two alternative designs for a 3×3 structure in three blocks of three units.

(a)   Block

	I	II	III
	$p_0 q_0$	$p_0 q_1$	$p_0 q_2$
	$p_1 q_1$	$p_1 q_2$	$p_1 q_0$
	$p_2 q_2$	$p_2 q_0$	$p_2 q_1$

(b)   Block

	I	II	III
	$p_0 q_0$	$p_0 q_1$	$p_0 q_2$
	$p_1 q_2$	$p_1 q_0$	$p_1 q_1$
	$p_2 q_1$	$p_2 q_2$	$p_2 q_0$

while, in design 15.11(*b*), block differences are equivalent to comparisons between

$$J_1 = p_0 q_0 + p_1 q_2 + p_2 q_1,$$
$$J_2 = p_0 q_1 + p_1 q_0 + p_2 q_2,$$

and

$$J_3 = p_0 q_2 + p_1 q_1 + p_2 q_0.$$

Any comparison between the three I groups of combinations is orthogonal to any comparison between the three J groups of combinations, since each I group contains one combination from each J group. Similarly, any comparison between the I groups, or between the J groups, is orthogonal to main effect comparisons of P or of Q, and hence I group comparisons and J group comparisons are orthogonal components of the PQ interaction. The orthogonality of the I group comparisons and the J group comparisons means that two orthogonal I group comparisons and two orthogonal J group comparisons constitute a set of four orthogonal comparisons, all of which are PQ interaction contrasts and, since there can be only four orthogonal PQ interaction comparisons, it follows that the set of I and J group comparisons contains complete information about the PQ interaction effects. However, as noted in Section 15.2, the I and J comparisons are not apparently related in any way to the PQ interaction effects suggested in Chapter 13 as being practically relevant.

To see how the I and J comparisons can be related to practical forms of PQ interaction effects, consider a particular PQ effect, from Chapter 13,

$$P'Q' = \frac{1}{2}(p_2 - p_0)(q_2 - q_0)$$

and two general contrasts from the I group and the J group

$$C_1 = \beta_1 I_1 + \beta_2 I_2 + \beta_3 I_3, \quad \text{where } \beta_1 + \beta_2 + \beta_3 = 0,$$

and

$$C_2 = \gamma_1 J_1 + \gamma_2 J_2 + \gamma_3 J_3, \quad \text{where } \gamma_1 + \gamma_2 + \gamma_3 = 0.$$

C_1, C_2 and $C_3 = C_1 + C_2$ can be represented as combinations of $p_i q_j$ as shown in Figure 15.12. For C_3 to be identical with $P'Q'$, we require, in addition to the general restrictions for contrasts,

$$\beta_1 + \beta_2 + \beta_3 = 0, \quad \gamma_1 + \gamma_2 + \gamma_3 = 0,$$

Figure 15.12. Representation of two contrasts of the I and J subsets and the conditions for a combination of contrasts to estimate P'Q'.

Contrast C_1

	P: 0	1	2
Q 0	β_1	β_3	β_2
1	β_2	β_1	β_3
2	β_3	β_2	β_1

Contrast C_2

	P: 0	1	2
Q 0	γ_1	γ_2	γ_3
1	γ_2	γ_3	γ_1
2	γ_3	γ_1	γ_2

$C_3 = C_1 + C_2$

	P: 0	1	2
Q 0	$\beta_1 + \gamma_1$	$\beta_3 + \gamma_2$	$\beta_2 + \gamma_3$
1	$\beta_2 + \gamma_2$	$\beta_1 + \gamma_3$	$\beta_3 + \gamma_1$
2	$\beta_3 + \gamma_3$	$\beta_2 + \gamma_1$	$\beta_1 + \gamma_2$

$P'Q'$

	P: 0	1	2
Q 0	$\frac{1}{2}$	0	$-\frac{1}{2}$
1	0	0	0
2	$-\frac{1}{2}$	0	$\frac{1}{2}$

the additional restrictions

$$\beta_1 + \gamma_1 = \frac{1}{2} \qquad \beta_3 + \gamma_3 = \frac{1}{2}$$

$$\beta_1 + \gamma_2 = \frac{1}{2} \qquad \beta_2 + \gamma_3 = \frac{1}{2} \qquad \beta_1 + \gamma_3 = 0.$$

$$\beta_2 + \gamma_2 = 0 \qquad \beta_3 + \gamma_2 = 0$$

$$\beta_2 + \gamma_1 = 0 \qquad \beta_3 + \gamma_1 = 0$$

The solution is

$$(\beta_1, \beta_2, \beta_3) = \left(\frac{2}{6}, -\frac{1}{6}, -\frac{1}{6}\right)$$

$$(\gamma_1, \gamma_2, \gamma_3) = \left(\frac{1}{6}, \frac{1}{6}, -\frac{2}{6}\right).$$

Similar identification of the other usual PQ interaction effects,

$$P''Q' = \frac{1}{4}(p_2 - 2p_1 + p_0)(q_2 - q_0)$$

$$P'Q'' = \frac{1}{4}(p_2 - p_0)(q_2 - 2q_1 + q_0)$$

$$P''Q'' = \frac{1}{8}(p_2 - 2p_1 + p_0)(q_2 - 2q_1 + q_0)$$

with sums of contrasts of the I and J groups is possible, and the construction of the appropriate contrasts is left to the reader.

The practically sensible form of partial confounding for the 3^2 treatment structure in two replicates, each containing three blocks of three units, is to confound the J comparisons in one replicate and the I comparisons in the other replicate. That is, use Figure 15.11(a) for the first replicate and Figure 15.11(b) for the second. The complete, two replicate, design is shown in Figure 15.13. In practice, four or even six replicates might be used repeating the initial two replicates.

Figure 15.13. Two replicates of a 3×3 structure in six blocks of three units with partial confounding of the PQ interaction.

			Block		
I	II	III	IV	V	VI
$p_0 q_0$	$p_0 q_1$	$p_0 q_2$	$p_0 q_0$	$p_0 q_1$	$p_0 q_2$
$p_1 q_1$	$p_1 q_2$	$p_1 q_0$	$p_1 q_2$	$p_1 q_0$	$p_1 q_1$
$p_2 q_2$	$p_2 q_0$	$p_2 q_1$	$p_2 q_1$	$p_2 q_2$	$p_2 q_0$

Combinations of comparisons between (I_1, I_2, I_3) and between (J_1, J_2, J_3) can provide estimates of the more usual form of interaction effects. Similarly, sums of squares calculated from the totals for (I_1, I_2, I_3) and for (J_1, J_2, J_3) provide the interaction SS for the analysis of variance of the design shown in Figure 15.13. We have already remarked that the set of four comparisons, two between I's and two between J's, comprises a complete set of interaction effects for the 4 df of the PQ interaction, and correspondingly the interaction SS can be calculated from the set of four comparisons, or equivalently from the two sets of totals for the I's and the J's. The argument can be supported by an algebraic proof based on the results of Section 4.7, and is illustrated in the following example.

Example 15.10

Suppose that a complete 3^2 factorial in r replicates, with no confounding, gives the treatment totals shown in Table 15.10.

Table 15.10.

	p_0	p_1	p_2	Total
q_0	20	25	30	75
q_1	24	28	30	82
q_2	28	30	30	88
Total	72	83	90	245

The simple calculation of main effects and interaction SS is

$$\text{SS(P)} = (72^2 + 83^2 + 90^2)/(3r) - 245^2/(9r) = 54.89/r,$$
$$\text{SS(Q)} = (75^2 + 82^2 + 88^2)/(3r) - 245^2/(9r) = 28.22/r,$$
$$\text{total treatment SS} = (20^2 + 25^2 + \ldots + 30^2)/r - 245^2/(9r) = 99.56/r,$$
$$\text{interaction SS} = (99.6 - 54.9 - 28.2)/r = 16.44/r.$$

The totals for the I's and the J's are

$$I_1 = 78, I_2 = 84, I_3 = 83,$$
$$J_1 = 80, J_2 = 79, J_3 = 86,$$
$$\text{SS(I)} = (78^2 + 84^2 + 83^3)/(3r) - 245^2/(9r) = 6.89/r,$$
$$\text{SS(J)} = (80^2 + 79^2 + 86^2)/(3r) - 245^2(9r) = 9.55/r,$$
$$\text{SS(I)} + \text{SS(J)} = 16.44/r = \text{interaction SS.}$$

The importance, for partially confounded experiments, of this alternative method of calculating the interaction SS is that, in the standard partially confounded design of Figure 15.13, while the normal form of interaction effects are confounded, the I and J comparisons are each unconfounded in one replicate. In the first replicate (blocks I, II, III) the I comparisons are confounded with blocks, but the J comparisons are not, so that the SS(J) can be calculated in the first replicate. Similarly, SS(I) can be calculated in the second replicate (blocks IV, V, VI), in which J comparisons are confounded. The PQ interaction SS is therefore calculated in two halves, each on 2 df from the two separate replicates, and the overall interaction SS is assessed by combining SS(I) + SS(J) on 4 df. The F-test for significance of the PQ interaction will be less powerful than in an experiment where the interaction is not confounded, being based effectively on a single replicate compared with the two replicates for the main effects. The analysis of variance structure is outlined in Table 15.11. Note that, while the two replicate partially confounded design has been used for illustration, the fact that there are only 4 df for the estimation of σ^2 makes the design unsuitable for practical use, and the four replicate design which provides 16 df for error would usually be required.

Table 15.11.

Source	SS calculation	df
Blocks	usual	5
A main effect	totals from both replicates	2
B main effect	totals from both replicates	2
I comparison	totals from second replicate 2	
J comparison	totals from first replicate 2	
Interaction	$SS_2(I) + SS_1(J)$	4
Error		4
Total	usual	17

The other design for factors with three levels, which utilises the ideas of partial confounding, is for three factors in blocks of nine units. The principles are exactly the same as for the two-factor design. As remarked in Section 15.2 the 27 combinations can be arranged in three blocks of nine, so that all main effects and two-factor interactions are not confounded. The four different designs which were summarised in Figure 15.7 each confound a part of the three-factor interaction. Since the three-factor interaction has 8 df, and a replicate in three blocks necessarily confounds two single degrees of freedom comparisons, it should not be surprising to find

(i) that the four different designs of Figure 15.7 each confound a different pair of comparisons;

(ii) that the four pairs of comparisons are all orthogonal and together constitute a complete subdivision of the 8 df for the three-factor interaction, and

(iii) that a four-replicate design can be used which provides information on each pair of comparisons in three out of four replicates.

Corresponding to the I, J comparison for the 3^2 design the comparisons for the 3^3 designs are traditionally referred to as W, X, Y and Z. Details of the four-replicate design and the calculations for the analysis of variance are given in other books, notably Cochran and Cox (1957). They will not be repeated here for three reasons. Such details add nothing further to the principles which are clearly displayed in the 3^2 design. If it is desired to use the four-replicate 3^3 design in blocks of nine, the analysis of variance should be done by computer and a modern computer program will give the correct analysis without special manipulation. However, the third, and most compelling reason for not giving details of the calculations, is that the design is not practically relevant. Four replicates will almost always be excessive. Two or three replicates will usually be found to give the most appropriate level of precision. In addition, it will rarely be necessary to examine the three-factor interaction. Interactions of three factors rarely provide useful information. If a two-replicate design is used, then some information on the three-factor interaction is available, though not in a symmetrical fashion and this can be extracted by the use of a suitable computer package analysis.

There is one further aspect of partial confounding that merits discussion, though the theory for this is not yet properly developed. We have discussed the use of different confounding systems, each in a complete replicate. However, it is possible to use different confounding systems in different fractions of a single replicate. Consider again the confounding of a 2^4 experiment in blocks of four units. If treatment levels are allocated sequentially, then all main effects are kept unconfounded, and one two-factor interaction is inevitably confounded between the first two blocks. Each block will include all four combinations of P and Q. The levels of R are added so that PQR is identical with the comparison between the first two blocks and the last two blocks. When the levels of S are allocated to the first two blocks one two-factor interaction involving S, say RS, must be confounded with the difference between the first two blocks. However, when the allocation of treatments in blocks III and IV is considered, there is no necessity to confound the RS interaction again. If, instead, the QS interaction is confounded the resulting design is shown in Figure 15.14. The design shown in Figure 15.14 provides full information on the four main effects and four of the two-factor interactions, and half information on the other two interactions, RS and QS.

Figure 15.14. Partial confounding of different two-factor interactions in two half replicates of a $2 \times 2 \times 2 \times 2$ structure.

Block

I	II	III	IV
$p_0 q_0 r_0 s_0$	$p_0 q_0 r_0 s_1$	$p_0 q_0 r_1 s_0$	$p_0 q_0 r_1 s_1$
$p_0 q_1 r_1 s_1$	$p_0 q_1 r_1 s_0$	$p_0 q_1 r_0 s_1$	$p_0 q_1 r_0 s_0$
$p_1 q_0 r_1 s_1$	$p_1 q_0 r_1 s_0$	$p_1 q_0 r_0 s_0$	$p_1 q_0 r_0 s_1$
$p_1 q_0 r_0 s_0$	$p_1 q_1 r_0 s_1$	$p_1 q_1 r_1 s_1$	$p_1 q_1 r_1 s_0$

To illustrate further the use of partial confounding within different fractions of a replicate, consider three alternatives for the design of a 2^4 experiment using two replicates with eight blocks of four units each. In the first design one two-factor interaction, RS, is completely confounded in both replicates, so that there is no information on that interaction, but full information on other main effects and two-factor interactions. Alternatively, a different two-factor interaction may be confounded in different replicates, RS in replicate one, and QS in replicate two. In the third design we could partially confound RS between the first two blocks, QS between blocks III and IV, PR between blocks V and VI, and QR between blocks VII and VIII, the resulting design being shown in Figure 15.15. The resulting information on different treatment effects from the different

Figure 15.15. Partial confounding in four half replicates of a $2 \times 2 \times 2 \times 2$ structure.

Block

I	II	III	IV
$p_0 q_0 r_0 s_0$	$p_0 q_0 r_0 s_1$	$p_0 q_0 r_1 s_0$	$p_0 q_0 r_1 s_1$
$p_0 q_1 r_1 s_1$	$p_0 q_1 r_1 s_0$	$p_0 q_1 r_0 s_1$	$p_0 q_1 r_0 s_0$
$p_1 q_0 r_1 s_1$	$p_1 q_0 r_1 s_0$	$p_1 q_0 r_0 s_0$	$p_1 q_0 r_0 s_1$
$p_1 q_1 r_0 s_0$	$p_1 q_1 r_0 s_1$	$p_1 q_1 r_1 s_1$	$p_1 q_1 r_1 s_0$

V	VI	VII	VIII
$p_0 q_0 r_0 s_0$	$p_0 q_0 r_1 s_0$	$p_0 q_0 r_0 s_1$	$p_0 q_0 r_1 s_1$
$p_0 q_1 r_0 s_1$	$p_0 q_1 r_1 s_1$	$p_0 q_1 r_1 s_0$	$p_0 q_1 r_0 s_1$
$p_1 q_0 r_1 s_1$	$p_1 q_0 r_0 s_1$	$p_1 q_0 r_0 s_0$	$p_1 q_0 r_1 s_1$
$p_1 q_1 r_1 s_0$	$p_1 q_1 r_0 s_0$	$p_1 q_1 r_1 s_1$	$p_1 q_1 r_0 s_1$

Table 15.12.

	Complete confounding (%)	Partial confounding in each replicate (%)	Partial confounding in half replicates (%)
P	100	100	100
Q	100	100	100
R	100	100	100
S	100	100	100
PQ	100	100	100
PR	100	100	75
PS	100	100	100
QR	100	100	75
QS	100	50	75
RS	—	50	75

designs is summarised in Table 15.12. The choice between these alternative designs, like the choice of which interactions to confound, depends on the prior assessment of the relative importance of the different effects. If the experimenter is quite certain that the RS interaction is of no importance, then the first, completely confounded, design would be appropriate. However, in general, it seems likely that spreading the information evenly over different interactions of a similar order will often prove attractive, in which case the third alternative should be chosen.

15.5 The split unit design as an example of confounding

The structure of a confounded experiment is essentially the same as that of a split unit design, although the terminologies that have evolved for the two types of design are different. In a confounded design, there are one or more replicates, each replicate including a single observation for each treatment combination. Each replicate is divided into blocks, each block containing the same number of units. The choice of treatment combinations to be allocated to each block of units determines which treatment effects are estimated in terms of difference between units within blocks, and which will be estimated in terms of differences between blocks, described as confounded with block differences.

Split unit designs include several blocks, each block including one observation for each treatment combination. Each block is divided into main units and levels of one or more factors are applied to main units. Each main unit contains the same number of split units and the levels or level combinations for the remaining factors are allocated to the split units within each main unit. The choice of factors to be applied to main units and to split units determines which treatment effects will be estimated in terms of differences between split units within main units and which will be estimated in terms of differences between main units.

The equivalence of the two design structures is summarised as follows:

split unit designs		*confounded designs*
blocks	≡	replicates
main units	≡	blocks
split units	≡	units.

Recognising split unit designs as a special case of confounding can improve understanding of both types of design. The inefficiency of split unit designs derived from the fact that the comparisons of main unit treatments are generally much less precise than the comparisons of split unit treatments. However, consideration of confounded designs shows that, if some treatment comparisons must be estimated more precisely than others, there is no necessity for the less precise comparisons to be main effects, as in the split unit designs. In the development of confounded designs the less precise comparisons are chosen to be interactions of several factors. Thus, except when split unit designs are used because they are required for practical reasons, we should not only prefer the fully randomised factorial to the split unit design, but should consider whether the confounded design, losing information on the multi-factor interactions, should not be preferred to both.

For example, consider the choice of design for the three two-level factors, P, Q and R. We might decide that R is of less importance and consider a split unit design with R as a main unit factor and P and Q as split unit factors. Alternatively we could use a confounded design with PQR confounded between main units (or blocks in confounding terminology). To prefer the split unit design it is necessary to argue that the three-factor interaction PQR is a more important effect than the main effect of R. In practice this argument would be difficult to believe, particularly since the interpretation of interaction effects through tables of means requires the main effect estimates.

One aspect of confounded designs which we have so far neglected and which the contrast with split unit designs emphasises is the information about treatment effects which is available from comparisons between blocks. In the split unit design, although the precision of main unit treatment effects is reduced, the information about the effects of main unit factors is important. In the confounded design, differences between blocks can provide estimates of those interaction effects which are confounded between blocks. To see how much information may be recovered from

Figure 15.16. Three replicates of a $2 \times 2 \times 2$ structure in six blocks of four units.

Block

I	II	III	IV	V	VI
$P_1 q_0 r_0$	$P_0 q_1 r_1$	$P_1 q_1 r_1$	$P_1 q_0 r_1$	$P_0 q_0 r_0$	$P_1 q_0 r_0$
$P_0 q_1 r_0$	$P_0 q_0 r_0$	$P_0 q_0 r_1$	$P_1 q_1 r_0$	$P_0 q_1 r_1$	$P_0 q_0 r_1$
$P_1 q_1 r_1$	$P_1 q_1 r_0$	$P_1 q_0 r_0$	$P_0 q_1 r_1$	$P_1 q_0 r_1$	$P_0 q_1 r_0$
$P_0 q_0 r_1$	$P_1 q_0 r_1$	$P_0 q_1 r_0$	$P_0 q_0 r_0$	$P_1 q_1 r_0$	$P_1 q_1 r_1$

inter-block information, we reconsider Example 15.4 with three two-level factors in blocks of four units.

Typically three or four replicates using 24 or 32 total observations would be used. Suppose 24 observations are to be used, with six blocks of four, as shown in Figure 15.16. Then the estimate of the three-factor interaction, PQR, will be given by the contrast of block totals

$$(B_1 - B_2 + B_3 - B_4 + B_6 - B_5)$$

with a multiplying factor of $1/12$ to convert the estimate to a per unit basis. This estimate is simply the average of three differences between block means and the precision of this estimate will depend on the variance of block differences. The situation is identical with that for three pairs of observations, each pair being for two treatments $T_1(= +PQR)$ and $T_2(= -PQR)$. The analysis of variance of block totals will be as shown in Table 15.13.

Table 15.13.

Source	df
PQR	1
Replicates (pairs)	2
Block error	2
Total block variation	5

If the pairing of blocks into replicates is not done in any systematic way, then the between replicate variation is pooled with the block error to give a block error sum of squares, based on 4 df.

The model implicit in this analysis may be derived from the model for the unconfounded design in three blocks of eight units

$$y_{ijkl} = \mu + b_i + P_j + Q_k + R_l + (PQ)_{jk} + (PR)_{jl} + (QR)_{kl} + (PQR)_{jkl} + \varepsilon_{ijkl}.$$

When using six blocks of four units, the algebraic appearance of the model remains the same, but only four of the possible (j, k, l) combinations occur for each i value. The sets of four are chosen so that all possible combinations of any two of (j, k, l) do occur with each i, but the $(PQR)_{jkl}$ terms occurring with i are such that, in each block, the four combinations are either

$$\{(0, 0, 0), (0, 1, 1), (1, 0, 1) \text{ and } (1, 1, 0)\}$$

or

$$\{(1, 0, 0), (0, 1, 0), (0, 0, 1) \text{ and } (1, 1, 1)\}.$$

However, the definition of $(PQR)_{jkl}$ implies that

$$\sum (PQR)_{jkl} = 0,$$

where the summation is over any one suffix, so that all the combinations in the first group are equal, and so are those of the second group, the two common values being equal and opposite. Hence, all those combinations in the first group may be written as $-PQR$, and all those in the second as $+PQR$. The block totals may be expressed in the form

$$Y_{i...} = 4\mu + 4b_i \pm 4PQR + \sum_{jkl} \varepsilon_{ijkl},$$

where the sign of PQR depends on the value of i, being $+$ for $i = 1, 3, 6$ and $-$ for $i = 2, 4, 5$. For this model, estimates of μ and PQR may be obtained by rewriting the model

$$Y_{i...} = 4\mu \pm 4PQR + \eta_i$$

to obtain not only the obvious estimate

$$PQR = (Y_{1...} - Y_{2...} + Y_{3...} - Y_{4...} + Y_{6...} - Y_{5...})/24$$

but also the analysis of variance structure as follows:

	SS	df
PQR	$24(PQR)^2$	1
Residual	S_r	4
Total	$\sum Y_{i..}^2 - Y_{...}^2/6$	5

A variance for the estimate of PQR is obtained from the residual mean squares $(S_r/4)$ which provides an estimate of $\text{Var}(\eta)$.

The parallel between split unit designs in which main effects are estimated from comparisons between main units, and other confounded designs, in which high-order interactions are estimated from comparisons between blocks, emphasises two aspects of design. First, that in both designs information about treatment effects occurs at two levels of variation, and the information at both, or in more complex designs all, levels of variation should be considered. Second, that whenever information is split between different levels of variation, there is a choice of how the information is distributed between different treatment effects.

15.6 Confounding for general block size and factor levels

The discussion of confounding has been concerned solely with those designs in which factor levels or combinations of factor levels appear equally frequently in each block. Consequently only block sizes which are products of the numbers of levels of the various factors have been considered for confounded experiments. This restriction may seem inappropriate, given the emphasis in Chapter 7 on always seeking the natural block size. Inevitably the theory and the recognition of suitable patterns for confounding are less advanced for the more general confounding situation and the philosophy of general confounding is presented only briefly mainly through examples.

If block size is not related to the numbers of levels for the several factors, the estimation of main effects and interactions of different factors will not be orthogonal to blocks. Consequently, the estimation of different factorial effects will no longer be orthogonal between different effects, even if all treatments are replicated equally over the whole experiment. Hence, estimates of effects will depend on the set of effects which are fitted, and the sums of squares attributed to different effects will depend on the order of fitting. The scale of variation caused by different orders of fitting is demonstrated in the following example.

Example 15.11

The design shown in Figure 15.17 includes three replicates of a 2^3 treatment set arranged in eight blocks of three units each. The selection of treatment combinations in each block is such that for each factor both levels occur in each block, and main effects should therefore be well estimated. The data are shown in Table 15.14. The logic of main effects and interactions requires that main effects must be fitted before interaction effects involving corresponding factors. The possible orders of fitting are

Figure 15.17. Three replicates of a $2 \times 2 \times 2$ structure in eight blocks of three units.

				Block			
I	II	III	IV	V	VI	VII	VIII
$p_0 q_0 r_0$	$p_0 q_0 r_0$	$p_0 q_0 r_0$	$p_0 q_0 r_1$	$p_0 q_0 r_1$	$p_0 q_0 r_1$	$p_0 q_1 r_0$	$p_0 q_1 r_0$
$p_0 q_1 r_1$	$p_1 q_0 r_1$	$p_1 q_0 r_0$	$p_0 q_1 r_1$	$p_0 q_1 r_0$	$p_1 q_0 r_0$	$p_0 q_1 r_1$	$p_1 q_0 r_1$
$p_1 q_1 r_0$	$p_1 q_1 r_1$	$p_1 q_1 r_1$	$p_1 q_1 r_0$	$p_1 q_0 r_1$	$p_1 q_1 r_1$	$p_1 q_0 r_0$	$p_1 q_1 r_0$

Table 15.14.

	\multicolumn{8}{c}{Block}							
	I	II	III	IV	V	VI	VII	VIII
$p_0q_0r_0$	33.2	31.7	34.6					
$p_0q_0r_1$				44.1	43.2	40.1		
$p_0q_1r_0$					42.7		46.6	43.1
$p_0q_1r_1$	40.7			46.4			48.2	
$p_1q_0r_0$			43.0			47.0	51.3	
$p_1q_0r_1$		44.1			49.0			50.3
$p_1q_1r_0$	48.6			55.7				53.0
$p_1q_1r_1$		45.4	49.1			49.4		

numerous and sensible orders include:

$$\{P, Q, R\}, \quad [PQ, PR, QR], \quad PQR,$$

where the brackets indicate that the terms within a bracket may be fitted in any order,

$$\{P, Q,\}, \quad PQ, R, \quad \{PR, QR,\}, \quad PQR$$
$$\{P, R,\}, \quad PR, Q, \quad \{PQ, QR,\}, \quad PQR,$$

and

$$\{Q, R,\}, \quad QR, P, \quad \{PQ, PR,\}, \quad PQR.$$

Analysis of variance for some of the various orders are shown in Table 15.15 and this gives some idea of the variation in apparent information according to the order.

Although the degree of non-orthogonality for this design might be thought to be considerable the changes in the sums of squares for effects, as the order of fitting is varied, are not substantial in relative terms so that the interpretation is not likely to be altered by the particular order, or orders,

Table 15.15. *Sums of squares for effects fitted in six different orders.*

Blocks	265.5										
P	446.5	Q	88.0	R	19.6	Q	88.0	R	19.6	P	446.5
Q	65.3	R	14.9	P	446.6	P	423.8	Q	63.3	R	19.7
R	15.4	P	424.3	PR	2.2	PQ	0.1	QR	20.3	Q	61.0
PQ	0.4	QR	7.7	Q	66.3	R	15.8	P	411.8	PR	7.5
PR	7.4	PQ	0.0	QR	8.1	QR	7.4	PR	7.7	PQ	0.3
QR	7.8	PR	7.8	PQ	0.1	PR	7.8	PQ	0.1	QR	7.8
PQR	0.1										
Residual	6.5										

chosen. It is particularly notable that the information about the strength of the interaction effects hardly varies at all.

When all terms are included in the model, the variances of the seven estimates of effects and their covariances and correlations can be estimated in terms of the error variance, s^2, which is 0.7176 for this data, and these are shown in Table 15.16. The variances can be compared with the variances from the three replicate randomised block design to examine the efficiency of the design. The efficiencies tabulated in Table 15.16 are calculated as the ratio of variance of effect for randomised complete block design to variance of effect for incomplete block design (Figure 15.17), assuming that the variance, σ^2, is the same for both designs. This assumption is likely to be considerably wrong for the same reasons that led to the use of an incomplete block design in the first place. From Table 15.16 it can be seen that the efficiencies for main effect estimation are between 0.82 and 0.87, but that the efficiencies for interactions are lower, and unevenly so, suggesting that a design with a more even spread of interaction information could be found. If the use of the small blocks reduces σ^2 by 20% then Figure 15.17 is a sensible design for main effect estimation. If the reduction of σ^2 is 50% then Figure 15.17 is a good design for all effects. The incomplete block design has the further disadvantage of some correlations between estimates of effects, though in this case those correlations are small. However, although the three-unit blocks do have some disadvantages, there will always be situations where the use of the reduced block size will clearly produce a more precise experiment.

What general principles are necessary to guide the selection of appropriate designs for particular block sizes and treatment structures? Essentially, they are those developed at the beginning of this chapter and earlier, in Chapter 7, during the discussion of designs for comparing treatments in general block structures. For the 2^3 experiment just considered, the design

Table 15.16. *Variances and efficiencies for the seven effects for the design of Figure 15.17.*

Effect	Variance	Efficiency
P	0.1418	0.84
Q	0.1452	0.82
R	0.1370	0.87
PQ	0.1615	0.74
PR	0.1883	0.64
QR	0.2290	0.52
PQR	0.1885	0.63

principles are simply expressed, since each of the seven effects is a simple difference between two sets of treatment combinations. Therefore, in each block of three units, we should try to include three combinations such that for each effect the combinations include both positive and negative combinations. The correlations of estimates of different effects are determined by the interrelationship of effect estimates over the eight different blocks but, provided the joint pattern of + and − terms for pairs of different effects is varied as much as possible between blocks, then the correlation should be minimised. Precise methods for optimising the design relative to a criterion defined in terms of the relative importance of the seven different effects are not yet fully developed, but I believe that it is possible to design good designs essentially by 'trial and error' provided a general computer program is available to assess the efficiency of each possible design.

To complete this chapter we consider a range of examples of designs for experiments given

(i) a particular factorial structure,

(ii) a pattern of 'natural' block sizes.

The general philosophy of block–treatment designs from Chapter 7 is applied to each effect in turn, usually starting with main effects, and then two-factor interactions. The levels, or combinations of levels, should be 'spread evenly' over blocks so as to make the number of concurrences of pairs of levels, or of pairs of combinations, as nearly equal as is possible, for all pairs of levels or all pairs of combinations. Where the interaction of two factors is judged to be of particular interest, then all combinations of levels of the two factors should be treated as equally important in the sense of Chapter 7. It would not be realistic to treat interactions as *more* important than the related main effects because the interpretation of interactions requires the estimation of main effects. If block sizes vary then, of course, we should consider the sum of concurrences inversely weighted by block size.

Example 15.12
In the second example at the beginning of this chapter the treatment structure is three levels of molasses (M) × two levels of energy (E). The unit structure has 24 blocks (sheep) with two units per block. The possibility of a systematic difference between the first and second observation across all blocks is a complicating factor, but with 24 sheep and six treatment combinations it is obviously easy to ensure that each treatment occurs four times in first position and four times in the second position.

Figure 15.18. Eight replicates of a 2×3 structure in an array of 24 sheep $\times$ two periods.

Sheep

Period	I	II	III	IV	V	VI	VII	VIII
1	e_1m_1	e_2m_1	e_1m_0	e_2m_0	e_1m_0	e_2m_1	e_2m_0	e_1m_1
2	e_2m_0	e_1m_2	e_2m_1	e_1m_2	e_2m_2	e_1m_0	e_1m_1	e_2m_2
	IX	X	XI	XII	XIII	XIV	XV	XVI
1	e_1m_2	e_2m_2	e_1m_2	e_2m_2	e_1m_0	e_1m_2	e_2m_2	e_2m_2
2	e_2m_1	e_1m_0	e_2m_0	e_1m_1	e_2m_2	e_2m_1	e_1m_1	e_1m_0
	XVII	XVIII	XIX	XX	XXI	XXII	XXIII	XXIV
1	e_2m_0	e_1m_1	e_1m_1	e_2m_1	e_2m_1	e_1m_0	e_1m_2	e_2m_0
2	e_1m_1	e_2m_0	e_2m_2	e_1m_2	e_1m_0	e_2m_1	e_2m_0	e_1m_1

Clearly each sheep receives two treatments, which must include both energy levels, and two of the three molasses levels. There are six pairs of combinations which can be constructed:

$$(e_1m_0, e_2m_1), (e_1m_0, e_2m_2), (e_1m_1, e_2m_2)$$
$$(e_1m_1, e_2m_0), (e_1m_2, e_2m_0), (e_1m_2, e_2m_1).$$

Each pair can be applied to the sheep in either order, and two replicates of each of the two orders of each of the six pairs gives the design in Figure 15.18 which clearly arranges treatment effect comparisons to be orthogonal to time periods. Inevitably in the real life experiment everything did not go according to plan, and one sheep died leaving only 23 blocks and a non-orthogonal design. There was, however, no difficulty in calculating the analysis of variance and estimates of main effects and interaction effects from within sheep variation, allowing for a systematic difference between the two time periods.

Example 15.13

In the first example from the beginning of this chapter the original design included all nine combinations of between row spacing (S) and within row density (D) in each of the four blocks. The experimenters' original suggestion for including two varieties of cassava (C) and two varieties of cowpea (P) was to apply each of the four combinations of varieties to an entire block, providing information on the two-factor interaction between spacing or density and variety only, with no information on the main

Figure 15.19. Design for a $3 \times 3 \times 2 \times 2$ structure in four blocks of nine units.

Block

	I	II	III	IV
$s_1 d_1$	$c_1 p_1$	$c_1 p_2$	$c_2 p_1$	$c_2 p_2$
$s_1 d_2$	$c_2 p_2$	$c_2 p_1$	$c_1 p_2$	$c_1 p_1$
$s_1 d_3$	$c_1 p_2$	$c_1 p_1$	$c_2 p_2$	$c_2 p_1$
$s_2 d_1$	$c_2 p_2$	$c_2 p_1$	$c_1 p_2$	$c_1 p_1$
$s_2 d_2$	$c_1 p_1$	$c_1 p_2$	$c_2 p_1$	$c_2 p_2$
$s_2 d_3$	$c_2 p_1$	$c_2 p_2$	$c_1 p_1$	$c_1 p_2$
$s_3 d_1$	$c_1 p_1$	$c_1 p_2$	$c_2 p_1$	$c_2 p_2$
$s_3 d_2$	$c_2 p_1$	$c_2 p_2$	$c_1 p_1$	$c_1 p_2$
$s_3 d_3$	$c_1 p_2$	$c_1 p_1$	$c_2 p_2$	$c_2 p_1$

effects of varieties. This would be a form of split plot design without main plot replication.

To obtain within block information on the main effects for the variety factors and interactions the design requirements should be

(i) each cassava variety four or five times in each block,

(ii) each cowpea variety four or five times in each block,

(iii) each CD combination at least once in each block,

(iv) as (iii) for CS, PD, PS combinations,

(v) each CP combination at least twice each block.

One possible design, constructed very rapidly, is shown in Figure 15.19; the reader is invited to try to improve on this.

Example 15.14

In Chapter 1 we introduced the experimental problem of a microbiologist who wanted to investigate a set of treatments making up a $5 \times 4 \times 2 \times 2$ factorial structure but who was restricted to sets of at most 40 observations per day. The simplest approach to this design problem, recognising that 40 is simply 5×8 and that this factorisation fits the treatment structure rather neatly, is to divide the 16 possible combinations of levels for the last three factors into two sets of eight combinations retaining estimation for main effects and two-factor interactions for these three factors. The five levels of the first factor are then allocated three to one set of eight combinations and two to the other set to give a set of 40 observations for a day, a possible example for two days being shown in Figure 15.20 (not randomised).

Figure 15.20. Design for a $5 \times 4 \times 2 \times 2$ structure in two days of 40 observations.

Day 1

$P_1q_1r_1s_1 \quad P_2q_1r_1s_2 \quad P_3q_1r_1s_1 \quad P_4q_1r_1s_2 \quad P_5q_1r_1s_1$

$P_1q_1r_2s_2 \quad P_2q_1r_2s_1 \quad P_3q_1r_2s_2 \quad P_4q_1r_2s_1 \quad P_5q_1r_2s_2$

$P_1q_2r_1s_2 \quad P_2q_2r_1s_1 \quad P_3q_2r_1s_2 \quad P_4q_2r_1s_1 \quad P_5q_2r_1s_2$

$P_1q_2r_2s_1 \quad P_2q_2r_2s_2 \quad P_3q_2r_2s_1 \quad P_4q_2r_2s_2 \quad P_5q_2r_2s_1$

$P_1q_3r_1s_2 \quad P_2q_3r_1s_1 \quad P_3q_3r_1s_2 \quad P_4q_3r_1s_1 \quad P_5q_3r_1s_2$

$P_1q_3r_2s_1 \quad P_2q_3r_2s_2 \quad P_3q_3r_2s_1 \quad P_4q_3r_2s_2 \quad P_5q_3r_2s_1$

$P_1q_4r_1s_1 \quad P_2q_3r_1s_2 \quad P_3q_4r_1s_1 \quad P_4q_4r_1s_2 \quad P_5q_4r_1s_1$

$P_1q_4r_2s_2 \quad P_2q_4r_2s_1 \quad P_3q_4r_2s_2 \quad P_4q_4r_2s_1 \quad P_5q_4r_2s_2$

Day 2

$P_1q_1r_1s_2 \quad P_2q_1r_1s_1 \quad P_3q_1r_1s_2 \quad P_3q_1r_1s_1 \quad P_5q_1r_1s_2$

$P_1q_1r_2s_1 \quad P_2q_1r_2s_2 \quad P_3q_1r_2s_1 \quad P_4q_1r_2s_2 \quad P_5q_1r_2s_1$

$P_1q_2r_1s_1 \quad P_2q_2r_1s_2 \quad P_3q_2r_1s_1 \quad P_4q_2r_1s_2 \quad P_5q_2r_1s_1$

$P_1q_2r_2s_2 \quad P_2q_2r_2s_1 \quad P_3q_2r_2s_2 \quad P_4q_2r_2s_1 \quad P_5q_2r_2s_2$

$P_1q_3r_1s_1 \quad P_2q_3r_1s_2 \quad P_3q_3r_1s_1 \quad P_4q_3r_1s_2 \quad P_5q_3r_1s_1$

$P_1q_3r_2s_2 \quad P_2q_3r_2s_1 \quad P_3q_3r_2s_2 \quad P_4q_3r_2s_1 \quad P_5q_3r_2s_2$

$P_1q_4r_1s_2 \quad P_2q_4r_1s_1 \quad P_3q_4r_1s_2 \quad P_4q_4r_1s_1 \quad P_5q_4r_1s_2$

$P_1q_4r_2s_1 \quad P_2q_4r_2s_2 \quad P_3q_4r_2s_1 \quad P_4q_4r_2s_2 \quad P_5q_4r_2s_1$

Example 15.15

An experiment on exudates from plants involved sections of individual leaves being treated with chemicals and placed (and observed for several days) in a quarter of a petri dish. The basic experimental unit is a section of a leaf in a quarter of a petri dish. The petri dishes are kept in piles of four. Four plastic bags, each containing four dishes, each dish divided into four quarters, provide a total of 64 experimental units. The plastic bags may be assumed to be different providing one blocking system. A second system of blocking is suggested as the position of each dish within the pile (top, middle, bottom). A lower level of blocking is the individual petri dish within which the four quarters should be similar.

The treatments proposed are three varieties × five chemical treatments of the plants. The design requirements would be

(i) in each dish all three varieties should occur (one twice),

(ii) in each dish four of the five chemicals should occur,

Figure 15.21. Design for a 5×3 structure in four sets of four blocks of four units.

Layer		Bag		
	I	II	III	IV
	$c_1 v_1$	$c_1 v_2$	$c_1 v_3$	$c_1 v_2$
Top	$c_2 v_2$	$c_2 v_3$	$c_2 v_1$	$c_3 v_2$
	$c_3 v_3$	$c_3 v_1$	$c_4 v_2$	$c_4 v_1$
	$c_4 v_3$	$c_5 v_2$	$c_5 v_1$	$c_5 v_3$
	$c_2 v_3$	$c_1 v_1$	$c_1 v_2$	$c_1 v_3$
Upper	$c_3 v_2$	$c_2 v_2$	$c_2 v_1$	$c_2 v_1$
Middle	$c_4 v_1$	$c_4 v_2$	$c_3 v_1$	$c_3 v_3$
	$c_5 v_1$	$c_5 v_3$	$c_4 v_3$	$c_5 v_2$
	$c_1 v_2$	$c_1 v_3$	$c_2 v_2$	$c_1 v_1$
Lower	$c_3 v_1$	$c_2 v_1$	$c_3 v_3$	$c_2 v_3$
Middle	$c_4 v_3$	$c_3 v_2$	$c_4 v_1$	$c_4 v_2$
	$c_5 v_2$	$c_4 v_3$	$c_5 v_3$	$c_5 v_1$
	$c_1 v_3$	$c_1 v_2$	$c_1 v_1$	$c_2 v_2$
	$c_2 v_1$	$c_3 v_3$	$c_2 v_3$	$c_3 v_1$
Bottom	$c_4 v_2$	$c_4 v_1$	$c_3 v_2$	$c_4 v_3$
	$c_5 v_3$	$c_5 v_1$	$c_5 v_2$	$c_5 v_1$

(iii) in each bag all 15 combinations should occur at least once,

(iv) at each layer all 15 combinations should occur at least once.

A possible design is shown in Figure 15.21. Again the reader is invited to try to improve this design.

Example 15.16

In Chapter 7 we considered the problem of comparing nine treatments using a block structure of 18 blocks with two units per block. This problem actually arose in the context of an experiment in which the nine treatments had a 3×3 factorial structure and the design question required allocation of pairs of treatments to blocks to maximise information about important treatment contrasts. A comprehensive investigation of the design possibilities and the choice of an optimal design is a substantial piece of research and the problem is not discussed in depth here. Instead we compare two alternative designs which illustrate well the principles of this chapter.

With nine treatments there are 36 possible pairs of treatments, and the block structure requires that at most only 18 different pairs can be used in an experiment. Two possible experimental designs with clearly different philosophies might seem appropriate. Either

(*a*) each block contains a pair of treatments with a common level for one of the factors, or

(*b*) each block contains a pair of treatments with no common level of either factor.

There are exactly 18 pairs of treatments of each type so that each experiment can contain 18 different blocks. Using a link to represent a pair of observations in a block, the two designs are represented in Figures 15.22(a) and (b). A simple question is which is the better design for estimating main effects. The reader is urged to answer before reading further.

Figure 15.22. Two alternative designs for a 3×3 structure in 18 blocks of two units; (a) each block contains a common level for one of the factors, (b) each block pair of treatments has no common factor level.

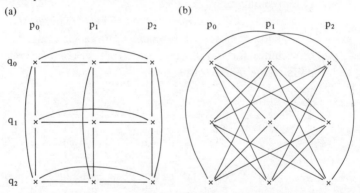

The answer is that the efficiencies of main effect and interaction estimates, relative to the design with four randomised blocks of nine units each, are given in Table 15.17. Thus the second design gives twice as much information about main effects as about interactions while the reverse is true for the first design. The reason why the second design is better for main effects is that in each block the levels of a factor are different and therefore each block contributes to the information on the main effect of the factor. Because the main effects are more efficiently estimated in design 15.22(b), the interactions must be correspondingly less efficiently estimated.

Table 15.17.

	Main effect	Interaction
Design 15.22 (*a*)	37.5%	75.0%
Design 15.22 (*b*)	75.0%	37.5%

Exercises 15

(1) Construct sensible designs for a $2 \times 3 \times 3$ factorial structure, assuming main effects are most important, followed by two-factor interactions, and explaining the advantages of your designs, for:
　　(i) six blocks of six units,
　　(ii) four blocks of nine units,
　　(iii) three blocks of 12 units,
　　(iv) nine blocks of four units,
　　(v) seven blocks of five units.
You may wish to assess any of your designs by using a statistical package (e.g. GLIM) to calculate variances of effects using a dummy analysis of variance.

(2) The data in Table 15.18 are taken from a 3^2 experiment on lettuce emergence, confounded in blocks of three. Compute the analysis of variance and obtain standard errors for comparisons between pairs of the nine treatments.

Table 15.18.

Replicate 1	block I	$a_0b_0$41	$a_1b_1$21	$a_2b_2$13
	block II	$a_0b_1$54	$a_1b_2$25	$a_2b_0$20
	block III	$a_0b_2$44	$a_1b_0$61	$a_2b_1$24
Replicate 2	block IV	$a_0b_0$52	$a_1b_2$30	$a_2b_1$24
	block V	$a_0b_1$44	$a_1b_0$39	$a_2b_2$26
	block VI	$a_0b_2$47	$a_1b_1$32	$a_2b_0$38

(3) The analysis of variance and table of means given in Table 15.19 comes from an experiment to investigate the effects of six spacing treatments combined with presence/absence of nitrogen and potassium fertilisers on maize. The basic experimental design was a 6×6 Latin square with one row missing to compare the six treatments (A, B, C, D, E, F). Each plot was split in two and the two split plots were allocated either the pair of treatments (n_0k_0, n_1k_1) or the pair (n_0k_1, n_1k_0). The first pair was used for all plots

Table 15.19(a). *Means.*

	A	B	C	D	E	F	
n_0k_0	21.8	27.2	31.6	22.2	23.2	21.4	24.6
n_0k_1	26.4	34.0	31.8	25.2	24.6	23.4	27.4
n_1k_0	28.8	32.6	31.0	29.8	24.6	30.8	29.6
n_1k_1	44.4	46.2	43.6	46.6	41.0	43.2	44.2
	30.35	35.0	34.5	30.95	28.35	29.7	

(b). *Analysis of variance.*

Source	SS	df	MS
Rows	968		
Columns (ign. spacings)	2721		
NK	2007		
Residual	714		
Spacings (elim. columns)	3055		
Error	1161		
Total	7905		
N	7020		
K	4629		
N × spacing	657		
K × spacing	132		
Error	490		
Total	20 833		

in columns 1, 3 and 4 and the second pair for plots in columns 2, 5 and 6 and the NK interaction is therefore confounded with columns.

Complete the analysis of variance, and summarise the results. (Note that the efficiency of the incomplete Latin square is 24/25. The means are adjusted for column effects.)

(4) Construct a design to compare six treatments in nine blocks of four units each, with no treatment occurring twice in a block when the treatment set is a 2×3 factorial, avoiding any confounding of the main effect for the two-level factor and with minimal confounding for the main effect of the three-level factor. Obtain variances of least squares estimates of main effect differences.

16

Some mathematical theory for confounding and fractional replication

16.0 *Preliminary examples*

(*a*) An industrial experiment is to be planned to investigate the effects of varying seven factors in a chemical process. Eight treatment combinations can be tested using the same batch of basic material. Eight different batches of material will be available, and it is expected that there may be substantial differences in output for sample units from the different batches. If it is decided to use two levels of each factor how shall the 64 treatment combinations to be included in the experiment be chosen, and how shall they be allocated to the eight blocks, or batches?

(*b*) An experiment on competition between grass and legume species is to be planned to investigate the effects and pairwise interactions of eight factors. The eight factors are:

(1) two different cutting regimes,

(2) two different grass species,

(3) two different legume species,

(4) two levels of phosphate,

(5) two levels of potassium,

(6) two levels of nitrogen,

(7) root competition between grass and legume allowed or restricted,

(8) shoot competition between grass and legume allowed or restricted.

The total number of experimental units that can be managed is 64 (or possibly 128). It is believed that the experimental units should be grouped into blocks of 16 (or possibly eight). A complication is that it may be necessary to have factor (1) and possibly some other factors applied to groups of units, necessitating a split unit design. What are the most sensible designs that should be considered and how much further information should the experimenter be asked for?

(c) An experiment on absorption of sugar by rabbits is to be designed to compare eight experimental treatments. The treatments have a 2^3 factorial structure, the three factors being

(i) concentrations of sugar in the diet,

(ii) chemical forms of sugar,

(iii) with or without an additive.

The experimental resources available for the experiment are five rabbits, each of which can be observed during five successive time periods. The design problem is therefore to allocate a 2^3 treatment structure within a 5×5 Latin square, recognising the implied requirement for maximum precision of main effects and minimum precision for the three-factor interaction.

16.1 The negative approach to confounding

In Chapter 15 the intuitive ideas of confounding, partial confounding and fractional replication were explored, and the thesis was developed that designing appropriate experiments for many combinations of block size and factorial structure is straightforward. In the present chapter mathematical theory is developed to provide support for that thesis, not only by demonstrating why the methods of design construction used in the previous chapter should be effective, but also providing the means to construct designs when the size of the treatment set becomes so large that the intuitive approach becomes more difficult to organise.

Whereas the intuitive approach in the previous chapter involves the construction of designs to allow all those effects which are regarded as important to be estimated without confounding, the classical, mathematical, approach to confounding is in terms of the effects which are confounded with blocks, and which therefore cannot be estimated. If, in an experiment with a number of blocks, certain effects are selected to be confounded with blocks, then the crucial mathematical theory identifies which other effects are also confounded, and hence which other effects are not confounded.

We illustrate this negative approach to confounding with a very simple example, which is almost invariably used in any presentation of the mathematical theory of confounding. Consider a 2^4 factorial structure with factors P, Q, R and S. Suppose it is initially decided that the experiment should be designed in blocks of eight units. Of all the possible treatment effects (four main effects, six two-factor interactions, four three-factor interactions and one four-factor interaction) it would seem reasonable to decide that the four-factor interaction is most easily ignored. Since the 15 treatment effects are orthogonal, if the treatment combinations are allocated so that the four-factor interaction is identical with the difference between blocks then the other 14 effects will be orthogonal to blocks and therefore not confounded. The four-factor interaction is

$$PQRS \propto (p-1)(q-1)(r-1)(s-1)$$
$$= \{pqrs + pq + pr + ps + qr + qs + rs + (1)\}$$
$$- (pqr + pqs + prs + qrs + p + q + r + s).$$

the familiar pattern of positive 'even' combinations and negative 'odd' combinations identified in Chapter 13. If in each replicate the even combinations are allocated to one block and the odd combinations to the other, then the PQRS effect is confounded with blocks in each replicate, and all other effects will be unconfounded. The design is shown in Figure 16.1(a), and it can quickly be verified that all other treatment effects are unconfounded, and that this is the same design as that in Figure 15.5(b), produced by the direct allocation approach.

Suppose, after further thought, it is decided that blocks of eight are too large, and that blocks of four must be used. The PQRS effect is already

Figure 16.1. Dividing a $2 \times 2 \times 2 \times 2$ structure (a) into two blocks confounding PQRS, (b) into four blocks confounding PQR.

(a)   Block

I	II
(1)	p
pq	q
pr	r
ps	s
qr	pqr
qs	pqs
rs	prs
pqrs	qrs

(b)   Block

I_1	I_2	II_1	II_2
ps	pq	p	s
qs	pr	q	pqs
rs	qr	r	prs
pqrs	(1)	pqr	qrs

confounded. The most obvious candidate for a second confounded effect is one of the three-factor interactions. Suppose it is decided to confound PQR in addition to PQRS. Now PQR is also defined as the difference between even and odd combinations:

$$PQR \propto (p-1)(q-1)(r-1)(s+1)$$
$$= (pqrs + pqr + ps + qs + rs + p + q + r)$$
$$- \{pqs + prs + qrs + pq + pr + qr + s + (1)\}.$$

If, from each of the blocks of Figure 16.1(a), two blocks are constructed, one containing even combinations w.r.t. PQR and the other containing odd combinations w.r.t. PQR, the four blocks shown in Figure 16.1(b) are obtained. However, since there are four blocks, there must be three contrasts between blocks and hence there should be three effects confounded with block comparisons (or possibly more than three effects not unconfounded). Inspection of the four blocks of Figure 16.1(b) reveals that the other confounded effect is the main effect of S, blocks I and IV having all the combinations with the upper level of S, and blocks II and III all the combinations with the lower level of S. Clearly a design in which the main effect of one factor is confounded is most unlikely to be suitable. It is necessary to devise a method of construction of designs, by selecting the effects to be confounded, which avoids confounding important effects.

The fundamental principle on which the theory of choosing confounded designs is based was established by Fisher. In the next section this theory is developed for two-level factors and in subsequent sections we consider factors with more than two levels.

16.2   Confounding theory for 2^n factorial structures

If, in an experiment for n factors, each with two levels, the 2^n treatment combinations are divided into 2^m blocks each having 2^{n-m} treatment combinations, then the subdivision into blocks can be conceived sequentially. The 2^n combinations are first split into two halves with one effect, X, confounded between the two halves; then the halves are split into quarters with a further effect, Y, confounded; then the quarters into eighths with a further effect, Z, confounded, and so on, until after m splits there are 2^m blocks each with 2^{n-m} combinations. In this procedure, m effects are selected to be confounded in the m splits. Since there are 2^m blocks, there are $(2^m - 1)$ block contrasts, and the same number of confounded effects. For a given set of chosen confounded effects, the complete set of confounded effects is determined by the following result.

Fisher's multiple confounding rule

If, in a replicate of a 2^n factorial, two effects, X and Y, are both confounded, then the generalised interaction, $Z(=XY)$, is also confounded, where the set of factors involved in the interaction, Z, includes all factors involved in either X or Y, but not in both.

Thus, if X and Y are the interactions PQRS and PQR, then $Z(=XY)$ will be S. If X and Y are PQS and PRTU, then $Z(=XY)$ will be QRSTU.

The validity of the rule may be demonstrated as follows: if the effect X is confounded with blocks, then the blocks are split into two groups, $\{X_1\}$ and $\{X_2\}$, such that the blocks in $\{X_1\}$ contain only treatment combinations which are even for X; that is, with an even number of factors from X having their upper level. Correspondingly, the blocks in $\{X_2\}$ contain only treatment combinations odd for X, that is with an odd number of factors from X having their upper level. If Y is confounded, there is a similar division of blocks into two groups $\{Y_1\}$ which has even combinations, and $\{Y_2\}$, which has odd combinations.

Consider the four groups caused as a result of the two divisions and, in particular, the blocks which occur in both $\{X_1\}$ and $\{Y_1\}$. All treatment combinations in these blocks are even for X and even for Y; that is, for each treatment combination, the number of factors from X at their upper level is even, and the number of factors from Y at their upper level is even. It follows that, for these combinations, the number of upper level factors in X or Y, but not both, must also be even, since it is equal to the sum of two even numbers minus twice the number of common upper level factors. Hence, all treatment combinations in both $\{X_1\}$ and $\{Y_1\}$ are even for Z.

Similarly, treatment combinations in $\{X_2\}$ and $\{Y_2\}$ are even for Z, and those in $\{X_1\}$ and $\{Y_2\}$ or in $\{X_2\}$ and $\{Y_1\}$ are odd for Z. To express the result more formally we use two mathematical concepts. First the idea of modulo arithmetic, where the expression

$$a = b(c)$$

means that a is the remainder when b is divided by c. The numbers a, b and c are integers and clearly a is an integer between 0 and $(c-1)$. Thus, for example,

$$7(2) = 1, \ 10(3) = 1, \ 10(5) = 0.$$

The numbers, b, such that $b(2) = 0$ are the even numbers and those for which $b(2) = 1$ are the odd numbers.

Secondly we use the set theory concepts of union and intersection. For two sets, $\{A\}$ and $\{B\}$ the union $\{A \cup B\}$ comprises all elements which are in A or B or both: the interaction $\{A \cap B\}$ comprises all elements which are in both A and B. Hence the set of elements which are in exactly one of $\{A\}$

and {B} is

$$\{A \cup B\} - \{A \cap B\}.$$

Formally, if two effects X and Y are to be confounded, and if for any treatment combination, U(X) is the number of factors involved in the effect X which have their upper level in the treatment combination then

for combinations in $\{X_1\}$ $U(X) = 0(2)$ (= even),
for combinations in $\{X_2\}$ $U(X) = 1(2)$ (= odd)
similarly, in $\{Y_1\}$ $U(Y) = 0(2)$
       in $\{Y_2\}$ $U(Y) = 1(2).$

But

$$U(Z) = U(X \cup Y) - U(X \cap Y)$$

and

$$U(X \cup Y) = U(X) + U(Y) - U(X \cap Y)$$

hence

$$U(Z) = U(X) + U(Y) - 2U(X \cap Y).$$

Since $2U(X \cap Y) = 0(2)$ for any value of $U(X \cap Y)$ it follows that

in $\{X_1 Y_1\}$ $U(Z) = 0(2).$

Similar arrangements lead to the corresponding results:

in $\{X_1 Y_2\}$ $U(Z) = 1(2),$
in $\{X_2 Y_1\}$ $U(Z) = 1(2),$
in $\{X_2 Y_2\}$ $U(Z) = 0(2).$

Hence, the interaction effect Z is such that the blocks in $Z_1 = \{X_1, Y_1\} + \{X_2, Y_2\}$ are 'even' for Z, and the blocks in $Z_2 = \{X_1, Y_2\} + \{X_2, Y_1\}$, are 'odd' for Z. Fisher's rule enables us to identify immediately all confounded effects when more than one effect is confounded. If three effects, W, X, Y, are chosen initially to be confounded, then the set of confounded effects will be W, X, Y, WX, WY, XY and WXY. The pattern of the limitations on confounding is quickly found by considering examples:

(i) 2^4 in four blocks of four:

confound PQRS and PQR $\Rightarrow$ S confounded;
confound PQR and PQS $\Rightarrow$ RS confounded.

This latter design was found in Figure 15.5(c).

(ii) 2^5 in four blocks of eight:

confound PQRST and PQRS $\Rightarrow$ T confounded,
confound PQRS and PQRT $\Rightarrow$ ST confounded,
confound PQRS and PRT $\Rightarrow$ QST confounded

(see Figure 15.5(d).

(iii) 2^5 in eight blocks of four:

confound PQRST, PQRS, QRST⇒T, P, PT, QRS also confounded,
confound PQRS, PRT, PQR   ⇒QST, S, QT, PRST also
confounded,
confound PQRS, PRT, RST   ⇒QST, PQT, PS, QR also
confounded.

(iv) 2^6 in eight blocks of eight:

confound PQRSTU, PQR, PST ⇒STU, QRU, QRST, PU also
confounded,
confound PQRS, PQTU, PRT   ⇒RSTU, QST, QRU, PSU also
confounded.

In each of these examples, we have deliberately included a confounding system including the highest possible factor interaction, and also the optimal confounding system in the sense of avoiding confounding main effects as a first priority, and two-factor interactions as a second priority. It can be seen that confounding the highest possible interaction is never optimal.

The guidelines for optimal systems of confounding three effects can be simply deduced by observing that no factor can be included in all three effects X, Y and Z$=$XY. Therefore, if three effects are to be confounded, each factor will be included in at most two of the confounded effects. The complete omission of any factor from all three confounded effects must involve unnecessarily low order interactions. Hence, in the set of three confounded effects, all factors must appear twice, and the most efficient confounding system will require confounded effects involving n_1, n_2 and n_3 factors, respectively, where n_1, n_2 and n_3 are the most nearly equal partition of 2^n into three components.

More generally, for a 2^n set of treatment combinations, in 2^m blocks, there will be $2^m - 1$ confounded effects, and each factor should occur 2^{m-1} times. The optimal confounding system should be sought by considering the most nearly equal partition of $n(2^{m-1})$ into $2^m - 1$ components. Thus, for $n=6$ and $m=3$, we have to divide 24 into seven components, and the best conceivable solution would be $(4, 4, 4, 3, 3, 3, 3)$, as was actually achieved in example (iv).

It may be helpful to think of an 'average' number of factors in the set of confounded interactions which will be

$$\frac{n2^{m-1}}{2^m-1},$$

which, if m is 3 or more, is approximately $n/2$. Hence, if a replicate is divided into a large number of blocks, the confounding system should be such that most of the confounded effects are interactions of about half the factors in the experiment. If higher order interactions are confounded, this necessitates confounding also interactions of fewer factors. For $m=1$, of course, the higher order interaction can be confounded. For $m=2$, the 'average' number of factors in the confounded effects will be $2n/3$, and it is always possible to achieve the most nearly equal partition of $2n$ into three components.

The selection of the set of confounded effects defines the design uniquely, but there remains the task of identifying the treatment combinations in each block. Fisher's rule for determining the set of confounded effects implies the pattern of 'even' and 'odd' treatment combinations in the different blocks. There must be one block, known as the principal block, in which all treatment combinations are even for all the confounded effects. That is, for every treatment combination in the principal block and for each confounded effect the number of factors involved in the confounded effect having an upper level in the treatment combination will be even:

$$U(X) = 0(2)$$
$$U(Y) = 0(2)$$
$$. . . . .$$

Each treatment combination is defined by the factors for which the upper level is used in the combination. If two treatment combinations occur in the principal block, then so also will the product combination defined by the factors having an upper level for either of the combinations, but not for both.

Any combination not in the principal block can be used as the seed of another block. The new block will contain the new treatment combination and all product combinations formed from the principal block combinations and the new combination. Since the principal block combinations are even for all confounded effects, all the other combinations in a block with this new combination will have the same 'even' or 'odd' class as the new combination for each confounded effect.

Example 16.1

2^6 in eight blocks of eight. The seven confounded effects will be interactions such that the total of the number of factors per interaction will be 24, and a possible set of confounded effects is

PQRS, PQTU, PRT, RSTU, QST, QRU, PSU.

The combinations which are even with respect to each of these seven effects are those which are even for the first three (the others are

Figure 16.2. A 2^6 design in eight blocks of eight units confounding PQRS, PQTU, PRT.

Block

I	II	III	IV	V	VI	VII	VIII
Principal	xp	xq	xpq	xr	xpr	xqr	xpqr
(1)	p	q	pq	r	pr	qr	pqr
pst	st	pqst	qst	prst	rst	pqrst	qrst
qsu	pqsu	su	psu	qrsu	pqrsu	rsu	prsu
rstu	prstu	qrstu	pqrstu	stu	pstu	qstu	pqstu
pqtu	qtu	ptu	tu	pqrtu	qrtu	prtu	rtu
pru	ru	pqru	qru	pu	u	pqu	qu
qrt	pqrt	rt	prt	qt	pqt	t	pt
pqrs	qrs	prs	rs	pqs	qs	ps	s

generalised interactions of the first three) and may be identified as pst, qsu, rstu and their products pqtu, pru, qrt and pqrs, together with the combination of lowest levels of all factors, (1).

The other blocks are constructed using any combination not already included and forming products with the principal block combinations. The resulting design is shown in Figure 16.2. Note that for any non-confounded three-factor interaction effect, such as PQR, all eight combinations ((1), p, q, r, pq, pr, qr, and pqr) will occur in each of the eight blocks.

As for all other confounded experiments where effects are either completely confounded or not confounded, the analysis of variance will include sums of squares for all important effects, except those which are confounded. For this example, the structure of the analysis of variance is as shown in Table 16.1.

Table 16.1.

Source	df
Blocks	7
Main effects	6
Two-factor interactions	15
Three-factor interactions (except for PRT, QST, PSU, QRU)	16
Error	19
Total	63

16.3 Confounding theory for other factorial structures; dummy factors

All theory for confounding in factorial structures other than the 2^n involves either adapting the ideas of 2^n confounding or generalising those ideas in a mathematically neat, but practically unattractive, fashion. The adaptation of the 2^n confounding ideas to factors with other numbers of levels is achieved by defining dummy factors each with two levels, such that the combinations of levels of these dummy factors correspond to the actual levels of the real factors, although the dummy factors and their levels have no sensible practical interpretation. A four-level factor, P, may be represented by two two-level dummy factors, P' and P'', as follows:

$$p'_0 + p''_0 \Rightarrow p_0$$
$$p'_0 + p''_1 \Rightarrow p_1$$
$$p'_1 + p''_0 \Rightarrow p_2 \qquad (16.1)$$
$$p'_1 + p''_1 \Rightarrow p_3$$

The design of a 4^n experiment with confounding in blocks is achieved by first determining the set of confounded effects, and the allocation of treatment combination to blocks, in terms of the dummy factors, and then translating the combinations of levels of dummy factors to levels of the real factors through the definition (16.1). The choice of which interactions of the dummy factors to confound must, of course, be primarily based on the real factors and must involve consideration of the translation from the dummy factors to the genuine factors. The main effects and interactions of P' and P'' are all main effects of the genuine factor P since

$$P' \quad \propto (p_3 + p_2 - p_1 - p_0)$$
$$P'' \quad \propto (p_3 - p_2 + p_1 - p_0)$$
$$P'P'' \propto (p_3 - p_2 - p_1 + p_0).$$

Hence the interaction of dummy factors $P'P''$ constitutes a main effect comparison of P. However, the interaction of dummy factors $P'Q''$ is an interaction comparison for the two factors P and Q.

Example 16.2

To demonstrate the necessary steps consider a complete 4^3 experiment to be designed in eight blocks of eight units each. The corresponding dummy factor design has six two-level dummy factors in eight blocks of eight, for which an appropriate design has been constructed in the previous example. Let the four-level factors be A, B, C and the corresponding dummy factors, A', A'', B', B'', C', C''. In Example 16.1, the set of confounded interactions was

(PQRS, PQTU, PRT, RSTU, QST, QRU, PSU).

We would like to allocate $(A', A'', B', B'', C', C'')$ to (P, Q, R, S, T, U) so that each of the seven confounded effects involves a dummy factor from each of the three genuine factors. This is not possible and the best that can be achieved is to have four of the confounded effects involving a dummy factor from each genuine factor, and the other three involving dummy factors from only two genuine factors. A typical solution is $P = A'$, $Q = B''$, $R = B'$, $S = A''$, $T = C'$, $U = C''$, giving confounded effects $(A'A''B'B''$, $A'B''C'C''$, $A'B'C'$, $B'A''C'C''$, $B''A''C'$, $B''B'C''$, $A'A''C'')$. The confounded effects, expressed in terms of the levels of the genuine factors, are

$$A'A''B'B'' = (a_3 - a_2 - a_1 + a_0)(b_3 - b_2 - b_1 + b_0)(c_3 + c_2 + c_1 + c_0)$$
$$A'B''C'C'' = (a_3 + a_2 - a_1 - a_0)(b_3 - b_2 + b_1 - b_0)(c_3 - c_2 - c_1 + c_0)$$
$$A'B'C' = (a_3 + a_2 - a_1 - a_0)(b_3 + b_2 - b_1 - b_0)(c_3 + c_2 - c_1 - c_0)$$
$$A'B''C'C'' = (a_3 - a_2 + a_1 - a_0)(b_3 + b_2 - b_1 - b_0)(c_3 - c_2 - c_1 + c_0)$$
$$A''B''C' = (a_3 - a_2 + a_1 - a_0)(b_3 - b_2 + b_1 - b_0)(c_3 + c_2 - c_1 - c_0)$$
$$B'B''C'' = (a_3 + a_2 + a_1 + a_0)(b_3 - b_2 - b_1 + b_0)(c_3 - c_2 + c_1 - c_0)$$
$$A'A''C'' = (a_3 - a_2 - a_1 + a_0)(b_3 + b_2 + b_1 + b_0)(c_3 - c_2 + c_1 - c_0).$$

Notice that $A'A''$ is recognisable as the quadratic effect of factor A, assuming equally spaced quantitative levels. The treatment combinations are translated into levels of A, B, C by

$$ps \Rightarrow a_3, \quad p \Rightarrow a_2, \quad s \Rightarrow a_1, \quad (1) \Rightarrow a_0$$
$$qr \Rightarrow b_3, \quad r \Rightarrow b_2, \quad q \Rightarrow b_1, \quad (1) \Rightarrow b_0$$
$$tu \Rightarrow c_3, \quad t \Rightarrow c_2, \quad u \Rightarrow c_1, \quad (1) \Rightarrow c_0.$$

The resulting design is shown in Figure 16.3, with the combinations in the same order in each block as those in Figure 16.2. In the analysis, the sums of squares for each two-factor interaction are derived from only eight of the usual nine degrees of freedom, because of the confounded component of each two-factor interaction. The actual calculation of sums of squares can be done either by calculating sums of squares for each of the eight non-confounded component effects, or much more simply by using a general analysis program which recognises the loss, through confounding, of the ninth component. The recognition that the confounded effect of the AB interaction is the quadratic × quadratic would suggest that calculating the sums of squares for other particular components of that effect such as linear × linear and linear × quadratic would be useful. More generally the fact that only one out of the nine degrees of freedom for each two-factor interaction is confounded enables the more important contrasts of the

Figure 16.3. A $4 \times 4 \times 4$ design in eight blocks of eight units.

Block

I	II	III	IV	V	VI	VII	VIII
$a_0 b_0 c_0$	$a_2 b_0 c_0$	$a_0 b_1 c_0$	$a_2 b_1 c_0$	$a_0 b_2 c_0$	$a_2 b_2 c_0$	$a_0 b_3 c_0$	$a_2 b_3 c_0$
$a_3 b_0 c_2$	$a_1 b_0 c_2$	$a_3 b_1 c_2$	$a_1 b_1 c_2$	$a_3 b_2 c_2$	$a_1 b_2 c_2$	$a_3 b_3 c_2$	$a_1 b_3 c_2$
$a_1 b_1 c_1$	$a_3 b_1 c_1$	$a_1 b_0 c_1$	$a_3 b_0 c_1$	$a_2 b_3 c_1$	$a_3 b_3 c_1$	$a_1 b_2 c_1$	$a_3 b_2 c_1$
$a_1 b_2 c_3$	$a_3 b_2 c_3$	$a_1 b_3 c_3$	$a_3 b_3 c_3$	$a_2 b_0 c_3$	$a_3 b_0 c_3$	$a_1 b_1 c_3$	$a_3 b_1 c_3$
$a_2 b_1 c_3$	$a_0 b_1 c_3$	$a_2 b_0 c_3$	$a_0 b_0 c_3$	$a_1 b_3 c_3$	$a_0 b_3 c_3$	$a_2 b_2 c_3$	$a_0 b_2 c_3$
$a_2 b_2 c_2$	$a_0 b_2 c_2$	$a_2 b_3 c_2$	$a_0 b_3 c_2$	$a_1 b_0 c_2$	$a_0 b_0 c_2$	$a_2 b_1 c_2$	$a_0 b_1 c_2$
$a_0 b_3 c_1$	$a_2 b_3 c_1$	$a_0 b_2 c_1$	$a_2 b_2 c_1$	$a_0 b_1 c_1$	$a_2 b_1 c_1$	$a_0 b_0 c_1$	$a_2 b_0 c_1$
$a_3 b_3 c_0$	$a_1 b_3 c_0$	$a_3 b_2 c_0$	$a_1 b_2 c_0$	$a_3 b_1 c_0$	$a_1 b_1 c_0$	$a_3 b_0 c_0$	$a_1 b_0 c_0$

interaction to be arranged to be unconfounded. The structure of the analysis of variance is shown in Table 16.2.

Table 16.2.

Source	df
Blocks	7
A	3
B	3
C	3
AB	8
AC	8
BC	8
Error	23
Total	63

Designs for mixtures of two- and four-level factors can be constructed similarly by using dummy two-level factors for the four-level factors.

To use the 2^n confounding systems for factors with three levels, it is necessary to accept unequal replication of the three levels. The presumption that this is a major disadvantage reflects the tendency to regard equal replication as inevitable without questioning whether all effects are equally important. Certainly it will sometimes be true that one level of a three-level factor should be regarded as more important than the other two and should be replicated more frequently. Again two dummy factors, P′ and P″, with two levels, are used to represent the levels of the genuine factor, P, as follows:

$$p_0' + p_0'' \Rightarrow p_0$$
$$p_0' + p_1'' \Rightarrow p_1$$
$$p_1' + p_0'' \Rightarrow p_1$$
$$p_1' + p_1'' \Rightarrow p_2.$$

The choice of the level (p_1) to be duplicated is for the experimenter.

The translation from dummy factor levels to real factor levels has two other important differences from that for four-level factors. There is a symmetry of the two dummy factors in their relationship to the genuine factor, and the interpretation of the main effects and interaction of the dummy factors is directly relevant to the main effects of the genuine factor. The main effect of P' is

$$P' \propto (p_1' - p_0')(p_1'' + p_0'') \equiv p_2 - p_1 + p_1 - p_0 = (p_2 - p_0).$$

Similarly, $P'' \propto (p_2 - p_0)$, while the interaction is

$$P'P'' \propto (p_1' - p_0')(p_1'' - p_0'') = (p_2 - p_1 - p_1 + p_0) = (p_2 - 2p_1 + p_0).$$

Both P' and P'' provide estimates of the $(p_2 - p_0)$ main effect of P which, for quantitative levels, is the linear effect of P. The interaction $P'P''$ provides an estimate of the quadratic effect of P. More generally as noted in Chapter 13 the pair of contrasts (i) between two levels, and (ii) between that pair and the third is, practically, almost the only useful form of contrast for three levels. The obvious advantage of having two estimates of the linear main effect of P is that it is possible to confound one of the dummy factor main effects and still have an estimate of the linear main effect of P from the other dummy factor main effect. This advantage is tempered by the reduced precision of each dummy factor effect as an estimator of the linear main effect of P. In fact, the duplication of the middle level changes the entire pattern of precision. To examine this change of precision pattern, consider two sets of 12 observations. The first consists of four replicates of a single three-level factor; the second consists of three replicates of the 2×2 dummy factors, equivalent to three replicates of p_0 and p_2, and six replicates of p_1. Consider the estimates of the linear and quadratic effects of P:

(a) P_L estimated by $(\bar{p}_2 - \bar{p}_0)$ variance $2\sigma^2/4 = \sigma^2/2$,

   P_Q estimated by $1/2(\bar{p}_2 - 2\bar{p}_1 + \bar{p}_0)$ variance $1/4(6\sigma^2/4) = 3\sigma^2/8$.

(b) P_L estimated by $2(\bar{p}_1' - \bar{p}_0')$ variance $4(2\sigma^2/6) = 4\sigma^2/3$,

   also estimated by $2(\bar{p}_1'' - \bar{p}_0'')$ variance $4(2\sigma^2/6) = 4\sigma^2/3$,

   P_Q estimated by $1/2(\bar{p}_1' - \bar{p}_0')(\bar{p}_1'' - \bar{p}_0'')$ variance $1/4(4\sigma^2/3) = \sigma^2/3$.

If both dummy factor main effects are used to estimate P_L, the variance of the combined estimate is $2\sigma^2/3$, compared with $\sigma^2/2$ for the equal replication design, an efficiency of 3/4. Correspondingly, the estimation of P_Q is more efficient with unequal replication by a factor of 9/8. If only one of the dummy factor main effects is used to estimate P_Q, then the efficiency relative to the equal replication design is 3/8. Although the estimation of the linear main effect has been discussed in terms of using one, or both, dummy main effect estimates, in practice of course we would not calculate the two separate estimates but would calculate, directly, a single estimate. Thus, the estimation of P_L is relatively better in the equal replicate design, but the estimation of P_Q is relatively better in the unequal replicate design. This obviously has some relevance to the subsequent consideration of alternative forms of confounding system.

The benefit of retaining some information about the linear effect, even if one dummy factor main effect is confounded, is not likely to be a usable advantage for main effects where the linear effect is already less precise than the quadratic because of the unequal replication. But, when interactions of the real factors are used as confounded effects, then the loss of one estimate of a linear × linear interaction effect can be tolerated because there are three other estimates. The design of Example 16.1 for a 2^6 experiment confounded in eight blocks of eight units can be used again to illustrate the use of dummy factors for confounding 3^n experiments.

Example 16.3

Consider the 3^3 design using dummy factors A′, A″, B′, B″, C′, C″. The six dummy factors must be equated to the six factors of Example 16.1 so that the confounded effects

PQRS, PQTU, PRT, RSTU, QST, QRU, PSU

represent effects where loss is not unacceptable. From the similar exercise in Example 16.2, some of the confounded effects must be components of two-factor interactions. A sensible solution is

$$P = A', \; Q = B', \; R = C', \; S = A'', \; T = C'', \; U = B'',$$

giving confounded effects

A′B′C′A″, A′B′C″B″, A′C′C″, C′A″C″B″, B′A″C″, B′C′B″, A′A″B″.

The third, sixth and seventh confounded effects would give estimates of the linear A × quadratic C, linear C × quadratic B, and linear B × quadratic A effects, but in each case there is a second estimator of the same effect (A″C′C″, B′B″C″ and A′A″B′) which is not confounded.

The allocation of treatment combinations to blocks is determined by translation of the dummy factor levels to the real three-level factor levels:

$$ps \Rightarrow a_2, \quad p \Rightarrow a_1, \quad s \Rightarrow a_1, \quad (1) \Rightarrow a_0$$
$$qu \Rightarrow b_2, \quad q \Rightarrow b_1, \quad u \Rightarrow b_1, \quad (1) \Rightarrow b_0$$
$$rt \Rightarrow c_2, \quad r \Rightarrow c_1, \quad t \Rightarrow c_1, \quad (1) \Rightarrow c_0.$$

The resulting design is shown in Figure 16.4. All components of all main effects and interactions of the real factors can be estimated, but the precision of different estimates will vary both because of the unequal replication and because of the loss of some dummy factor effects through confounding. The structure of the analysis of variance is inevitably very similar to that for an experiment without confounding (see Table 16.3).

The standard errors for the various linear and quadratic main effects and two-factor interactions can be obtained simply from a general analysis program, and are given in Table 16.4. For comparison with the unconfounded, equal replicate design, the standard errors for the same effects in

Figure 16.4. A $3 \times 3 \times 3$ design in eight blocks of eight units.

Block

I	II	III	IV	V	VI	VII	VIII
$a_0b_0c_0$	$a_1b_0c_0$	$a_0b_1c_0$	$a_1b_1c_0$	$a_0b_0c_1$	$a_1b_0c_1$	$a_0b_1c_1$	$a_1b_1c_1$
$a_2b_0c_1$	$a_1b_0c_1$	$a_2b_1c_1$	$a_1b_1c_1$	$a_2b_0c_2$	$a_1b_0c_2$	$a_2b_1c_2$	$a_1b_1c_2$
$a_1b_2c_0$	$a_2b_2c_0$	$a_1b_1c_0$	$a_2b_1c_0$	$a_1b_2c_1$	$a_2b_2c_1$	$a_1b_1c_1$	$a_2b_1c_1$
$a_1b_1c_2$	$a_2b_1c_2$	$a_1b_2c_2$	$a_2b_2c_2$	$a_1b_1c_1$	$a_2b_1c_1$	$a_1b_2c_1$	$a_2b_2c_1$
$a_1b_2c_1$	$a_0b_2c_1$	$a_1b_1c_1$	$a_0b_1c_1$	$a_1b_2c_2$	$a_0b_2c_2$	$a_1b_1c_2$	$a_0b_1c_2$
$a_1b_1c_1$	$a_0b_1c_1$	$a_1b_2c_1$	$a_0b_2c_1$	$a_1b_1c_0$	$a_0b_1c_0$	$a_1b_2c_0$	$a_0b_2c_0$
$a_0b_1c_2$	$a_1b_1c_2$	$a_0b_0c_2$	$a_1b_0c_2$	$a_0b_1c_1$	$a_1b_1c_1$	$a_0b_0c_1$	$a_1b_0c_1$
$a_2b_1c_1$	$a_1b_1c_1$	$a_2b_0c_1$	$a_1b_0c_1$	$a_2b_1c_0$	$a_1b_1c_0$	$a_2b_0c_0$	$a_1b_0c_0$

Table 16.3.

Source	df
Blocks	7
Three main effects	6
Three two-factor interactions	12
Three-factor interactions	8
Error	40
Total	63

Table 16.4.

Effect	SE for confounded design	SE for full block design $(\times \sqrt{(27/64)})$
A linear	$\sigma\sqrt{(1/8)} = 0.35\sigma$	$\sigma\sqrt{(3/32)} = 0.31\sigma$
A quadratic	$\sigma/4 = 0.25\sigma$	$\sigma\sqrt{(9/128)} = 0.27\sigma$
AB lin × lin	$\sigma = 1.00\sigma$	$\sigma(3/4) = 0.75\sigma$
AB lin × quad	$\sigma\sqrt{(1/2)} = 0.71\sigma$	$\sigma\sqrt{(27/64)} = 0.65\sigma$
AB quad × lin	$\sigma = 1.00\sigma$	$\sigma\sqrt{(27/64)} = 0.65\sigma$

the latter design have been calculated and are also given in Table 16.4, but with a multiplying factor of $\sqrt{(27/64)}$ to adjust for the different numbers of units used in the two designs. The comparison of the two designs is difficult because not only are the relative precisions of the two experiments different for different comparisons, but the value of σ^2 will be different for the two designs, since the smaller block size of eight should give a much smaller σ^2 than the larger block size of 27. A choice between the two designs must therefore involve not only the consideration of the relative importance of different effects, but also the benefit of using smaller blocks.

The use of dummy factors to represent three-level factors also extends to factorial structures with a mixture of two- and three-level factors. For example, an experiment for the 36 combinations of $A(3) \times B(3) \times C(2) \times D(2)$ confounded in blocks of eight units could be constructed from the 2^6 design of Example 16.1 by using dummy two-level factors A′, A″, B′, B″ to represent the three-level factors A and B with duplication of the middle level, and defining the translation form (PQRSTU) to (A′A″B′B″CD) and thence to (ABCD).

For 3^n designs, there are also other theoretical approaches to constructing designs which do not require the use of duplicated levels. These are discussed in the next section, but have the disadvantage that the confounding structure is not related to the practically useful form of effects for three-level factors. From the viewpoint of requiring practically useful designs with confounding in terms of easily interpretable effects and with simple analyses, there is no ideal theoretical philosophy for the confounded 3^n experiment. For factors with quantitative levels other criteria for comparing designs are discussed in Chapter 18.

16.4 Confounding for 3^n

The classical approach to confounding for 3^n factorial structures extends the ideas of 2^n confounding in a different manner from the dummy factor approach. The basis of the 2^n confounding systems was the division of treatment combinations into two sets, characterised, with reference to a particular effect, as 'even' and 'odd'. More formally, the treatment combinations in the two sets, X_0 and X_1, are such that $U(X)$, the number of factors in effect X having their upper level in the treatment combination, must be

$$U(X) = 0(2) \text{ for all treatment combinations, X, in } X_0$$

and

$$U(X) = 1(2) \text{ for all treatment combinations in } X_1.$$

With three levels, any equal subdivision of treatment combinations will be into three sets, X_0, X_1 and X_2. The characterisation of the three sets must now involve all three levels and, instead of a simple count of factors with upper levels in treatment combinations, $U(X)$ is defined as a weighted sum of levels for the factors included in the effect, X. The combinations in the three treatment sets, X_0, X_1 and X_2, will be those for which $U(X) = 0(3)$, 1(3) and 2(3), respectively.

The actual subdivision of combinations into three sets is determined by the form of the weighted sum of levels $U(X)$. The general form is

$$\alpha i + \beta j + \gamma k + \dots,$$

where $i, j, k, \dots$ represent the levels of the different factors for a treatment combination $(p_i q_j r_k \dots)$ and $\alpha, \beta, \gamma, \dots$ represent the weights.

Consider the various possible subdivisions for two factors. There are nine treatment combinations,

$$p_0 q_0, \ p_0 q_1, \ p_0 q_2, \ p_1 q_0, \ p_1 q_1, \ p_1 q_2, \ p_2 q_0, \ p_2 q_1, \ p_2 q_2.$$

(1) If $\alpha = 1$, $\beta = 0$, $U(X) = i$.

Treatment combinations are allocated to sets according as

$$i = 0(3), \ i = 1(3) \text{ or } i = 2(3).$$

In other words,

set 1 includes all combinations with p_0,
set 2 includes all combinations with p_1,
set 3 includes all combinations with p_2.

(2) If $\alpha = 2$, $\beta = 0$, $U(X) = 2i$.

This levels to exactly the same allocation except that set 2 has the p_2 combinations and set 3 the p_1 combinations. As a basis for a confounded design the difference is immaterial.

(3) If $\alpha = 0$, $\beta = 1$, $U(X) = j$.

Now the allocations are based on q_0 (set 1), q_1 (set 2), q_2 (set 3).

(4) If $\alpha = 0$, $\beta = 2$, $U(X) = 2j$.

Again this provides the same allocation as (3).

(5) If $\alpha = 1$, $\beta = 1$, $U(X) = i + j$.

Treatment combinations are allocated to sets based on the sum of i and j. This gives

set 1	$p_0 q_0$,	$p_1 q_2$,	$p_2 q_1$,
set 2	$p_0 q_1$,	$p_1 q_0$,	$q_2 q_2$,
set 3	$p_0 q_2$,	$p_1 q_1$,	$p_2 q_0$.

(6) If $\alpha = 1$, $\beta = 2$, $U(X) = i + 2j$.

Now it is $i + 2j$ which determines the allocation

set 1	$p_0 q_0$,	$p_1 q_1$,	$p_2 q_2$,
set 2	$p_0 q_2$,	$p_1 q_0$,	$p_2 q_1$,
set 3	$p_0 q_1$,	$p_1 q_2$,	$p_2 q_0$.

(7) If $\alpha = 2$, $\beta = 1$, $U(X) = 2i + j$.

This provides the same allocation as (6).

(8) If $\alpha = 2$, $\beta = 2$, $U(X) = 2i + 2j = 2(i + j)$.

This, perhaps more obviously, provides the same allocation as (5).

The four alternative patterns of allocation can be interpreted in terms of the treatment effects as follows:

(1) or (2) confounds the main effect of P between sets

(3) or (4) confounds the main effect of Q between sets.

In (5), (6), (7), (8) main effects are clearly not confounded and hence the confounded components are part of the PQ interaction. It should also be noted that the allocations (5) and (6) are precisely the two designs constructed in Section 15.2 (Figure 15.6).

To distinguish the four subdivisions which lead to part of the PQ interaction effect being confounded, the subdivisions are identified in the form $P^\alpha Q^\beta$. Thus subdivision (5) is $P^1 Q^1$ or PQ, and (6) is $P^1 Q^2$ or PQ^2. As already noted, (7), which is $P^2 Q^1$ or $P^2 Q$, gives the same allocation as PQ^2 (6), and (8), which is $P^2 Q^2$, is identical with PQ (5). Comparisons between totals for the three sets for any subdivision are contrasts for the corresponding effect. For either PQ or PQ^2 the comparisons between totals are orthogonal to main effect comparisons for P and for Q and must therefore be interaction effect contrasts. The comparisons between $(PQ)_0$, $(PQ)_1$ and $(PQ)_2$ are thus interaction effect contrasts (which we have met previously as contrasts between J_1, J_2, J_3 in Section 15.4). Similarly, comparisons between $(PQ^2)_0$, $(PQ^2)_1$ and $(PQ^2)_2$ are interaction effect contrasts (equivalent to those between I_1, I_2, I_3 in Section 15.4).

Although the contrasts between the three sets (I_1, I_2, I_3) or (J_1, J_2, J_3) are not directly practically meaningful, they can, as noted in Section 15.4, be used to estimate effects of practical interest. More importantly, they provide a basis for constructing confounding systems with several effects confounded, in the same way that the 'even' and 'odd' division did for 2^n treatment structures. The general statement providing for the implications of multiple confounding is exactly the same.

Fisher's multiple confounding rule

If, in a replicate of a 3^n factorial, two effects X and Y are both confounded, then the generalised interaction, $Z = XY$, is also confounded. The generalised interaction of two effects

$$X = P^\alpha Q^\beta R^\gamma \ldots \text{ and } Y = P^{\alpha'} Q^{\beta'} R^{\gamma'} \ldots$$

is

$$Z = P^{\alpha''} Q^{\beta''} R^{\gamma''} \ldots,$$

where

$$\alpha'' = \alpha + \alpha'(3)$$
$$\beta'' = \beta + \beta'(3).$$

The argument for this result is essentially the same as that for the earlier result for two-level factors.

Confounding X requires that the treatment combinations are split into three groups, X_0, X_1, X_2, such that $\alpha i + \beta j + \ldots \equiv 0$, 1, or 2(3), and each block must be wholly contained in one of X_0, X_1 and X_2. Similarly, confounding Y defines three groups of treatment combinations, Y_0, Y_1, Y_2, such that $\alpha' i + \beta' j + \ldots \equiv 0$, 1, or 2(3). When both X and Y are confounded, each block is wholly contained in one of the nine sets deriving from all combinations of (X_0, X_1, X_2) with (Y_0, Y_1, Y_2).

Consider now the sets (Z_0, Z_1, Z_2) defined to include treatment combinations such that

$$\alpha'' i + \beta'' j + \ldots \equiv 0, 1 \text{ or } 2(3).$$

Now

$$\alpha'' = \alpha + \alpha'(3) \text{ etc.}$$

Hence

$$\alpha'' i + \beta'' j + \ldots \equiv (\alpha + \alpha') i + (\beta + \beta') j + \ldots (3)$$
$$\equiv (\alpha i + \beta j + \ldots) + (\alpha' i + \beta' j + \ldots) + \ldots (3)$$

and Z_0 will consist of treatment combinations such that either

$$\alpha i + \beta j + \ldots = 0(3) \text{ and } \alpha' i + \beta' j + \ldots = 0(3)$$

or

$$\alpha i + \beta j + \ldots = 1(3) \text{ and } \alpha' i + \beta' j + \ldots = 2(3)$$

or

$$\alpha i + \beta j + \ldots = 2(3) \text{ and } \alpha' i + \beta' j + \ldots = 1(3).$$

That is Z_0 will consist of combinations in either

(X_0 *and* Y_0) or (X_1 *and* Y_2) or (X_2 *and* Y_1).

Similarly, Z_1 and Z_2 will consist of three of the nine sets defined by combinations of (X_0, X_1, X_2) and (Y_0, Y_1, Y_2). Hence, each block of treatment combinations, which must be wholly contained in one of the nine combinations of (X_0, X_1, X_2) with (Y_0, Y_1, Y_2), must be wholly contained in one of Z_0, Z_1, Z_2. This is sufficient to demonstrate that Z is confounded. It should, however, be noted that Z_0, Z_1 and Z_2 will be equal sized groups since, for each set of levels ($j, k, \ldots$) for Q, R, $\ldots$, there is one level of P for which the value of i satisfies

$$\alpha''i + \beta''j + \ldots \equiv 0(3),$$

one value of i for the corresponding equations with 0 replaced by 1 and one value for the equations with 0 replaced by 2.

Further, since the effects X and XX or X^2 define the same division into three sets of treatment combinations, the simultaneous confounding of X (and X^2) and Y (and Y^2) implies the additional confounding not only of $Z = XY$, but also of $W = X^2Y$, $V = XY^2$ and $U = X^2Y^2$. Since confounding two effects, X and Y, implies nine blocks, and each confounded effect corresponds to 2 df, it is inevitable that the deliberate confounding of two effects will imply a total of eight confounded effects.

The development of suitable systems of confounded effects is rather easier for 3^n experiments than for 2^n, essentially because the confounded component interactions comprise more than a single effect. However, just as in 2^n confounding, when we confound more than one effect, we should expect to be unable to restrict the confounded effects to the highest possible order interactions. Some examples will show this more clearly:

(1) Three factors, 27 combinations in nine blocks of three units each. If PQR and PQR^2 (or any other pair of three-factor interaction components) are confounded, a main effect must also be confounded, the other confounded effects being

$$(PQR)(PQR^2) = P^2Q^2 \equiv PQ$$

and

$$(PQR)(PQR^2)^2 = R^2 \equiv R.$$

If PQ and PR are confounded then so are

$$(PQ)(PR) = P^2QR \equiv PQ^2R^2$$

and

$$(PQ)(PR)^2 = QR^2.$$

(2) Four factors, 81 combinations in nine blocks of nine units each. If a pair of four-factor interaction components are confounded this implies

the additional confounding of a main effect or two-factor interaction. If PQRS and PQ^2RS^2 are confounded then so are

$$(PQRS)(PQ^2RS^2) = P^2R^2 = PR$$

and

$$(PQRS)(PQ^2RS^2)^2 = Q^2S^2 = QS.$$

If two three-factor interaction component effects are confounded, then the other effects confounded can also be three-factor interaction component effects. If PQR and PR^2S are chosen to be confounded, then the other confounded effects are

$$(PQR)(PR^2S) = P^2QS \equiv PQ^2S^2$$

and

$$(PQR)(PR^2S)^2 = QRS^2.$$

This latter example illustrates the benefit of three-level factors instead of two-level factors for confounding. With four two-level factors, confounding two three-factor interactions implies that a two-factor interaction must be confounded. With three-level factors the existence of both PQR and P^2QR allows sufficient extra alternatives that we can avoid confounding any two-factor interaction.

The definition of the confounding system provides a unique design. The actual composition of treatment combinations in blocks is determined in exactly the same manner as for 2^n confounded designs. First, construct the principal block containing only treatment combinations in the first set of the three sets from each confounded effect. The required combinations $(p_iq_jr_k\ldots)$ satisfy

$$\alpha i + \beta j + \ldots = 0(3)$$
$$\alpha' i + \beta' j + \ldots = 0(3)$$
$$\ldots\ldots\ldots\ldots\ldots\ldots$$

for as many independent confounded effects as are defined. For example (2), the combinations would satisfy

$$i + j + k \equiv 0(3)$$

and

$$i + 2k + l \equiv 0(3).$$

Since PR is not confounded all (i, k) combinations must occur in each block. For each (i, k) combination, there will be unique values of j and l. The principal block will therefore be

$$p_0q_0r_0s_0 \quad p_0q_2r_1s_1 \quad p_0q_1r_2s_2$$
$$p_1q_2r_0s_2 \quad p_1q_1r_1s_0 \quad p_1q_0r_2s_1$$
$$p_2q_1r_0s_1 \quad p_2q_0r_1s_2 \quad p_2q_2r_2s_0.$$

Figure 16.5. A $3 \times 3 \times 3 \times 3$ design in nine blocks of nine units.

Block

I	II	III	IV	V
$p_0 q_0 r_0 s_0$	$p_0 q_1 r_0 s_0$	$p_0 q_2 r_0 s_0$	$p_0 q_0 r_0 s_1$	$p_0 q_0 r_0 s_2$
$p_0 q_2 r_1 s_1$	$p_0 q_0 r_1 s_1$	$p_0 q_1 r_1 s_1$	$p_0 q_2 r_1 s_2$	$p_0 q_2 r_1 s_0$
$p_0 q_1 r_2 s_2$	$p_0 q_2 r_2 s_2$	$p_0 q_0 r_2 s_2$	$p_0 q_1 r_2 s_0$	$p_0 q_1 r_2 s_1$
$p_1 q_2 r_0 s_2$	$p_1 q_0 r_0 s_2$	$p_1 q_1 r_0 s_2$	$p_1 q_2 r_0 s_0$	$p_1 q_2 r_0 s_1$
$p_1 q_1 r_1 s_0$	$p_1 q_2 r_1 s_0$	$p_1 q_0 r_1 s_0$	$p_1 q_1 r_1 s_1$	$p_1 q_1 r_1 s_2$
$p_1 q_0 r_2 s_1$	$p_1 q_1 r_2 s_1$	$p_1 q_2 r_2 s_1$	$p_1 q_0 r_2 s_2$	$p_0 q_0 r_2 s_0$
$p_2 q_1 r_0 s_1$	$p_2 q_2 r_0 s_1$	$p_2 q_0 r_0 s_1$	$p_2 q_1 r_0 s_2$	$p_2 q_1 r_0 s_0$
$p_2 q_0 r_1 s_2$	$p_2 q_1 r_1 s_2$	$p_2 q_2 r_1 s_2$	$p_2 q_0 r_1 s_0$	$p_2 q_0 r_1 s_1$
$p_2 q_2 r_2 s_0$	$p_2 q_0 r_2 s_0$	$p_2 q_1 r_2 s_0$	$p_2 q_2 r_2 s_1$	$p_2 q_2 r_2 s_2$

VI	VII	VIII	IX
$p_0 q_1 r_0 s_1$	$p_0 q_1 r_0 s_2$	$p_0 q_2 r_0 s_1$	$p_0 q_2 r_0 s_2$
$p_0 q_0 r_1 s_2$	$p_0 q_0 r_1 s_0$	$p_0 q_1 r_1 s_2$	$p_0 q_1 r_1 s_0$
$p_0 q_2 r_2 s_0$	$p_0 q_2 r_2 s_1$	$p_0 q_0 r_2 s_0$	$p_0 q_0 r_2 s_1$
$p_1 q_0 r_0 s_0$	$p_1 q_0 r_0 s_1$	$p_1 q_1 r_0 s_0$	$p_1 q_1 r_0 s_1$
$p_1 q_2 r_1 s_1$	$p_1 q_2 r_1 s_2$	$p_1 q_0 r_1 s_1$	$p_1 q_0 r_1 s_2$
$p_1 q_1 r_2 s_2$	$p_1 q_1 r_2 s_0$	$p_1 q_2 r_2 s_2$	$p_1 q_2 r_2 s_0$
$p_2 q_2 r_0 s_2$	$p_2 q_2 r_0 s_0$	$p_2 q_0 r_0 s_2$	$p_2 q_0 r_0 s_0$
$p_2 q_1 r_1 s_0$	$p_2 q_1 r_1 s_1$	$p_2 q_2 r_1 s_0$	$p_2 q_2 r_1 s_1$
$p_2 q_0 r_2 s_1$	$p_2 q_0 r_2 s_2$	$p_2 q_1 r_2 s_1$	$p_2 q_1 r_2 s_2$

The other blocks will satisfy other combinations of

$$i+j+k=0, 1 \text{ or } 2(3),$$

and

$$i+2k+l=0, 1 \text{ or } 2(3),$$

and these are generated most easily by cyclic rotation of the levels of j ($0 \rightarrow 1 \rightarrow 2 \rightarrow 0$) or of l, or both. The resulting design is shown in Figure 16.5.

16.5 Fractional replication

When the use of a design including only a fraction of all the possible treatment combinations was considered in Section 13.7, the requirements of such fractional replicate designs were specified in terms of being able to estimate the important effects. For each effect to be estimated it was assumed that all the combinations of levels for all factors included in the effect should occur equally frequently. Thus, for a main effect of a factor, all levels of that factor must occur equally frequently. For a two-factor interaction all combinations of levels for these two factors should occur equally frequently. This concept is essentially the same as that for allocating treatment combinations to blocks in confounded designs, and a fractional replicate design may be derived as a single block from a confounded experiment. The more formal development of the principles of confounding provides a system for identifying aliases of effects in fractional replication designs.

Suppose the effects which are confounded to produce the block which is then used for the fractional replicate design include an effect X. Then all combinations in the block will have the same relationship to X; that is, for a 2^n experiment, either all combinations in the block will be even with respect to X, or all combinations will be odd; for a 3^n experiment, the combinations will all be in one of the three sets defined by the effect X. An effect, X, used to define a fractional replicate in this way is called a *defining contrast*. Now consider the estimation of another effect, Y, from the fractional replicate which has been constructed as a single block from an experiment in which X has been confounded. The Y effect contrast, or contrasts, will be calculated from the totals for the combinations in the different sets defined by Y (evens and odds for 2^n; three sets, Y_0, Y_1, Y_2 for 3^n). But these sets of combinations will be identical with the sets of combinations defined by the effect, Z, which is the generalised interaction of X and Y. Hence, the estimate(s) of the effect(s) Z will be indistinguishable from those of Y or, in the earlier terminology, Z will be an alias of Y. The practical effect of this aliasing rule is shown through two examples:

Example 16.4

(1) In a 2^5 experiment, the half replicate defined by the five-factor interaction effect PQRST is used, so that the treatment combinations included in the experiment are:

(1)	pq	pr	qr
ps	qs	rs	pqrs
pt	qt	rt	pqrt
st	pqst	prst	qrst

The aliases of main effects and interactions are:

P $=$ QRST, Q $=$ PRST, R $=$ PQST, S $=$ PQRT, T $=$ PQRS

$$PQ = RST, \quad PR = QST, \quad PS = QRT, \quad PT = QRS, \quad QR = PST,$$
$$QS = PRT, \quad QT = PRS, \quad RS = PQT, \quad RT = PQS, \quad ST = PQR$$

which can be confirmed by formally writing out the various effects including only those combinations included in the experiment. Thus the full definitions of P and QRST are

$$16\,P = (p-1)(q+1)(r+1)(s+1)(t+1)$$
$$= [pqrst + pqr + pqs + pqt + prs + prt + pst + p]$$
$$+ [pqrs + pqrt + pqst + prst + pq + pr + ps + pt]$$
$$- [qrst + qr + qs + qt + rs + rt + st + (1)]$$
$$- [qrs + qrt + qst + rst + q + r + s + t]$$

and

$$16\,QRST = (p+1)(q-1)(r-1)(s-1)(t-1)$$
$$= [pqrst + pqr + pqs + pqt + prs + prt + pst + p]$$
$$- [pqrs + pqrt + pqst + prst + pq + pr + ps + pt]$$
$$+ [qrst + qr + qs + qt + rs + rt + st + (1)$$
$$- [qrs + qrt + qst + rst + q + r + s + t].$$

If the four groups of terms enclosed in brackets in the expressions above are abbreviated to

$$\{pqrst\}, \{pqrs\}, \{qrst\} \text{ and } \{qrs\}$$

then the two effects can be written

$$16\,P = (pqrst) + (pqrs) - (qrst) - (qrs)$$
$$16\,QRST = (pqrst) - (pqrs) + (qrst) - (qrs).$$

In the half replicate only the combinations (pqrs) and (qrst) are included. Hence the estimate of P is identical with the estimate of $-QRST$. If the other half replicate consisting of the combinations (pqrst) and (qrs) is also included in the experiment, we can estimate both P and QRST; if only this second half replicate is used, then P and QRST are again indistinguishable. If the definitions of P and QRST are written once more

$$16\,P = \quad (pqrs) - (qrst) \; + \; (pqrst) - (qrs) \; = \; 8L + 8K$$
$$16\,QRST = -[(pqrs) - (qrst)] + [(pqrst) - (qrs)] = -8L + 8K,$$

where

$$8L = (pqrs) - (qrst) \text{ and } 8K = (pqrst) - (qrs)$$

then the first half replicate provides an estimate L for P or for $-QRST$, while the other half replicate provides an estimate K for P or for $+QRST$. If the full replicate is used then $(L+K)/2$ would provide the estimate for P while $(K-L)/2$ would provide the estimate for QRST.

Example 16.5

(2) In a 3^4 experiment the one-third replicate defined by the component PQR^2S of the four-factor interaction is used. The treatment combinations to be used will be all those $p_i q_j r_k s_l$ such that

$$U(X) = i + j + 2k + l = 0(3).$$

Each of the 27 combinations of (i, j, k) will occur exactly once with the level of l determined by the $U(X)$ criterion. The treatment combinations are therefore those shown in Figure 16.6. The aliases of the effects or component effects will be:

$$
\begin{array}{llll}
P & \equiv P^2QR^2S \equiv QR^2S & Q & \equiv PQ^2R^2S \equiv PR^2S \\
R & \equiv PQS \quad\;\; \equiv PQRS & S & \equiv PQR^2S^2 \equiv PQR^2 \\
PQ \equiv P^2Q^2R^2S \equiv R^2S & & PQ^2 \equiv P^2R^2S \quad\; \equiv Q^2R^2S \\
PR \equiv P^2QS \quad\;\; \equiv QRS & & PR^2 \equiv P^2QRS \;\; \equiv QS \\
PS \equiv P^2QR^2S^2 \equiv QR^2 & & PS^2 \equiv P^2QR^2 \quad\; \equiv QR^2S^2 \\
QR \equiv PQ^2S \quad\;\; \equiv PRS & & QR^2 \equiv PQ^2RS \;\; \equiv PS \\
QS \equiv PQ^2R^2S^2 \equiv PR^2 & & QS^2 \equiv PQ^2R^2 \quad\; \equiv PR^2S^2 \\
RS \equiv PQS \quad\;\; \equiv PQR & & RS^2 \equiv PQ \quad\qquad \equiv PQRS^2.
\end{array}
$$

Thus, half of each two-factor interaction term is aliased with a half of another two-factor interaction. And, of course, because the effects used in 3^n confounding do not have a direct practical interpretation, this means that this design can fail to produce useful information about the two-factor interactions. Two methods might seem to be available to overcome this problem. One would be to try using a component effect of the three-factor interaction as the defining contrast for the one-third replicate. This,

Figure 16.6. A one-third replicate of a 3^4 structure defined by the contrast PQR^2S.

$$
\begin{array}{lll}
p_0 q_0 r_0 s_0 & p_0 q_0 r_1 s_1 & p_0 q_0 r_2 s_2 \\
p_0 q_1 r_0 s_2 & p_0 q_1 r_1 s_0 & p_0 q_1 r_2 s_1 \\
p_0 q_2 r_0 s_1 & p_0 q_2 r_1 s_2 & p_0 q_2 r_2 s_0 \\
p_1 q_0 r_0 s_2 & p_1 q_0 r_1 s_0 & p_1 q_0 r_2 s_1 \\
p_1 q_1 r_0 s_1 & p_1 q_1 r_1 s_2 & p_1 q_1 r_2 s_0 \\
p_1 q_2 r_0 s_0 & p_1 q_2 r_1 s_1 & p_1 q_2 r_2 s_2 \\
p_2 q_0 r_0 s_1 & p_2 q_0 r_1 s_2 & p_2 q_0 r_2 s_0 \\
p_2 q_1 r_0 s_0 & p_2 q_1 r_1 s_1 & p_2 q_1 r_2 s_2 \\
p_2 q_1 r_0 s_2 & p_2 q_2 r_1 s_0 & p_2 q_2 r_2 s_1
\end{array}
$$

however, merely leads to some main effects having components of two-factor interaction as aliases. The other alternative accepts the design structure as given, and assumes that the sum of squares due to main effects and the half components of each two-factor interaction, which are not ambiguous, will indicate which interactions are likely to be more substantial. The ambiguous effects can then be interpreted as caused by the more likely of the two aliases, and the interpretation can proceed accordingly. This approach may not be totally successful, but can lead to a partial interpretation of the underlying situation. Some of the practical aspects of interpretation have been illustrated in the example in Section 13.7.

Fractional replicate designs for mixed levels can also be constructed, the philosophy for the construction being essentially the same as for the confounding designs at the beginning of Chapter 15.

Example 16.6
Suppose we are considering a $2^3 \times 3^2$ treatment structure, and wish to use less than 72 observations. The most likely sizes smaller than 72 are 36 or 48 observations. The latter can be constructed most simply by considering the design for 24 observations and using the remaining 48 observations, since any contrasts which are aliased in the 24 observations will also be aliased in the 48 observations.

The design with 36 observations can clearly include all combinations of levels from four of the factors Q(2), R(2), S(3) and T(3), and the crucial stage of the design is the choice of the level of P(2) to combine with each qrst combination. Previous experience with confounding suggests that it should be possible to choose the levels of P to include all combinations of levels of
  (i) PQST
  (ii) PRST
and to include each of the PQR level combinations four or five times. The resulting set of treatment combinations for the design is shown in Figure 16.7. In the confounded design with two blocks of 36 combinations the effects which would have been partly confounded are PQR and PQRST. In the analysis of the confounded design, the PQR effect could be estimated by a non-orthogonal analysis. In the same manner the aliasing of $P \equiv QR$, $Q \equiv PR$ and $R \equiv PQ$ is partial, and both effects from each pair can be estimated from a non-orthogonal analysis. The effects P and QR, for example, are estimated with a correlation of 4/36 because the degree of departure from equal replication of all PQR combinations amounts to four observations in 36. This may be seen directly from the

Figure 16.7. A half replicate of a $2^3 \times 3^2$ structure.

$$
\begin{array}{llll}
p_0q_0r_0s_0t_0 & p_1q_0r_1s_0t_0 & p_1q_1r_0s_0t_0 & p_0q_1r_1s_0t_0 \\
p_1q_0r_0s_0t_1 & p_0q_0r_1s_0t_1 & p_0q_1r_0s_0t_1 & p_1q_1r_1s_0t_1 \\
p_0q_0r_0s_0t_2 & p_1q_0r_1s_0t_2 & p_1q_1r_0s_0t_2 & p_0q_1r_1s_0t_2 \\
p_1q_0r_0s_1t_0 & p_0q_0r_1s_1t_0 & p_0q_1r_0s_1t_0 & p_1q_1r_1s_1t_0 \\
p_0q_0r_0s_1t_1 & p_1q_0r_1s_1t_1 & p_1q_1r_0s_1t_1 & p_0q_1r_1s_1t_1 \\
p_1q_0r_0s_1t_2 & p_0q_0r_1s_1t_2 & p_0q_1r_0s_1t_2 & p_1q_1r_1s_1t_2 \\
p_0q_0r_0s_2t_0 & p_1q_0r_1s_2t_0 & p_1q_1r_0s_2t_0 & p_0q_1r_1s_2t_0 \\
p_1q_0r_0s_2t_1 & p_0q_0r_1s_2t_1 & p_0q_1r_0s_2t_1 & p_1q_1r_1s_2t_1 \\
p_0q_0r_0s_2t_2 & p_1q_0r_1s_2t_2 & p_1q_1r_0s_2t_2 & p_0q_1r_1s_2t_2 \\
\end{array}
$$

least squares equations for the effects P and QR:

$$36\hat{P} - \quad 4\hat{Q}R = (\text{total for } p_1) - (\text{total for } p_0)$$
$$-4\hat{P} + 36\hat{Q}R = (\text{total for } q_0r_0 \text{ and } q_1r_1) - (\text{total for } q_0r_1 \text{ and } q_1r_0).$$

16.6   Confounding in fractional replicates

The examples of fractional replication discussed so far have all been weak in one practical respect, and this is the absence of blocking. We have argued previously that it is rare for there to be no appropriate blocking structure among large sets of observations, and the examples of 27 and 36 sets of observations come in the category of large sets of observations.

The use of blocking within a fractional replicate causes no major conceptual problems. The intuitive approach of Chapter 15 is effective for treatment structures in which the total number of treatments is not large. For the important main effects and interactions each block of the fractional replicate must include equal replication of levels for each factor, and of combinations of levels for each important interaction.

Using the more formal approach of this chapter there are three distinct stages:

(i) Choose the fractional replicate without consideration of the later confounding choices. The fraction enabling the maximum number of relevant effects to be estimated with unimportant aliases should always be chosen.

(ii) Identify the group of effects which must be estimated together with the aliases of all such effects.

(iii) Choose a set of confounded effects such that none is an alias of an important effect and so that no generalised interaction of confounded effects is an alias of an important effect.

The choice of confounded effects is therefore complicated by the number of implications of each possible choice. However, the design problem in practical situations is not as complex as might appear from a general discussion, since the choices are always limited, and a careful listing of the possibilities can usually be simply achieved. Once again the general philosophy of the two approaches is illustrated through a series of examples as follows.

Example 16.7

For a 2^8 treatment structure 64 observations (a quarter replicate) are available and the block size is chosen to be 16 units. An initial consideration of the degrees of freedom structure for the analysis of variance reveals that the main effects and two-factor interactions account for 8 and 28 df respectively, leaving a further 27 df. After allowing for block and error degrees of freedom there are unlikely to be many degrees of freedom available for three-factor interactions, and it is probably sensible at this stage to abandon interest in three-factor interactions. With so many factors it is probably easier to design the experiment through the more formal approach.

First choose as the three defining contrasts the highest order interaction effects possible:

> PQRST, STUVW and the generalised interaction PQRUVW.

Examination of aliases for a subset of effects gives

$$
\begin{array}{llll}
\text{P} & \equiv \text{QRST} & \equiv \text{PSTUVW} & \equiv \text{QRUVW} \\
\text{S} & \equiv \text{PQRST} & \equiv \text{TUVW} & \equiv \text{PQRSUVW} \\
\text{U} & \equiv \text{PQRSTU} & \equiv \text{STVW} & \equiv \text{PQRVW} \\
\text{PQ} & \equiv \text{RST} & \equiv \text{PQSTUVW} & \equiv \text{RUVW} \\
\text{PS} & \equiv \text{QRT} & \equiv \text{PTUVW} & \equiv \text{QRSUVW} \\
\text{PU} & \equiv \text{QRSTU} & \equiv \text{PSTVW} & \equiv \text{QRVW} \\
\text{ST} & \equiv \text{PQR} & \equiv \text{UVW} & \equiv \text{PQRSTUVW.}
\end{array}
$$

All other main effects and two-factor interactions show the same patterns as one of these. These patterns reflect the structure of the defining contrasts which split the factors into three groups

> PQR; ST; UVW,

such that factors in a group have the same aliasing structure. That is, the aliases for the main effect of Q, like those for P, include a four-factor, a five-factor and a six-factor interaction; the aliases for the two-factor interaction QT (or SW) will include a three-factor, a five-factor and a six-factor interaction.

Table 16.5.

df	
8	Main effects
28	Two-factor interactions
18	Three-factor interactions in the pattern PSU
9	Three-factor interactions in the pattern PQU
63	Total

The confounded effects can be three-factor interactions or higher order interactions. We therefore consider the alias structure for three-factor interactions keeping the group structure for the factors in mind.

$$P \quad S \ U \equiv QR \quad T \ U \equiv P \quad T \quad VW \equiv QR \ S \ VW$$
$$PQ \quad U \equiv \quad R \ ST \ U \equiv PQ \ ST \ VW \equiv \quad R \quad VW$$

The first form of three-factor interaction includes one factor from each group. There are $3 \times 2 \times 3 = 18$ such three-factor interactions and they therefore account for 18 of the remaining 27 df. The second form of three-factor interaction has two factors from the first group with one from the third, and includes as an alias a similar three-factor interaction with the groups reversed. There are 3×3 such interactions and they account for the remaining 9 df. All the degrees of freedom in the analysis of variance are now identified (see Table 16.5). The confounding problem requires the choice of two of these 27 three-factor interactions such that the generalised interaction of the two is also an alias of a member of the 27. We observe that one of the aliases for the second group is of the form PQ ST VW, that is two factors from each group, and deduce that two three-factor interactions from the first group with no letter in common will satisfy all the requirements.

The design is therefore defined as follows:

defining contrasts	PQRST, STUV (implies PQRUVW)
confounded effects	$PSU \equiv QRTU \equiv PTVW \equiv QRSVW$
	$QTV \equiv PRSV \equiv QSUW \equiv PRTUW$
implying	$PQSTUV \equiv RUV \equiv PQW \equiv RSTW$

The principal block must be even w.r.t. all defining contrasts and confounded effects, and the resulting design is given in Figure 16.8.

Example 16.8
Consider again the one-third replicate of the 3^4, constructed in Example 16.5. Obviously blocks of 27 are too large and three blocks of nine units

Figure 16.8. A quarter replicate of a 2^8 structure in four blocks of 16 units.

Block

I	II	III	IV
(1)	pq	pr	qr
rsu	pqrsu	psu	qsu
qtw	ptw	pqrtw	rtw
qrstuw	prstuw	pqstuw	stuw
pqst	st	qrst	prst
pqrtu	rtu	qtu	ptu
psw	qsw	rsw	pqrsw
pruw	qruw	uw	pquw
pquv	uv	qruv	pruv
pqrsv	rsv	qsv	psv
ptuvw	qtuvw	rtuvw	pqrtuvw
prstvw	qrstvw	stvw	pqstvw
stuv	pqstuv	prstuv	qrstuv
rtv	pqrtv	ptv	qtv
qsuvw	psuvw	pqrsuvw	rsuvw
qrvw	prvw	pqvw	vw

seems the simplest blocking structure. The degrees of freedom structure for the analysis of variance gives

	df
P	2
Q	2
R	2
S	2
Remainder	18

The aliasing patterns are listed in the earlier example and the nine effects, $PQ(\equiv R^2S)$, PQ^2, PR, $PR^2(\equiv QS)$, $PS(\equiv QR^2)$, PS^2, QR, QS^2 and RS account for the remaining 18 df. The confounding effect must be chosen from one of these. If RS is chosen, then the 27 treatment combinations will be arranged in three blocks defined by

I (r_0s_0, r_1s_2, r_2s_1)
II (r_0s_1, r_1s_0, r_2s_2)
III (r_0s_2, r_1s_1, r_2s_0),

giving the design shown in Figure 16.9. Direct construction of the design by first allocating levels for P and Q to the three blocks of nine and then allocating the levels of R, followed by levels of S, would inevitably lead to the same design or to the design with R and S levels interchanged.

Figure 16.9. A one-third replicate of a 3^4 structure in three blocks of nine units.

Block

I	II	III
$p_0 q_0 r_0 s_0$	$p_0 q_0 r_2 s_2$	$p_0 q_0 r_1 s_1$
$p_0 q_1 r_2 s_1$	$p_0 q_1 r_1 s_0$	$p_0 q_1 r_0 s_2$
$p_0 q_2 r_1 s_2$	$p_0 q_2 r_0 s_1$	$p_0 q_2 r_2 s_0$
$p_1 q_0 r_2 s_1$	$p_1 q_0 r_1 s_0$	$p_1 q_0 r_0 s_2$
$p_1 q_1 r_1 s_2$	$p_1 q_1 r_0 s_1$	$p_1 q_1 r_2 s_0$
$p_1 q_2 r_0 s_0$	$p_1 q_2 r_2 s_2$	$p_1 q_2 r_1 s_1$
$p_2 q_0 r_1 s_2$	$p_2 q_0 r_0 s_1$	$p_2 q_0 r_2 s_0$
$p_2 q_1 r_0 s_0$	$p_2 q_1 r_2 s_2$	$p_2 q_1 r_1 s_1$
$p_2 q_2 r_2 s_1$	$p_2 q_2 r_1 s_0$	$p_2 q_2 r_0 s_2$

Example 16.9

Finally, consider again the design problem of Example 16.6, the half replicate of the $2^3 \times 3^2$ for which three blocks of 12 units might be an appropriate blocking structure. This revised design problem re-emphasises a result encountered previously in both approaches to confounding. When the restrictions are changed it is not usually appropriate to modify the original design but to consider the new problem directly. If we attempt to allocate the 36 combinations in the optimal half replicate of Example 16.6 to three blocks of 12 by allocating all QRS combinations to each block, adding the T levels so as to avoid confounding QRT and minimising the confounding of ST, then the distribution of levels of P between blocks is extremely uneven (ten:two split in one block). The half replicate of the $2^3 \times 3^2$ in three blocks of 12 must therefore be constructed directly.

Plainly all main effects can be allocated to be unconfounded, and it would seem possible to try to keep the two-factor interactions PQ, PR, PS, PT, QR, QS, QT, RS, RT unconfounded. In constructing the design there is no point in avoiding confounding three-factor interactions between the three blocks because, as was pointed out in the earlier example, there are not enough degrees of freedom to estimate three-factor interactions. However, all combinations of levels for each three-factor interaction should be included as nearly equally in the total 36 combinations as is possible to avoid introducing aliases of important effects for other important effects.

Two alternative designs are shown in Figure 16.10. The choice between

Figure 16.10. Two alternative designs for a half replicate of a $2^3 \times 3^2$ structure in three blocks of 12 units.

(a)

Block		
I	II	III
$p_0 q_0 r_0 s_0 t_0$	$p_1 q_0 r_0 s_0 t_1$	$p_0 q_0 r_0 s_0 t_2$
$p_1 q_0 r_0 s_1 t_1$	$p_0 q_0 r_0 s_1 t_2$	$p_1 q_0 r_0 s_1 t_0$
$p_0 q_0 r_0 s_2 t_2$	$p_1 q_0 r_0 s_2 t_0$	$p_0 q_0 r_0 s_2 t_1$
$p_1 q_0 r_1 s_0 t_0$	$p_0 q_0 r_1 s_0 t_1$	$p_1 q_0 r_1 s_0 t_2$
$p_0 q_0 r_1 s_1 t_1$	$p_1 q_0 r_1 s_1 t_2$	$p_0 q_0 r_1 s_1 t_0$
$p_1 q_0 r_1 s_2 t_2$	$p_0 q_0 r_1 s_2 t_0$	$p_1 q_0 r_1 s_2 t_1$
$p_1 q_1 r_0 s_0 t_0$	$p_0 q_1 r_0 s_0 t_1$	$p_1 q_1 r_0 s_0 t_2$
$p_0 q_1 r_0 s_1 t_1$	$p_1 q_1 r_0 s_1 t_2$	$p_0 q_1 r_0 s_1 t_0$
$p_1 q_1 r_0 s_2 t_2$	$p_0 q_1 r_0 s_2 t_0$	$p_1 q_1 r_0 s_2 t_1$
$p_0 q_1 r_1 s_0 t_0$	$p_1 q_1 r_1 s_0 t_1$	$p_0 q_1 r_1 s_0 t_2$
$p_1 q_1 r_1 s_1 t_1$	$p_0 q_1 r_1 s_1 t_2$	$p_1 q_1 r_1 s_1 t_0$
$p_0 q_1 r_1 s_2 t_1$	$p_1 q_1 r_1 s_2 t_0$	$p_0 q_1 r_1 s_2 t_1$

(b)

Block		
I	II	III
$p_0 q_0 r_0 s_0 t_0$	$p_0 q_0 r_0 s_0 t_1$	$p_0 q_0 r_0 s_0 t_2$
$p_1 q_0 r_0 s_1 t_1$	$p_1 q_0 r_0 s_1 t_2$	$p_1 q_0 r_0 s_1 t_0$
$p_0 q_0 r_0 s_2 t_2$	$p_0 q_0 r_0 s_2 t_0$	$p_0 q_0 r_0 s_2 t_1$
$p_1 q_0 r_1 s_0 t_1$	$p_1 q_0 r_1 s_0 t_2$	$p_1 q_0 r_1 s_0 t_0$
$p_0 q_0 r_1 s_1 t_2$	$p_0 q_0 r_1 s_1 t_0$	$p_0 q_0 r_1 s_1 t_1$
$p_1 q_0 r_1 s_2 t_0$	$p_1 q_0 r_1 s_2 t_1$	$p_1 q_0 r_1 s_2 t_2$
$p_1 q_1 r_0 s_0 t_2$	$p_1 q_1 r_0 s_0 t_0$	$p_1 q_1 r_0 s_0 t_1$
$p_0 q_1 r_0 s_1 t_0$	$p_0 q_1 r_0 s_1 t_1$	$p_0 q_1 r_0 s_1 t_2$
$p_1 q_1 r_0 s_2 t_1$	$p_1 q_1 r_0 s_2 t_2$	$p_1 q_1 r_0 s_2 t_0$
$p_0 q_1 r_1 s_0 t_0$	$p_0 q_1 r_1 s_0 t_1$	$p_0 q_1 r_1 s_0 t_2$
$p_1 q_1 r_1 s_1 t_1$	$p_1 q_1 r_1 s_1 t_2$	$p_1 q_1 r_1 s_1 t_0$
$p_0 q_1 r_1 s_2 t_2$	$p_0 q_1 r_1 s_2 t_0$	$p_0 q_1 r_1 s_2 t_1$

them depends on the relative importance of obtaining some information on ST, and of correlation between the estimates of P and QR. The first design, shown in Figure 16.10(a), is a different half replicate from that used earlier, but it still includes all combinations of PQST, of PRST and of QRST and the partial defining contrasts are still PQR and PQRST. The ST interaction has 2 df completely confounded between blocks and the partial aliasing of P with QR (and of course Q with PR and R with PQ) occurs to the same extent as in Example 16.6.

The second alternative shown in Figure 16.10(b) allows some information on the ST interaction which is partially confounded with blocks at the cost of increasing the correlation between $\hat{P}$ and $\hat{QR}$.

16.7   Confounding in row and column designs

Confounding has been discussed so far only in the context of a single blocking classification. However, the principles of confounding can be applied when there are two orthogonal systems of blocks, which, as in Chapter 8, will be referred to as rows and columns. For the simpler forms of factorial structure, it will usually be most appropriate to use the mathematical approach of this chapter. The more intuitive approach of the previous chapter provides the more suitable means of controlling confounding for less simply structured problems. The philosophy of confounding in row and column designs is presented through four examples.

The general theory of confounding for two blocking classifications is less developed than for a single blocking structure. The standard mathematical theory of confounding in a single blocking classification is sufficient for single replicates of factorial structures which have equal numbers of levels for all factors, using a rectangular array both of whose dimensions are powers of the number of levels per factor. This is, of course, a very restricted set of situations, and the results are not of much practical importance, but they do illuminate the general pattern of such problems. Two confounding systems are defined, one for the rows and one for the columns, such that the two sets of confounded effects are entirely distinct. The allocation of treatment combinations to rows and to columns in accordance with the double confounding requirements provides a uniquely defined design.

Example 16.10

If a 2^5 structure is to be investigated using a unit structure in four rows × eight columns, then a column confounding system of seven effects and a separate row confounding system of three effects are required.

Figure 16.11. Division of 2^5 combinations into eight columns confounding PQR, PRS and PST.

I	II	III	IV	V	VI	VII	VIII
(1)	p	q	r	s	t	qr	qt
prt	rt	pqrt	pt	prst	pr	pqt	pqr
qrst	pqrst	rst	qst	qrt	qrs	st	rs
pqs	qs	ps	pqrs	pq	pqst	prs	pst

Suitable sets of effects could be

PQR, PRS, PST, QS, QRST, RT, PQT,

for columns, and

PQRS, PRT and QST

for rows. The treatment combinations in the eight columns are quickly derived as shown in Figure 16.11. Because each of the row confounded effects is orthogonal to all of the column confounded effects, the combinations in each column must be rearrangeable into four rows such that combinations are

(1) even w.r.t. PQRS and PRT,
(2) even w.r.t. PQRS and odd w.r.t. PRT,
(3) odd w.r.t. PQRS and even w.r.t. PRT,
(4) odd w.r.t. PQRS and PRT.

The resulting unrandomised design is shown in Figure 16.12. The randomisation procedure would use permutations of rows and columns as in Chapter 9. A consequence of the method of construction with the combination (1) in row 1 column 1 is that the element in row i and column j is the 'product' of the two combinations in row 1 and column j, and row i and column 1, where the 'product' is generalised in the same sense as the generalised interaction of multiple confounding theory.

In the analysis of variance, effects confounded with rows or columns cannot be estimated. The analysis of variance structure for the design is

Figure 16.12. A 2^5 structure arranged in four rows × eight columns.

Column

	1	2	3	4	5	6	7	8
Row 1	(1)	qs	rst	pqrs	qrt	pr	pqt	pst
2	prt	pqrst	ps	qst	pq	t	qr	rs
3	qrst	rt	q	pt	s	pqst	prs	pqr
4	pqs	p	pqrt	r	prst	qrs	st	qt

Table 16.6.

Source	df
Rows	3
Columns	7
Main effects	5
Two-factor interactions (not QS and RT)	8
Error	8
Total	31

therefore as shown in Table 16.6. Designs with more than one replicate can be constructed by combining several single replicate designs with partial confounding. However, if the column (or row) effects might be expected to be consistent across rectangles, then it would be preferable not to employ partial confounding for *both* rows and columns and this creates additional problems.

Example 16.11
Consider first a design in two replicates with each replicate treated completely separately. In the design shown in Figure 16.13 the first

Figure 16.13. Two replicates of a 2^5 structure arranged in four rows × eight columns with different confounding systems.

Replicate 1

	Column							
	1	2	3	4	5	6	7	8
Row 1	(1)	qs	rst	pqrs	qrt	pr	pqt	pst
2	prt	pqrst	ps	qst	pq	t	qr	rs
3	qrst	rt	q	pt	s	pqst	prs	pqr
4	pqs	p	pqrt	r	prst	qrs	st	qt

Replicate 2

	Column							
	9	10	11	12	13	14	15	16
Row 5	(1)	ps	qt	rst	pqst	prt	qrs	pqr
6	prs	r	pqrst	pt	qrt	st	pq	qs
7	qst	pqt	s	qr	p	pqrs	rt	prst
8	pqrt	qrst	pr	pqs	rs	q	pst	t

Table 16.7.

Source	Based on	df
Replicates		1
Columns (within replicates)		14
Rows (within replicates)		6
Main effects	both replicates	5
QS, RT	2nd replicate	2
PR, QT	1st replicate	2
Other two-factor interactions	both replicates	6
PQS, QRS, QRT, RST	1st replicate	4
PQR, PQT, PRT, QST	2nd replicate	4
Error		19
Total		63

replicate uses the same confounding systems as in Example 16.10,

columns	PQR	PRS	PST	QS	RT	QRST	PQT
rows	PQRS	PRT	QST.				

The second replicate, has confounding systems:

columns	PQS	QRS	PST	PR	QT	PQRT	RST
rows	PRS	QRT	PQST.				

Because the design involves partial confounding, some of the sums of squares in the analysis of variance are calculated from only one of the two replicates. The use of two replicates allows information on some of the three-factor interactions to be calculated if that is considered useful. The structure of the analysis of variance is shown in Table 16.7.

Example 16.12

Now suppose that the two replicates are to be arranged within a unit structure of eight rows × eight columns. It would be possible to use the same column confounding system (PQR, PRS, PST, QS, QRST, RT, PQT) with one row confounding system (PQRS, PRT, QST) for the first replicate in the first four rows and a different row confounding system (PRST, PQS, QRT) in the last four rows. However, this still leaves the two-factor interactions QS and RT totally confounded with columns, even though with eight units in each column it would seem reasonable to hope that it would not be necessary to confound any two-factor interactions.

To confound groups of only three effects between columns and between rows, assuming that this is possible because each column or row contains

Figure 16.14. Basic confounding system for two replicates of 2^5 structure in eight rows × eight columns.

PQR	even	even	odd	odd
PST	even	odd	even	odd

PRT	QST				
even	even	$(1)_{pqt}$	pr_{qrt}	$pqrs_{rst}$	qs_{pst}
even	odd	$qrst_{prs}$	$pqst_s$	pt_q	rt_{pqr}
odd	even	st_{pqs}	$prst_{qrs}$	$pqrt_r$	qt_p
odd	odd	qr_{prt}	pq_t	ps_{qst}	rs_{pqrst}

eight combinations, we may proceed as follows. Suppose PQR, PST and QRST are to be confounded between columns, and PRT, QST and PQRS with rows, then the design is not fully defined since each system has only four classes. The allocation defined by the confounding systems is shown in Figure 16.14. Each 'row' and 'column' combination contains two treatment combinations. If the full 8 × 8 unit structure is divided into four quarters, each 4 × 4, then the allocation in Figure 16.14 can be split into two 4 × 4 structures. The obvious way of splitting the pairs is on the even/odd value for the five-factor interaction PQRST. Thus the 'even' treatment combination of each pair is allocated to the quarters in the north-west and south-east corners. After randomising the orders of rows and of columns the resulting design is shown in Figure 16.15.

The confounding method of choosing three effects to confound with rows, three to confound with columns and one to confound with quarters

Figure 16.15. Final design for 2^5 in eight rows × eight columns.

Column

		1	2	3	4	5	6	7	8
Row	1	prt	t	ps	qr	qst	pq	pqrst	rs
	2	pqt	qrt	pqrs	(1)	rst	pr	pst	qs
	3	qrst	pqst	q	prs	pt	s	rt	pqr
	4	pqs	qrs	pqrt	st	r	prst	p	qt
	5	prs	s	pt	qrst	q	pqst	pqr	rt
	6	qr	pq	qst	prt	ps	t	rs	pqrst
	7	(1)	pr	rst	pqt	pqrs	qrt	qs	pst
	8	st	prst	r	pqs	pqrt	qrs	qt	p

Table 16.8.

Source	df
Rows	7
Columns	7
Main effects	5
Two-factor interactions	10
Non-confounded three-factor interactions	6
Error	28
Total	63

(before randomisation) implies that there is a further confounding of effects with half rows in quarters, or half columns in quarters. The effects confounded in this manner include main effects $P(= PQRST \times QRST)$ and $T(= PQRST \times PQRS)$ and the results can still be seen after randomisation. In each column t always occurs in the first three rows and the last row (columns 1, 2, 5 and 7), or in rows 4 to 7 (columns 3, 4, 6, 8). However, if the additive model for the row and column design correctly represents the blocking structure then this 'confounding' is not practically relevant because the 'confounding blocks' are not represented by terms in the assumed model. (This is one design where randomisation theory would diverge from the infinite population model theory. Because the randomisation procedure confounds effects with semi-columns × rows or with semi-rows × columns, the effects confounded with these 'blocks' should be assessed relative to the variation between semi-rows within columns or semi-columns within rows, requiring two further levels of variation within the analysis of variance, similar to the criss-cross design. If this is felt to be important then a better choice than PQRST for confounding between quarters would be PQT.) The analysis of variance structure for the design of Figure 16.15 is shown in Table 16.8.

Example 16.13
The final example, quoted at the beginning of the chapter, is a design problem of a more general form, encountered during consultancy, and belongs in spirit more to Chapter 15, though it involves confounding both between rows and between columns. The experimental problem is to investigate absorption by rabbits of sugar occurring in various forms and amounts in the rabbits' diets. The experimental treatments selected have a 2^3 structure, being two concentrations (C) × two forms (F) × presence or absence of an additive (A). The experimental units are short periods of

three weeks for each rabbit. Five rabbits are available for the experiment and each may be treated and observed for the same five time periods. There will definitely be consistent differences between rabbits, and there may be consistent differences between the five time periods. The design problem is to arrange a 2^3 treatment structure within a 5×5 row and column structure. Main effects should be most efficiently estimated (least affected by row or column effects) but two-factor interactions are also of interest.

From the general results for row and column designs in Chapter 8, we know that precision of main effect comparisons depends on the frequency of occurrence, together within a row or column, of the two levels to be compared. Therefore we seek to construct a design such that, for each main effect, the two levels shall occur most nearly equally in each row or column. In other words each row and each column should include two or three replicates of each level for each of the three factors. A design which satisfies these requirements is shown in Figure 16.16. It is not known whether this design is, in any sense, optimal, but it appears to be sensible, and illustrates the possibility of constructing particular designs to satisfy particular requirements.

By maximising the 'equality' of occurrence for main effect levels, interaction contents may be less even. For example, the CF interaction comparison $(cf+(1))-(c+f)$ has $+$ and $-$ terms allocated in rows in the proportions $(\frac{3}{2}, \frac{1}{4}, \frac{3}{2}, \frac{3}{2}$ and $\frac{2}{3})$, while the proportions in the columns are $(\frac{3}{2}, \frac{3}{2}, \frac{3}{2}, \frac{2}{3}$ and $\frac{2}{3})$. These proportions are obviously not very disparate so that the design will provide considerable information about two-factor interactions.

All treatment effects are not orthogonal to rows or columns so the order of fitting is important. The analysis of variance structure is shown in Table 16.9.

Figure 16.16. A 2^3 design in five rows × five columns.

		Rabbit			
		1	2	3	4	5
Time period	1	(1)	acf	a	f	c
	2	ac	f	cf	c	af
	3	cf	a	ac	af	(1)
	4	af	ac	(1)	acf	cf
	5	acf	c	f	a	ac

Table 16.9.

Source	df
Rows	4
Columns	4
C	1
F	1
A	1
CF	1
CA	1
FA	1
CFA	1
Error	9
Total	24

Exercises 16

(1) An experimenter is working on the effects of various chemicals on fungal growths on leaves. The natural experimental unit is a half leaf, and he can either use a leaf (=two half leaves) as a block, or use a pair of leaves (=four half leaves) as a block. Clearly the random variance will be larger in the latter case. The experimenter proposes to examine the effects of five chemicals using a 2^5 factorial structure, with the two levels for each chemical being presence or absence. He is interested in the main effect of each chemical and also in whether there are any two-factor interactions.

Devise suitable confounding schemes for a single replicate of a 2^5 factorial

(i) in eight blocks of four units,

(ii) in 16 blocks of two units, avoiding confounding any main effect,

(iii) in 16 blocks of two units confounding one main effect, and as few two-factor interactions as possible.

If sufficient material is available for five replicates, explain why (ii) is undesirable. Also discuss possible partial confounding schemes in five replicates for (i) and (iii), and explain how you would choose between schemes based on (i) and (iii) in terms of the random variances σ_1^2 for (i) and σ_2^2 for (iii).

(2) A factorial experiment with five two-level factors A, B, C, D and E is arranged in blocks of eight plots per replication, certain interactions being confounded with block differences. The same interactions are confounded in each replication. The treatment combinations tested in one of the blocks are

a, b, cd, ce, ade, bde, abcd, abce,

where, for example, the symbol ade denotes the combination of factors A, D and E at their upper levels, and B and C at their lower levels. Which interactions are confounded?

(3) Determine which effects can be estimated and which are confounded with blocks in each of the following designs:

(i) Three factors, N (four levels), T (four levels) and S (two levels), see Figure 16.17.

Figure 16.17. A $4 \times 4 \times 2$ design confounded in four blocks of eight units each.

Block

I	II	III	IV
$n_0 t_0 s_1$	$n_0 t_0 s_0$	$n_0 t_1 s_0$	$n_0 t_1 s_1$
$n_0 t_3 s_0$	$n_0 t_3 s_1$	$n_0 t_2 s_1$	$n_0 t_2 s_0$
$n_1 t_1 s_0$	$n_1 t_1 s_1$	$n_1 t_0 s_1$	$n_1 t_0 s_0$
$n_1 t_2 s_1$	$n_1 t_2 s_0$	$n_1 t_3 s_0$	$n_1 t_3 s_1$
$n_2 t_1 s_1$	$n_2 t_1 s_0$	$n_2 t_0 s_0$	$n_2 t_0 s_1$
$n_2 t_2 s_0$	$n_2 t_2 s_1$	$n_2 t_3 s_1$	$n_2 t_3 s_0$
$n_3 t_0 s_0$	$n_3 t_0 s_1$	$n_3 t_1 s_1$	$n_3 t_1 s_0$
$n_3 t_3 s_1$	$n_3 t_3 s_0$	$n_3 t_2 s_0$	$n_3 t_2 s_1$

(ii) Seven factors, A, B, C, D, E, F and G, with two levels each in four blocks of 32 units. One block (of the four) is shown in Figure 16.18.

Figure 16.18. One block of a 2^7 design in four blocks of 32 units each.

fg	bf	af	abfg
e	beg	aeg	abe
dg	bd	ad	abdg
def	bdefg	adefg	abdef
cg	bc	ac	abcg
cef	bcefg	acefg	abcef
cdfg	bcdf	acdf	abdefg
cde	bcdeg	acdeg	abcde

(4) A single replicate of a 3^3 experiment on the effects of three levels of nitrogen, three levels of phosphorus and three levels of potash on germination of lettuce seedlings gave the results as in Figure 16.19 and Table 16.10.

Figure 16.19. Design and data for a single replicate of a 3^3 experiment.

Block

I	II	III
$n_0 p_0 k_0$ 41	$n_0 p_0 k_2$ 40	$n_0 p_0 k_1$ 53
$n_0 p_1 k_2$ 11	$n_0 p_1 k_1$ 38	$n_0 p_1 k_0$ 54
$n_0 p_2 k_1$ 30	$n_0 p_2 k_0$ 44	$n_0 p_2 k_2$ 41
$n_1 p_0 k_1$ 11	$n_1 p_0 k_0$ 61	$n_1 p_0 k_2$ 48
$n_1 p_1 k_0$ 21	$n_1 p_1 k_2$ 39	$n_1 p_1 k_1$ 46
$n_1 p_2 k_2$ 11	$n_1 p_2 k_1$ 20	$n_1 p_2 k_0$ 25
$n_2 p_0 k_2$ 12	$n_2 p_0 k_1$ 42	$n_2 p_0 k_0$ 35
$n_2 p_1 k_1$ 21	$n_2 p_1 k_0$ 24	$n_2 p_1 k_2$ 37
$n_2 p_2 k_0$ 13	$n_2 p_2 k_2$ 46	$n_2 p_2 k_1$ 58
171	354	394

Table 16.10. *Two-way treatment totals.*

	n_0	n_1	n_2	p_0	p_1	p_2	Total
k_0	139	107	69	134	99	82	315
k_1	121	77	121	106	105	108	319
k_2	92	98	95	100	87	98	285
p_0	134	120	86	340	291	288	
p_1	103	106	82				
p_2	115	56	117				
Total	352	282	285				

$\sum y^2 = 37425$

Calculate the analysis of variance and report your conclusions.

(5) You are asked to design an experiment (on wheat) to investigate the effects of water stress or irrigation during each of three periods (three two-level factors, P_1, P_2, P_3), four rates of nitrogen application (four-level factor N) and two types of nitrogen (two-level factor, T). The experimenter has 64 plots available, and

considers the maximum block size is 16 plots. Construct a single replicate of the $2^4 \times 4$ experiment in four blocks, bearing in mind the experimenter's preference (in order of decreasing importance):

(a) no main effects to be confounded,

(b) all comparisons between the eight combinations of water stress and irrigation to be unconfounded,

(c) at least 12 df left for error,

(d) all two-factor interactions to be unconfounded,

(e) three- and four-factor interactions involving two or three of the water stress factors (P's) to be unconfounded.

Outline the analysis of your design.

(6) In a 2^6 experiment with factors A, B, C, D, E and F all three-factor and higher interactions may be assumed negligible as may the two-factor interactions involving either A or C. Construct a quarter replicate which allows main effects and other two-factor interactions to be separated. Suggest a suitable blocking scheme for a quarter replicate in two blocks of eight units.

(7) Figure 16.20 gives the field plan in standard notation for a half replicate of a 2^6 design arranged in four blocks of eight. What is the defining contrast for this half replicate? What interactions are confounded with blocks?

Figure 16.20. Half-replicate of a 2^6 design in four blocks of eight units each.

Block I				Block II			
abce	bf	cdef	af	abcf	cd	ae	bdef
abcdef	ad	bd	ce	adef	cf	be	abcd

Block III				Block IV			
abde	ef	ac	bc	bcef	acef	acde	(1)
de	acdf	bcdf	abef	ab	bcde	df	abdf

Construct the analysis of variance for this design (you may assume that the three-, four-, five- and six-factor interactions are zero). Show clearly which treatment contrasts are used to estimate the residual variance, and how the three-factor interactions are accounted for in the analysis.

(8) An experiment is to be designed to compare different recipes for a cake mixture. Three treatment factors are proposed:

A = level of a milk product   (two levels),
B = level of a starch product   (two levels),
C = level of an emulsifier   (three levels).

Each of the 12 cake mix combinations is to be replicated four times. On each of two days cakes could be baked at each of 12 different times, and at each time two cakes could be baked, one in the left side of the oven, the other in the right side. There may be consistent differences between the two sides, and also between the 12 times within a day. The unit structure is therefore 12 rows (times) × two columns (sides) repeated on two days. How do you allocate the 12 treatment combinations so that the three main effects and the AB interaction are all estimated as efficiently as possible?

17

Quantitative factors and response functions

17.0   *Preliminary examples*

(*a*) A research student was proposing an extensive investigation of the decay of seed germination over time. The time interval over which the change of germination rate was to be examined was two years. Seeds would be treated in various different ways, and seeds from many different sources would be used. The major design questions were how often to test the germination rate during the two years, and how many seeds to use on each test occasion. The student, when questioned about the pattern of decay with time, was adamant that all previous data showed clearly that the relationship of germination rate with time was linear. For each combination of seed source and treatment, about 2000 seeds would be available. How should the 2000 seeds be sampled over the two-year period?

(*b*) In a rather similar investigation into the decline of strength of weldings used for oil rigs, again looking at the pattern over a long period of stress, the proposed design on which comments were requested was that which was regarded by the engineers concerned as too obvious to require statistical advice. As in the seed germination investigation, it was believed absolutely that the relationship of strength with level of stress was linear. The only analysis which was contemplated was to fit a straight line regression of the variable measuring strength on the length of time for which the stress was applied. The 'obvious' design proposed by the engineers was to use equal replication for eight, equally spaced, durations of stress. Can the despised statistician offer any improvement?

17.1 The use of response functions in the analysis of data

When an experiment involves levels of a quantitative factor, it is essential to examine the pattern of response of the observed variable to the changing stimulus of the factor. It is almost always appropriate, as a first step, to plot mean yields against the factor level. This will often be only a first step, and frequently we should seek to draw inferences about the pattern of response by fitting an appropriate response function, and interpreting the response in terms of the values of the parameters of the response function. The analysis may be taken further, with the comparison of fitted response functions for several levels of other factors included in the treatment structure.

First we consider the set of available and appropriate response functions. There are different levels of appropriateness. There is always a level in the modelling of a biological process at which any simple algebraic model is inadequate to represent the true biology. At the other extreme, there are models which essentially attempt no more than a crude smoothing of the data to provide a very simple summary of the data. Between these two extremes there is the main area of the development of algebraic models which are judged by whether their behaviour is compatible with general or specific patterns of response believed to hold for the biological system being investigated, and by whether the models fit the data, using statistical assessment of the fit.

The simplest response model used is the straight line

$$y = \alpha + \beta x.$$

This is hardly ever considered as a practically realistic model in spite of the two examples at the beginning of this chapter, but is sometimes useful for calculating an average rate of increase of y with x over a limited range of x values for which the linear relationship is approximately true.

The simplest, and historically the most used, curved relationship is the quadratic

$$y = \alpha + \beta x + \gamma x^2.$$

The disadvantages of this model are that it is too restrictive to be realistic. Specifically, if extrapolated beyond the range of the data, it rapidly gives impossible values. For the most common response pattern with β positive and γ negative, and a maximum value of y at $x = -\beta/2\gamma$, negative values of y are implied when x differs from this value by more than $(\beta^2 - 4\alpha\gamma)^{\frac{1}{2}}/2\gamma$. A second disadvantage is the symmetry of the quadratic about the optimum. It will rarely be credible in a practical situation that the mechanisms governing the increase of y to its maximum value will be mirrored by the mechanisms controlling the decrease beyond the maximum. The third disadvantage of the quadratic curve is that it does not include an asymptotic form of response; that is a response in which y increases (or decreases) monotonically but at a gradually reducing rate so that, as x increases, y tends to a limiting value.

The symmetry disadvantages can be overcome by considering powers of x, and possibly of y. To obtain curves which also satisfy the other objections, we have to use models which are more complex in a fundamental way than those considered so far. The straight line, and the quadratic, with or without pre-specified powers, are all linear response functions in the sense that the parameters occur in a linear form, though if a power is treated as a parameter to be estimated the model is no longer linear. This linearity of parameters is important because it leads to a very simple form of estimation using the general linear model ideas of Chapter 4. This ease of estimation was extremely important before computers were available, and, although non-linear models were used, it was necessary to devise clever and complex methods of fitting the models to data. Consequently, the quadratic response was used frequently when not appropriate because of the difficulty of fitting a more appropriate model.

There are now fewer computational difficulties in the use of non-linear models, and the quadratic model should retreat to a less prominent position, being mainly used as a smoothing curve of rather greater flexibility than the straight line. The two major families of non-linear curves which provide greater flexibility in modelling response patterns are based on exponential and on reciprocal relationships, respectively.

First, the exponential relationships. These have been used extensively for asymptotic responses. The simplest form is

$$y = A(1 - e^{-kx}),$$

the Mitscherlich response used extensively for the response of crop yield to varying fertiliser level. A corresponding form for decay curves in chemical and biological processes, or in medical survival studies, is

$$y = Ae^{-kx}.$$

Numerous extensions of these models have been used for particular situations. A general form including both the previous particular cases is

$$y = A + be^{-kx}.$$

More exotic forms of exponential relationships include

$$y = xAe^{kx}$$

used by Carmer and Jackobs (1965), the Gompertz curve

$$y = y_0 A^{B^x}$$

used extensively in growth studies and, perhaps the most exotic of all,

$$y = A - B \exp(1 - kx^\theta)$$

used by Reid (1972) to describe the response of ryegrass swards to a very wide range of nitrogen levels.

Reciprocal relationships, of which the simplest form is

$$y = \beta_0 + \beta_1(x - x_0)^{-1},$$

have been used in many areas of research, and have been justified on theoretical grounds in some cases of biological research, notably in kinetic relationships in enzyme chemistry (the Michaelis–Menten equation) and in crop density–yield relations in agronomy (the 'law of constant final yield'; Hozumi, Asahira and Kira, 1956). Reciprocal relationships have been generalised to form the class of inverse polynomials

$$x/y = \text{polynomial in } x$$

by Nelder (1966). Particular cases which have been found useful are the inverse linear

$$1/y = \beta_0/x + \beta_1,$$

an asymptotic curve much used in yield–density relations (after Hozumi *et al.*) and the inverse quadratic

$$1/y = \beta_0/x + \beta_1 + \beta_2 x,$$

which gives a non-symmetric curve with an optimum (at $x = (\beta_0/\beta_2)^{\frac{1}{2}}$). An alternative form of non-symmetric optimum curve is

$$1/y^\theta = \beta_0/x + \beta_1,$$

which has been justified, on theoretical biological grounds, for yield–density relations for agronomic studies.

One other group of relationships which continues to receive attention (in spite of determined efforts by *some* statisticians) is the 'broken stick' model in which the response is modelled by two straight lines

$$y = \alpha_0 + \beta_0 x \qquad\qquad x \leqslant x_1$$
$$y = \alpha_0 + \beta_0 x_1 + \beta_1(x - x_1) \qquad x \geqslant x_1,$$

where x_1 is a fourth parameter.

17.2   Design objectives

Fitting a response model to experimental data may be appropriate for various different reasons. Correspondingly, we can define several different objectives for model fitting. As the objective of the fitting varies, so will the choice of treatment levels (the design) which best satisfies the objective. Various authors have produced lists of possible objectives, notably Box and Draper (1974). To identify which objectives are most important, we must re-examine our motives in fitting response models. These may include:

(a) supplying a summary of data for the purpose of presentation,

(b) examining the appropriateness of a postulated model,

(c) estimating the parameters or functions of parameters of a well established response function to provide biological or chemical information,

(d) providing a smoothed version of the data to give predictions of response over a range of levels of the stimulus factor,

(e) as a step in identifying optimal levels of a set of stimulus factors.

Of these, (a) has no implications for design and, of the other four, we can separate (b) and (c), which are strongly related to a particular model or models, from (d) and (e), which are primarily related to prediction of the true response, the models in (d) and (e) being used as a means to retaining most of the data information in a concise form, but with no implications of interest in the parameters individually. We consider (d) and (e) in the next chapter, which is concerned with the exploration of response surfaces representing the response to several variables.

We can further subdivide (c) and can relate (b) to (c). Most of the models contemplated for fitting to data will have between two and four parameters. While it is possible to construct models with more parameters, it is extremely difficult to obtain a sufficiently informative set of data to be able to make inferences about more than four parameters. For many data sets an attempt to fit a model with four parameters leads to the conclusion that the estimated correlation between the fitted values of two of the parameters is so high that the data do not contain independent information about both parameters, but only about some combination of them.

The major division in experimental design for parameter estimation is between choosing a design to provide maximum information about a single parameter or a single combined function of parameters, or a design to provide maximum information about the set of parameters as a whole. Algebraic results are more easily available (in general) for the latter, but the danger of treating all parameters as being of equal importance is that information on the more important parameters may be markedly reduced.

The examination of the appropriateness of the fitted model implies that there is some idea of how the model might be inappropriate. This in turn implies an alternative model and the need to discriminate between the original and alternative models. If the alternative model is essentially the original model with an additional parameter, then the problem of optimal discrimination between the two models is equivalent to optimising the estimation of the additional parameter. If the alternative model does not include the original model as a special case, then discrimination cannot be reduced to a particular example of estimation. A formal approach to discrimination is often unfruitful because, if two alternative curves are both believed to provide a realistic description of the pattern of response which is expected, then the differences between the curves are likely to be relatively slight. This is illustrated in Figure 17.1(a), in which two similar asymptotic curves, one exponential and the other a linear inverse polynomial, are superimposed. It is clear that very precise data will be needed to discriminate between the two models.

A further illustration of the similarity of alternative models for response functions is provided in Figure 17.1(b). Three models which have been

Figure 17.1. Demonstration of similarity of curves. (a) Exponential and linear inverse polynomial, (b) three models (1), (2) and (3) in text, used by Sparrow (1979).

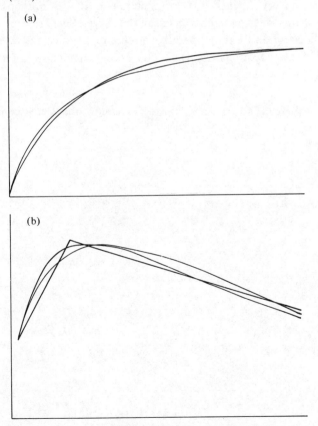

used for describing fertiliser response curves, in which the data show a maximum, are displayed. The three models are

(1) $y = (\beta_0 + \beta_1 x)/(1 + \beta_2 x + \beta_3 x^2)$,

(2) $y = (\beta_0 + \beta_1 x + \beta_2 x^2)/(1 + \beta_3 x)$,

(3) a pair of straight lines (the broken stick model of the previous section).

These three curves were included in a wide range of models which were compared by Sparrow (1979) for fitting a set of 83 experimental sets of data on the response of spring barley to nitrogen fertiliser with nine levels of nitrogen in each data set. A typical set of data is shown in Figure 17.2. The three curves considered here gave very similar patterns of fit, the average residual mean squares being 30, 32 and 34 ((kg/ha)$^2 \times 10$) for the three models. Sparrow concludes that all three curves 'represent the response of spring barley to nitrogen with some degree of success' and 'the three models differed little in their average predicted optimal yields', and even after such a very extensive set of experiments it was not possible to distinguish any clear preference between the curves. Such an experience is extremely common from comparative studies of response curves.

The next three sections of this chapter examine the choice of levels for

Figure 17.2. Response of spring barley to nitrogen: data from Sparrow (1979).

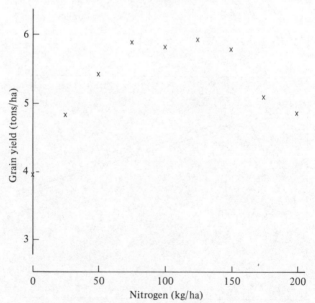

specific parameter estimation, for general parameter estimation and for discrimination. The penultimate section is concerned with the practically important question of considering several criteria and examining the robustness of different optimal designs to alternative criteria. Finally a different experimental approach to investigating response functions is considered.

17.3   Specific parameter estimation

The simplest response function is a straight line relationship

$$y = \alpha + \beta x,$$

and the simplest design question concerns the choice of values of x to optimise the estimation of β, the slope of the relationship. To minimise the variance of $\hat{\beta}$, we must minimise

$$\sigma^2 \bigg/ \sum_i (x_i - \bar{x})^2,$$

which requires choosing the x's to maximise

$$S = \sum_i (x_i - \bar{x})^2.$$

In discussing maximisation, it is assumed that the total number of observations is fixed, since, without this restriction, precision can always be improved by using more observations. It is immediately apparent first that S may be increased by increasing the range of x values, and second that, for a fixed range, S is maximised by taking half the permitted total observations at the upper end of the permissible range of x values and the other half at the lower end of the permitted range.

This result illustrates immediately the practical difficulties implicit in considering criteria for optimal design. An experimenter may be initially confident that the expected response in his experiment is a straight line relationship. Often when presented with the resulting optimal design requiring observations at only two levels of x, the experimenter's confidence in the straightness of the relationship will waver, and a compromise design will be selected. Nonetheless, the study of optimal designs for specific criteria is valuable because it indicates the ideal design under extreme assumptions, and provides a well defined basis from which a compromise solution may be determined.

The optimal design for the simple criterion of optimising the estimation of the slope of a straight line relationship is easily obtained. There are a number of other interesting criteria for parameters or functions of parameters of the simple response functions discussed in the first section of this chapter. These include:

(a) for the quadratic model, $y = \alpha + \beta x + \gamma x^2$, the estimation of either the linear or quadratic parameter, β or γ;

(b) for the same quadratic model, the estimation of the position of the maximum yield, $x_m = -\beta/2\gamma$, the criterion being either the minimisation of the variance of the estimate of x_m or the maximisation of the expected yield at the predicted optimum;

(c) for the exponential model, $y = A(1 - e^{-kx})$, the estimation of either the asymptote parameter A or the curvature parameter k;

(d) for the inverse linear model, $y = x(\beta_0 + \beta_1 x)^{-1}$, the estimation of either of the two parameters β_0 or β_1;

(e) for the inverse quadratic model, $y = x(\beta_0 + \beta_1 x + \beta_2 x^2)^{-1}$, the estimate of the position of the optimum, $x = (\beta_0/\beta_1)^{\frac{1}{2}}$;

(f) for any of the models (c), (d) or (e), the estimation of the x value for which the slope of the relationship is equal to some value, C, corresponding to the cost per unit of the input variable x.

Apart from (a), for which the algebraic argument follows exactly the same pattern as that for the slope of a straight line relationship, no exact algebraic results are possible for any of these criteria, since all involve either non-linear parameters or non-linear combinations of parameters. Such results as are available are based on approximate or asymptotic variances of parameters or parameter combinations, and the possible approaches are illustrated through two examples.

Example 17.1

Suppose, for the quadratic model,

$$y = \alpha + \beta x + \gamma x^2$$

with $\gamma < 0$, the interest is in estimating the value of x for which y is a maximum,

$$x_m = -\beta/2\gamma.$$

The obvious estimate is the ratio of the least squares estimates of β and γ:

$$\hat{x}_m = -\hat{\beta}/2\hat{\gamma},$$

and the variance–covariance matrix of $(\hat{\alpha}, \hat{\beta}, \hat{\gamma})$ can be used to obtain an approximate, asymptotic, variance for $\hat{x}_m$.

Ideally, we would consider a completely general set of n observations, $x_1, x_2, \ldots x_n$, but the problem is then too general for a simple analytic solution. The set of possible designs is therefore restricted to those with $(k + 1)$ equally spaced levels of x, spread between zero and an upper limit d, each level being replicated r times. The values of k, d and r may be chosen subject to an overall limit on the total number of observations,

$$(k+1)r = n.$$

The approximate variance of $\hat{x}_m$ is

$$V(\hat{x}_m) \simeq (\sigma^2/4\gamma^4)\{\gamma^2 \text{ Var}(\hat{\beta}) + \beta^2 \text{ Var}(\hat{\gamma}) - 2\beta\gamma \text{ Cov}(\hat{\beta}, \hat{\gamma})\},$$

where the variance–covariance matrix of $(\hat{\alpha}, \hat{\beta}, \hat{\gamma})$ is

$$\sigma^2 \begin{bmatrix} n & \sum x & \sum x^2 \\ \sum x & \sum x^2 & \sum x^3 \\ \sum x^2 & \sum x^3 & \sum x^4 \end{bmatrix}^{-1}.$$

Inserting the relevant algebraic sums for $\sum x^i$ and simplifying we obtain

$$V(\hat{x}_m) \simeq [3k\sigma^2/\{n(k+2)\gamma^2 d^2\}]$$
$$\times [1 + \{15k^2/(k-1)(k+3)\}(1 - 2x_m/d)^2].$$

Considering this variance as a function of d and k, we see first that the terms in k occur as $k/(k+2)$ or $k^2/(k-1)(k+3)$, both of which are increasing functions of k, provided k is at least two (and of course $k=1$ is of no interest, since we must estimate three parameters, α, β and γ). Hence, the approximate variance is minimised by taking $k=2$, that is by using just three observations, each replicated $n/3$ times. With $k=2$, the approximate variance reduces to

$$V(\hat{x}_m) \simeq 3\sigma^2/(2n\gamma^2 d^2)\{1 + 12(1 - 2x_m/d)^2\}.$$

The absolute minimum is obtained by taking d infinitely large. However, investigation shows that there is a local minimum when

$$d = \{24/(9 + \sqrt{3})\}x_m \simeq 2.24x_m,$$

and, in fact, the approximate variance can be reduced below the variance for this value of d only by taking d larger than $4.6x_m$. The conclusions from this example are:

(i) Only three different x values should be used when the objective is to fit a three-parameter quadratic response function. To use more reduces the efficiency of estimation of parameters.

(ii) The range of values over which observations should be taken will depend on a rough guess of the position of the optimum. Without such a guess, no proper choice can be made.

(iii) The optimal pattern for the three observations using equally spaced observations is to use $(0, x, 2x)$, where x is taken to be about the guessed optimal position, with a tendency to choose x higher than the guessed position, rather than lower.

Note that, if we wish to consider a range of values from (c to d), where c is a non-zero lower limit, then, because a change of origin leaves a quadratic function as a quadratic function, the value of d should be chosen so that the midpoint of the (c, d) interval is slightly above the guessed optimum. However, in the original formulation of the approximate variance, d^2 is replaced by $(d - c)^2$, showing that the variance is minimised by choosing c as small as possible, which, in most practical circumstances, is zero.

In practice, experimenters are extremely reluctant to use values as high as twice their guessed optimum, and the statistician will have to argue strongly to make the range of observations sufficiently wide to provide efficient estimates. The experimenters' reluctance is, in part, correct, since it stems from an unexpressed doubt about the validity of the quadratic response function (or, for that matter, any specified response function) over a wide range. Therefore, the choice of design must reflect a compromise between pressures. A wide range of x values is required to reduce the variance of parameter estimates. But the range of x values should be restricted so that the assumed response model is a good approximation.

Example 17.2

The second example shows a method of solution through computational rather than analytic methods. For most practically relevant questions of design for response functions, the criterion of primary interest is a variance of either a non-linear parameter or a non-linear combination of linear parameters. Therefore, the variance to be minimised is an approximate, asymptotic variance, and a crucial component in the choice of the factor levels is the assumption that the approximation is good, and that it is equally good for all sets of x observations. The adequacy of the approximation can be assessed through simulation, and unfortunately it appears that in some particular cases the usual first-order approximation may become inadequate for precisely those designs (choice of x values) which minimise the approximate variance.

The use of simulation to check results derived from an approximate variance function suggests the direct use of simulation. If the efficiency of any design is to be assessed by the simulated variance of the estimate of the parameter criterion, then the optimality criterion reduces to a search procedure. We require to find the design for which the simulation variance of the estimate of the parametric combination is minimised. Since the simulation variance is not exact, we are searching for the optimum of a function which we cannot observe exactly, but only with error. The size of the simulation sample may be varied to reflect the precision required. Thus, when the design is some way from the optimal design, only a small sample is required, but, near the optimal design, larger samples are needed to obtain more precise information about the true variance function. In an asymptotic model such as the exponential

$$y = A(1 - e^{-kx})$$

the estimation of the optimum position is not necessary, since y increases monotonically with x. However, if the input to the system, x, has a non-zero cost, then the interest is in the value of x for which dy/dx is equal to the cost of x. The economic yield to be maximised is

$$y = A(1 - e^{-kx}) - cx,$$

and the optimum value of x is given by

$$dy/dx = Ake^{-kx} - c = 0,$$

that is,

$$x_m = -(1/k) \log (kA/c).$$

In a study of optimal designs for the exponential model, Francis (1978) considered various criteria, and in particular examined designs for estimating the value x_m. Francis compared the precision of estimates of x_m using an approximate formula for the asymptotic variance of x_m and also performing a simulation study. The study included various sets of parameter values, and the results considered here are for $k = 0.1$, $A = 200$ and $c = 0.9$. Twenty simulations were performed for each of 16 design pairs of (x_1, x_2) values using ten observations at x_1 and ten at x_2 with $\sigma^2 = \frac{1}{6}$. The results for the approximate asymptotic variance and the observed simulation variance are shown in Table 17.1(a) and (b). Various patterns emerge from these results. It is clear that the approximate asymptotic variance gives an underestimate of the sampling variance. However, the general

Table 17.1(a). *Approximate asymptotic variance for x_m.*

	x_1/x_2			
	0.2	0.4	0.6	0.8
$x_2 = 12.5$	0.0307	0.0231	0.0382	0.1367
$x_2 = 25$	0.0048	0.0055	0.0125	0.0590
$x_2 = 37.5$	0.0026	0.0044	0.0140	0.0890
$x_2 = 50$	0.0022	0.0056	0.0258	0.2278

(b). *Simulation variances for x_m.*

$x_2 = 12.5$	0.0412	0.0288	0.0330	0.1780
$x_2 = 25$	0.0073	0.0092	0.0168	0.0886
$x_2 = 37.5$	0.0050	0.0048	0.0141	0.0957
$x_2 = 50$	0.0030	0.0052	0.0466	0.2480

pattern of optimality is similar for the two criteria with the optimal x_1/x_2 ratio decreasing as x_2 increases. The results suggest that the optimal x_1/x_2 ratio is higher for the simulated variances than for the asymptotic variances. To establish whether the asymptotic variance formula gives an accurate estimate of the optimal x_1/x_2 values for any x_2 level, a more extensive simulation study in the neighbourhood of the optimum would be needed. However, a reasonable estimate of the optimal ratio x_1/x_2 can be obtained from the investigation of either the asymptotic variance or the simulation variance. When x_2 is small, about 12.5, the optimal ratio is about 0.4, and when x_2 is large, about 50, the optimal ratio diminishes to a bit less than 0.2.

17.4 Optimal design theory

In linear models, the information about the precision of parameters is contained in the variance–covariance matrix of the parameter estimates. The variance of a single parameter or the variance of a linear combination of parameters, such as $\alpha + 3\beta - 2\gamma$, may be derived by evaluating that variance directly from the variance–covariance matrix. However, interest is not always confined to a single criterion, and there will often be interest in several aspects of the parameters of the model.

One approach to this diversity of interest is to assume that all parameters of the model are of equal interest, and to consider how to estimate the parameters as a whole as efficiently as possible. The variance–covariance matrix of the set of parameters provides a possible measure. In the same way that a confidence interval for a single parameter is calculated from the variance for the parameter estimate, a confidence region for a set of parameters may be derived from the variance–covariance matrix of the parameters, and the volume of the region is measured by the determinant of the variance–covariance matrix. The smaller the volume of the region, the smaller the 'uncertainty' about the parameters and, correspondingly, the greater the precision of estimation of the parameters as a whole.

The minimisation of the confidence region can be achieved by minimising the determinant of the variance–covariance matrix, referred to as the generalised variance, or equivalently by maximising the determinant of the least squares matrix, $|A'A| = |C|$. The two simplest, and only important, one-dimensional response functions with linear parameters, for which the generalised variance may be examined, are the linear and quadratic polynomials.

For $y = \alpha + \beta x$, the least squares matrix is

$$C = \begin{bmatrix} n & \sum x \\ \sum x & \sum x^2 \end{bmatrix}$$

and the generalised variance is
$$G = \sigma^2 / \{n \sum x^2 - (\sum x)^2\}.$$
It may be noted that minimising G is equivalent to minimising the variance of β, which we considered in the previous section. This equivalence of the generalised variance and a specific criterion variance is exceptional.

For the quadratic model
$$y = \alpha + \beta x + \gamma x^2,$$
the least squares matrix is
$$C = \begin{bmatrix} n & \sum x & \sum x^2 \\ \sum x & \sum x^2 & \sum x^3 \\ \sum x^2 & \sum x^3 & \sum x^4 \end{bmatrix}$$

and the generalised variance is
$$G = \sigma^2 / (S_0 S_2 S_4 + 2 S_1 S_2 S_3 - S_0 S_3^2 - S_1^2 S_4 - S_2^3),$$
where $S_i = \sum x^i$. The variances of $\hat{\beta}$ and $\hat{\gamma}$ are
$$\mathrm{Var}(\hat{\beta}) = G(S_0 S_4 - S_2^2)$$
and
$$\mathrm{Var}(\hat{\gamma}) = G(S_0 S_2 - S_1^2),$$
and clearly both provide criteria different from that of the generalised variance.

There is a large amount of general theory deriving from studies of the generalised variance criterion. It can be shown that designs which are optimal in the generalised variance sense are also optimal according to related and equivalent criteria, in particular a criterion involving the average length of confidence intervals for all individual parameters. There are other forms of optimality which may be considered as alternatives to G. These all consider the parameters symmetrically but express the precisions of parameters in different combined forms.

Although the concept of generalised variance is defined exactly only for linear models, there is an equivalent concept for non-linear models. This is defined as the determinant of the matrix of approximate, or asymptotic, variances and covariances derivable from the second differentials of the sum of squared deviations of yield from its expected value, $\sum \{y - \mathrm{E}(y)\}^2$, or equivalently from the likelihood, when normal distribution assumptions are made. The adequacy of the approximation may be assessed by simulation in exactly the same way that is necessary for a specific parameter criterion.

A further general comment about the generalised variance criterion is that its use is not restricted to quantitative factors. The concept of the overall precision of a set of parameters is applicable to any design problem since the determinant of the variance–covariance matrix is interpretable as the volume of the multi-dimensional confidence region of the parameters.

There are, for the response function models which are the subject matter of this chapter, two important conclusions which emerge from consideration of the generalised variance criterion. First, and most important, for any linear or non-linear model, the optimal design in the generalised variance sense never requires more distinct observations than there are parameters in the assumed model. This supports the patterns observed in the investigations of specific parameter criteria. For the estimation of the slope of a straight line response, only two observation values were required; for the estimation of the optimum position in the quadratic, only three observation values were required. In general, for a p-parameter response function, and the associated p-dimensional generalised variance criterion, the optimal design concentrates information on p observation values, the use of more than p observation points dissipating the available information. The loss of information through the use of more than p observations is considered further in the penultimate section of this chapter.

The second aspect of the generalised variance criterion that is important is that it is a general criterion, and consequently must provide less information than is possible about specific parameters, or combinations of parameters. There can be no absolute rule for the choice between general design optimality and specific design optimality. In general, the more specific the interest in a particular parameter, the less appropriate will be the use of a general criterion. But also, for more complex criteria involving several parameters, the generalised variance criterion which reflects the joint precision of all parameters may be expected to become more relvant.

17.5   Discrimination

The main concern of an experimenter presented with the optimal design for estimating the slope of a linear regression, that is half the observations at each end of the permissible range, is the absence of observations at intermediate points, and in particular at the centre of the range. This concern arises quite correctly from a need to check the assumed model. More generally, there is a class of design criteria deriving from the desire to discriminate between models. Often in biological research there are alternative models used to represent the response to a range of stimuli. The desire to compare alternative models, and to design an experiment to determine which model is appropriate, or, more reasonably, which is not appropriate, is a proper stimulus for an experiment.

There are, therefore, two reasons for considering discrimination as a criterion for choosing a design: either as a subsidiary criterion, necessary before the estimation which is the primary objective can be confidently applied, or as the primary criterion in establishing appropriate classes of models. The former is relatively simple, and is equivalent to a different estimation criterion. It was explained in Section 17.2 that the latter is, practically, extremely difficult, and it is not discussed further.

The testing of an assumed model is most simply represented as a comparison between the assumed model and a more complicated, but related, model. In this situation, the more complicated model will include the assumed model as a special case. Usually there will be a single additional parameter in the more complicated model. For example, the natural expression of the doubt about the relevance of a linear regression model, given that there is good reason for believing that the linear regression model might be relevant, is to include a quadratic term, γx^2, in the model. The check of the adequacy of the linear regression may be achieved by considering the sum of squares for fitting the quadratic term, in addition to the linear regression. In general, the adequacy of an assumed model may be checked by assessing the improvement in the fit due to an additional parameter and, when the additional parameter is an extra, linear, term in the model, the appropriate criterion is the extra SS for fitting that term.

However, the criterion of maximising the sum of squares due to fitting an additional parameter is equivalent to the criterion of minimising the variance of the estimate of the extra parameter. For linear models, this can be seen immediately from the results of Chapter 4. The sum of squares for fitting a single additional parameter θ_2, is

$$\hat{\theta}_2'(C^{22})^{-1}\hat{\theta}_2.$$

The variance of the estimate of θ_2 is $\sigma^2 C^{22}$, where C^{22} is the diagonal element corresponding to the parameter θ_2, in the variance matrix for the fitted parameters. The logic clearly applies to non-linear models also, though no corresponding exact result exists.

Example 17.3
Consider the discrimination against a quadratic model as an alternative to a linear regression model. The terminology is the same as that used in previous sections. The expected value of the sum of squares for fitting γ,

allowing for the fitting of α and β, is

$$E(\hat{\gamma}^2)\{(S_0 S_2 S_4 - S_1^2 S_4 - S_0 S_3^2 + 2 S_1 S_2 S_3 - S_2^3)/(S_0 S_2 - S_1^2)\}.$$

Equivalently, the variance of $\hat{\gamma}$ is

$$\mathrm{Var}(\hat{\gamma}) = (S_0 S_2 - S_1^2)/(S_0 S_2 S_4 - S_1^2 S_4 - S_2^3 + 2 S_1 S_2 S_3 - S_0 S_3^2).$$

Suppose observations are taken at three equally spaced values $(0, x, 2x)$, with n_1, n_2 and n_3 replications, respectively, then

$$\mathrm{Var}(\hat{\gamma}) = (n_1 n_2 + 4 n_1 n_3 + n_2 n_3)/(4 n_1 n_2 n_3 x^4).$$

It is rapidly verified that, given a fixed total, $n = n_1 + n_2 + n_3$, the variance is minimised by taking

$$n_1 = n_3 = n/4, \quad n_2 = n/2.$$

This may be recognised as sensible by considering that the quadratic parameter measures the difference between the response at the middle of the range and the average response at the two ends of the range. The resources are equally split between the middle and the ends.

17.6   Designs for competing criteria

If an experiment has a single objective then it is comparatively easy for the statistician to determine the best design in the sense of satisfying that objective. The objective may be specific or general. The principle of how to find an optimal design for any single criterion is hopefully clear from the examples in the previous sections, though considerable computation may be necessary to achieve the design.

However, experimenters rarely restrict themselves to a single criterion but seek to answer several questions. Statisticians may regret this tendency but should also recognise that it is not only natural but also scientifically proper. The practical design problem therefore requires consideration of more than one objective, and it is necessary to examine the suboptimality of designs across a range of objectives.

Example 17.4

The principles of suboptimality are illustrated for a trivial example in which the response model is a straight line. The philosophy is the same for other response models but the details are less clear. The generalised variance optimal design is identical with that for optimising the variance of the slope estimate. The variances for four criteria for each of eight three-point designs with all points lying within the range $(0, 10)$ are shown in Table 17.2. The results for individual criteria are fairly obvious. For the slope (or equivalently the general) criterion, the optimal design requires observations only at the two ends of the range. For the intercept criterion

Table 17.2.

Observations			Criteria			
			Var($\hat{\alpha}$)	Var($\hat{\beta}$)	Var($\hat{\alpha}+5\hat{\beta}$)	Var($\hat{\alpha}+8\hat{\beta}$)
0	0	10	0.500	0.015	0.375	0.660
0	10	10	1.000	0.015	0.375	0.360
0	5	10	0.833	0.020	0.333	0.385
2	5	8	1.722	0.056	0.333	0.833
6	8	10	8.333	0.125	1.458	0.333
5	9	10	4.905	0.071	0.976	0.333
0	0	2	0.500	0.375	7.375	20.500
0	1	5	0.619	0.071	0.976	3.286

two observations must be at zero, and, for each criterion concerned with prediction at a particular point, the optimal design requires the mean of the three design points to be at the point for which prediction is required. If we try to select a single design from those considered in Table 17.2 which minimises the maximum ratio of variance/minimum variance for the set of four criteria considered in this example, then the design with equally spaced values spanning the complete range should be chosen. The optimal design for the general criterion is not much worse, but the crucial distinction is between designs spanning the possible range and those with a reduced range. Each of the latter is very poor for at least one criterion.

However, the more important aspect of this simple example is the degree of suboptimality for one criterion when a design chosen for a different criterion is used. There is no design which is optimal for any of the four criteria, which does not give an inefficiency of at least a factor of two for at least one other criterion. Therefore, before accepting a design which is optimal for one criterion, it is necessary to consider the efficiency of the design for other criteria of interest. It will rarely be appropriate to use a design which does not cover the possible range of values.

Example 17.5
Consider again the problem of the student investigating the decay rate of seed germination with time. The optimal design, assuming a linear regression model and the criterion of minimising the variance of the slope, is to test 1000 seeds at the beginning of the two-year period and 1000 at the end of the two years. This proposed design was greeted with some dismay by the student, and even after some discussion about why the design was optimal he was not prepared to accept the design although he understood the reasonableness of the design concept. Essentially he needed protection

Table 17.3. *Efficiency of estimation of linear slope and of discrimination power against a quadratic effect for various designs.*

Design observations					Variance efficiency	Power efficiency
-1	-0.5	0	0.5	1.0	$S_0 S_2/2000^2$	$(S_0 S_4 - S_2^2)/(500 S_0)$
1000	0	0	0	1000	1.00	0.00
900	0	200	0	900	0.90	0.36
800	0	400	0	800	0.80	0.64
700	0	600	0	700	0.70	0.84
600	0	800	0	600	0.60	0.96
500	0	1000	0	500	0.50	1.00
400	400	400	400	400	0.50	0.70
600	300	200	300	600	0.675	0.65
200	400	800	400	200	0.30	0.54
800	100	200	100	800	0.825	0.50

against the failure of the linearity assumption. From consideration of the optimal design to discriminate between the straight line and the simplest alternative, the quadratic, we know that this would require 1000 seeds tested at the middle of the two-year period with 500 at each end. For estimating the slope, this design, which is optimal for discrimination, is only 50% efficient. Essentially, the two criteria are completely incompatible. The efficiencies (i) for estimating the slope compared with the optimal (1000, 0, 1000), and (ii) for the power of discrimination compared with the optimal (500, 1000, 500), are shown in Table 17.3.

The design finally chosen involved a compromise, as all practical designs must. Seeds were tested at six-monthly intervals, with 800 at the beginning and end, 200 at the middle (one year) and 100 each after six and 18 months. This design is 82.5% efficient for estimating the slope, it gives a reasonable power for detection of the non-linearity (which turned out not to be detectable), and it allowed the student to be 'doing something' throughout his research period!

Little work has yet been done on compromise designs for particular problems but I believe that there is sufficient evidence to identify some general principles for the practical selection of levels.

(1) It is essential to know what kind of model is intended for the analysis of results, and to be able to define approximately the range of levels of x for which the model should be appropriate.

(2) The levels chosen should cover the maximal range of levels of x, for which the assumed model is believed to be appropriate.

(3) The number of levels should be the same as the number of parameters in the intended model, or one more. For any single criterion all optimal designs appear to need no more levels than the number of parameters in the model, and this is of course the minimum number of levels necessary to fit the model. The extra level provides both for a conflict of interests between the competing criteria, and some protection against major failure of the model. The argument for dispersing the fixed total resources among many excess levels to detect any of a large number of possible discontinuities of the model is faulty because there will not be adequate precision at any point to detect a discontinuity, while the dispersal of resources reduces precision for the major criteria.

In practice, as mentioned earlier, models with more than four parameters for a single stimulus variable are not informative, and the number of levels for a quantitative factor should almost invariably be three or four. In general these levels should be replicated equally, though this recommendation will not be appropriate for compromise designs when the relative importance of the two criteria is clearly unequal, as in Example 17.5.

17.7 Systematic designs

There is one form of experimentation where the advice to use a small number of levels may be ignored. In some agricultural crop experiments the requirement that the conditions of growth for the harvested crop shall be representative of the conditions for large-scale production of the crop for a particular treatment may lead to harvesting only a small part at the centre of each experimental plot, the surrounding area being a guard area. In the particular case of experiments to compare different crop spacings, the harvested proportion of an experiment to compare four densities in a randomised block design may be as low as 40%, as shown in Figure 17.3(a). Systematic designs have been used for such experiments, following the ideas of Nelder (1962) and have also been used for several other types of treatment, including fertilisers and crop protection chemicals.

The concept of a systematic design is that over a series of subplots the level of the quantitative factor varies systematically but slowly (e.g. 10% change) through the sequence of subplots. This form of design has already been discussed briefly in Section 14.6. Each particular treatment level is then surrounded by treatment levels differing only slightly, and therefore

Figure 17.3. Efficient use of area with a systematic design: harvested area is shown within rectangles for (a) four spacings in randomised design, (b) a systematic design.

(a)

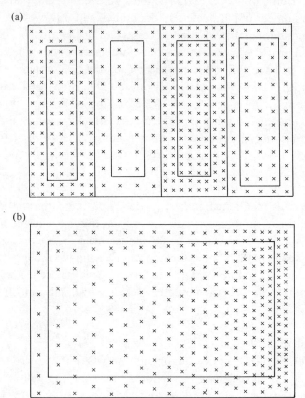

(b)

guard areas are unnecessary. Examples of a Nelder fan design and a row design for systematic density variation are shown in Figure 17.4. When a crop would normally be planted on a rectangular lattice, the planting arrangement about individual plants is distorted by the systematic arrangement, as can be seen in Figure 17.4(b). The fan design retains a constant planting arrangement while the density is varied. Fans are rather awkward to arrange in the field but they have successfully been used in many experiments particularly for tree crops. For comparison with the randomised block design the harvested area for a systematic design is shown in Figure 17.3(b), and may be as much as 80%. The inefficiency arising from using many levels must be weighed against the greater proportion of the experimental resources used in the harvested data.

The analysis for systematic designs has already been discussed in Section 14.6. Because the treatment levels are not randomised there may

Figure 17.4. Systematic designs for changing plant density. (a) Parallel rows, (b) fan design.

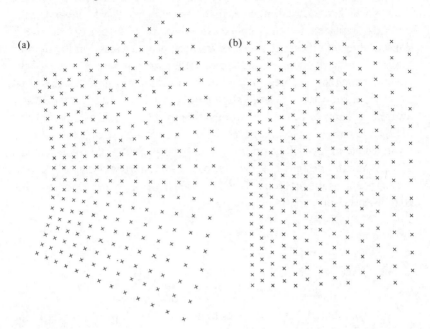

be doubts about the validity of an analysis of variance. However, for most systematic designs the analysis of variance would not be the most appropriate initial step in the analysis. Instead the fitting of a response function to represent the effect of the systematic factor for each set of data should usually be used. The subsequent analysis is based on comparisons of parameters of the fitted response functions. Such comparisons may be between replicates of the systematic plot or between different treatments applied to the different whole plots within which the systematic treatment level variation occurs.

In general then, the use of systematic treatment plots will occur within an experiment in which there are other treatment factors which are applied to whole plots. The systematic treatment then resembles a split plot factor, except that rather than comparing individual split plot treatments the fitted response function represents the information from each main (or whole) plot. The allocation of the other treatment to whole plots, the replication blocking and randomisation of whole treatment plots is exactly the same as for ordinary, non-systematic experiments. In addition, the direction of the systematic change of the treatment level within each whole plot can, and should, be randomly selected.

Systematic designs are an extremely useful addition to the practical statisticians library of designs. They are not the same as the systematic designs which were used much earlier this century and which were, quite properly, replaced by the randomised designs of Fisher. Their use is limited, chiefly in the early stages of research programmes, but it would be foolish to discard them simply because they do not satisfy either of the general principles that treatment allocation should be randomised, or that the numbers of treatment levels should not exceed four. Many statisticians, aware of the importance of randomisation in the analysis of experimental data, have been slow to accept the advantages of the use of systematically arranged factor levels. The opposition to the use of systematic components of experimental designs is, I believe, based on a narrow view of experimentation, and of the methods of analysis which are appropriate for drawing conclusions from experimental data.

Exercises 17

(1) How inefficient was the proposed design in Section 17.0(a) using eight equally spaced levels to estimate the slope of a straight line relationship?

(2) Assuming a quadratic response function, $y = \alpha + \beta x + \gamma x^2$, for an experiment in which y is observed once at each of three levels, 0, x and 1, where $0 < x < 1$, find the variance–covariance matrix of the least squares estimates of the parameters α, β and γ. Find

   (a) the algebraic form of the generalised variance; and the variances of the estimates of:

   (b) γ,

   (c) the slope of the response at $x = 0.5$,

   (d) the change of y between $x = 0$ and $x = 0.5$,

   (e) the y value at $x = 1$.

Discuss and explain the results, obtaining optimum values of x where these exist.

(3) Assuming the model $y_i = \sigma + \beta_1 x_{1i} + \beta_2 x_{2i} + \varepsilon_i$, n_1 observations are taken at $x_1 = x_2 = 0$, n_2 at $(1, 1)$ and n_3 at $(2, 0)$. Obtain expressions for the variances and covariances of $\hat{\beta}_1$ and $\hat{\beta}_2$ and hence show that for

$x_1 + x_2 + x_3 = N$ (fixed),

   (a) β_1 is estimated most precisely if $n_2 = 0$, $n_1 = n_3 = N/2$,

   (b) β_2 is estimated most precisely if $n_2 = N/2$, $n_1 = n_3 = N/4$,

(c) the sum of variances is least when $n_1 = n_3 = N/(\sqrt{2}+2)$ and

$$n_2 = \frac{\sqrt{2N}}{\sqrt{2}+2},$$

(d) $\hat{\beta}_1$ and $\hat{\beta}_2$ are independent only if $n_1 = n_3$.

(4) In an experiment to investigate the effect of temperature on the period of pupation of a certain species of butterfly, a total of $4n$ pupae are to be studied at three different temperatures t_0, t_0+1 and t_0+2. Equal numbers of pupae, n, are incubated at temperatures t_0 and t_0+1, whilst the remaining $2n$ are incubated at temperature t_0+2. Assuming that the relationship between time of pupation y and temperature t is of the form

$$y = \alpha + \beta(t-t_0) + \gamma(t-t_0)^2 + \varepsilon,$$

where the ε are independent normally distributed errors with zero mean and common (unknown) variance σ^2, show that the least squares estimates of β and γ are given by

$$\hat{\beta} = 2\bar{y}_1 - \bar{y}_0 - \bar{y}_2$$
$$\hat{\gamma} = \frac{1}{2}(\bar{y}_2 - \bar{y}_1),$$

where $\bar{y}_0$, $\bar{y}_1$ and $\bar{y}_2$ are the averages of the observations at temperatures t_0, t_0+1 and t_0+2, respectively.

If $\gamma > 0$, show that the minimum expected value of y is at a temperature $t_m = t_0 - \beta/2\gamma$, and that the variance of the estimator $\hat{t}_m = t_0 - \hat{\beta}/2\hat{\gamma}$ is approximately

$$\sigma^2(\beta^2 + 6\gamma\beta + 11\gamma)^2/(8n\gamma^4).$$

18

Response surface exploration

18.0 Preliminary examples

(a) A series of experiments on nitrogen fertilisation for winter wheat is to be planned. A single design will be repeated at each of three locations. It is decided that two blocks of 24 plots each can be available at each location and the principle design question concerns the choice of the 24 treatments. Nitrogen can be applied at three times (early, normal and late). The most important factor is expected to be the total amount of nitrogen applied and this should vary over the range from zero to 320 kg/ha. The secondary factors are the timing of application and it is accepted that up to 80 kg/ha could be applied early and up to 80 kg/ha late. Early application is believed to be more effective than late application.
(b) An experiment to investigate the effect of the levels of two additives on the quality of a cake production process (Bailey, 1982). The resources available are 25 experimental units using the same five ovens on each of five days. The blocking structure must therefore be a 5×5 row and column design. The design choices concern the treatment levels and their replication, and particularly the pattern of allocation of treatment combinations to experimental units. There is a particular interest in the linear × linear interaction term and the linear and quadratic main effects of each factor must also be regarded as important.

18.1 General estimation objectives

In the previous chapter, we considered parametric response functions for individual treatment factors. In this chapter, we always consider several factors and, in contrast to the philosophy of the previous chapter, the interest is in the overall response surface, rather than the response to individual factors. Some specific characteristics of the surface are of special interest, but the general pattern of response is also important.

A specific characteristic which is often important is the optimum position; that is, the combination of factor levels for which the expected response is a maximum. The response to be maximised may be a simple yield or output, or it may be modified to allow for costs, but the overall principle is unchanged. Essentially, we are searching for the highest point on an unknown surface without assuming an established parametric model. In the immediate region of the optimum, the surface may be approximated by a simple model, the simplest sensible model being a second-degree polynomial, and such a model is often assumed, implicitly or explicitly. However, until the position of the optimum has been clearly identified it may not be known whether the planned experiment is in the neighbourhood of the optimum.

A second characteristic of response surface exploration is that we are interested in predicting yield, usually over quite a wide range of factor levels, and usually in a region about the optimum. Since prior information about the form of the surface is not available, interest may lie in any direction from the point at which experimentation is commenced. Thus, the axes of the factors, whose values are being varied to provide the exploration of the surface, do not necessarily define more interesting directions than other directions. Generally, the design of an experiment to explore the surface will cover a region of the surface and will have a more-or-less clearly defined centre which should correspond to the position which is believed, *before the experiment*, to be of primary interest. This may be an estimated position of the optimum, or economic optimum from some previous experiment, or simply a combination of 'standard' levels.

In constructing designs to explore response surfaces, we should not forget other ideas in experimental design, and one particularly important idea which is extremely relevant is that of blocking. In experiments to investigate response surfaces, as in other forms of experimentation, it is usually necessary to have replication of observations, and this should lead to consideration of blocking the experimental units. In addition, the set of combinations of factor levels for which the response is to be measured may be quite extensive and, for a sensible grouping of units into blocks, it will frequently not be possible to include all the different experimental treatments in a single block or even in two blocks. The choice of allocation of treatments to blocks has to be made efficient in terms of the particular form of analysis to be employed.

Another related aspect of design concerns the possibility of sequential experimentation. For investigations of industrial chemical processes, it may be sensible to plan each individual observation in the knowledge of the results from all preceding observations. Alternatively, it may be appropriate to take small groups of observations sequentially, with each group being designed to take account of the results of previous groups. Again, when groups of observations are planned, the form of analysis to be applied must be considered to ensure that the group is of sufficient size to enable the analysis to be performed. Sequential ideas of experimentation can be important, even when observations must be acquired in large groups separated by long periods, as in agricultural crop trials.

Before these aspects of the design of response surface experiments are considered in detail, we must consider the form of analysis. The interest in the general pattern of the surface, with the intent of predicting yield over the region explored by the design, and probably with particular interest in the position of the optimum, implies the fitting of a multivariable response equation. In addition to the estimates of the parameters of the response surface, a measure of the lack of fit of the surface should be obtainable. This will require both an estimate of pure error and the use of more factor level combinations than there are parameters to be estimated for the response surface equation. The choice of experimental treatment combinations should provide lack-of-fit information sensitive to those departures from the form of the proposed response equation which might be thought more possible, so that the lack-of-fit test should be reasonably powerful.

The most frequently used response surface equation is the general second-degree polynomial:

$$y = \beta_0 + \sum_i \beta_i x_i + \sum_i \beta_{ii} x_i^2 + \sum_i \sum_{j \neq i} \beta_{ij} x_i x_j. \tag{18.1}$$

The advantages of this model are that it is extremely simple to fit, it is familiar to most experimenters in a variety of contexts, and it has a simply defined optimum, given by the set of equations derived from the derivatives of y w.r.t. each x_i:

$$2\beta_{ii} x_i + \sum_{j \neq i} \beta_{ij} x_j = -\beta_i$$

for each value of i.

The practical use of the fitted second-degree polynomial and the associated equations for estimating the values of x_i required for the maximum yield is more difficult than it might appear. Clearly if there is a simple well-defined optimum and the observations are exactly those given by the quadratic model, then the fitted surface will be identical with the true model and the optimum position will be estimated exactly. However, the second-degree polynomial is only an approximation to the true model and observations always include a random component due to the particular experimental units. The set of random unit effects can very easily lead to a set of observations showing a different pattern from the assumed model. Not only will there be shifts of position, but frequently the fitted surface which summarises the observations will not have a simple optimum. When this happens the solution to the above equations will not define the position of the maximum but some other point where all the derivatives are zero.

In the one-dimensional second-degree response function the only practically interesting alternative to a maximum is a minimum. However, when there are two or more factors the random variation involved in the observations frequently leads to other apparent patterns. The simplest is the saddle point in two dimensions for which the surface should be envisaged from two perpendicular directions. In one, each vertical slice of the surface will show a maximum, in the other a minimum, hence the 'saddle'. A contour diagram of a saddle point surface is shown in Figure 18.1. In more than two dimensions, there are modifications of the simple saddle point, and the larger the number of dimensions, the greater is the probability of some form of saddle point. A detailed example of fitting a second-degree response surface with saddle point problems is considered in Section 18.5.

Although the second-degree polynomial is the most frequently used model, it is not the only one, nor is it always the best. As computational

Figure 18.1. Contours for a saddle point surface.

facilities improve alternative families of response surfaces, which do not have the restrictive characteristics of second-degree polynomials discussed in Section 17.1, should become more frequently used. Other functions which have been used are the generalised exponential

$$y = A(1 - e^{-\sum_i k_i x_i})$$

and the general inverse second-degree polynomial. This latter is written in the form:

$$y^{-1} = P(x_1, x_2, \ldots, x_x) \Big/ \prod_{i=1}^{k} x_i,$$

where $P(x_1, \ldots, x_x)$ is the general second-degree polynomial, or, alternatively,

$$y^{-1} = \beta_0 + \sum_i \beta_i x_i + \sum_i \gamma_i x_i^{-1} + \text{cross product terms.}$$

18.2   Some response surface designs based on factorial treatment structures

Before considering specific requirements of response surface designs, we review some of the available treatment structures, with particular emphasis on the use of factorial structures. The first two designs which might be considered for response surface investigation are the 2^n and 3^n factorial structures.

The 2^n structure does not allow estimation of quadratic terms in the second-degree polynomial because, with only two levels of a factor, there can be no information on the change of slope. Therefore, the 2^n structure will provide information only on the gradient of the surface in all possible directions. For more than four variables, the 2^n structure uses many more combinations than are necessary for the estimation of the linear surface equation. If quadratic terms are omitted, then the polynomial response equation takes the form:

$$y = \beta_0 + \sum_i \beta_i x_i + \sum_i \sum_{j \neq i} \beta_{ij} x_i x_j.$$

The pattern of increase of number of treatment combinations (2^n), and the number of parameters to be fitted $\{1 + (n^2 + n)/2\}$, with the number of factors, n, is shown in Table 18.1, which shows clearly the wastefulness of the 2^n series in terms of the information per treatment combination. In spite of this inefficiency, and the restriction to information about gradients in all possible directions, the 2^n experiment does offer some considerable benefits in the initial stages of an investigation of a response surface. If many factors are being used, fractional replicates may be used since the model to be fitted includes only main effect and two-factor interaction

Table 18.1.

n	2	3	4	5	6
Combinations	4	8	16	32	64
Parameters	4	7	11	16	22

terms. Thus, any fractional replicate in which all main effects and two-factor interactions are not aliased with each other may be used. A half replicate becomes appropriate when $n=5$, and a quarter replicate when $n=8$. The minimum number of combinations required for estimation of all two-factor interactions, using the smallest feasible fractional replicate, may be compared with the number of parameters (see Table 18.2).

Table 18.2.

n	2	3	4	5	6	7	8
Replicate	1	1	1	$\frac{1}{2}$	$\frac{1}{2}$	$\frac{1}{2}$	$\frac{1}{4}$
Combinations	4	8	16	16	32	64	64
Parameters	4	7	11	16	22	29	37

Although 2^n structures, using fractional replicates for large n, are suitable for initial investigations, the lack of information about quadratic effects makes the structures impractical for later experiments closer to the optimum, where the curvature becomes more important than linear trends. The quadratic parameters can be estimated from the 3^n structure and the equation to be fitted will be the full quadratic model, equation (18.1). However, the 3^n experimental structures suffer even more severely from the increasing size of the number of combinations compared with the number of parameters. The combinations and parameter numbers are shown in Table 18.3. Clearly, therefore, for $n=3$ or more, the simple 3^n structure is extremely inefficient.

The first designs that were developed to replace the 3^n structures were called composite designs because they were made up from several components. The initial component was a 2^n structure (or fractional replicate of a 2^n). The two levels were interpreted as being on either side of a central value for each factor, the combination of central values being

Table 18.3.

n	2	3	4	5
Combinations	9	27	81	243
Parameters	6	10	15	21

thought of as the origin, or centre, of the design. The use of observations at the centre, although it increases the number of levels for each factor to three, is not sufficient to estimate all the quadratic coefficients. Indeed, when only a centre point is added to the 2^n structure, then all quadratic coefficients will be equal since they are estimated as the difference between the yield at the centre and the average of the yields at the two extreme levels. Therefore, additional points are added to the design on each axis (or equivalently for each factor). The three components of the design, levels being defined relative to a zero level at the origin, are thus:

(i) 2^n all combinations of ± 1 for each factor,

(ii) origin observations $(0, 0, 0, \ldots, 0)$

(iii) axial observations $(\pm \alpha, 0, 0, \ldots, 0)$

$$(0, \pm \alpha, 0, \ldots, 0)$$

$$\vdots$$

$$(0, 0, 0, \ldots, \pm \alpha).$$

The relationship between the number of observations for a central composite design, and the number of parameters, for various numbers of factors, n, is shown in Table 18.4, the numbers of combinations in brackets

Table 18.4.

n	2	3	4	5	6
Combinations	9	15	25	43(27)	77(45)
Parameters	6	10	15	21	28

being for half replicates of the basic 2^n component. The choice of the value of α is available to the experimenter. Usually α will be chosen to be between 1 and 1.5. In calculating the numbers of observations we have included only a single observation at the origin. There are sometimes advantages in using more observations at the origin than at other combinations.

The central composite designs satisfy the general requirements of response surface designs which are

(i) that the parameters of the model to be fitted can be estimated,

(ii) that the number of treatment combinations is not allowed to become too large, and

(iii) the observations are spread fairly evenly over the region within which information about the surface is required.

The last characteristic is very important, but is difficult to appreciate because it involves an assessment of the spread of points in n-dimensional

Figure 18.2. Two-factor designs. (a) Factorial, (b) central composite.

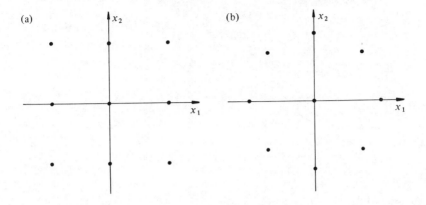

space. The situation for two factors provides only a partial insight into the problem and the three-factor case is already difficult to represent.

In the case for two factors, the designs for the 3^n and for a standard composite design are shown in Figure 18.2(a) and (b), the value of α being chosen to be $\sqrt{2}$ (for reasons which will be discussed in the next section). It can be seen that the difference between the designs is that the 'corner' points in Figure 18.2(a) are relatively far away from the centre of the design, in contrast with the more even spread of distances from the centre in Figure 18.2(b). The choice between the designs depends on the relative interest in the different regions or directions.

For three factors the 3^3 factorial and the classical central composite design are represented in Figure 18.3(a) and (b). The major differences which should be perceived in the two diagrams are, first the non-directional effect of the composite design, and second the absence in the second design of observations with exactly two out of the three factors differing from the origin values. The selection of components for inclusion in a composite design is somewhat arbitrary. This can be seen by considering the third design in Figure 18.3(c), which is composed solely of the points in Figure 18.3(a), which were omitted when forming Figure 18.3(b), together with the origin point. The cuboctahedron design of Figure 18.3(c) is an entirely sensible alternative design to consider for investigating a response surface. Much of the discussion in the later sections of this chapter will centre on the classical central composite designs described earlier in this section. They have been used in practical experimentation and are known to provide practically useful information. However, it is important to recognise that they are certainly not the only appropriate form of design.

Figure 18.3. Three-factor designs. (a) Factorial, (b) central composite, (c) cuboctahedron.

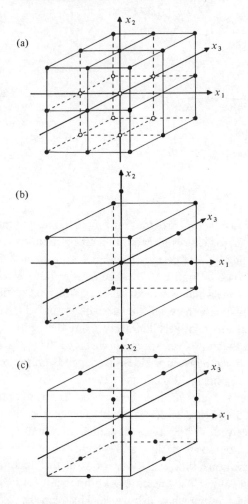

Various modifications of the basic central composite designs have been suggested and used, particularly in fertiliser response investigation. These usually reflect particular knowledge about the pattern of response. A very simple example is the situation where it is known that the requirements for two fertilisers are closely and positively related. Then it will be sensible to define two new treatment factors, the first being the sum of the two original factors (possibly an unequally weighted sum), and the second the ratio of the two original factors. If it is believed that the critical ratio of the two fertilisers is $3:2$ for $X_1:X_2$ then an appropriate design for a 3^2 factorial is that shown in Figure 18.4(a), the corresponding central composite design

Figure 18.4. Two-factor designs with $(X_1 \times X_2)$ and (X_1/X_2) as factors. (a) Factorial, (b) central composite.

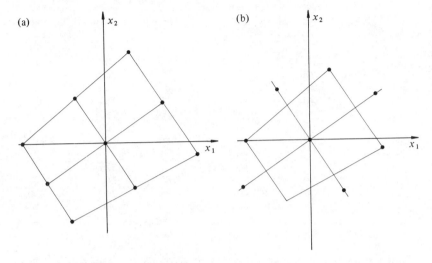

being shown in Figure 18.4(b).

An alternative family of designs to the central composite designs developed for fertiliser trials by Rojas (1963, 1972) are the San Cristobal designs. These all include the set of 2^n factorial combinations. In addition the type 1 San Cristobal designs have axial points asymmetrically placed at -1 and α, where α is larger than 1 and generally larger than the α values used in the corresponding central composite. The type 2 designs are even more asymmetrical with axial points only in the positive direction. The two types of design are illustrated in Figure 18.5(a) and (b).

Figure 18.5. San Cristobal designs. (a) Type 1, (b) Type 2.

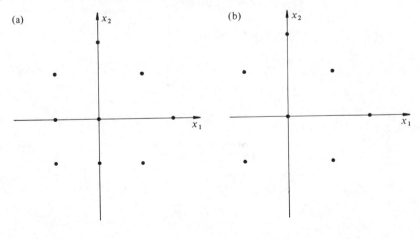

Table 18.5. *Combination of early, normal and late nitrogen levels chosen for a winter wheat experiment.*

Early	Normal	Late	Early	Normal	Late
0	0	0	0	240	0
0	80	0	40	200	0
40	40	0	0	200	40
80	0	0	40	160	40
0	40	40	80	160	0
			80	120	40
0	160	0	40	120	80
40	120	0	80	80	80
0	120	40			
40	80	40	0	320	0
80	80	0	40	240	40
80	40	40	80	240	0
80	0	80	80	160	80

For a final illustration of the philosophy of selecting a design to match the practical requirements, we return to the example explained at the beginning of this chapter. Five levels of total nitrogen (0, 80, 160, 240 and 320) were chosen. The numbers of treatment combinations at these levels were (1, 4, 7, 8, 4) and the total set of treatment combinations selected are shown in Table 18.5. A diagrammatic representation of the early and late levels for each total level is shown in Figure 18.6.

Figure 18.6. Combinations of levels for early and late nitrogen applications within five levels of total nitrogen (0, 80, 160, 240, 320).

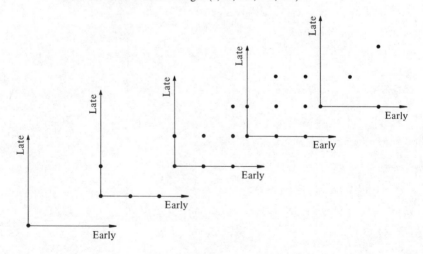

18.3 Prediction, rotatability and testing fit

In the previous section alternative designs have been discussed in a general fashion, emphasising the need for a subjective approach to choosing a design. It would be wrong, however, to imply that there are no technical considerations which might influence choice. Two criteria which can provide further information are concerned with predictions of yield, and with testing the fit of the assumed model. For any particular design, the variance of the predicted yield at a point $(x_1, x_2, \ldots)$ on the surface may be calculated. This prediction variance can be plotted to display the variation of precision for prediction over the region of interest. The plots in Figure 18.7(a) and (b) are for the 3^2 and central composite designs shown in Figure 18.2(a) and (b). The plots show the contours of prediction variance, with the minimum variance represented by the lowest contour

Figure 18.7. Contours for variance ($\hat{y}$). (a) 3^2, (b) central composite, (c) central composite with three origin points, (d) central composite with five origin points.

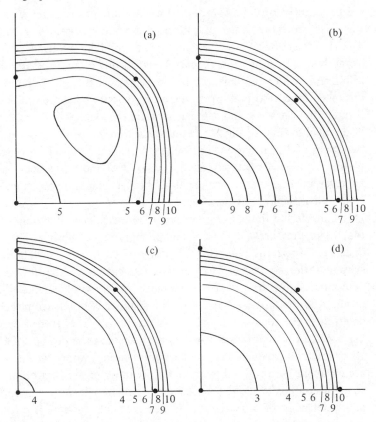

value. Only one-quarter of each design is shown since the other quarters are exact repetitions. For the 3^2 factorial the prediction is most precise in areas towards the centre of each quarter, though the variance is not much larger at the centre of the complete design. For the central composite design the contours are circular about the centre of the whole design, and the most precise prediction is obtained at about three-quarters of the distance from the centre to the outer ring of observations. A further control over the prediction variance can be achieved through variation of the number of observations at the origin, and, in Figure 18.7(c) and (d), the effects on prediction variance of increasing the replication at the origin from one in Figure 18.7(b) to three in Figure 18.7(c) and five in Figure 18.7(d) are shown. To make the prediction variances comparable in all four diagrams the variances in Figure 18.7(c) and (d) have been scaled to allow for the greater number of observations in the designs with additional centre replication. The effect of extra centre replication inevitably improves precision near the centre with some corresponding loss of precision towards the outside of the design.

In Figure 18.7(b), the contours of prediction variance are circular, and this arises from the choice of α in the composite design of Figure 18.2(b). The characteristic of circular contours for prediction variance is termed 'rotatability'. For the general central composite design for n factors, using a full replicate of the basic 2^n, it can be shown that the value of α, the distance of the axial point from the origin, should be chosen to be $2^{n/4}$ for rotatability. For $n = 2$ and 4, the distance of the axial point from the origin is identical to that for the 2^n treatment combinations. For $n = 3$ and 5, the two types of distance are very similar, but as n increases further the axial points are placed relatively further away. For most practical purposes, however, the observations can be thought of as placed on an approximately spherical shell about the origin.

In considering the variation of prediction variances, it should be noted that there are three important components of the design. The relative distances of various points of the design from the origin determine the shape of the contours. The relative replication of the origin points determines how the prediction variance varies with distance from the origin. And the range of the design (the distance of points from the origin) determines the region over which an acceptably low level of prediction variance can be maintained. Just as in the estimation designs of Chapter 17, where for a particular assumed response model more efficient estimates are obtained by using only a minimum number of different observations, so in the composite design there is no real benefit of introducing both inner and outer rings of observations. In any particular direction through the origin, the information about the quadratic response is essentially provided by three values, two at the circumference of the shell and one at the origin, and all necessary changes in patterns of prediction variances can be achieved through controlling the origin replication and the distance of the observations from the origin.

Another important general characteristic of response surface designs is the capacity to test the fit of the model. In central composite designs, this is achieved by comparing the lack of fit SS (with degrees of freedom equal to the difference between the number of distinct observation points and the number of fitted parameters) with the pure error SS provided either by overall replication of the design observations, or by replication of origin observations. When $n = 2$ or 3, there will almost always be several replications of the complete design, but for $n = 5$ or 6, the design may be restricted to a single replicate. However, for $n = 5$ or 6, the origin replication will usually be quite substantial, to achieve reasonable prediction variances in the neighbourhood of the origin, and this replication also permits a test of fit.

The test of fit using a variance ratio statistic from the lack-of-fit and pure error mean squares is of a very general nature. If any particular alternative form of model is suspected, then it should be possible to choose a design to optimise discrimination against the particular alternatives. In practice, this is rarely considered, and the general test is used.

18.4 Blocking and orthogonality

As in all experiments, the design of a response surface experiment must involve consideration of the natural variation amongst the experimental units. The composite designs most commonly used for response surface exploration involve quite large numbers of observations, and the fitted response equation will include a large number of parameters estimated from a complex set of equations. The need for efficient blocking in response surface designs has sometimes been given little importance, with the result that large blocks are often used in practice. When blocking has been considered, the requirement of orthogonal estimation of parameters has been given heavy weight because of the need to be able to fit the response equation with relatively simple calculations. The influence of modern computing facilities on blocking designs for response surface experiments appears to have been slight. It is clear from general consideration of blocking principles and the interpretation of mildly non-orthogonal effects that more efficient designs can be achieved through the use of non-orthogonal blocking.

The principles behind the choice of treatment allocation to blocks are essentially similar to those of Chapters 7 and 15. Ideally the comparisons of interest should be orthogonal to block differences. Where this ideal cannot be achieved, the level of non-orthogonality should be reduced to the minimum possible. To illustrate this general philosophy, we consider the two-factor composite design in detail, and then examine briefly a range of combinations of treatment set size and block size.

The two-factor composite design contains nine observations. Suppose that it is decided to duplicate the origin observation, giving ten observations per replicate. Suppose further that it is felt that blocks of ten observations will give an unnecessarily large value of σ^2, and it is decided to use blocks of five observations. Consider the model to be fitted:

$$y = \beta_0 + \beta_1 x_1 + \beta_2 x_2 + \beta_{11} x_1^2 + \beta_{22} x_2^2 + \beta_{12} x_1 x_2$$

and the ten observations:

$$y_1 = (0, 0), \ y_2 = (0, 0), \ y_3 = (1, 1), \ y_4 = (1, -1),$$
$$y_5 = (-1, 1), \ y_6 = (-1, -1), \ y_7 = (\sqrt{2}, 0), \ y_8 = (-\sqrt{2}, 0),$$
$$y_9 = (0, \sqrt{2}), \ y_{10} = (0, -\sqrt{2}),$$

The least squares estimates of the parameters, ignoring block effects, are:

$$\hat{\beta}_0 = \frac{1}{2}(y_1 + y_2)$$

$$\hat{\beta}_1 = \frac{1}{8}\{y_3 + y_4 - y_5 - y_6 + \sqrt{2}(y_7 - y_8)\}$$

$$\hat{\beta}_2 = \frac{1}{8}\{y_3 - y_4 + y_5 - y_6 + \sqrt{2}(y_9 - y_{10})\}$$

$$\hat{\beta}_{11} = \frac{1}{32}\{-8(y_1 + y_2) + 2(y_3 + y_4 + y_5 + y_6) + 6(y_7 + y_8)$$
$$-2(y_9 + y_{10})\}$$

$$\hat{\beta}_{22} = \frac{1}{32}\{-8(y_1 + y_2) + 2(y_3 + y_4 + y_5 + y_6) - 2(y_7 + y_8)$$
$$+6(y_9 + y_{10})\}$$

$$\hat{\beta}_{12} = \frac{1}{4}(y_3 - y_4 - y_5 + y_6).$$

It is now apparent that splitting the treatments into two blocks:

$$I = (y_1, y_3, y_4, y_5, y_6)$$

and

$$II(y_2, y_7, y_8, y_9, y_{10})$$

has no effect on the parameter estimates, because if block effects b_1, $b_2 (= -b_1)$ are included in the model the expected values of the parameter estimates do not involve b_1 and b_2. In this example, then, the treatments can be allocated to blocks so that all parameters are estimated orthogonally to the blocks. In passing, it is worth noting that the parameters β_0, β_{11}, β_{22} are not mutually orthogonal, the correlations of the estimates being $(4/7)^{1/2}$ for β_0 and β_{11} and $(3/7)^{1/2}$ for β_{11} and β_{22}. Note also that, for any other required pattern of block sizes, it is not possible to achieve treatment allocation so that estimation of parameters is orthogonal to blocks. However, if blocks of four were required so that a double replicate would occupy five blocks then there are various choices of block allocations such that block effects have little influence on parameter estimates. For example, a basic set of five blocks

(y_1, y_2, y_7, y_8),
(y_1, y_2, y_9, y_{10}),
(y_3, y_4, y_5, y_6) twice,
(y_7, y_8, y_9, y_{10})

provides orthogonal estimation for β_1, β_2, and β_{12}. An alternative set of five blocks

(y_1, y_3, y_7, y_8),
(y_1, y_4, y_6, y_7),
(y_2, y_3, y_6, y_9),
(y_2, y_5, y_9, y_{10}),
(y_5, y_4, y_8, y_{10})

leaves all estimates non-orthogonal to blocks but at mild levels of non-orthogonality.

In general, when a second-degree polynomial equation is fitted to data from a composite design, some consistent patterns in the estimation of parameters can be identified:

(i) the $\hat{\beta}_0$ estimate involves origin observations,

(ii) the $\hat{\beta}_i$ estimates involve two sets of comparisons, one for the 2^n component of the design, and one for the axial points,

(iii) the $\hat{\beta}_{ii}$ estimates involve all combinations, with the dominant comparisons being those between the axial points and the origin points,

(iv) the $\hat{\beta}_{ij}$ estimates involve comparisons only within the 2^n component of the design.

Hence, any orthogonal blocking system should keep the 2^n component of the design in blocks separate from the axial component. By examining the numbers of treatment combinations in the two components, sensible structures for blocking composite designs of various sizes can be devel-

Table 18.6.

n	2	3	4	5	6
2^n component	4	8	16	32(16)	64(32)
Axial	4	6	8	10	12
Origin points	2	3	4	5	6

oped. The sizes of the different components of the design are shown in Table 18.6. The numbers of origin observations suggested is quite arbitrary and will be determined by the requirements for the prediction variance. Usually, there should be an origin observation in each block.

To attempt to tabulate all possible reasonable designs is outside the philosophy of this book. Instead we make some suggestions for $n = 3$ and $n = 4$ and hope that the general principles will be clear.

$n = 3$. Two blocks of between seven and nine observations keeping the 2^3 and axial group in separate blocks would be very suitable. If smaller blocks are required, then the 2^3 can be split into two groups of four, confounding the three-factor interaction which is not used in estimating the second-degree polynomial. The axial points can be split into pairs.

$n = 4$. Three blocks of eight combinations each with an added origin point is ideal. The four-factor interaction in the 2^4 component may be confounded. Blocks of four involve confounding one two-factor interaction (though the fractional partial confounding ideas of Section 15.4 may be appropriate here). The possibilities of (*a*) confounding the 2^4, (*b*) splitting pairs of axial points off the axial group, and (*c*) origin points, allow almost any block size to be catered for. Thus,

Block size $= 4$. Confound the 2^n in four blocks of four. Four further blocks each containing two axial points + two origin points.

Block size $= 5$. 2^4 confounded in four blocks of four with an origin point in each block. Two further blocks each with two pairs of axial points plus an origin point.

Block size $= 6$. 2^4 confounded in four blocks of four, two blocks also including an axial pair, and two including two origin points. A fifth block with two axial pairs plus two origin points.

Block size $= 7$. 2^4 confounded in four blocks of four with each block also containing an axial point pair and an origin point.

18.5   Sequential experimentation

It is comparatively rare for an experiment devised to investigate a response surface to be performed in isolation. Instead each experiment will be one of a sequence of experiments. There are several different patterns of sequential experimentation, ranging from sequences of single observations to a series of linked experiments devised to satisfy separate but related objectives. The choice of the most appropriate pattern of sequential experimentation will depend partly on the objectives of the investigation, but also on the particular nature of the research. Thus, in an investigation into the operation of an industrial chemical process, observations of the process under varied conditions may be made singly or in small batches, with no restrictions on the time interval between successive sets of observations. In contrast, the investigation into optimal fertiliser levels for an agricultural crop is limited by the need to grow the crop at a particular time in each growing season, and hence there will be a gap of a year between successive experiments. These two extreme forms of sequential experimentation are discussed here in some detail, but it must be remembered that various intermediate forms are also possible.

First, consider the investigation of the response of an industrial chemical process to variation of the levels of several input factors. The experimentation will occur within an industrial operation and this will influence the choice of design levels, since there will be a requirement that the deviation from optimal (but unknown) conditions must be limited. The fundamental ideas of designing experiments to investigate the operation of an industrial process, within the normal operation of that process, are known as EVOP, or evolutionary operation, and stem from original work by Box (1957) and Box and Youle (1955). Box's own definition required that 'a process should be run so as to generate product plus information on how to improve the product'. Clearly, the optimal operation of the process is a dominant consideration, but the requirements for information are wider than that single objective.

The philosophy of EVOP is dominated by the use of 2^n factorial designs with added centre points. In the context of a continuing process, it is desirable that sets of observations be made in as short a time period as possible. Consequently, an EVOP experiment is often quite small. EVOP designs for investigating many factors employ fractional replicates for the estimation of main effects, with some two-factor interactions.

The 2^n design with added centre points permits only the estimation of linear main effects, and two-factor linear × linear interactions, together with information from the centre points about the overall level of curvature of the surface. As mentioned in Section 18.2, there can be no information about the separate curvature effects of different treatment factors, since the yield for the middle level of a factor is the same for all factors. This is illustrated in Figure 18.8 for a four-factor experiment using

Figure 18.8. Lack of information about curvature with only one observation with origin levels.

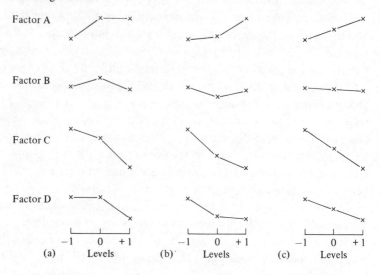

a complete 2^4 design with added centre points. The mean yields for the two main levels of each factor vary, although *their average remains constant* and the yield at the centre level remains unchanged, with the result that the patterns of curvature for the four factors must be very similar, showing an apparent maximum for each factor (Figure 18.8(a)), or an apparent minimum for each factor (Figure 18.8(b)), or an apparent linear trend for each factor (Figure 18.8(c)). (The representations for the four factors are displaced vertically for visual clarity.) It is therefore important in interpreting the results of such an experiment to avoid any attempt to interpret the curvature of the response pattern for any single factor. The important information provided by the centre points concerns only the overall curvature of the response surface, that is the amount by which the centre point yield differs from the average yield of the 2^n observations. In the context of a sequence of experiments, the difference of the centre point yield from the mean yield of the other observations, when compared to the linear trend effects, provides a measure of the nearness to the optimum of the surface. Larger centre point deviations relative to the linear trend differences generally tend to indicate that the design is closer to the optimum.

The EVOP designs of '2^n + centre' can be extended to include axial points, thus giving composite designs, if the design is believed to be sufficiently close to the optimum to warrant estimation of quadratic coefficients.

For a second illustration of the sequential experimental investigation of a response surface, consider the response of an agricultural crop to variation of the levels of several nutrient factors. The gap between successive experiments is a year, and each experiment will require adequate replication for the results of the experiment to be analysable and interpretable in isolation from other experiments in the sequence. Unlike the EVOP form of experimentation, it is quite probable that some factors will disappear from the later experiments on the grounds that there is no detectable effect of the factors. Initial experiments will almost inevitably include large numbers of factors, since the potential waste from ignoring a factor initially and introducing it subsequently, with the possibility of discovering that it changes the pattern of other factor responses through two-factor interactions, is considerable. There is a natural progression from initial experiments involving only two levels of each factor, possibly as in EVOP with centre points, through three levels to composite designs as the region of interest becomes better defined. The region of interest for composite designs should never become very small. There are two reasons for this. The level of variability of yields within agricultural experiments is always quite high, and for precise information about the form of the response a wide range of levels is necessary for each factor. In addition, the wide fluctuation of environmental conditions, and related fluctuations of costs imply that the region of interest cannot be narrowly identified, and that information on predicted yield is required over a wide range to allow predictions under varying price and cost structures to be evaluated.

The analysis of each experiment in a sequence of agricultural experiments should be intensive since the limitation to one experimental stage per year allows considerable time for analysis, and it is important to obtain all possible information from each experiment. Provided each factor has three levels, a response function should always be fitted to the results. Usually a quadratic polynomial will be adequate to summarise the pattern of response, though, for a wider range of levels, an inverse polynomial or exponential function may be more appropriate. The saddle point pattern, mentioned in Section 18.1, is a very real one when three or more factors are involved. It is comparatively rare to find a set of data for which the attempt to estimate the position of the optimum provides estimates of the factor levels for an optimum within the range of the experiment, without some modification of the analysis. This problem and some ways of overcoming the problem will be discussed further in the following section.

The major objective in most forms of sequential response surface exploration is the determination of optimal conditions. Much has been written about algorithms for seeking the maxima or minima of functions, and it is important to identify clearly the characteristics of such algorithms which are relevant to the problem of seeking the optimum of an experimental response surface. The development of optimum seeking algorithms has been dominated by consideration of functions without error. In the investigation of response surfaces, observations always include error, the size of the error depending on the subject under investigation. Some of the philosophy of function optimisation methods can be relevant to the search for an optimum of a response surface. However, there will always be a need to adapt a function optimisation procedure for optimum searching for observations with error to allow for the uncertainty about predictions. One fundamental problem with designs for optimum seeking is that, like function minimisation algorithms, the searching philosophy must change as the search proceeds. Initially, the requirement is for identification of the direction in which the search should move. Latterly, the requirement is for estimation of the optimum within a region. For the initial stages, estimates of curvature effects are irrelevant, and to include observations designed to provide such estimates is wasteful of resources. Later, the absence of information about curvature can lead to the search moving away from the region of the optimum. However, identifying when the search should change from the initial to the final form is difficult because the scale of the response pattern is not known.

18.6   Analysis of response surface experimental data

The two examples discussed in this section are chosen to illustrate first the problems that can arise when fitting models to response surface data, and second the use of inverse polynomial models. It should not be assumed from the fact that the discussion of saddle point problems in the first example is in the context of a second-degree polynomial model that such problems are confined to these models. They can occur for any form of model, though it might be hoped that they will be less severe for models with a more empirically credible pattern of behaviour.

The problems in estimating maxima of fitted response surfaces often arise from imprecise estimates of the $\beta_{ij}x_ix_j$ product terms in the fitted quadratic model. One approach to this problem is to multiply the β_{ij} coefficients by a factor of $k(<1)$, reducing the effects of the cross-product terms. Alternatively, or additionally, some β_{ij} may be set to zero, particularly choosing for such omission those terms whose β_{ij} are not large compared with their standard errors. The philosophy of this kind of amended analysis is in the spirit of 'ridge regression'. Essentially, while not knowing anything very precisely about the individual parameters of the response function, we can be confident that the surface being investigated does have a maximum, probably not very distant from the region over which the experiment is conducted. This rather vague qualitative information should be allowed to influence the analysis and interpretation of the data. A further general point is that, although the derivatives of the fitted surfaces may not provide estimates of a maximum, the fitted surface may still give a very good approximation to the true surface.

Example 18.1

We consider a three-factor response surface for which the true response is

$$y = 30 + x_1 - x_2 - 3x_3 - x_1^2 - 2x_2^2 - 3x_3^2 + 2x_1x_2 + 3x_1x_3 - 4x_2x_3,$$

and a central composite design is used with 15 observations at

$$(0, 0, 0), (\pm 1, \pm 1, \pm 1), (\pm 1.5, 0, 0), (0, \pm 1.5, 0), (0, 0, \pm 1.5),$$

each replicated twice. The simulated mean yields using simple additive error terms are as shown in Table 18.7. The least squares equations for the parameters of the fitted second-degree model are

$$12.5\hat{\beta}_1 = -6.5$$
$$12.5\hat{\beta}_2 = 6.5$$
$$12.5\hat{\beta}_3 = -44.00$$
$$15\hat{\beta}_0 + 12.5(\hat{\beta}_{11} + \hat{\beta}_{22} + \hat{\beta}_{33}) = 401.00$$
$$12.5\hat{\beta}_0 + 18.125\hat{\beta}_{11} + 8\hat{\beta}_{22} + 8\hat{\beta}_{33} = 331.25$$

Table 18.7. *Observations from a three-factor composite design.*

x_1	Factor levels x_2	x_3	Yield
0	0	0	30
−1	−1	−1	31
−1	−1	+1	25
−1	+1	−1	37
−1	+1	+1	14
+1	−1	−1	15
+1	−1	+1	25
+1	+1	−1	33
+1	+1	+1	23
−1.5	0	0	27
+1.5	0	0	30
0	−1.5	0	29
0	+1.5	0	26
0	0	−1.5	33
0	0	+1.5	23

$$12.5\hat{\beta}_0 + 8\hat{\beta}_{11} + 18.125\hat{\beta}_{22} + 8\hat{\beta}_{33} = 326.75$$
$$12.5\hat{\beta}_0 + 8\hat{\beta}_{11} + 8\hat{\beta}_{22} + 18.125\hat{\beta}_{33} = 329.00$$
$$8\hat{\beta}_{12} = 21.00$$
$$8\hat{\beta}_{13} = 29.00$$
$$8\hat{\beta}_{23} = -37.00.$$

The solutions to these equations are

$$\hat{\beta}_0 = 31.23$$

$\hat{\beta}_1 = -0.52$	$\hat{\beta}_{11} = -1.58$	$\hat{\beta}_{12} = 2.62$
$\hat{\beta}_2 = 0.52$	$\hat{\beta}_{22} = -2.02$	$\hat{\beta}_{13} = 3.62$
$\hat{\beta}_3 = -3.52$	$\hat{\beta}_{33} = -1.80$	$\hat{\beta}_{23} = -4.62.$

The equations for the estimation of the factor levels for which the derivatives are zero are

$$3.16x_1 - 2.62x_2 - 3.62x_3 = -0.52$$
$$-2.62x_1 + 4.04x_2 + 4.62x_3 = 0.52$$
$$-3.62x_1 + 4.62x_2 + 3.60x_3 = -3.52$$

It may be immediately obvious to the reader that the large value of $\hat{\beta}_{23}$ makes it inevitable that the solution to these equations will not correspond to a point of maximum yield. In fact the solutions are

$$x_1 = +0.79, \quad x_2 = -1.805, \quad x_3 = 2.14,$$

for which the value predicted by the fitted equation is 26.79. Since this is clearly less than the fitted value of 31.23 at $(x_1 = 0, x_2 = 0, x_3 = 0)$ it cannot be a maximum.

The true maximum (from the true response model) is at the combination

$$x_1 = -1, \quad x_2 = 0.75, \quad x_3 = -1.5,$$

and the prediction from the fitted model for this point is 39.32 showing that the predictions of the fitted model are not totally inappropriate (the true model gives a yield of 31.38 at its optimum).

This particular example is a very severe one because all three cross-product terms in the fitted model have the 'wrong' sign. Modifying the relative sizes of the quadratic and cross-product terms might not be expected to provide a satisfactory solution. However, if we examine the extreme modification in which the cross-product terms are set to zero, the estimates

$$x_1 = -0.16, \quad x_2 = 0.13, \quad x_3 = -0.98$$

are obtained, for which the predicted yield from our full fitted model would be 33.98 (or 32.87 if the cross-product terms are omitted from the fitted model). This at least gives some indication of the directions in which the factor levels should be modified to obtain higher yields. The factor most requiring change is x_3, which should be reduced.

Further insight can be gained by calculating the predicted values from the fitted model at all 27 combinations of the three levels $(-1, 0, 1)$ for the three factors. These fitted values are given in Table 18.8 with the corresponding values for the true model. In view of the failure of the analysis of the fitted model to provide estimates of the optimum position, the general agreement between the fitted and true models is, at least qualitatively, extremely good.

The five highest predicted values from the fitted model suggest that the yield is maximised in the region around those five points

$$(0, 1, -1), (-1, 1, -1), (-1, 0, -1), (1, 1, -1) \text{ and } (0, 0, -1).$$

Table 18.8. *Predicted values from fitted and response functions.*

x_1	x_2	x_3	Fitted model	True model
-1	-1	-1	30.97	28
-1	-1	0	30.25	29
-1	-1	1	25.93	24
-1	0	-1	35.51	31
-1	0	0	30.17	28
-1	0	1	20.23	19
-1	1	-1	36.01	30
-1	1	0	26.05	23
-1	1	1	12.49	10
0	-1	-1	25.79	25
0	-1	0	28.69	29
0	-1	1	27.99	27
0	0	-1	32.95	30
0	0	0	31.23	30
0	0	1	25.91	24
0	1	-1	36.07	31
0	1	0	29.73	27
0	1	1	19.79	17
1	-1	-1	17.45	20
1	-1	0	23.97	27
1	-1	1	26.89	28
1	0	-1	27.23	27
1	0	0	29.13	30
1	0	1	27.43	27
1	1	-1	32.97	30
1	1	0	30.25	29
1	1	1	23.93	22

Table 18.9.

	$x_2 = -1$	$x_2 = 0$	$x_2 = 1$
$x_1 = -1$	30.97	35.51	36.01
$x_1 = 0$	25.79	32.95	36.07
$x_1 = 1$	17.45	17.45	32.97

The subtable of predicted values for $x_3 = -1$ is shown in Table 18.9. For the true model the values corresponding to the five highest values from the fitted model are five of the seven true highest values.

The moral of this example which I believe applies widely is that the fitted surface will usually give an accurate overall representation of the pattern of the true surface. However, the imprecision of individual parameter estimates will often lead to useless conclusions when functions are calculated from the parameters except, of course, when the functions are the set of predicted values for the fitted surface. When rather fewer parameters are fitted then functions of parameters may be less useless.

In the sequential context the importance of examining the fitted response surface and basing the next experiment on an overall assessment of the surface rather than a single predicted optimum, even in the rare cases where such a prediction is available, is obvious. For this example it is clear that the range of levels for x_3 needs to be changed for the next experiment, probably to $(-2, -1, 0)$, and that the interactions of x_3 with x_1 and with x_2 would suggest smaller shifts for x_1 and x_2 but with a wider spread for each factor. Thus the new sets of levels might be (in terms of the old levels)

$$x_1 (-2, -0.75, 0.5)$$
$$x_2 (-0.5, 0.75, 2)$$
$$x_3 (-2, -1, 0).$$

Example 18.2

In the winter wheat experiment already mentioned twice in this chapter, there were 24 treatments in two blocks of 24 plots each. The experiment was repeated for three years at each of three farm sites, Home Farm, Upton Grey and Easton Manor. The mean yields, in tonnes per hectare from the three sites for one year, are given in Table 18.10. After examination of the data for the pairs of replicates and of the general pattern of response it was decided to fit models of the form

$$1/y = \beta_0 + \beta_1(z+c)^{-1} + \beta_2(z+c)^{-2} + \beta_3(x_E+d)^{-1} + \beta_4(x_L+f)^{-1},$$

Table 18.10.

| Nitrogen (kg/ha) | | | | | Mean yields (tonnes/ha) | | |
Total	Early	Normal	Late		Easton Manor	Home Farm	Upton Grey
0	0	0	0		5.20	3.86	5.06
80	0	80	0		8.20	6.60	7.40
80	40	40	0		8.10	7.07	6.84
80	80	0	0		8.34	7.56	7.11
80	0	40	40		8.44	6.00	6.95
160	0	160	0		8.25	8.44	8.12
160	0	120	40		8.38	7.61	7.68
160	40	120	0		9.12	9.07	8.91
160	40	80	40		9.94	8.92	8.38
160	80	80	0		9.58	8.95	9.00
160	80	40	40		8.78	9.08	8.88
160	80	0	80		9.32	8.74	8.74
240	0	240	0		8.00	8.26	8.30
240	0	200	40		9.33	8.36	8.28
240	40	200	0		9.23	9.36	9.01
240	40	160	40		9.86	9.22	8.02
240	40	120	80		10.00	9.08	8.85
240	80	160	0		9.82	9.52	8.04
240	80	120	40		9.30	9.48	9.30
240	80	80	80		10.01	9.42	9.18
320	0	320	0		9.60	8.68	7.52
320	40	240	40		9.94	9.36	9.07
320	80	240	0		9.66	9.70	9.06
320	80	160	80		9.95	9.52	9.25

where z, x_E and x_L are the levels of total nitrogen, early nitrogen and late nitrogen, respectively. The model includes eight parameters, three of which (c, d, f) represent base levels of nitrogen available in the soil, and are inherently non-linear parameters. The pattern of replicate variation suggested strongly that variation of $1/y$ was homogeneous. The models were therefore fitted by multiple regression with five linear and three non-linear parameters using the mean of the $1/y$ values from the two replicates as the dependent variable.

The estimation of the non-linear parameters was pursued by grid search methods to determine a minimum residual sums of squares. In fact, it was found that the complete model contained more parameters than could be estimated properly and models including both β_1 and β_2 were not fitted.

Some of the grid search results for the residual SS for four models

(1) model $1/y = \beta_0 + \beta_1(z+c)^{-1}$,

(2) model $1/y = \beta_0 + \beta_1(z+c)^{-1} + \beta_3(x_E+d)^{-1}$,

(3) model $1/y = \beta_0 + \beta_1(z+c)^{-1} + \beta_3(x_E+d)^{-1} + \beta_4(x_L+f)^{-1}$,

(4) model $1/y = \beta_0 + \beta_2(z+c)^{-2} + \beta_3(x_E+d)^{-1} + \beta_4(x_L+f)^{-1}$

are tabulated in Table 18.11 for the Home Farm data set. It is clear that the parameters d and f are very poorly determined. Similar results are obtained for the other two data sets and it was decided to set $d = f = 20$. The residual SS were then examined for a range of values of c for revised

Table 18.11(a). *Model (1).*

c	10	20	30	40	50	60
RSS	364	238	180	163	172	194

(b). *Model (2).*

d	\multicolumn{5}{c}{c}				
	20	30	40	50	60
5	111	65	55	64	85
10	110	65	54	64	85
20	109	64	54	64	85
30	109	65	54	64	85
40	110	65	55	65	86

(c). *Model (3).*

f	$c=$		30			40			50	
	$d=$	10	20	30	10	20	30	10	20	30
10		59	58	59	46	46	46	54	54	55
30		59	58	59	46	45	46	54	53	54
40		59	58	59	46	45	46	54	53	54

(d). *Model (4).*

d	\multicolumn{7}{c}{c}						
	60	70	80	90	100	110	160
10	88	64	51	46	48	56	67
20	88	64	51	46	48	56	67
30	88	64	51	46	49	56	68

Table 18.12(a). *Model (2a)*.

c	Home Farm	Easton Manor	Upton Grey
20	110	67	178
30	64	65	154
40	54	71	143
50	64	81	140
60	85	93	141

(b). *Model (2b)*.

30	58	58	145
40	46	65	135
50	54	75	133

(c). *Model (2c)*.

60	88	67	166
70	64	66	154
80	51	68	145
90	46	71	139
100	48	77	135
110	56	83	134
120	67	90	134

models

(2a) model $1/y = \beta_0 + \beta_1(z+c)^{-1} + \beta_3(x_E+20)^{-1}$,

(3a) model $1/y = \beta_0 + \beta_1(z+c)^{-1} + \beta_3(x_E+20)^{-1} + \beta_4(x_L+20)^{-1}$,

(4a) model $1/y = \beta_0 + \beta_2(z+c)^{-2} + \beta_3(x_E+20)^{-1}$,

with the results for the three sites as shown in Table 18.12. Testing the hypothesis of invariance of c values over the three sites gives the analysis of residual SS for model (4a), see Table 18.13, where the degrees of freedom are based on the numbers of linear and non-linear parameters fitted and

Table 18.13.

	SS	df
Separate c values	246	59
Single c value	256	61
Difference	10	2

are only approximate. The variation of c is clearly non-significant, assuming that the approximate F-test for the ratio of the mean squares is at all appropriate. Similar results are obtained for the other two models.

Comparison of the different models for the three sites shows that the inclusion of β_4 is never significant. The model with the quadratic term for total nitrogen is slightly superior to that with a linear term and the fitted models selected are

Home Farm: $\quad 1/y = 0.094 + 1267(z + 90)^{-2} + 0.183(x_E + 10)^{-1}$

Easton Manor: $1/y = 0.094 + 682(z + 90)^{-2} + 0.099(x_E + 10)^{-1}$

Upton Grey: $\quad 1/y = 0.105 + 653(z + 90)^{-2} + 0.111(x_E + 10)^{-1}$.

The residual mean squares from these fits may be compared with the error mean squares from the analysis of variance for the two blocks × 24 treatment designs giving the results in Table 18.14. Since the dependent variable used for the model fitting is the mean of two replicates the residual mean square should be approximately half of the corresponding error mean square, and with the possible exception of the Upton Grey data the models clearly provide an adequate fit. The general fit for the Home Farm data is confirmed in Figure 18.9, though there is a suggestion that the fitted inverse polynomials bend too gently.

Figure 18.9. Fitted models and mean yields for winter wheat data for various levels of early and late nitrogen: ● (0, 0), ○ (0, 40), △ (40, 0), ▲ (40, 40 or 80), ■ (80, 0), □ (80, 40 or 80).

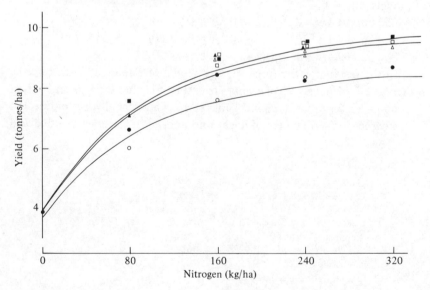

Table 18.14.

| | Model fitting | | | Analysis of variance |
	RSS	df	RMS	EMS
Home Farm	46	20.3	2.3	4.3
Easton Manor	71	20.3	3.5	6.3
Upton Grey	139	20.3	6.8	9.8

18.7 Experiments with mixtures

In some response surface experiments the factors which may be varied in the experiment are inevitably not independent, because an increase in the level of one factor necessitates an overall reduction in the total of the levels of the other factors. Many chemical processes, including production of edible materials, are based on mixing together several components. Experimentation may be required to investigate the effects of changing the proportions of different components. But since the total amount of material used in a process is fixed, the levels of the various factors to be investigated must sum to a fixed total.

The general principles developed in this chapter and the previous one apply also to experiments on the response functions and surfaces when the factors are components in a mixture. However, the technical details do require new formulation. The subject of mixture experiments has an extensive literature and there is a review of the subject in the book by Cornell (1981).

The fundamental difference between mixture experiments and experiments in which all factors may be varied independently is most clear when only two or three factors are considered. Suppose there are two factors in an investigation into the nutritional benefit of different types of bread. Families may be classified by the proportion of bread they consume which is made from white flour (factor A) or the proportion made from wholemeal flour (factor B). If different diets based on 25%, 50%, 75% and 100% white bread are to be compared, then it follows that in these diets there will be 75%, 50%, 25% and 0% wholemeal bread. The two factors are exactly related, and varying the levels of one inevitably produces a corresponding variation in the levels of the other. Effectively there is only one factor for variation.

Suppose, however, that bread is classified into three classes: white flour (W), partially refined flour (PR), wholemeal flour (WM). We might now consider ten treatments:

	W:	PR:	WM:
(1)	25%;	0%;	75%;
(2)	50%;	0%;	50%;
(3)	75%;	0%;	25%;
(4)	100%;	0%;	0%;
(5)	25%;	25%;	50%;
(6)	25%;	50%;	25%;
(7)	25%;	75%;	0%;
(8)	50%;	25%;	25%;
(9)	50%;	50%;	0%;
(10)	75%;	25%;	0%.

Now changing the level of one factor does not inevitably change the level of another, (e.g. (1)→(2) does not change PR, (2)→(5) does not change WM) but the levels of the three factors are interrelated. The advantage of the choice of the ten treatments listed may be seen by examining Figure 18.10(a). The triangle represents the set of all possible diets in terms of the subdivision into W, PR or WM. With the restriction of including at least 25% W, the set of observations are spread evenly over the remaining area, in agreement with the general philosophy of factorial experimentation. If we attempt to impose a complete factorial structure for two factors, say W and WM, then the maximum levels of the two factors cannot total more than 100%. The form of experiment that would result from a complete factorial of, for example, 25%, 40%, 55% W, with 0%, 15%, 30%, 45% WM

Figure 18.10. Mixture designs. (a) Simplex design with at least 25% W, (b) factorial for W and WM factors. W = white flour, WM = wholemeal flour, PR = partially refined flour.

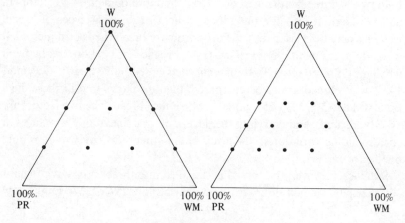

is represented in Figure 18.10(b) and plainly covers a much more restricted set of mixtures than the design of Figure 18.10(a).

The design in Figure 18.10(a) is described as a simplex design and is the form of design widely used for experiments on mixtures. It is not necessary for the levels of factors to vary widely, or even similarly. Some of the possible variations of form of simplex are shown in Figure 18.11. The concept of simplex designs generalises immediately to four or more factors.

Figure 18.11. Possible alternative simplex designs.

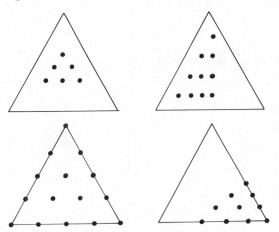

A very simple simplex in five factors with three levels each, and a more extensive design in four factors with four levels each, are tabulated in Table 18.15. In general, a simplex design with n factors and r levels of each factor results in

$$\frac{(n+r-2)!}{(r-1)!(n-1)!}$$

combinations.

As in response surface designs without restrictions on the sets of levels of factors it will often be found useful to add centre points to simplex designs. Of the three examples previously considered only the bread example already includes a centre point observation (50% W, 25% PR, 25% WM). Unless the number of factors is one less than the number of levels per factor then a centre point will not occur. For the designs of Table 18.15 we might add centre points:

(24, 19, 19, 19, 19) for Table 18.15(a),

and

(25, 25, 25, 25) for Table 18.15(b).

Table 18.15(a). *Simplex design for five factors with three levels each.*

A	B	C	D	E	A	B	C	D	E
20	15	15	15	35	30	15	15	15	25
20	15	15	25	25	30	15	15	25	15
20	15	15	35	15	30	15	25	15	15
20	15	35	15	15	30	25	15	15	15
20	15	25	15	25	40	15	15	15	15
20	15	25	25	15					
20	35	15	15	15					
20	25	25	15	15					
20	25	15	25	15					
20	25	15	15	25					

(b). *Simplex design for four factors with four levels each.*

A	B	C	D	A	B	C	D
10	10	10	70	30	10	10	50
10	10	30	50	30	10	30	30
10	10	50	30	30	10	50	10
10	10	70	10	30	30	10	30
10	30	10	50	30	30	30	10
10	30	30	30	30	50	10	10
10	30	50	10	50	10	10	30
10	50	10	30	50	10	30	10
10	50	30	10	50	30	10	10
10	70	10	10	70	10	10	10

Such centre points may be replicated more frequently than other combinations to concentrate the predictive precision towards the centre of the design.

As with factorial structures it is important to emphasise that it is not necessary to use a complete simplex design. The example of nitrogen applications for winter wheat can be viewed as containing a set of mixture designs at different levels of total nitrogen, with incomplete simplices at each level. This representation of the winter wheat experiment is displayed in Figure 18.12 with the levels of total nitrogen represented by different digits ($1 = 80$, $2 = 160$, $3 = 240$, $4 = 320$).

In the same way that the interdependence of factor levels leads to the replacement of the complete factorial structure by a simplex design symmetric in the factors, the response function models used for mixture designs must also differ from those for factorial structured treatments. The

Figure 18.12. Mixture design representation of the winter wheat trial.

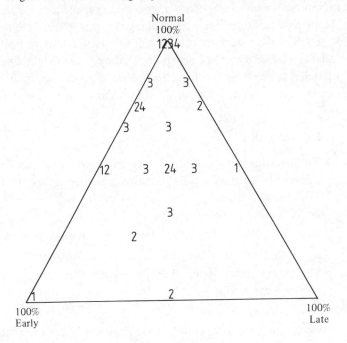

simple linear response model in three variables is

$$y = \beta_0 + \beta_1 x_1 + \beta_2 x_2 + \beta_3 x_3.$$

Because the factor levels are interrelated,

$$x_1 + x_2 + x_3 = 1.$$

One variable could be expressed in terms of the other two and the model written in terms of only two variables. However, this destroys the symmetry of the model, and a simple form retaining symmetry is obtained by rewriting the model in the form

$$y = \beta_0(x_1 + x_2 + x_3) + \beta_1 x_1 + \beta_2 x_2 + \beta_3 x_3$$
$$= \gamma_1 x_1 + \gamma_2 x_2 + \gamma_3 x_3,$$

where $\gamma_i = \beta_0 + \beta_i$. A similar reformulation for the second-degree model leads to

$$y = \gamma_1 x_1 + \gamma_2 x_2 + \gamma_3 x_3 + \gamma_{12} x_1 x_2 + \gamma_{13} x_1 x_3 + \gamma_{23} x_2 x_3,$$

and extensions to third-degree polynomials or the use of reciprocal terms can also be used.

By their nature, mixture designs are mainly concerned with the more general exploration of response surfaces discussed in this chapter rather than the particular parameter estimation of the previous chapter. The other important ideas of this chapter, concerning sequential experimentation, blocking of observations, controlling the prediction precision and testing the fit of the model are all relevant to mixture designs. For details the reader is referred again to Cornell (1981), though careful thought about the principles considered in this chapter should make the modifications required for application to mixture designs clear.

Exercises 18

(1) To investigate the effect of two fertilisers, nitrogen (N) and phosphate (P), on the yield of maize, an experiment consisting of 13 observations was performed using the following coded pairs of levels (x_1 and x_2) of N and P:

$$x_1 \quad -1 \ -1 \quad 1 \ 1 \ -\sqrt{2} \ \sqrt{2} \ 0 \quad \ \ 0 \quad 0\ 0\ 0\ 0\ 0$$
$$x_2 \quad -1 \quad 1 \ -1 \ 1 \ 0 \quad \ \ 0 \quad -\sqrt{2} \ \sqrt{2} \ 0\ 0\ 0\ 0\ 0.$$

The experimenter assumed that the yield of the ith observation may be represented by:

$$y_i = \beta_0 + \beta_1 x_{1i} + \beta_2 x_{2i} + \beta_{11} x_{1i}^2 + \beta_{12} x_{1i} x_{2i} + \beta_{22} x_{2i}^2 + \varepsilon_i,$$

where the ε_i are independent and identically distributed with zero mean and variance σ^2.

Find the least squares estimators of the coefficients of the model, and show that the estimated yield for any combination of levels (x_1, x_2) has variance

$$\sigma^2 \left(\frac{1}{5} - \frac{3\rho^2}{40} + \frac{23\rho^4}{160} \right),$$

where $\rho^2 = x_1^2 + x_2^2$. Comment on the implications of this result.

(2) In the investigation of a three-factor response surface, a cuboctahedron design is used, consisting of the 12 points $(0, \pm1, \pm1)$, $(\pm1, 0, \pm1)$, $(\pm1, \pm1, 0)$, plus four observations at the centre point $(0, 0, 0)$. Show how to fit the second-order surface:

$$Y = \beta_0 + \sum_{i=1}^{3} \beta_i x_i + \sum_{i=1}^{3} \sum_{j=1}^{3} \beta_{ij} x_i x_j.$$

Obtain variances of the estimates of β's and indicate which estimators are correlated. What test of the adequacy of the second-order fit is available?

(3) The expected value of a response depends linearly on factor values $x_1, \ldots, x_k$, the errors of the responses being independently distributed with constant variance, σ^2. Prove that, whatever

design is used with n observations, the variance of the estimated response at the centroid of the design points is σ^2/n.

Table 18.16.

Design	Design points	No. of observations
A	$(0, -1)$	$n/8$
	$(1, 1)$	$3n/8$
	$(-1, 1)$	$3n/8$
B	$(1, -1)$	$n/8$
	$(-1, -1)$	$n/8$
	$(1, 1)$	$3n/8$
	$(-1, 1)$	$3n/8$

With $k = 2$, the designs A and B shown in Table 18.16 both have their centroids at $(0, \frac{1}{2})$. Verify directly that the estimated response at this point has variance σ^2/n. Make a thorough comparison of the designs with respect to other properties that would be relevant.

(4) A design is required for an experiment on three quantitative treatment factors, T_1, at five equally spaced levels, and T_2 and T_3 each at two levels, in two blocks of ten plots each. The equation

$$E(y) = \beta_0 + \beta_1 x_1 + \beta_{11} x_1^2 + \beta_{111} x_1^3 + \beta_{1111} x_1^4 + \beta_2 x_2 + \beta_3 x_3$$
$$+ \beta_{12} x_1 x_2 + \beta_{13} x_1 x_3 + \beta_{23} x_2 x_3 + \beta_{123} x_1 x_2 x_3$$

is to be fitted to the results; x_1, x_2, x_3 are the levels of the three factors. Construct a design confounding β_{23} as little as possible and leaving all other effects in the equation unconfounded.

Calculate the loss of information on β_{23}.

Determine which effects, if any, will be biased by variation in the quadratic effect of T_1 between different levels of T_2 and T_3. (The orthogonal polynomial quadratic coefficients for five equally spaced levels are 2, -1, -2, -1, 2.)

PART IV

CODA

/

0

Designing useful experiments

0.0   ***Some more real problems***
Many of the problems considered in previous chapters have
occurred during statistical consultancy sessions and are appropriate to motivate this chapter. Typical examples are the peppers
experiment in Chapter 7 and the rabbits experiment in Chapter
16. Three more problems occurred in quick succession within
three weeks in early 1984 and illustrate a wide range of practical
design problems.

(*a*) Problem 1: the moving cups. In assaying the chemical concentration of liquid samples, a set of about 24 samples can be
automatically assayed in a single batch. The samples are placed in
cups which are held in position on a circular disc, and the disc
rotates so that each cup in turn appears beneath the assay
machinery. It is required to investigate changes in the concentration of chemicals over time, and therefore the 24 samples are
assayed over a time period of between one and two hours. Two
shapes of cup are available, and two sizes of sample are possible
within each cup. Three chemicals, sodium (Na), chlorine (Cl) and
potassium (K) are of interest, and it is proposed to have two levels
(zero and some) for each chemical. The cups may be covered
during the run of 24 samples until each is sampled, but the covers,
if used, must be used for all 24 samples in the run. Given that the
change of chemical concentration with time is the primary
interest, and that the experimenter wishes to examine the effects
on this change of varying cup shape, sample size, cover, Na, Cl
and K levels, how should an experiment consisting of eight to ten
complete runs, of 24 samples each, be planned?

(*b*) Problem 2: mowing and species. The experiment is to examine the effect on germination of weed seeds of different mowing treatments for grass, within which the weed seeds are spread. A set of 15 mowing treatments is proposed, including various combinations of mower type, height of blades, mowing frequency, and collection of clippings (a subset of a $3 \times 2 \times 2 \times 2$ factorial structure). There are ten weed species whose germination is to be examined. The minimum practical plot size for a mowing treatment is $1.5 \, m \times 1 \, m$ and, within that plot, there is space for two areas within which seeds of two different species can be broadcast. Given that the primary objective is the comparison of seed germination rates for different mowing treatments, both overall and also for each weed species, with comparison of germination rates for the different weed species a secondary objective, how should the experiment be designed? The total resources available allow five complete replicates of the 15×10 combinations (750 mini-plots). What size plot for each mowing treatment should be used, how many weed species per plot, and how should the weed species be chosen for each plot?

(*c*) Problem 3: recognising daddies. This is a design problem only in retrospect, but it is a design problem nonetheless, since before the analysis can be performed the design must be identified. Each of 32 young children were asked to draw two pictures of a person. Each picture was then scored for various aspects of anatomical accuracy. Each child was asked to draw (i) a man, and (ii) their daddy; the drawing was to use either (i) coloured crayons, or (ii) black pencil on white paper; and the order of the two requests could be (i) daddy first, and man second or (ii) man first and daddy second. Each of the four possible patterns:

(i) 1st, man, colour;   2nd, daddy, black and white
(ii) 1st, man, black and white;   2nd, daddy, colour
(iii) 1st, daddy, colour;   2nd, man, black and white
(iv) 1st, daddy, black and white;   2nd, man, colour

was used with each pattern given to eight children. Given the set of two scores for each of 32 children, how should the effects of man/daddy, colour/black and white, and order be estimated, together with their interactions? And how does the analysis alter when, as actually happened for one set of scores, three of the 32 children failed to produce drawings?

0.1 Design principles or practical design

Any book on experimental design develops the concepts on which the subject is based in a logical, systematic manner so that the principles can be clearly understood. In doing so, the author inevitably produces a sequence of examples, which are essential to the orderly progress of the subject, but which, inevitably, offer themselves as a set of recipes of available designs. I have tried in this book to put a greater emphasis on principles and on the intuitive application of those principles, with rather fewer standard design examples, but the book still contains a large number of potential recipes. The problem with this seemingly inevitable format is that it is not appropriate to the statistician, or experimenter, when they are considering how to design an experiment to satisfy a particular set of practical conditions.

The designer of an actual experiment is required to produce a design appropriate to a very particular set of circumstances. But, except where the designer is very experienced, he or she will not be able to assess the entire spectrum of design ideas, and make decisions about the design, from the basis of a comprehensive knowledge of how all the principles of design might relate to this particular problem. Instead 'designing an experiment' has very frequently been interpreted as picking a design recipe which hopefully fits the particular circumstances and, if necessary, making compromises in the objectives of the experiment and the structure of the experimental material in order to fit the problem into the recipe's requirements. As evidence of this tendency to Procustean design, it is only necessary to observe the very large proportion of experiments, perhaps as high as 85%, which take the form of randomised complete block designs. If the principles of blocking are being correctly applied then we have to accept that in every case (*a*) the natural block structure of the experimental units consists of blocks of equal size, and (*b*) the natural block size which gives the most efficient comparison of treatments happens to be exactly the same as the number of treatments required for the objectives of the experiment. This is a level of coincidence that any statistician should find hard to accept.

Ideally, a book about designing experiments should start from the problems which the experimenter has to solve, and should develop the subject from the viewpoint of the needs of the experimenter. Now this pattern patently cannot be used since, without an understanding of the basic principles, the solution of problems would proceed in a disorganised and unintelligible fashion. The standard format of books on design is inevitable, and therefore correct. But it is not sufficient. A book must also present the other point of view, and it is the purpose of this chapter to try to describe how experiments should be designed. It is Chapter 0 because it should come first, but it is not placed first because it requires all the other chapters, or at least some parts of them, to precede it. The stimulus for this chapter comes from a dinner conversation with Richard Jarrett, to whom I am much indebted.

The process of designing an experiment is described from the viewpoint of a statistician being fully involved in the design of an experiment for a research scientist. I believe that this is the only efficient approach to designing experiments. No non-statistician has enough knowledge, without expert statistical advice, to design the best possible experiment on all occasions. And, unless the statistician is prepared to be deeply involved in the work of the research scientists and to have extensive discussion with the experimenter he will not always be able to give fully relevant advice.

The need for an experiment arises from a question or set of questions to which the research scientist wants to find answers. The questions will imply the form of experimental unit which is to be used, though there may often be some flexibility in the precise definition of an experimental unit. Thus, questions about drugs purporting to cure cancer in humans have, at some stage, to be answered by trying the drugs on individual men and women: a person is a unit! Different pasture systems must be tried on substantial areas of land, and the systems assessed either by harvesting the grass or by grazing the pasture with animals. The experimental unit could be a defined area of pasture, or it could be the area of pasture plus a group of animals grazing the pasture. To answer questions about optimal conditions for a chemical process, the experimental unit will require the operation of the process on a particular machine for a specified length of time, the length of time being chosen by the experimenter.

Having recognised the questions to which the experiment must give answers and the form of experimental unit to be used, then the next two stages are taken in parallel. First, the available resources of experimental units must be assessed. Second, the formulation of experimental treatments must be made to fit the questions. As has been emphasised previously, the separation of these two aspects of design is essential. The third, and final, stage consists of combining the two components, units and treatments, to produce the experimental design.

0.2   Resources and experimental units

The experimenter, having considered the general objective of the experiment, has identified the natural experimental unit. There are then two aspects of the set of units which must be considered. The first is the total amount of resources that can be made available for the experiment. Sometimes there is a clear and simple limit. The number of animals available for a nutrition experiment may be restricted to a set of, say, 25, in the required condition at the time of experiment; there may be a further limitation that the animals are available for 16 weeks only. Sometimes the limit is not so clearly defined, but it may be assumed by those responsible for allocating resources that the number of units for a particular type of experiment could not be greater than, say, 100. For example, in a chemical industrial experiment in which the unit is a day's production, it might be assumed that no experiment could exceed two months, because of the restrictions on normal practice imposed by experimental conditions. A further type of limitation is that imposed by the back-up facilities available for chemical analysis of samples from the experimental units, and by storage space for material awaiting analysis. Whatever the cause of the restrictions on resources, it is important to recognise the limits from the outset of the planning of the experiment, so that the choices available to experimenter and statistician are clearly defined.

The second consideration is the structure of the experimental units. Structure can be identified in two forms: the inevitable structure and the suspected structure. Inevitable structure may arise from the choice of units. If it is decided that the unit should be a period of time for a particular animal, and the same pattern of periods is to be used for each of the available animals, then there will be a two-dimensional structure of animals × periods. Alternatively if each patient tests a sequence of different drugs, without control of the times at which the drugs are taken, then there is a one-dimensional structure of units within patients. It is important in recognising structure to distinguish between the double classification crossed structure, in which the pattern of units for an animal or patient is identical for each subject, and the simple single classification structure, in which the unit classification is not consistent across subjects.

Suspected structure is concerned with using the practical experience of the experimenter and statistician to try to anticipate patterns of variation in the output or analysis of the units. Thus, in animal or medical trials, there are many characteristics of the animal or patient which could be recorded before the experiment, and which might be thought to be relevant to the future performance of the animal or patient; age, sex, size, condition and a wide range of variables measuring previous performance or environment might all be candidates for defining suspected structure relevant to the design of the experiment. In experiments with field crops, the basis for suspected structure becomes less well defined, and plots may be expected to produce more or less similar yields according to geographical closeness, in addition to similarities arising from physically defined variables such as moisture and water levels, height, slope, and so on. Potential characteristics defining suspected structure for other forms of experimentation have been discussed in Chapters 2 and 7.

In preparing the definition of the set of units to be available for the experiment, it is essential that the initial identification of structure be made *without consideration of the detailed choice of treatments to be used.* The clear identification of the structure of the available units is probably the major factor in determining the precision of the final experiment, and unnecessary compromises introduced into the structure of the experiment by anticipating the treatment allocation stage may make the experiment less precise.

The considerations of structure may result in various degrees of definition of the unit structure to carry forward to the treatment allocation stage. Thus possible definitions of structure could include:

(i) Forty-eight units available in a 16×3 cross-classification structure, the 16 super-units being themselves grouped into four groups of three and one of four.

(ii) Between 40 and 60 units which will be grouped into three major groups of between ten and 25 units, with a further desirable restriction that at most five units can be managed in a similar way at the same time, so that groups of up to five units should be defined within the major groups.

(iii) Seventy-five units available in 14 groups containing, respectively, eight, eight, seven, six, six, six, six, five, five, four, four, four, three, three units.

(iv) Up to 60 units available with a desirable group structure in which each group should contain at most eight units, though nine or even ten units might just be acceptable.

(v) About 80 units likely to be available (spread over a considerable time period) with classification into groups defined by four major factors giving 16 groups, the size of which might be expected, in advance, to vary between 0 and 15.

Any less specific description of the information about resources and structure of the available units will usually indicate that the experimenter and statistician have not thought sufficiently about the units to be used in the experiment.

This discussion of suspected structure has been mainly in terms of only a single grouping of units, but there will often be more than one possible basis of blocking. The experimental units may therefore be identified within a two-dimensional structure, with each unit being classified into groups in each of two separate classifications, each of which is comprehensive in that all units are classified in each classification. In many situations, it will not be clear how many blocking classifications are sensible. What can always be achieved, however, is a list of possible classification systems, and a rough ordering of likely importance, ranging from essential to almost certainly negligible. In most practical situations one or two blocking classifications will appear to be clearly important, and sometimes there will be perhaps one or two further classifications that might be important. At the conclusion of the consideration of the structure of the experimental units, the units will usually have been classified in a single set of blocks or a double blocking classification (as discussed in Chapter 8). A blocking classification system may be based on a combination of blocking factors if there is no reason to believe that differences between levels of one blocking factor will be consistent over different levels of another blocking factor. The use of a double blocking classification implies such consistency. It will rarely be appropriate to use no blocking system or to use a treble or higher-dimensional blocking system. Information on classifications which are not used to provide blocking structure should always be recorded for possible later use as covariates. The information about multiple blocking patterns should be identifiable in the same way which we have illustrated for a single blocking system.

0.3   Treatments and detailed objectives

In Chapter 12, we discussed several types of treatment structure and the objectives implied by each type. We now approach the problem from the more valid viewpoint of considering the objectives first, and then deducing the treatments. Scientifically, this latter approach is obviously correct, and it must be said that it is astonishing how many scientists arrive at a statistician's office for discussions of experimental design or, more frequently, for analysis of experimental data, with well-defined treatments, but with no clear idea of the questions for which the treatments should provide answers. A reasonable test of whether the experimental treatments have been chosen appropriately to answer well-defined questions is whether the detailed form of analysis of treatment variation is clearly defined. If the experimenter does not know how to analyse and interpret the treatment variation, then he or she has not properly considered the objectives of the experiment.

So how should the objectives of an experiment be identified? First, it is important to realise that experiments are rarely isolated events, but are usually part of an overall research structure. Individual experiments must be planned in the context of an overall structure, and this means not only that each experiment must respond to the results of previous experiments, but also that the prospect of future experiments should not be ignored in the planning of the current experiment. Overall, research structures are broadly classifiable into two classes; in the first, the objective is to seek an optimum form of operation of a system; the second is concerned with developing an understanding of the mechanics of a system; a third stream of research experiments arises from requirements to satisfy regulations concerning testing of materials.

The search for an optimal mode of operation is more technological than scientific, and can be characterised by a reasonably well-defined objective, and by definition of the ranges of the components of the system, or factors, which can be used. Thus, in considering the growth of a mixture of two crops, maize and beans, simultaneously on the same piece of land, the research project should allow for variation of the spatial pattern of arrangement of the two crops, nutrient provision, management of the crops, particularly timing and control of pests, and varieties or genotypes of the crops. A breeding programme might be attached to the research project or it may be that only varieties previously available are to be considered. There may be limits on the costs of material inputs or management materials to make the investigation relevant to the farmers for whom the research findings are intended. A quite different example would be in the examination of the effects of drugs, for example alcohol, on performance of human beings. Here, it is not so much an optimum but a reasonable limit that is the primary interest. The research will be limited to some particular aspects of performance, and this limitation will have implications for the set of conditions under which performance is observed. For example, we may be interested in the effect of alcohol on performance of drivers, and this would eliminate consideration of persons below the minimum age for driving, and would also eliminate from consideration individuals who were unable to drive or who were restricted in possible alcohol intake. We might wish to consider the variation of forms and amounts of alcohol, of condition of the persons receiving the alcohol, and the environment within which the driving was to be performed.

In any experiment within an optimum seeking research programme, the objectives for each particular experiment must be considered. A major question will usually be how many aspects of variation, or factors, to consider simultaneously in an experiment. This will depend on the stage of the experiment within the research programme. At the beginning of the programme, it is important to include many factors, because failure to investigate the interaction of factors at an early stage can easily lead to wasting resources through pursuing too narrow a line of research before eventually checking on the effects of other factors. It is difficult to conceive of an investigation with too many factors at an early stage. The number of levels per factor will normally be small at early stages, because only broad patterns of response can be examined.

In development stages of research programmes, experiments will predominantly be concerned with between two and four factors, with other factors included only to provide checks against changes of interaction patterns. The number of levels may increase but, unless it is proposed to consider very complex models in the analysis of the data, it is probable that three or four levels chosen to optimise the information required about the effect of each factor will be adequate for most purposes. The systematic designs of Chapters 14 and 17 depart from this pattern in order to use the resources more efficiently. The majority of experiments involving quantitative factors will be concerned directly with the estimation of the optimal levels of the factors and three levels, sufficiently widely spaced to allow precise estimation of the optimum, will usually be sufficient.

Late in the research programme, the major concern will be to provide overall information on the optimal conditions, and on the deviations from the optimum, and here the multi-factor designs for estimating response surfaces will often be appropriate, with many factors included, partly to provide simultaneous information about the effects of all important factors, but also to continue to provide some checks on interaction information.

In summary then, in technological experiments the important decisions about treatments concern the number of factors and numbers of levels. The factorial structure is the natural one from which to commence consideration of experimental treatments, but increasingly, as the knowledge of the research programme expands, complete factorial structures will be found to include combinations known to be ineffective, and these should be omitted. In general, experimenters have not sufficiently accepted the requirement for using many factors, particularly in the early stages of investigations, and this leads to much inefficient use of resources. The numbers of levels should usually be two (earlier) or three (later) except for response surface designs, which, because they relate to the whole surface and not to individual factors, may have up to five levels.

In research programmes whose primary purpose is to develop understanding of underlying mechanisms, the objectives are different, and the experiments will be correspondingly different. One major difference will be in the pattern of observations made for each unit. For the technological experiment, there will usually be a single measure of 'yield' or 'output', possibly with some subsidiary measures. In the scientific experiment, there will often be a series of measurements, often spread over time, and this will tend to reduce the total number of units that can be managed. It may be necessary to increase the number of levels of a quantitative factor to allow for uncertainty about the appropriate form of response model. Nonetheless, it is difficult to believe that large numbers of levels are particularly desirable, since this leads to a diffusion of information. It is of little use having information about very many levels, to guard against the occurrence of a sudden change which may occur at any level, because the precision of information at each level is likely to be so poor that detection of any particular sudden change is improbable.

This discussion may seem too much concerned with general good sense that the principles of use to an experimenter may be very opaque. More specifically then the questions an experimenter should ask are.

(1) Where does the present experiment occur in the overall research programme? Are the implications of previous experiments well understood? What are the questions arising from previous research that should be answered in this experiment? And how much future experimentation should be expected?

(2) What are the precise questions that the experiment should answer? There will usually be several questions. Various possible forms of question are:

(i) does the difference between A and B cause a change in yield? How big is this change?

(ii) Is this difference affected by other changes?

(iii) What is the yield response to change in the stimulus level of a factor?

(iv) Which of a range of alternative procedures produces the highest yield?

In most experiments, the total number of questions asked should correspond to between 10 and 50 df. The idea of relating degrees of freedom to questions arises from the analysis to be performed on the experimental results, but it is essential that this is considered at the design stage. A question of form (i) is a single degree of freedom question. Those of type (ii) may be single or multiple degrees of freedom, depending on the degree of complexity of interactions that is to be allowed. A single treatment factor may produce a single degree of freedom question of type (i), or more than one. For example, if an additional constituent is considered for a recipe, it may be added at either of two times. There are then two questions: does the constituent affect the finished product; and does the timing affect the product? In terms of treatment contrasts, these may be represented as $(-2, +1, +1)$, $(0, -1, +1)$.

For yield response to a quantitative factor, the number of degrees of freedom is equal to the number of parameters in the proposed response function. For multi-factor response functions, the number of degrees of freedom is equal to the number of parameters in the multi-variable response function. If a large number (n) of alternative procedures (synthesised drugs, breeding lines) are being compared, then $(n-1)$ degrees of freedom are implied.

0.4   The resource equation

A major principle of design arises from the consideration of degrees of freedom for treatment comparisons. The total information in an experiment involving N experimental units may be represented by the total variation based on $(N-1)$ degrees of freedom. In the general experimental situation, this total variation is divided into three components, each serving a different function:

(a) the treatment component, T, being the sum of degrees of freedom, corresponding to the questions to be asked;

(b) the blocking component, B, representing all the environment effects to be allowed for. Usually $(b-1)$ degrees of freedom for the b blocks, but other alternatives are $(r-1)+(c-1)$ for row and column designs, or more generally, all effects in multiple control or multiple covariate designs. It may also be desirable to allow a few degrees of freedom for unforeseen contingencies such as missing observations.

(c) The error component, E, being used to estimate s^2 for the purpose of calculating standard errors for treatment comparison, and, more generally, for inference about treatments.

We can express this division of information as *the resource equation* in terms of the degrees of freedom used in each component T, B and E:

$$T+B+E=N-1.$$

Now, to obtain a good estimate of error, it is necessary to have at least 10 df, and many statisticians would take 12 or 15 df as their preferred lower limit, although there are situations where statisticians have yielded to pressure from experimenters and agreed to a design with only 8 df. The basis for the decision is most simply seen by examining the 5% point of the t-distribution, and using sufficient degrees of freedom so that increasing the error degree of freedom makes very little difference to the significance point, and hence to the interpretation.

To return to the implications for design, it is necessary to have at least 10 df for error, to estimate σ^2 and, in choosing the number of treatment question degrees of freedom, we should not normally allow E to fall below 10. But equally, if E is allowed to be large, say greater than 20, then the experimenter is wasting resources by not asking enough questions. This is a failure of experiments which occurs much more frequently than allowing E to fall too low, and it is just as much a failure of design. The basic economic advantage of factorial experiments arises from asking many questions simultaneously and to insist, as many scientists do, on asking only one or two questions in each experiment, is to waste resources on a grand scale.

One final comment on this question of efficient use of resources. It is of course possible to keep the value of E in the region of 10 to 20 by doing several small experiments, rather than one rather larger experiment in which the total number of experimental units is much less than the sum of the units in the separate small experiments. This is inefficient because of the need to estimate σ^2 for each experiment. The experimenter should identify clearly the questions he wishes the experiment to cover, and he should also consider carefully if he is asking enough questions to use the experimental resources efficiently.

0.5   The marriage of resources and treatments

One of the recurring themes of this book has been that the available computing facilities have very substantially reduced the restrictions of possible designs, and the main focus of this modern freedom of design is at the stage of allocating treatments to units within a natural blocking structure. The reader should, by this stage, be expecting to be told that, having decided the natural blocking structure and the relevant treatment structure embodying all the desired questions, the allocation problem is simple. And he would be correct, because it is. Nevertheless it is necessary to consider how to combine the ideas of the previous chapters when solving a particular problem.

First, we should check whether there are, within the set of experimental treatments, restrictions on the possible size of units to which some factor levels can be applied. Is it necessary to consider split unit designs, and, if so, what are the restrictions imposed by the practical nature of the experiment? If some treatments require larger units than the proposed experimental unit size, then information about treatment comparisons will be divided between (at least) two strata. Each stratum should have an adequate error degree of freedom, and this could lead to using more factors as main plot factors than is strictly necessary. For example,

Table 0.1.

Main plot		Split plot	
Source	df	Source	df
Blocks	2	Q	1
P	1	R	1
Error	2	S	1
		two-factor Xns^2	6
		three-factor Xns	4
		four-factor Xns	1
		error	28

suppose a 2^4 treatment structure is proposed, with three replicates, in three blocks of 16 units. If one factor, P, requires larger units than the basic unit (of which there are 48), then the natural design is to use two main plots in each of three blocks. Each main plot will be split into eight split plots for the other eight treatment combinations giving an analysis of variance structure as shown in Table 0.1. This design provides virtually no information about the precision of estimation of the effect of P. Now, suppose that, although the levels of P require units bigger than the basic unit, a main plot of four basic units is adequate for a level of P. Then better information about P can be obtained by having four main plots within each block, and four split plots within each main plot. To use such a design, other treatments must be allocated to main plots, and the unthinking split plot user might decide to allocate factor Q also to main plots, giving an experiment and analysis of variance structure as shown in Table 0.2. However, the reader of Chapters 15 and 16 should argue that the effects, in addition to P, allocated to main plots should be those of least

Table 0.2.

Main plot		Split plot	
Source	df	Source	df
Blocks	2	R	1
P	1	S	1
Q	1	RS	1
PQ	1	other two-factor Xns	4
Error	6	three- and four-factor Xns	5
Err		error	24

interest, since main plots are likely to be less precise. Therefore, the other effects to be allocated to main plots should be the three-factor interaction QRS, and the interaction with P, PQRS. The design is then as shown in Figure 0.1, in which block I has not been randomised to display structure more clearly, but blocks II and III have been randomised. The analysis of variance structure is as shown in Table 0.3. Usually it would be decided, before the experiment, to 'pool' the sums of squares for QRS and PQRS with the main plot error to provide 8 df for error in the main plot analysis.

The purpose of this example is to illustrate that the ideas of different

Figure 0.1. Split unit design with P and QRS as main plot effects.

Block I

Main plot 1	Main plot 2	Main plot 3	Main plot 4
pqrs	p	(1)	qrs
pq	prs	qs	r
pr	pqs	rs	q
ps	pqr	qr	s

Block II

qs	ps	prs	s
(1)	pq	pqr	qrs
qr	pqrs	pqs	q
rs	pr	p	r

Block III

pqs	q	pqrs	rs
prs	r	ps	qr
p	qrs	pq	(1)
pqr	s	pr	qs

Table 0.3.

Main plot		Split plot	
Source	df	Source	df
Blocks	2	Q	1
P	1	R	1
QRS	1	S	1
PQRS	1	two-factor Xns	6
Error	6	other three-factor Xns	3
		error	24

chapters must be considered together. The necessity of a split plot design does not mean that the ideas of confounding should be ignored. Similarly, if an experiment has two levels of one factor, and a large number of levels of a second factor (testing breeding lines under two environments), and the two levels of the first factor must be applied to large units (e.g. of size = six basic units), the necessity of a split plot design does not require that all levels of the second factor must appear as split units within each main plot. Suppose there are 18 levels of the second factor, then the simple split plot design has two blocks each containing two main plots, each of which contains 18 split plots. Each main plot is now very large, and variation between split plots within main plots will be unnecessarily large. If it is felt that main plots should include no more than six split plots for the within main plot variation, σ^2, to be acceptably small and properly estimable, then the proper design will consist of two blocks, each block containing six main plots, three for level 1 of the first factor, and three for level 2; the levels of the second factor will be allocated in groups of six to the total of 12 main plots, so that the concurrence of pairs of levels is as even as possible, which requires that most pairs of levels occur once together with a few zero and double joint occurrences. A possible design is shown in Figure 0.2 (not randomised). Note that testing the 18 levels as split plot treatments implies that the design allocation of these 18 to main plots of six split plots each is resolvable within each set of three main plots with the same main plot factor level. The joint occurrences are therefore less even than would be achieved in a non-resolvable design.

Leaving the topic of split plot designs, remembering that the comments relevant to other designs apply in some measure to split plot designs also, we need to bring together the various ideas relating to small block sizes for large numbers of experimental treatments. The two situations discussed at length in previous chapters concern the allocation of treatments to small blocks, when all treatment comparisons are of equal importance (Chapters 7 and 8) and when there is a clear priority ordering for factorial effects with main effects more important than two-factor interactions, which in turn are more important than three-factor interactions (Chapters 15 and 16).

In practice, it will often be true that neither of these situations exactly fits the experimental requirements. The general situation is that there are a number of treatment comparison questions, of which there will be a subset felt to be rather more important than the rest. If the natural block size pattern does not allow all experimental treatments to occur in each block, then the following steps may be helpful:

(1) Decide if any subset of treatments are sufficiently important that replication of individual experimental treatments should vary.

Figure 0.2. Experimental plan for a 2×18 structure in two blocks each containing six main plots, each containing six split plots.

Block I

Main plot

1	2	3	4	5	6
$p_0 q_1$	$p_0 q_7$	$p_0 q_{13}$	$p_1 q_1$	$p_1 q_3$	$p_1 q_5$
$p_0 q_2$	$p_0 q_8$	$p_0 q_{14}$	$p_1 q_2$	$p_1 q_4$	$p_1 q_6$
$p_0 q_3$	$p_0 q_9$	$p_0 q_{15}$	$p_1 q_7$	$p_1 q_9$	$p_1 q_{11}$
$p_0 q_4$	$p_0 q_{10}$	$p_0 q_{16}$	$p_1 q_8$	$p_1 q_{10}$	$p_1 q_{12}$
$p_0 q_5$	$p_0 q_{11}$	$p_0 q_{17}$	$p_1 q_{13}$	$p_1 q_{15}$	$p_1 q_{17}$
$p_0 q_6$	$p_0 q_{12}$	$p_0 q_{18}$	$p_1 q_{14}$	$p_1 q_{16}$	$p_1 q_{18}$

Block II

Main plot

1	2	3	4	5	6
$p_0 q_1$	$p_0 q_2$	$p_0 q_3$	$p_1 q_1$	$p_1 q_2$	$p_1 q_3$
$p_0 q_4$	$p_0 q_5$	$p_0 q_6$	$p_1 q_6$	$p_1 q_4$	$p_1 q_5$
$p_0 q_7$	$p_0 q_9$	$p_0 q_8$	$p_1 q_{10}$	$p_1 q_8$	$p_1 q_7$
$p_0 q_{10}$	$p_0 q_{12}$	$p_0 q_{11}$	$p_1 q_{12}$	$p_1 q_9$	$p_1 q_{11}$
$p_0 q_{13}$	$p_0 q_{14}$	$p_0 q_{15}$	$p_1 q_{14}$	$p_1 q_{13}$	$p_1 q_{16}$
$p_0 q_{16}$	$p_0 q_{17}$	$p_0 q_{18}$	$p_1 q_{15}$	$p_1 q_{18}$	$p_1 q_{17}$

The relative replication of the experimental treatments should be determined to reflect this relative importance. A simple example of this is the use of a positive control treatment when all treatments are to be compared directly with the control. Here, the relative replication of the control should be between 1 and $\sqrt{n}$, where there are n treatments for comparison with the control. Alternatively, the control may be required only for general background information, and then the relative replication of the control may be less than one (i.e. only one or two control plots in the whole experiment). More complex situations might involve, say, a $2 \times 2 \times 3$ factorial structure plus six additional treatments consisting of a subset of the $2 \times 2 \times 3$ structure for a changed level of some new factor. The structure of the treatments might be as in Figure 0.3.

Figure 0.3. Treatment set of 18 from 24 'possible' combinations from a $2^3 \times 3$ structure.

		$p_0 q_0$	$p_0 q_1$	$p_1 q_0$	$p_1 q_1$
	r_0	x	x	x	x
s_0	r_1	x	x	x	x
	r_2	x	x	x	x
	r_0	x	x		
s_1	r_1	x	x		
	r_2			x	x

The experimenter then has to decide in terms of the important comparisons whether any or all of the s_1 combinations require duplication. To leave the replication of the s_1 combinations at the same level as the s_0 combinations means, for example, that the $(s_0 - s_1)$ comparison is less precise that those of $(p_0 - p_1)$ or $(q_0 - q_1)$.

Although the question of relative replication should be considered, I believe that in only a few cases will it be important to use differential replication. There are two reasons for this. It is often difficult to determine the relative precision required for different comparisons at all precisely. The achieved variation of precision (as measured by variances of treatment differences) is always relatively less than the variation of replication. And one side effect of unequal treatment replication is an overall reduction in precision. To illustrate this for a very simple three treatment set with a total of 24 observations, consider different replications of the treatments, and the standard errors of treatment differences as set out in Table 0.4. It can be seen that using an unequal

Table 0.4.

Replication			Standard error			
t_1	t_2	t_3	$t_1 - t_2$	$t_1 - t_3$	$t_2 - t_3$	$t_1 - (t_2 + t_3)/2$
8	8	8	0.5	0.5	0.5	0.43
12	6	6	0.5	0.5	0.57	0.41
16	4	4	0.56	0.56	0.71	0.43
12	8	4	0.46	0.57	0.61	0.41

replication pattern gives small improvements in precision for one or two comparisons, at the cost of a larger loss of precision for other comparisons. The greater concentration of resources on some comparisons is largely ineffective partly because of the general loss of efficiency resulting from the unequal replication. It is therefore usually sensible to assume equal replication of treatments in most cases.

(2) Decide which of the two basic models is more nearly relevant. Either (a) all treatment comparisons important or, (b) some treatment comparisons of negligible importance. The second situation will almost always occur when the treatments have factorial structure, with the unimportant effects being higher order interactions. For (a) the experimenter should consider whether there are comparisons which are more important and should make a list in order of decreasing importance. For (b), list the major comparisons, the negligible comparisons and an intermediate group which, while clearly not of negligible importance, are also less important than the major comparisons.

(3) For (2a), allocate treatments to units using the principles of making joint occurrence of treatments as nearly equal as possible but, where there must be some treatments with an extra replicate, or some pairs with an extra joint occurrence, choosing from the list of the more important comparisons.

(4) For (2b), allocate treatments to units, confounding effects of negligible importance with blocks and, if further effects must be partially confounded in different replicates, or fractions of replicates, choosing from the intermediate group of effect comparisons.

(5) After the treatment allocation, examine the actual precisions that are predicted for treatment comparisons, using a computer program to obtain an analysis of variance of dummy data, with σ^2 set equal to one to obtain the variance–covariance structure of treatment comparison estimates. Consider where the predicted variance pattern differs from that which might be desirable and try modifying the design by switching particular treatments or sets of treatments.

This last step is rarely attempted, but the facilities allowing the examination of the expected precision of comparisons over a range of alternative designs should now be possible with most computing facilities, and should become a regular part of the design process.

0.6   Three particular problems

We return now to the three problems posed at the beginning of this chapter.

Problem 1: The moving cups

The investigation of change of chemical concentration with time, within a run of a mechanical assay process, with additional factors of cup shape, sample size, use of covers and levels of Na, Cl and K.

Eight or ten runs of the assay process with 24 samples per run are available, and variation between runs is thought to be large compared with variation within runs. Therefore, the change with time must be estimated from within run differences. Although the change with time is probably well approximated by a linear regression, it was decided to use three times (three samples) for each combination of factors tested within a run, so that any curvature in the trend could be assessed. It was therefore decided that for each run the set of 24 samples should consist of three samples for each of eight combinations from the six factors additional to the time factor. There remained two main questions: how to arrange the three times × eight combinations within the 24 samples for each run, and how to choose the eight combinations for each run.

The objectives for the allocation of the three times × eight combinations are to optimise the precision of the estimation and comparison of the time change effects. If the three times × eight combinations are allocated randomly to the 24 samples, then the time interval from which the time effect is estimated for each combination will vary in an uncontrolled fashion (and could even be small for *all* combinations). Even if the 24 samples are split into three groups of eight, and the eight combinations allocated randomly within each group (early, middle, late), the precision of the regression effect will vary over combinations. The estimation and comparison of time-trend regression effects is made most precise by allocating the eight combinations randomly to the first eight samples, and repeating the order of combinations in the second and third eights.

The choice of eight combinations for each run requires the application of ideas of split units and confounding. The use of covers must be for a complete run or not at all. Therefore, covers is a main plot (run) factor. Should we choose two other main plot factors, and use three main plot and three split plot factors, or should we use four, or even five, factors within each run? Although the time repetition within each run is a complicating aspect of the design, it is completely irrelevant to the choice of combinations. To obtain maximum precision of main effects, and two-factor

Figure 0.4. Design plan for one run of the moving crops problem.

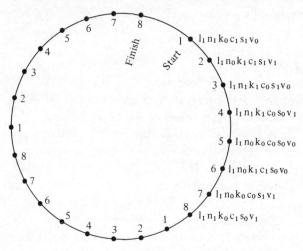

1 $\quad l_1 n_1 k_0 c_1 s_1 v_0$

2 $\quad l_1 n_0 k_1 c_1 s_1 v_1$

3 $\quad l_1 n_1 k_1 c_0 s_1 v_0$

4 $\quad l_1 n_1 k_1 c_0 s_0 v_1$

5 $\quad l_1 n_0 k_0 c_0 s_0 v_0$

6 $\quad l_1 n_0 k_1 c_1 s_0 v_0$

7 $\quad l_1 n_0 k_0 c_0 s_1 v_1$

8 $\quad l_1 n_1 k_0 c_1 s_0 v_1$

interactions of the five factors, and of the interactions of the time change effect with these factors, all five factors should be varied within each run, confounding two three-factor and one four-factor interaction. The confounded interactions actually selected were NaKCl, Na × shape × size, and the four-factor interaction, K × Cl × shape × size. The design pattern for one run is shown in Figure 0.4 and the allocation of combinations of levels of the six factors to the eight runs is shown in Figure 0.5 the factor labels used in the figure being N for Na, K for K, C for Cl, L for cover (lid), S for shape and V for volume (sample size).

The analysis of results is based in the linear and quadratic components of response to time (late − early; late + early − 2 (middle)). Each component is calculated for each of the eight combinations in each run. The analysis of variance structure is shown in Table 0.5.

Table 0.5.

	df
Covers	1
Main plot (run) error	6
Main effects (Na, K, Cl, S, V)	5
Two-factor interactions (including covers)	15
Three-factor interactions (excluding NaKCl, NaSV)	18
Split plot error	18
Total	63

Figure 0.5. Overall design for the moving crops problem.

Run 1	Run 2	Run 3
$l_1 n_1 k_0 c_1 s_1 v_0$	$l_0 n_1 k_1 c_0 s_0 v_1$	$l_0 n_0 k_1 c_0 s_0 v_0$
$l_1 n_0 k_1 c_1 s_1 v_1$	$l_0 n_0 k_1 c_1 s_0 v_0$	$l_0 n_0 k_0 c_1 s_1 v_1$
$l_1 n_1 k_1 c_0 s_1 v_0$	$l_0 n_1 k_1 c_0 s_1 v_0$	$l_0 n_1 k_1 c_1 s_0 v_1$
$l_1 n_1 k_1 c_0 s_0 v_1$	$l_0 n_1 k_0 c_1 s_0 v_1$	$l_0 n_1 k_0 c_0 s_1 v_0$
$l_1 n_0 k_0 c_0 s_0 t_0$	$l_0 n_0 k_0 c_0 s_0 v_0$	$l_0 n_1 k_0 c_0 s_0 v_1$
$l_1 n_0 k_1 c_1 s_0 v_0$	$l_0 n_1 k_0 c_1 s_1 v_0$	$l_0 n_0 k_0 c_1 s_0 v_0$
$l_1 n_0 k_0 c_0 s_1 v_1$	$l_0 n_0 k_0 c_0 s_1 v_1$	$l_0 n_0 k_1 c_0 s_1 v_1$
$l_1 n_1 k_0 c_1 s_0 v_1$	$l_0 n_0 k_1 c_1 s_1 v_1$	$l_0 n_1 k_1 c_1 s_1 v_0$

Run 4	Run 5	Run 6
$l_1 n_0 k_0 c_1 s_1 v_0$	$l_0 n_1 k_1 c_1 s_0 v_0$	$l_1 n_1 k_1 c_1 s_1 v_0$
$l_1 n_0 k_0 c_1 s_0 v_1$	$l_0 n_0 k_0 c_1 s_1 v_0$	$l_1 n_1 k_1 c_1 s_0 v_1$
$l_1 n_1 k_1 c_1 s_0 v_0$	$l_0 n_1 k_1 c_1 s_1 v_1$	$l_1 n_1 k_0 c_0 s_0 v_1$
$l_1 n_0 k_1 c_0 s_1 v_0$	$l_0 n_0 k_0 c_1 s_0 v_1$	$l_1 n_0 k_0 c_1 s_1 v_1$
$l_1 n_1 k_1 c_1 s_1 v_1$	$l_0 n_1 k_0 c_0 s_0 v_0$	$l_1 n_1 k_0 c_0 s_0 v_0$
$l_1 n_0 k_1 c_0 s_0 v_1$	$l_0 n_0 k_1 c_0 s_0 v_1$	$l_1 n_0 k_0 c_1 s_0 v_0$
$l_1 n_1 k_0 c_0 s_1 v_1$	$l_0 n_1 k_0 c_0 s_1 v_1$	$l_1 n_0 k_1 c_0 s_1 v_1$
$l_1 n_1 k_0 c_0 s_0 v_0$	$l_0 n_0 k_1 c_0 s_1 v_0$	$l_1 n_0 k_1 c_0 s_0 v_0$

Run 7	Run 8
$l_1 n_0 k_1 c_1 s_1 v_0$	$l_0 n_1 k_0 c_1 s_0 v_0$
$l_1 n_1 k_0 c_1 s_1 v_1$	$l_0 n_0 k_1 c_1 s_1 v_0$
$l_1 n_0 k_0 c_0 s_0 v_1$	$l_0 n_1 k_1 c_0 s_1 v_1$
$l_1 n_0 k_0 c_0 s_1 v_0$	$l_0 n_1 k_0 c_1 s_1 v_1$
$l_1 n_0 k_1 c_1 s_0 v_1$	$l_0 n_0 k_0 c_0 s_1 v_0$
$l_1 n_1 k_1 c_0 s_0 v_0$	$l_0 n_0 k_0 c_0 s_0 v_1$
$l_1 n_1 k_1 c_0 s_1 v_1$	$l_0 n_0 k_1 c_1 s_0 v_1$
$l_1 n_1 k_0 c_1 s_0 v_0$	$l_0 n_1 k_1 c_0 s_0 v_0$

Problem 2: Mowing and species

The primary interest was in the comparison of germination rates under the 15 mowing treatments, but the secondary interest, in the comparative germination of the ten weed species, was also a major objective of the experiment. No specific knowledge of the experimental area was available except that the variation between plots would be smaller for plots which were closer together.

Before proceeding further, it is necessary to ask why so many different treatments were needed in the experiment. When the proposed mowing treatments were examined, they were found to involve four factors, each with two or three levels. It was proposed to compare three mower types, two heights of cutting blades, two mowing frequencies, and the collection or non-collection of clippings. Of the potential 24 combinations, only three of the four combinations of height and frequency were feasible and one mower type did not permit collection of clippings; hence, the 15 combinations were the complete set of sensible factorial combinations and all were required to be able to assess main effects and interactions for each factor or combination of factors. The ten weed species resulted from an extensive selection process of the possible species and represent the major types of weeds occurring in grass. The statistician may often doubt whether as many species are needed as are often proposed in this kind of experiment, because of doubts whether the number of questions being asked really require as many as ten species, with 9 df for comparison. However, ultimately the statistician must accept the experimenter's decision on the number of species, after the statistician has made plain the benefits of concentration of resources on a smaller number of species to answer more specific questions more precisely.

The treatment structure consisted of 15 mowing treatments × ten weed species giving 150 combinations in a complete replicate. A mowing treatment required a minimum plot size of 1.5 m × 1 m. Each weed species required a plot of half that size. Three philosophies of design were discussed:

(a) Use large plots for each mowing treatment, sufficient to include all ten weed species. This is a split plot design with mowing treatments as main plots, species as split plots. It is clearly inappropriate because, in addition to the general disadvantages of split plot designs, the treatment comparisons of primary interest are applied to the main plots, and will therefore be the least precise.

(b) Divide the ten weed species into five pairs. Use 75 main plots for the 15 mowing treatments × five pairs, the two species in each pair being split plot treatments. This gives comparisons between mowing treatments as precisely as is achievable. Of the 45 comparisons between weed species, five (between the two species of each pair) are liable to be considerably more precise than the other 40. After consideration of the particular species, the experimenter concluded that it was not reasonable to select five pairs such that within pair comparisons were of considerably more importance than the remaining 40 comparisons.

(c) Still using 75 main plots in each replicate, five for each mowing treatment, use different pairings of species for each mowing. Choose pairings of weed species so that over the whole experiment the pairings of species would occur as evenly as possible. Since there were 45 pairs of species, which could be grouped in nine sets of five pairs, such that each species occurred once in each set, sets of five pairs could be allocated to the different mowing treatments so that each set was used once or twice in each replicate. This design pattern made the main effect comparison of species as even as possible. The precision of mowing treatment comparisons would be as for (b), except that there would be a small level of non-orthogonality of species and mowing treatment effects in the main plot level of the analysis.

Figure 0.6. Division of pairs of weed species into nine basic sets of five pairs each.

Set	1	AB	CD	EF	GH	IJ
	2	AC	BG	DI	EJ	FH
	3	AD	BH	CE	GI	FJ
	4	AE	BD	CF	GJ	HI
	5	AF	BJ	CH	DG	EI
	6	AG	BI	CJ	DF	EH
	7	AH	BC	DJ	EG	FI
	8	AI	BF	CG	DE	HJ
	9	AJ	BE	CI	DH	FG

Figure 0.7. Allocation of sets of weed species pairs to mowing treatments in each replicate.

Mowing treatment	I	II	III	IV	V
1	1	8	4	5	7
2	7	9	1	4	6
3	3	2	7	1	4
4	3	8	4	2	6
5	9	5	3	2	1
6	6	7	9	8	4
7	5	4	2	6	8
8	1	3	5	7	9
9	8	2	9	7	5
10	4	3	8	5	9
11	6	1	8	9	3
12	2	9	5	4	7
13	4	7	6	1	3
14	2	6	7	3	1
15	5	1	6	8	2

It was decided to use design philosophy (*c*). The experimental resources allowed five replicates. The identification of nine sets of five pairs, such that each set included all ten species and no pair was repeated in different sets, was a simple incomplete block design problem, and a solution is shown in Figure 0.6. It was decided to allocate sets to mowing treatments, in such a manner that

(i) for each mowing treatment, the sets were different in each of the five replicates,

(ii) in each replicate, the nine sets each occurred once or twice,

(iii) over the entire experiment, the nine sets each occurred eight or nine times. The allocation of sets to mowing treatments in the five replicates is shown in Figure 0.7.

Problem 3: recognising daddies

The problem was to identify the design, and hence the analysis, of the experiment. The pattern of the experiment is displayed in Figure 0.8. The

Figure 0.8. Structure of data for the 'recognising daddies' problem.

Children	Picture 1	Picture 2
1	Cd 1	Bm 2
2	Cd 1	Bm 2
.	.	.
.	.	.
.	.	.
8	Cd 1	Bm 2
9	Cm 1	Bd 2
10	Cm 1	Bd 2
.	.	.
.	.	.
.	.	.
16	Cm 1	Bd 2
17	Bd 1	Cm 2
18	Bd 1	Cm 2
.	.	.
.	.	.
.	.	.
24	Bd 1	Cm 2
25	Bm 1	Cd 2
26	Bm 1	Cd 2
.	.	.
.	.	.
32	Bm 1	Cd 2

set of eight possible observations consists of all possible combinations of three factors, each with two levels:

colour: C or B
daddy: d or m
order: 1 or 2.

The construction of the experiment ensures that each main effect is not influenced by differences between children. We may recognise children as blocks, in the design terminology of this book, and the two observations for each child as experimental units (within blocks). Examination of Figure 0.8 reveals not only that main effects are not confounded but that all three two-factor interactions are confounded between blocks (children), whereas the three-factor interaction is not confounded. The design is that which would have been constructed by methods of Chapters 15 or 16 if it had been determined that main effects must not be confounded. It seems likely that the design actually evolved along the philosophic lines of Chapter 15, ensuring that each child experienced each level of each factor (inevitable for order).

The analysis and interpretation of the results are of some interest, because they can be approached in various ways. The information, and hence the analysis, occurs at two levels: comparison within children and between children. In the analysis of variance structure for a complete set of 32 children, eight for each possible pair of treatments, all treatment effects are clearly orthogonal and the structure is shown in Table 0.6. The simplest method of calculating the analysis is to calculate the sum and

Table 0.6.

Source	df
Between children	31
colour × daddy	1
colour × order	1
daddy × order	1
'main unit' error	28
Within children	32
colour	1
daddy	1
order	1
colour × daddy × order	1
'split unit' error	28

difference of scores for each child (as in Section 13.3). There will be four types of sum and four types of difference corresponding to the four possible patterns of treatment combination pairs:

$$S_1 = (1, m, c) + (2, d, b)$$
$$S_2 = (1, m, b) + (2, d, c)$$
$$S_3 = (1, d, c) + (2, m, b)$$
$$S_4 = (1, d, b) + (2, m, c)$$

$$D_1 = (1, m, c) - (2, d, b)$$
$$D_2 = (1, m, b) - (2, d, c)$$
$$D_3 = (1, d, c) - (2, m, b)$$
$$D_4 = (1, d, b) - (2, m, c).$$

For each sum type and each difference type, the mean and variance are calculated. The effects are then estimated from the means, as follows:

$$\text{colour} = (c - b)(d + m)(1 + 2) = (\bar{D}_1 - \bar{D}_2 + \bar{D}_3 - \bar{D}_4)/4$$
$$\text{daddy} = (c + b)(d - m)(1 + 2) = (-\bar{D}_1 - \bar{D}_2 + \bar{D}_3 + \bar{D}_4)/4$$
$$\text{order} = (c + b)(d + m)(1 - 2) = (\bar{D}_1 + \bar{D}_2 + \bar{D}_3 + \bar{D}_4)/4$$

$$\text{CD} = (c - b)(d - m)(1 + 2) = (-\bar{S}_1 + \bar{S}_2 + \bar{S}_3 - \bar{S}_4)/4$$
$$\text{CO} = (c - b)(d + m)(1 - 2) = (\bar{S}_1 - \bar{S}_2 + \bar{S}_3 - \bar{S}_4)/4$$
$$\text{DO} = (c + b)(d - m)(1 - 2) = (-\bar{S}_1 - \bar{S}_2 + \bar{S}_3 + \bar{S}_4)/4$$

$$\text{CDO} = (c - b)(d - m)(1 - 2) = (-\bar{D}_1 + \bar{D}_2 + \bar{D}_3 - \bar{D}_4)/4.$$

The four sample variances for differences should be averaged to give s_d^2 and the three sample variances for sums should be averaged to give s_s^2. In each case, before pooling, the variances should be examined to check that pooling is not unreasonable. The variances of each effect can then be calculated as $s_d^2/32$ or $s_s^2/32$ for effects estimated in terms of differences, or sums, respectively.

If the numbers of children for each of the four patterns of treatment pairs are not equal, then the orthogonal analysis of variance cannot be calculated. However, the calculation of means and variances for sums and for differences for each of the four types is still the appropriate first step. Effects can be estimated from the four means as previously, and the pooled variance estimates can be calculated using weighted averages. The variance of each effect must reflect the numbers of children for each type, n_1, n_2, n_3, n_4, and will be

$$s_d^2(n_1^{-1} + n_2^{-1} + n_3^{-1} + n_4^{-1})/16.$$

Note the similarity of the analysis of this experiment to that for two-period, two-treatment, cross-over designs, discussed in Chapter 8.

Finally, to obtain a set of comparable estimates of mean performance for the eight experimental treatments, we calculate the mean effect

$$\text{mean} = (c+b)(d+m)(1+2) = (\bar{S}_1 + \bar{S}_2 + \bar{S}_3 + \bar{S}_4)/4$$

and use the inverse relation between treatment combinations and effects

$$
\begin{bmatrix} c\ d\ 1 \\ c\ d\ 2 \\ c\ m\ 1 \\ c\ m\ 2 \\ b\ d\ 1 \\ b\ d\ 2 \\ b\ m\ 1 \\ b\ m\ 2 \end{bmatrix} = (1/8)
\begin{bmatrix}
+1 & -1 & +1 & -1 & +1 & -1 & +1 & -1 \\
+1 & +1 & +1 & +1 & +1 & +1 & +1 & +1 \\
+1 & -1 & -1 & +1 & +1 & -1 & -1 & +1 \\
+1 & +1 & -1 & -1 & +1 & +1 & -1 & -1 \\
+1 & -1 & +1 & -1 & -1 & +1 & -1 & +1 \\
+1 & +1 & +1 & +1 & -1 & -1 & -1 & -1 \\
+1 & -1 & -1 & +1 & -1 & +1 & +1 & -1 \\
+1 & +1 & -1 & -1 & -1 & -1 & +1 & +1
\end{bmatrix}
\begin{bmatrix} \text{mean} \\ \text{order} \\ \text{daddy} \\ \text{DO} \\ \text{colour} \\ \text{CO} \\ \text{DC} \\ \text{CDO} \end{bmatrix}
$$

Since some of variances of the effects are calculated in terms of s_d^2 and some in terms of s_s^2, the variances of comparison of the treatment combination estimates will be combinations of these. The particular form of the variance will depend on the particular comparison and, of course, on the numbers of observations of each type.

0.7 The relevance of experimental results

Within research institutes and universities it is often accepted that the designed experiment is the natural medium through which to develop knowledge. However, particularly in the more technological subject disciplines, it is necessary to examine the relevance of experimental results to applications. The concept of a designed experiment involves the tight control of all those factors which are not to be investigated in the experiment. This implies the use of small, carefully monitored, units. The pursuit of precision implies ever more rigidly defined experimental procedures.

In contrast, the practical implications of experimental results, if they are to be useful outside the confines of the research establishment and the associated world of learned scientific journals, must be applied to a wide range of practical situations. In agricultural crops, results from an experimental set of small plots must be applied in large fields. For animals, diets tested successively on the same animals must be applied to large herds of animals. In medicine, drugs tested on a small sample of patients available to the research scientist, often in hospital or other unusual surroundings, will be applied to a large population of people who may be brought into contact with the drugs through various sources. In psychology, assessment of various forms of treatments using volunteer subjects may have implications for large populations of individuals.

How do we ensure that experimental results are relevant? The twin requirements of precision and relevance appear to be totally conflicting. If experimental units are chosen randomly from the population to which the results might be applied, then the experimental error variation could be enormous. If the precision of experimental results depends on careful overall control of experimental conditions, together with the use of blocking, then the only means of making the results more relevant is to allow those factors which may be expected to vary within the target population to be varied within the experiment or, more generally, within the experimental programme. From the viewpoint of the relevance of results, the use of blocking factors to improve precision offers a secondary benefit of making the investigation of relevance more possible. Suppose that, in a medical trial of several drugs, the blocking factors sex, age and previous clinical history are used for controlling variation within the set of available experimental units. Then the comparison between drugs is essentially an average of the comparisons within each block (or combination of sex, age and history), and the effects of sex, age and history are eliminated from the comparison of interest, the difference between drugs. However, in addition to averaging the estimates of drug difference obtained from the different blocks, the variation of those estimates may be investigated to assess whether the differences are consistent over the different blocks, that is, over the different ages, sexes and histories.

More generally, factors may be deliberately introduced into an experiment to try to establish the validity of the results for a wider range of situation. Additional factors may be used as treatments. For example, different crop genotypes may be introduced as a factor in a nutrient experiment, possibly confounding some high-order interactions between the additional factor and the previous factors to avoid increasing block size. Alternatively, and less effectively, additional factor levels may be applied to whole blocks.

A frequently used extension of the idea of applying factors to whole blocks is to repeat experiments in different years or at different sites. If repetition of experiments is to be used in this way to make the results more relevant, then it is important to realise that there are statistical principles which are important in the design of sets of experiments. Simply repeating experiments as frequently and as widely as possible at randomly selected locations is not an efficient use of resources. Ideally, the research scientist should consider the population to which the results of the experimental programme are to be applied and should seek to characterise the major differences within that population. Wherever possible, the characteristics should be translated into additional experimental factors to be included within the controlled experiment. Where this is not possible because the characteristics of major differences represent environmental variation, then locations with widely different levels of those characteristics should be chosen to be represented in the experimental programme, so as to provide maximum information about the influence of those character- istics. The general principles of choosing levels of factors, discussed in Chapter 17, apply again here. It is better to have experiments at fewer sites, but with greater precision per experiment, provided the chosen sites span the range of the important characteristics, than a much larger number of experiments, each less precise, covering a complete spectrum of variations of the characteristic.

Even in those situations where it is felt that the important characteristics of variation among sites cannot be simply identified, it is important to select sites deliberately and restrictively in terms of their known character- istics. The apparent objective advantage of selecting a set of sites randomly to 'represent' the target population in an undefined manner is almost invariably a pseudo-respectable device for wasting resources. It cannot be too strongly emphasised that randomisation, which within designed experiments plays a crucial role in providing valid results, has nothing to contribute in the planning and control of variation. Randomisation is irrelevant to the improvement of precision of individual experiments, and of experimental programmes; it is necessary purely to provide valid estimates of the precision that has been achieved, within each individual experiment.

0.8   Block × treatment and experiment × treatment interactions

The analysis and interpretation of experiments repeated over years or sites is an area of statistics that has often been felt to be complicated and difficult. While it is true that the results from such analyses are often disappointing, this is not due to any inherent difficulty of the ideas, but to the inevitable consequences of asking questions which can only be answered by repetition of experiments. Consider again the motivation for repeated experiments. We suppose that the comparisons, made as precisely as possible within a controlled experiment, may vary over other conditions, and we therefore repeat experiments in different conditions to examine this variation. We should therefore expect that the estimates of treatment differences will often be different between experi- ments. However, it is unrealistic to suppose that only the treatment differences will change as major changes to the environment of the experiment occur. Clearly, we should expect that the variation between experimental units, represented by the error variance σ^2, will also change.

Hence, it is important never to attempt an overall analysis of a set of experiments without first completing analyses of the individual experiments. The approach to the analysis of a set of experiments should assume that it is unlikely that a joint analysis will be valid. It may be possible to find a transformation of variable to a scale on which a joint analysis is valid, but it is probably most often true that interpretation of a set of experiments should be based on the individual analyses of results from the separate experiments.

However, when the error variance is credibly invariant over experiments, how should an overall analysis be constructed and how should the results be interpreted? Consider the case where we have a set of t treatments in b blocks in each of r repeats of the experiment. Each individual analysis structure is

blocks: $b-1$ df,
treatments: $t-1$ df,
error: $(b-1)(t-1)$ df.

When the results from different experiments are to be considered in a single analysis of variance, we must already have decided that it is reasonable to assume that the error variance is consistent over experience. There is usually no reason to expect block differences to be consistent over experiments so separate block effects must be estimated for each experiment. Variation due to treatment differences should be divided into two components; average differences over experiments, and experiment–treatment interactions. The combined analysis of variance structure will be as shown in Table 0.7. Much of the discussion about repeated experiments has concerned the basis for making treatment comparisons.

Table 0.7.

Source	df	MS
Between experiments	$r-1$	
Blocks within experiments	$r(b-1)$	
Treatment (main effect)	$t-1$	TMS
Experiment × treatment interaction	$(r-1)(t-1)$	ETMS
Pooled error	$r(b-1)(t-1)$	s^2
Total	$rbt-1$	

First, for the calculation of F-ratios, there is an apparent choice of calculating TMS/s^2 or $TMS/ETMS$ to test the size of the treatment differences, in addition to the calculation of $ETMS/s^2$ to test the extent to which treatment differences vary over the set of experiments. Second, the standard errors for comparing treatment means could be based either on s^2 or on the error treatment mean square. The logic of this second choice is identical to that of the choice of divisor in the F-ratio calculation.

There is no absolute statistical answer to the question of the proper ratio to test treatment effects. The two tests are different, and they provide answers to two different questions. Which question is appropriate must depend on the particular reasons for conducting the set of experiments. The statistician and experimenter must be quite clear what the different mean squares represent. The treatment mean square of course represents the average overall variation between treatments. The error treatment mean square represents the inconsistency of treatment effects between different experiments. The error mean square, s^2, represents the inconsistency of the treatment effects over the blocks within each experiment.

If the repeated experiments are randomly selected, then it seems appropriate that the treatment average effects should be compared with the variation of those effects over different experiments, since the general prediction objective concerns the treatment effects to be expected, and the variation over repeated experiments is the only available guidance to the precision of such predictions. However, if the repeated experiments were chosen deliberately to represent particular changes of conditions under which the treatment differences should be observed, then the sample of experiments is no longer an appropriate framework against which to measure treatment main effects. Both the treatment main effect and the interaction of treatments with the controlled differences between experiments should be assessed against the only random source of variation, that of plots within blocks. Note the importance of the concept of random variation. We may regard blocks as being randomly chosen, or may rely on the random allocation of treatments to units within each block to justify the use of s^2 as an appropriate measure for assessing treatment differences, either main effects or interactions. There is no additional formal randomisation across experiments, and so the use of error treatment mean squares as a measure for assessing treatment comparisons relies on the random nature of the selection of the set of experiments selected.

More complicated situations may occur, when the repetition of experiments may include both a systematically designed component and a random component, or when repetitions both over years and over sites are involved. The principles to be considered when choosing the mean squares for use either as denominators of F-ratios, or as the basis for standard errors, are universally relevant, and further detailed discussion of particular cases should not be necessary here.

When it is not legitimate to regard error variances as invariant over experiments, a combined analysis is not valid. In such cases, a sensible procedure is to base the interpretation of treatment differences on an analysis of effects. For each treatment effect, it is possible to consider the variation of the effect over experiments. For each experiment, the estimate of a particular treatment effect of importance may be calculated. If the set of experiments may now be regarded as a random sample, the values of the treatment effect estimate from each experiment can be viewed as a sample of estimates, and the mean and variance of the sample calculated. A mean treatment effect may be assessed using a standard error based on the variance of that effect over the sample of experiments.

As mentioned previously it is possible to take the analysis of effects further, and to calculate the effect in each block of each experiment. The analysis of variance of each particular treatment effect can then provide information about the variation of the effect between blocks within each experiment, and between experiments. In some examples, even though the experiment error variances are deemed to vary between experiments, the variance of a particular effect may appear consistent and the between block, within experiment variances of that effect may be pooled.

Discussion of experiment × treatment interactions leads naturally to considering whether block × treatment interactions are important. In general, experiments as usually designed contain no explicit information about block × treatment interactions. In a randomised complete block or incomplete block design, there is no repetition of treatments within a block, so that the block–treatment interaction variation cannot be distinguished from the error variation. If block–treatment interactions occur, then the usual linear model structure is inadequate, and the analysis of the experimental data must be reconsidered. This subject has been discussed briefly already in Chapter 11. It may also be argued that, if there are substantial block–treatment interactions, interpretation of experimental results is difficult, or even impossible, and even that there is little value in trying to examine block–treatment interactions.

Suppose, however, that we are concerned about block–treatment interactions; how do we try to estimate them? Extension of a model for a block–treatment design to include block–treatment interaction terms reveals the difficulty clearly. The model is

$$y_{ij} = \mu + b_i + t_i + (bt)_{ij} + \varepsilon_{ij}.$$

In the normal circumstances, where each block–treatment combination has at most a single observation, $(bt)_{ij}$ cannot be distinguished from ε_{ij}. A direct method of investigation is to alter the design of the experiment to include repetitions of treatments within blocks. This is rarely adopted, because it necessarily reduces the precision of treatment comparisons, since that precision depends essentially on the number of within block treatment differences, and each within block difference which is used to estimate the 'pure' error that is not contaminated by block–treatment interaction cannot be used to estimate treatment effects. However, it may sometimes be decided that obtaining separate estimates of σ^2, from pure error and from the block–treatment interaction mean square, is sufficiently valuable to outweigh the loss of treatment comparison precision.

An indirect method of examining block–treatment interaction is through division of the error sum of squares in a randomised block design in which the treatments have any kind of structure. The error SS is identical with the (block × treatment) SS. If the treatments have a structure so that the treatment SS may be divided into components, L_i, then there is a corresponding division of the error SS into components of the form

$$(block \times L_i) \text{ SS}.$$

This is the same idea as that of the analysis of effects and the use of analysis of sums and differences discussed in Section 13.3. Each (block × L_i) SS represents the variation of the treatment comparison for L_i over blocks. The existence of possible block–treatment interaction effects can be examined by testing the consistency of the set of (block × L_i) mean squares. If the simple additive block and treatment effect model is correct, then for any subdivision of the error sum of squares no substantial heterogeneity of variances should appear. Such an attempt to diagnose possible block–treatment effect interactions is not comprehensive since we can hardly consider all possible forms of treatment comparison. However, it does provide a further diagnostic test for the additivity assumed in the model and should be available in comprehensive packages for analysis of variance.

Exercises 0

(1) An agricultural research worker wishes to study the effect of an insecticide on the yield of a crop. He proposes to compare two methods of application (spray or powder – factor M), and two times of application (early or late – factor T). He can use about 50 plots in either of two ways:

(i) Regard absence and presence of insecticide as a third factor, X, in a formal 2^3 design on factors X, M, T in six randomised blocks of eight, although all combinations of M, T in the absence of insecticide are necessarily identical:

(ii) Use only one treatment without insecticide plus the four combinations of use of insecticide, to give a set of five treatments, and arrange in two 5×5 Latin squares.

Discuss critically the interesting and important contrasts between treatments for each design; using σ_1^2, σ_2^2, to represent variances per plot for (i), (ii), obtain the variance for each contrast. Hence indicate what guidance on the choice of design could be obtained from a fairly trustworthy guess at the ratio σ_1^2/σ_2^2. Identify clearly the circumstances in which one design appears to be a clear preference, and those in which the choice depends upon which aspect of the results is of chief concern to the investigator.

(2) Eighteen experimental units, which are arranged in three rows of six, are available for an experiment on three treatments A, B, C. Four designs are under consideration:

(i) completely randomised with six replications of each treatment;

(ii) completely randomised with eight replications of A, five of B and five of C;

(iii) six randomised blocks of three units each;

(iv) two 3×3 Latin squares, both of which have the same rows.

For each design show the form of the analysis of variance and indicate the calculations for the sums of squares. Compare the precisions with which treatment differences are estimated from the four designs.

If any of the four designs is wholly unsatisfactory, explain clearly why you would never use it. Indicate the merits of those not rejected and give an example of a situation in which each might be used.

(3) Ten animals are available for a trial to compare five diets for cattle. Two designs are considered. The first consists of a double Youden square with three treatment periods for each animal; there will be recovery periods between consecutive treatment periods which will hopefully eliminate residual effects. The second design uses two 5×5 Latin squares balanced for residual effects which must be allowed for because of the closeness of consecutive

Figure 0.9. Two alternative designs to compare five diets for cattle using ten animals with either (a) three periods or (b) five periods.

Animals

(a)		1	2	3	4	5	6	7	8	9	10
Time	1	A	B	C	D	E	A	B	C	D	E
	2	B	C	D	E	A	B	C	D	E	A
	3	C	D	E	A	B	D	E	A	B	C

(b)											
Time	1	A	B	C	D	E	A	B	C	D	E
	2	B	C	D	E	A	C	D	E	A	B
	3	D	E	A	B	C	B	C	D	E	A
	4	E	A	B	C	D	E	A	B	C	D
	5	C	D	E	A	B	D	E	A	B	C

treatment periods. Both designs are given in Figure 0.9. Obtain the standard error of estimated differences between direct treatment effects for each design, and discuss how an experimenter should choose between the two designs.

(4) In a consumer trial ten men are available to compare the effectiveness of five brands of razor blades. Each individual is supplied with two blades of different brands and identical instructions for face preparation. Each morning for a week he uses these two brands of blades, one brand for each side of his face. The smoothness of each shave is measured and the average value for each blade is used as a measure of the quality of the blade. Suggest an appropriate design for this experiment if the trial is to be completed over the one-week period. Give details of the analysis and state clearly any models you use and assumptions you make.

(5) It is proposed to investigate the effects of two chemicals (A, B) each at two concentrations $(A_1, A_2; B_1, B_2)$ on fungal growth on leaves. The basic experimental unit is a single leaf. It is believed that growth of the fungus is influenced by the age of leaf. The difference in growth for leaves of different ages is of some interest in itself, but it may also affect the differences between the chemical treatments. These effects are, however, of secondary interest compared with the comparisons of the chemicals. It is known that leaves from different plants may differ substantially; also leaves in a pair on a plant are likely to be particularly similar, and leaves of

a similar age will behave more similarly than those whose ages differ more substantially.

Three designs have been suggested as follows:

(i) From each of eight plants use four leaves of differing ages (the same ages for each plant). Regarding the 32 leaves as an eight plant × four age array, the four treatments $(A_1, A_2; B_1, B_2)$ are allocated so that each treatment appears twice for each leaf age and once for each plant.

(ii) From each of four plants use eight leaves (four old, four young) and apply the four treatments to leaves so that all eight combinations of chemical × concentration × age occur for each plant.

(iii) From each of four plants take four pairs of leaves (two old pairs, two young pairs). Allocate each chemical to one old pair and one young pair for each plant and allocate the two concentrations of each chemical to the two leaves of the pair.

Identify the appropriate form of analysis of variance for each design and discuss the relative merits of the three designs.

(6) A parallel line bioassay is carried out to compare seven distinct test preparations (A, B, C, D, E, F, G) simultaneously with a standard, S, using two doses of each and 32 subjects per preparation. Outline the partitioning of degrees of freedom for the analysis of freedom for the analysis of variance and mention any additional assumptions inherent in the usual analysis. Describe any changes in design and in the partitioning that would be appropriate to each of the following circumstances:

(i) Only 128 subjects can be treated on one day, and responsiveness may change markedly from day to day.

(ii) As (i), but also only eight combinations of preparation and dose level can be handled on one day.

(iii) As (i), but preparation G can be omitted. The 256 subjects are to be allocated approximately optimally for estimating the potency of each of A, B, ..., F, relative to S.

(iv) Four doses of each of the eight preparations are to be used instead of two, only 64 subjects can be treated per day, and only eight combinations of preparation and dose can be handled on each of four days.

(v) As (i), but time of treatment within a day may affect responsiveness. Only one person does the work and he requires about three minutes per subject.

(7) In an experiment to investigate the effects of nutrition level (N) and housing temperature (T) on the rate of growth of piglets, a factorial experiment is to be used, with litters of eight or nine piglets as blocks. A primary object of the experiment is to estimate the linear effects of nutrition (the rate at which yield increases with increasing nutrition level) and also the linear × linear interaction of N and T. Three designs are considered with different numbers of levels of a factor, and with possibly different ranges of levels of a factor (range = highest level − lowest level). The designs considered are

(i) A 2×2 factorial with factor level ranges N_2, T_2 in four blocks each containing two replicates of all four treatment combinations.

(ii) A 3×3 factorial with equally spaced levels over ranges N_3, T_3, in four blocks each containing a single replicate.

(iii) A 4×4 factorial with equally spaced levels over ranges N_4, T_4; two replicates each consisting of two blocks of eight units, each with the $N'N''T'T''$ interaction confounded (N', N'', $T'T''$ are dummy two-level factors for the four-level factors N, T).

For each design identify comparable estimates of the linear effect of nutrition and of the linear × linear interaction and find estimated variances of those estimates. For what values of N_2, T_2, N_3, T_3, N_4, T_4 are the estimates from the three designs equally precise?

Discuss briefly the experimenter's choice of design.

(8) The following comes from a letter requesting help for the design of an experiment:

'In the trial we are to study:

High and low oil silages fed at high and low levels, the silage to be treated with or without antioxidants, and diets are to be supplemented or not supplemented with limiting amino acids including tryptophan which is broken down in acid conditions.

In addition to these 16 treatment combinations there are to be four control treatments, being high and low oil fish meals fed at high and low levels. All control feeds will include antioxidants and will be supplemented with amino acids.

We have two sets of cages available for this work:

A 16 cage flat deck system to take up to ten chicks for three weeks and a 16 cage metabolism unit to take four chicks to three weeks. We plan to use both sets of cages and put four chicks per cage for two full experimental periods and one cage unit for a further experimental period.

I would appreciate any comments you have regarding this experimental design.'

How would you reply?

References

Bailey, R. A. (1982). The decomposition of treatment degrees of freedom in quantitative factorial experiments. *J. R. Stat. Soc. B* 44, 63–70.

Barnett, V. and Lewis, T. (1978). *Outliers in Statistical Data*. Wiley, Chichester.

Bartlett, M. S. (1978). Nearest neighbour models in the analysis of field experiments (with discussion). *J. R. Stat. Soc. B* 40, 147–74.

Besag, J. E. (1974). Spatial interaction and the statistical analysis of lattice systems (with discussion). *J. R. Stat. Soc. B* 36, 192–236.

Besag, J. E. and Kempton, R. A. (1986). Analysis of field experiments using spatial statistics. *Biometrics* 42, 231–51.

Bleasdale, J. K. A. (1967). Systematic designs for spacing experiments. *Exp. Agric.* 3, 73–85.

Box, G. E. P. (1957). Evolutionary operation: a method for increasing industrial productivity. *Appl. Stat.* 6, 81–101.

Box, G. E. P. and Cox, D. R. (1964). An analysis of transformations (with discussion). *J. R. Stat. Soc. B* 26, 211–52.

Box, G. E. P. and Draper, N. R. (1974). Robust designs. *M. R. C. Technical Summary Report No. 1420*, University of Wisconsin.

Box, G. E. P. and Youle, P. V. (1955). The exploration and exploitation of response surfaces: an example of the link between the fitted surface and the basic mechanism of the system. *Biometrics* 11, 287–322.

Bryan–Jones, J. and Finney, D. J. (1983). On an error in 'Instruction to Authors'. *Hort. Sci.* 18, 279–82.

Carmer, S. G. and Jackobs, J. A. (1965). An exponential model for predicting optimum plant density and maximum corn yield. *Agronomy J.* 57, 241–4.

Cleaver, T. J., Greenwood, D. J. and Wood, J. T. (1970). Systematically arranged fertiliser experiments. *J. Hort. Sci.* 45, 457–69.

Cochran, W. G. and Cox, G. M. (1957). *Experimental Designs*. Wiley, New York.

Cornell, J. A. (1981) *Experiments with Mixtures: Designs, Models and The Analysis of Mixture Data*. Wiley, New York.

Corsten, L. C. A. (1958). Vectors, a tool in statistical regression theory. *Mededelingen van de Landbouwhogeschool te Wageningen* 58, 1–92.

Cox, D. R. (1958). *Planning of Experiments*. Wiley, New York.

Dorfman, R. (1943). The detection of defective members of large populations. *Ann. Math. Stat.* 14, 436–40.

Dyke, G. V. and Shelley, C. F. (1976). Serial designs balances for effects of neighbours on both sides. *J. Agric. Sci.* 87, 303–5.

Fisher, R. A. and Yates, F. (1963). *Statistical Tables for Biological Agriculture and Medical Research.* Oliver and Boyd, Edinburgh.

Francis, L. (1978). Experimental designs for the two parameter exponential response curve. MSc Thesis, University of Reading.

Freeman, G. H. (1979). Some two-dimensional designs balanced for nearest neighbours. *J. R. Stat. Soc. B* 41, 88–95.

Green, P. J., Jennison, C. and Seheult. A. H. (1985). Analysis of field experiments by least squares smoothing. *J. R. Stat. Soc. B* 47, 299–315.

Gordon, T. and Foss, B. M. (1966). The role of stimulation in the delay of onset of crying in the new-born infant. *J. Exp. Psychol.* 16, 79–81.

Hills, M. and Armitage, P. (1979). Two-period cross-over clinical trial. *Br. J. Clin. Pharmacol.* 8, 7–20.

Hozumi, K., Asahira, T. and Kira, T. (1956). Intraspecific competition among higher plants: VI. Effect of some growth factors on the process of competition. *J. Inst. Polytech. Osaka Cy. Univ.* D7, 15–28.

Kempthorne, O. (1952). *The Design and Analysis of Experiments.* Wiley, New York.

Kempton, R. A. and Howes, C. W. (1981). The use of neighbouring plot values in the anaiysis of variety trials. *Appl. Stat.* 30, 59–70.

Kuipers, N. H. (1952). Variantie–analyse. *Statistica* 6, 149–94.

Maindonald, J. H. and Cox, N. R. (1984). Use of statistical evidence in some recent issues of DSIR agricultural journals. *N. Z. J. Agric.* 27, 597–610.

McCullagh, P. and Nelder, J. A. (1983). *Generalised Linear Models.* Chapman and Hall, London.

Morse, P. M. and Thompson, B. K. (1981). Presentation of experimental results. *Can. J. Plant Sci.* 61, 799–802.

Nelder, J. A. (1962). New kinds of systematic design for spacing experiments. *Biometrics* 18, 283–307.

Nelder, J. A. (1966). Inverse polynomials, a useful group of multi-factor response functions. *Biometrics* 22, 128–41.

O'Neill, R. and Wetherill, G. B. (1971). The present state of multiple comparison methods (with discussion). *J. R. Stat. Soc. B* 33, 218–41.

Patterson, H. D. and Silvey, V. (1980). Statutory and recommended list trials of crop varieties in the U. K. (with discussion). *J. R. Stat. Soc. A* 143, 219–52.

Patterson, H. D., Williams, E. R. and Hunter, E. A. (1978). Block designs for variety trials. *J. Agric. Sci.* 90, 395–400.

Pearce, S. C. (1963). The use and classification of non-orthogonal designs (with discussion). *J. R. Stat. Soc. B* 25, 353–77.

Pearce, S. C. (1975). Row-and-column designs. *Appl. Stat.* 24, 60–74.

Pocock, S. J. (1979). Allocation of patients to treatment in clinical trials. *Biometrics* 35, 183–97.

Rayner, A. A. (1967). *A first course in Biometry for Agriculture Students.* University of Natal Press, Pietermaritzburg, South Africa.

Reid, D. (1972). The effects of the long-term application of a wide range of nitrogen rates on the yields from perennial ryegrass swards with and without white clover. *J. Agric. Sci. (Camb.)* 79, 291–301.

Rojas, B. A. (1963). The San Cristobal design for fertiliser experiments. *Proceedings of the International Society of Sugar Cane Technologists,* Mauritius.

Rojas, B. A. (1972). The orthogonalised San Cristobal design. *Proceedings of the International Society of Sugar Cane Technologists,* Louisiana, USA.

Sobel, M. and Groll, P. A. (1959). Group testing to eliminate effectively all defectives in a binomial sample. *Bell Syst. Tech. J.* 38, 1179–252.

Sparrow, P. E. (1979). Nitrogen response curves of spring barley. *J. Agric. Sci. (Camb.)* 92, 307–17.

Wilkinson, G. N. (1970). A general recursive procedure for analysis of variance. *Biometrika* 57, 19–46

Wilkinson, G. N., Eckert, S. R., Hancock, T. W. and Mayo, O. (1983). Nearest neighbour (NN) analysis of field experiments (with discussion). *J. R. Stat. Soc. B* 45, 151–211.

Williams, R. M. (1952). Experimental designs for serially correlated observations. *Biometrika* 39, 151–67.

Index